This book presents a thorough and detailed description of the very successful Lund model of the dynamics of particle physics. The Lund model, inspired by quantum chromodynamics, has provided a very promising and pictorial approach to the dynamics of quark and gluon interactions. Starting with a brief reprise of basic concepts in relativity, the quantum mechanics of fields and particle physics, this book goes on to discuss the dynamics of the massless relativistic string, confinement, causality and relativistic covariance, Lund fragmentation processes, QED and QCD bremsstrahlung, multiplicities and particle-parton distributions. The book also explores the relationships between the Lund model and other models based on field theory (the Schwinger model, S-matrix models, lightcone algebra physics and variations of the parton model) or on statistical mechanics (the Feynman-Wilson gas, scaling, iterative cascade models).

The book will be of interest to experimental and theoretical particle physicists, and also to those working in other branches of physics who would like to develop a feel for these basic interactions. This title, first published in 1998, has been reissued as an Open Access publication on Cambridge Core.

CAMBRIDGE MONOGRAPHS ON PARTICLE PHYSICS,
NUCLEAR PHYSICS AND COSMOLOGY: 7
General Editors: T. Ericson, P. V. Landshoff

THE LUND MODEL

CAMBRIDGE MONOGRAPHS ON PARTICLE PHYSICS, NUCLEAR PHYSICS AND COSMOLOGY

The Lund Model

BO ANDERSSON
Lund University

CAMBRIDGE
UNIVERSITY PRESS

Shaftesbury Road, Cambridge CB2 8EA, United Kingdom

One Liberty Plaza, 20th Floor, New York, NY 10006, USA

477 Williamstown Road, Port Melbourne, VIC 3207, Australia

314–321, 3rd Floor, Plot 3, Splendor Forum, Jasola District Centre, New Delhi – 110025, India

103 Penang Road, #05-06/07, Visioncrest Commercial, Singapore 238467

Cambridge University Press is part of Cambridge University Press & Assessment, a department of the University of Cambridge.

We share the University's mission to contribute to society through the pursuit of education, learning and research at the highest international levels of excellence.

www.cambridge.org
Information on this title: www.cambridge.org/9781009401258

DOI: 10.1017/9781009401296

First published 1998
Reissued as OA 2023

A catalogue record for this publication is available from the British Library.

ISBN 978-1-009-40125-8 Hardback
ISBN 978-1-009-40128-9 Paperback

Contents

vii

Acknowledgements

This book is dedicated to the memory of my teacher, Gunnar Källén, who died much too young. He was the greatest teacher any man could have had. Those of us who enjoyed the vision of physics according to the way he described it, as a great adventure, have often asked what he would have said about the greatest adventure my generation will ever have, i.e. the confined field theory of QCD (which is so different from what we talked about in Lund in the 1960s!).

I would also like to thank my collaborators over the years, in particular the past and present members of the Lund High-Energy Theory Group. I will firstly mention Gösta Gustafson, who kindly has read most parts of this book and provided many useful remarks. He has been my principal collaborator for very many years and there is no way to thank him for all the joy and insight he has provided. Further, one of our recent students, Jari Häkkinen, has helped me both with my sometimes very temperamental computer and also with the figures. As there are very few 'free luncheons in life', as my American friends tell me, his help has been something that I will savour for all times. I have also been teased, kicked around and during the process learned a lot from my copy-editor, Susan Parkinson.

But even if I have been endowed with the best advisers, with regard to any remaining errors and omissions the buck rests with me!

1
Introduction

This book stems from lectures in different places and at different times. I would like to thank all those colleagues, graduate students and collaborators, who have patiently listened, commented upon and by insistent questioning given me insight into the physics described in this text.

You will find that the physics is described in a semi-classical language. I believe that my generation, the grandchildren of the wonderful generation that developed the tools of quantum mechanics, have largely learned to use semi-classical dynamical pictures while avoiding the quantum mechanical pitfalls. After having understood that the state density is different and that probabilities are not additive in quantum mechanics most of one's classical intuition can be used. I provide an example in Chapter 2 which shows that you can never fool Heisenberg's indeterminacy relations (i.e. position and conjugate momentum cannot be determined simultaneously with arbitrary precision). But you may choose your variables in such a way (rapidity and position for high-energy particles) that all the quantum mechanical rules are fulfilled and you may still transfer easily between the descriptions in terms of the different variable sets.

The material in the book has been chosen to stress the connections between different approaches to high-energy physics. The basic picture is nevertheless the one stemming from field theory as it is used in the Lund model. The Lund model has been successful in describing many of the dynamical features of multiparticle production because it contains so many relations to earlier and contemporary work, although often with very different dynamical starting points. I am very sorry that due to space limitations I have had to exclude many interesting and still-viable theoretical approaches to the physics of high-energy multiparticle production from this book.

It may at this point be useful to try to clarify what I mean by the Lund model in this book. There is some confusion because during the years many of the original contributors (and also people never working with the Lund Group) have provided a lot of material described as 'in accordance with the Lund model'. After chapters on *relativistic kinematics, field theory, renormalisation* and the *parton model*, all introduced to provide the notation as well as some useful formulas, I will consider *the Lund fragmentation model of quarks and gluons*.

This part of the Lund model (which was the first part produced and which, owing to lucky coincidences has not been changed very much over the years) makes use of the massless relativistic string as a model for the QCD color force fields. It provides a description of the transition from the partonic entities to the final-state observables in terms of the hadronic states. The model is described in detail in Chapters 6-15 and is implemented in the well-known Monte Carlo simulation program JETSET. The major achievements are

1 A consistent space-time and energy-momentum-space description leading to a unique (Markov) stochastic process for the breakup of the (string) field into hadrons. The process is described on the $(1+1)$-dimensional surface spanned by the string field during its periodic motion (and it is determined uniquely from the partons).

2 A highly nontrivial description of the partons, with the quarks (q-particles) and antiquarks ($\bar{q}$-particles) as endpoint excitations and the gluons (g) as internal excitations on the string field.

3 The breakup of the fields into 'new' $q\bar{q}$-pairs stems from a quantum mechanical tunnelling process. Although all the formulas of the model are derived in a semi-classical framework the final results can be interpreted within a consistent quantum mechanical scenario (and actually also within statistical mechanics, thereby providing the so-called Feynman-Wilson gas analogy).

4 It is possible within the model to account for the strong (transverse) polarisation effects observed and to describe more subtle quantum mechanical interference effects such as Bose-Einstein correlations.

There is secondly the *Lund dipole cascade model* (the DCM), which contains a description of the multiparton bremsstrahlung emissions in perturbative QCD, thereby providing the states for which the Lund fragmentation model may be applied. This is described in Chapters 16-18 and it is implemented in the ARIADNE Monte Carlo simulation program. A different approach, the *method of independent parton cascades*, has been implemented in the JETSET and, according to the Webber-Marchesini model, cf. Chapter 17, in the HERWIG Monte Carlo simulation programs.

There is finally (and this is a very recent advance) the *linked dipole chain model*, providing a description of the states occurring in deep inelastic scattering (DIS) events. I start with Chapter 19 on the 'ordinary' approach to DIS using the (double) leading-logarithm approximation as well as the results of approximating the matrix elements by the (major) lightcone singularities. The main problem is to describe the hadron structure functions, i.e. the partonic flux factors, stemming from the hadronic wave function, in accordance with perturbative QCD. The well-known Dokshitzer-Gribov-Lipatov-Altarelli-Parisi (DGLAP) equations are derived and also the considerations behind the Balitsky-Fadin-Kuraev-Lipatov (BFKL) mechanism. Finally I have included a section on the recently developed Ciafaloni-Catani-Marchesini-Fiorani (CCMF) model, which contains a very ambitious effort to re-sum the large-order contributions to the perturbative QCD diagrams.

The linked dipole chain (LDC) model, described in Chapter 20, is a generalisation and simplification of the results of the CCFM model and just as for CCFM it interpolates between the DGLAP and BFKL results for the structure functions. It provides a general framework to describe all kinds of deep inelastic scattering events (besides the 'ordinary' parton-probe events that occur in accordance with perturbative QCD and the Feynman parton model there are boson-gluon fusion events, which contribute a large part of the present HERA cross section, and Rutherford scattering between the resolved probe structure and the hadron structure).

In this way the Lund model contains one common general feature at every level of the description of QCD, i.e. the occurrence of dipoles:

- An excitation in the vacuum, e.g. from an e^+e^- annihilation event, produces a color $q\bar{q}$-dipole, which decays via gluon bremsstrahlung according to the dipole cascade model into a set of color dipoles, spanned between the partons. This is known as a 'timelike' cascade because the original large excitation mass decays into smaller and smaller dipole masses. The dipoles move apart thereby producing a force field similar to the modes of the massless relativistic string.

- Afterwards the string field breaks up into hadrons, 'the ultimate dipoles', produced in the Lund fragmentation model from a quark and antiquark from adjacent breakup vertices together with the field in between.

- When such a hadron is probed the states can again be described as a set of dipoles, according to the linked dipole chain model, spanned between the color-adjacent gluons emitted in the ensuing bremsstrahlung. This is known as a 'spacelike' cascade because it corresponds to probing the hadron wave function up towards larger and larger 'virtualities', i.e. more and more spacelike momentum transfers, $-q^2 = Q^2$ (smaller wavelengths $\lambda \sim 1/Q$). The interaction with the probe brings the whole chain on-shell and then the dipoles again decay via the dipole cascade model to smaller dipoles and finally into hadrons via the Lund fragmentation model.

At this point I would like to make two remarks. Firstly there is a duality between descriptions of perturbative QCD in terms of dipoles and in terms of gluonic excitations. The gluons correspond to pointlike excitations in the color field while the dipoles are the (field) 'links' between these points. In other words the color from one dipole meets the anticolor from the adjacent one at a gluon 'corner' (note that the color-8 gluons can be considered as a combination of $3\bar{3}$ color charges).

My second remark is that the only solvable confined field theory we know of, $(1 + 1)$-dimensional QED (the Schwinger model described in Chapter 6) is just a theory of dipoles. The Lagrangian of the original fermion–antifermion field interacting with the connecting electric field can be transformed into the Lagrangian of a free field, corresponding to a dipole density of massive quanta composed of such a pair and the adjoining field. It should be stressed, however, that it is not known whether confinement implies a dipole picture of the charges and the fields.

Hadronic interactions *per se* have been investigated during a longer timespan than any other parts of multiparticle dynamics, but we are still very far from a consistent and useful description. I have at different places introduced some features, e.g. the S-matrix and unitarity, which are so general that they must be part of any future theory. But I have owing to space limitations decided to exclude all specific models, although some of them, like Gribov's Reggeon theory, have beauty and generality sufficient to redeem even a partial study.

I have also generally avoided to include experimental material. It should be stressed that no phenomenological work is alive without the necessary experimental checks on the approach. There have been, however, a large number of investigations, reviews and comparisons with experimental data in all the conference proceedings of the last decade. They are all in agreement with the general approach of the book. I will as a further excuse make use of the following sentence, which occurs in many places and must have been invented for just this situation: 'New experimental material is also coming in at such a rapid rate that the book would date unnecessarily quickly by including only the presently available data'. I admire my experimental colleagues for the fact that it is a true statement!

But we should always keep in mind what Bacon has pointed out (this is a free translation of the credo of phenomenology): 'You have not learned anything by being in agreement with data, because there are always other possible explanations. But if you put forward an idea, calculate inside the framework in an honest way and find disagreements with experiments *then you have learned something*, i.e. that this approach is not taken by Nature'. Or as one of my friends enthusiastically said during a heated conference discussion: *We must dare to be wrong!*

I have used the units conventional in today's high-energy physics putting the velocity of light c and Planck's constant $\hbar$ equal to unity thereby making energy dimensions inverse to length dimensions. In that connection it is useful to remember that a transfer between energy and length units is with this convention provided by the rather precise approximation $1\,\mathrm{fm} \times 1\,\mathrm{GeV} \simeq 5$.

In order to keep the reference list reasonably short I have taken the liberty of omitting references to phenomena like the parton model, Wick's theorem, the Ward identity etc., which nowadays are all part of our common physics heritage. I may have overdone it and if so I apologise to the authors. I would like to mention that material included in the books

J.D. Jackson, *Classical Electrodynamics*, John Wiley & Sons

H. Goldstein, *Classical Mechanics*, Addison-Wesley

E. Merzbacher, *Quantum Mechanics*, John Wiley & Sons

is referred to by these authors' name only. There is evidently a set of equally useful basic text-books where you can find the same material, but it is impossible to be exhaustive. When it comes to quantum field theory the subject has still not matured to the extent of these text-books. A rather formal description (containing, however, many useful references) is given by C. Itzykson and J.B. Zuber, *Quantum Field Theory*, McGraw-Hill, 1980. For perturbative QCD there is a recent book, Yu.L. Dokshitzer, V.A. Khoze, A.H. Mueller and S.I. Troyan, Basics of Perturbative QCD, Editions Frontières, 1991, which is very good. An early reference to the Lund model (as of 1982) is *Phys. Rep.* **97** 31, 1983.

2

Relativistic kinematics, electromagnetic fields and the method of virtual quanta

The dynamics of the massless relativistic string (which we will meet at very many different places in this book) is a delightful theoretical laboratory to study the properties of the theory of special relativity. To make the book self-contained and also to define our notation we will briefly review in this chapter some properties of special relativity, in particular with respect to its implications for high-energy particle kinematics.

We will also review some properties of electromagnetic fields with particular emphasis on the features we are going to make use of later in the book. We will end with a description of the *interaction ability of an electrically charged particle*.

This is the first but not the last example in this book of the *law of the conservation of useful dynamics*. This says that every new generation of theoretical physicists tends to reinvent, reuse (and usually also rename) the most useful results of earlier generations. One reason is evidently that there are few situations where it is possible to find a closed mathematical expression for the solution to a dynamical problem.

Here our basic aim is to describe the interactions between charged particles which are moving with very large velocities (as they do in high-energy physics). As a charged particle interacts via its field the question can be reformulated into finding a way to describe the field of a charged particle which is moving very fast. To account for quantum mechanics we need a way to describe the quantum properties of the charged particle's field and this problem can be solved even at a semi-classical level. It is possible to obtain a closed formula for the *flux of the field quanta* in this case.

Fermi addressed the problem in the 1920s, Weizsäcker and Williams found the method independently of Fermi and each other in the 1930s. After that it became a standard tool in connection with QED in terms of the *method of virtual quanta*, the MVQ. Later again Feynman made use of it in order to introduce the *parton model*. We will discuss that model repeatedly in this book, but it is useful to see how 'partons' emerge even at the semi-classical level in electromagnetism.

2.1 The Lorentz boost

Michelson and Morley demonstrated that the velocity of light, c, is independent of the direction of a light beam. Einstein interpreted this finding to imply that the velocity c is independent of the relative motion of the light source and the detector.

We are not going to dwell upon the many basic questions that are raised by this interpretation but simply accept that it has profound implications with respect to measurements of events in space and time. The resulting predictions have been tested repeatedly and always been found to be true. In this section we will briefly consider some of these predictions.

I *The Lorentz boost*. Consider two observers A and B, moving with respect to each other. We will suppose that they have calibrated their watches and decided upon a common origin in space and time as well as the directions of the coordinate axes in space. The arrangement will be that they move along their common x-axis so that B has the velocity v with respect to A. We will for simplicity use units such that the velocity of light $c = 1$. Then an event (1) which for A occurs at the space-time coordinates

$$(1) \equiv (t_{1A}, x_{1A}, y_{1A}, z_{1A}) \tag{2.1}$$

will for B, in his system, seem to occur at the time and space coordinates (with the corresponding index B):

$$\begin{aligned}
t_{1B} &= \gamma(v)(t_{1A} - vx_{1A}) \\
x_{1B} &= \gamma(v)(x_{1A} - vt_{1A}) \\
y_{1B} &= y_{1A} \\
z_{1B} &= z_{1A}
\end{aligned} \tag{2.2}$$

This transformation is termed a *boost along the x-axis* and $\gamma(v) = 1/\sqrt{1 - v^2}$. The time- and the (*longitudinal*) x-coordinates get mixed by the transformation but the *transverse* coordinates, i.e. the y- and z-coordinates, are unaffected. Several boosts may be performed one after the other. It is easy to see that the final result does not depend upon the order and therefore the boosts along a single direction constitute a commutative (abelian) group.

More complex transformations also include rotations of the coordinate systems. Note that such rotations in general do not commute with each

other or with the boost transformations. This means that the outcome of
the total transformation depends upon the order in which each one of the
rotations and boosts is done.

II *The proper time*. The coordinate and time values are all differences
between the commonly agreed origin and the space-time point at
which event (1) occurs. They are all *relative coordinates*. A and B will
have different values for their measured t, x values for the event but
there is one combination which they will agree upon,

$$t_{1A}^2 - x_{1A}^2 = t_{1B}^2 - x_{1B}^2 \equiv \tau_1^2 \qquad (2.3)$$

The proper time of the event, τ_1, is evidently an *invariant* with respect
to all boosts along the x-axis. This means that it does not contain any
reference to the relative velocity of the observers along the x-axis.

The proper time is the value a watch would show if it started out from
the origin (i.e. at $t = 0, x = 0$) in A's system and moved away with velocity
$v_A = x_{1A}/t_{1A}$. Then it will arrive at x_{1A} at time t_{1A}, just when the event
(1) occurs. To see this imagine that observer B had chosen the velocity
$v = v_A$. It is therefore the time obtained in the *rest frame* of the watch.
This is the frame in which both events occur at the same place, the space
origin (make use of the second line in Eq. (2.2)!).

IIIA *Time dilation*. The observer A will conclude that the time difference
in his system that corresponds to the proper time τ_1 would be (make
use of the first line of Eq. (2.2)!)

$$t_{1A} = \frac{\tau_1}{\sqrt{1 - v_A^2}} \qquad (2.4)$$

This means that to A it will seem that the time difference is larger,
i.e. it will seem as if time is passing more slowly in the watch rest
system. This effect is called time dilation.

This is a noticeable effect for the fast-moving fragments of a collision
between cosmic ray elements and the atoms of the upper atmosphere.
There are e.g. the μ-particles, very short-lived when we produce them
basically at rest, in the laboratory on earth. The lifetime of a μ-particle is
around 2×10^{-6} seconds. Therefore even if it was moving with the velocity
of light it would only be able to cover about 600 metres!

Nevertheless the produced μ-particles survive a sufficiently long time to
be able to go all the way from the top of the atmosphere down to earth,
where we can find them in abundance.

To understand this effect we note that the decay time is related to the
properties of the particle in its rest frame while the 'survival time' we

observe is the time it will take a fast-moving particle (with velocity close to c) to move the distance δ from the top of the atmosphere (at a height of around 2×10^4 meters) to the observation point on earth. According to Eq. (2.4) this survival time is much longer and therefore many of the μ-particles survive to reach the ground.

IIIB *Lorentz contraction.* There is a corresponding effect for distances, which is called Lorentz contraction. For the surviving μ-particles, the distance δ, which to us is about 2×10^4 meters, will seem to be at most the 600 metres mentioned above. Considered from the rest system of the μ-particle the distance δ_{rest} is the length that the earth and its atmosphere moves towards it during its lifetime! From the Eq. (2.4) we conclude for the Lorentz contraction effect

$$\delta_{rest} = \delta \sqrt{1 - v^2} \tag{2.5}$$

IV *Covariance.* The scalar product of two ordinary vectors $\mathbf{a} \cdot \mathbf{b}$, written in terms of the coordinates as $a_x b_x + a_y b_y + a_z b_z$, is an invariant with respect to rotations. It is possible to write the invariant τ_1^2 as a (generalised) scalar product. The quantity

$$(1)(2) = t_1 t_2 - x_1 x_2 - y_1 y_2 - z_1 z_2 \tag{2.6}$$

will be invariant with respect to the general Lorentz transformations (i.e. boosts and rotations in any order) if the coordinates and times of the events (1) and (2) transform with respect to Lorentz boosts as in Eq. (2.2) (and (**1**) $\equiv (x_1, y_1, z_1)$ and similarly (**2**) transform as ordinary vectors under rotation).

Such quantities as (1) in Eq. (2.1) are called *four-vectors*. They transform as vectors with respect to the Lorentz transformations, in particular as in Eq. (2.2) for boosts along an axis. Besides the invariants, in the same way called *scalars* under the Lorentz transformations, and the four-vectors it is possible to define *four-tensors* (the electromagnetic field tensor is an example of such a quantity).

All these quantities are said to be *covariant*: they transform in a linear way with respect to the Lorentz transformations, i.e. the corresponding quantities in different Lorentz frames are related by means of linear equations.

V *The transformation of the velocity.* As an example of a quantity with more complex properties with respect to the Lorentz transformations we consider the velocity. We have already mentioned the velocity v_A measured in A's system. From B's point of view the corresponding velocity will be (use both the first and the second line of Eq. (2.2)!)

$$v_B = \frac{v_A - v}{1 - v_A v} \tag{2.7}$$

It is not difficult to show that if the velocities v_A, v do not exceed $c = 1$ then the velocity v_B will have the same property.

VI The *energy-momentum four-vector*. The classical (Newtonian) definition of momentum is the mass (m) times the velocity (v_p) of the
particle. But from Eq. (2.7) it is obvious that the transformation
properties of the velocity are complex under a Lorentz boost. In
order to generalise the definition of momentum Einstein made use
of the proper time of the particle motion in the following way.

The velocity of the particle is defined in terms of its trajectory $\mathbf{r}(t)$ (i.e.
its space position $\mathbf{r}$ labelled by means of the time t) as

$$\mathbf{v}_p = \frac{d\mathbf{r}}{dt} \tag{2.8}$$

For every (massive) particle it is possible to imagine a rest frame in which
the particle is always at the (space) origin. In this way it is possible to
define the proper time τ for the particle's motion; it is the time in this, the
particle's rest system.

Considered from any other Lorentz frame the proper time τ will be
related to the 'ordinary' time t by means of the differential equation

$$d\tau = dt\sqrt{1 - \mathbf{v}_p^2} \tag{2.9}$$

according to Eqs. (2.3), (2.4).

The proper time $\tau(t)$ defined in this way is unique as soon as proper
boundary conditions are given for the differential equation. (Its functional
dependence upon the time t will in general be different in different Lorentz
frames, however.)

We conclude that the corresponding *four-velocity* u defined by

$$u \equiv \left(\frac{dt}{d\tau}, \frac{d\mathbf{r}}{d\tau} \right) = \gamma(\mathbf{v}_p)(1, \mathbf{v}_p) \tag{2.10}$$

will transform covariantly as a vector under the Lorentz transformations.
(The third line of Eq. (2.10) is obtained from the differential equation
(2.9).) Note that the corresponding invariant $uu = u^2$ has the value $u^2 = 1$.
Einstein defined the *four-momentum* p of a particle as

$$p = (e, \mathbf{p}) = mu = m\gamma(\mathbf{v}_p)(1, \mathbf{v}_p) \tag{2.11}$$

The space components $\mathbf{p}$ (from now on the *momentum*) of this four-
momentum (which we sometimes will call the *energy-momentum vector*)

have the property that for small velocities $|\mathbf{v}_p| \equiv |v_p|$ (which should be interpreted to mean $|v_p| \ll c$, of course) they coincide with the classical momentum components.

The 'extra' component $e = m\gamma(\mathbf{v}_p)$ can be identified with the energy of the particle because for small velocities we obtain by expanding the square root

$$m\gamma(\mathbf{v}_p) \simeq m + m\mathbf{v}_p^2/2 \qquad (2.12)$$

The second term corresponds to the well-known expression for the kinetic energy of a (nonrelativistic) particle. The first term, the rest energy, corresponds to the famous Einstein conclusion that the mass content of a particle is related to a stored energy, e_s

$$e_s = mc^2 \qquad (2.13)$$

The ordinary vector velocity $\mathbf{v}_p$ can according to Eq. (2.11) be expressed as

$$\mathbf{v}_p = \frac{\mathbf{p}}{e} \qquad (2.14)$$

2.2   Particle kinematics

The invariance equation for the energy-momentum vector $p = (e, \mathbf{p})$, if we consider a particle moving along a fixed direction $\mathbf{p} = p\mathbf{n}$, described by the unit vector n is

$$e^2 - p^2 = m^2 \qquad (2.15)$$

This means that the energy (which always is positive for a particle) can be expressed as $e = \sqrt{p^2 + m^2}$.

VII *The rapidity variable.* According to Eq. (2.15) a particle with a fixed mass has a four-momentum which lies on a hyperbola in the ep-plane. It is possible to introduce a hyperbolic angle y_p to describe any particular point on the hyperbola:

$$\begin{aligned} e &= m \cosh y_p \\ p &= m \sinh y_p \end{aligned} \qquad (2.16)$$

This hyperbolic angle is called the rapidity, and we note from the relationship between (e, p) and the ordinary velocity v_p in Eq. (2.11) that

$$v_p = \tanh y_p \simeq y_p \qquad (2.17)$$

with the last line valid for small values of v_p and y_p. We also note that $\gamma(v_p) = \cosh y_p$.

For a Lorentz boost along the direction **n** we obtain, using the first two lines of Eq. (2.2), with a boost velocity $v = \tanh y$ and using the notation (e_B, p_B) for the energy-momentum components in the new frame,

$$e_B = \gamma(v)(e - vp)$$
$$= m(\cosh y_p \cosh y - \sinh y_p \sinh y) = m \cosh(y_p - y)$$
$$p_B = \gamma(v)(p - vE) \tag{2.18}$$
$$= m(\sinh y_p \cosh y - \cosh y_p \sinh y) = m \sinh(y_p - y)$$

This means that Lorentz boosts along **n** will move us along the hyperbola of Eq. (2.15). In particular any value of the energy-momentum can be obtained by a suitable boost from the rest system $y_p = 0$. In other words the *rapidity variable is additive*.

This also comes out of the relation for adding ordinary velocities, Eq. (2.7), if we express the velocities in terms of rapidities:

$$v_B \equiv \tanh y_B = \frac{v_A - v}{1 - v_A v} = \tanh(y_A - y) \tag{2.19}$$

If the rapidity is expressed in terms of the corresponding velocity v we obtain

$$y = \frac{1}{2} \ln \left(\frac{1+v}{1-v} \right) = \frac{1}{2} \ln \left(\frac{e+p}{e-p} \right) \tag{2.20}$$

It often occurs that in a given dynamical situation there may be a direction which is of particular importance. It is then useful to describe the particles under investigation in terms of their rapidities defined with respect to that direction (even if some or all of the particles move in somewhat different directions). This corresponds to using the velocity component, v_ℓ, along that (longitudinal) direction; we then obtain

$$y_\ell \equiv \frac{1}{2} \ln \left(\frac{1+v_\ell}{1-v_\ell} \right) = \frac{1}{2} \ln \left(\frac{e+p_\ell}{e-p_\ell} \right) \tag{2.21}$$

with p_ℓ the corresponding momentum component.

VIII *The lightcone components.* It is often useful to describe the energy-momentum vector with respect to the direction **n** in terms of the components

$$p_+ = e + p = m \exp y_p, \quad p_- = e - p = m \exp(-y_p) \tag{2.22}$$

For a boost with rapidity y along n these quantities transform as

$$p_+ \to p_+ \exp(-y), \quad p_- \to p_- \exp y \tag{2.23}$$

It is of course natural that their product is a constant, equal to the invariant in Eq. (2.15). For the case in Eq. (2.21) one defines the lightcone components $(e \pm p_\ell)$. They can then be described with respect to the rapidity y_ℓ in the same way as in Eq. (2.22) except that the mass m is exchanged for the *transverse mass* m_t. This quantity is defined by

$$m_t = \sqrt{m^2 + \mathbf{p}_t^2} \qquad (2.24)$$

in terms of the *transverse momentum vector* $\mathbf{p}_t$, corresponding to the two components of the momentum that are transverse to the chosen longitudinal direction.

We will at this point briefly consider Heisenberg's indeterminacy relations and indicate that although the position and the conjugate momentum of a particle cannot be determined simultaneously *it is possible to determine the rapidity and the position for a high-energy particle simultaneously with any degree of exactness*, [66].

The indeterminacy relations mean that owing to the commutation relation

$$[p, x] = -i \qquad (2.25)$$

it is necessary that the width of a wave-packet in position x, Δx, is related to the corresponding width in momentum p, Δp by

$$\Delta x \Delta p \geq 1/2 \qquad (2.26)$$

Merzbacher shows, by defining the mean and the width in the state with the wave function ψ as

$$\langle x \rangle = \int dx \psi^*(x) x \psi(x)$$
$$(\Delta x)^2 = \left\langle (x - \langle x \rangle)^2 \right\rangle = \int dx \psi^*(x)(x - \langle x \rangle)^2 \psi(x) \qquad (2.27)$$

with a similar relationship for p that there is a single kind of state, the Gaussian wave packet, for which Eq. (2.26) is an equality.

We can rewrite Eq. (2.26) in the following way for a particle with energy-momentum (e, p) with rapidity according to Eq. (2.16):

$$\Delta x \frac{\Delta p}{e} \equiv \Delta x \Delta y \geq \frac{1}{2e} \to 0 \qquad (2.28)$$

when e is very large. Note that Eq. (2.16) implies that $dp/e = dy$.

Relation (2.28) is shown for a free particle, in [66], by actual construction of the necessary wave-packets. It implies that, although you can never fool Heisenberg, you are allowed to choose your variables in such a way that quantum mechanical effects can be small or negligible.

As you will find in connection with the Lund model, when we are concerned with the longitudinal dynamics we shall use the freedom to present semi-classical pictures, in which we go between coordinate- and rapidity-space descriptions. This cannot be done in the same cavalier way in connection with the transverse dynamics, because transverse momenta are in general very limited in size in high-energy physics.

2.3   Timelike, lightlike and spacelike vectors in Minkowski space

Up to now we have neglected the fact that the invariant size of a four-vector, like the squared proper time in Eq. (2.3), is not positive definite as is the corresponding length of an ordinary vector. This means that it is possible to find space-time points for which the proper time squared is vanishing or negative.

In both these cases the interpretation of proper time discussed above is no longer valid. There is no (proper) Lorentz frame that is a rest frame for an observer, in which both the start (at the origin) and the event itself occur at the same point in space.

Those points for which the proper-time interpretation is valid are called *timelike* and we note that they fulfil

$$|t_{1A}| > |\mathbf{r}_{1A}| \equiv \sqrt{\mathbf{r}_{1A}^2} \qquad (2.29)$$

This is evidently a Lorentz-covariant definition.

All energy-momentum vectors for massive particles are also in the same way called timelike.

1   *Lightlike four-vectors*

In the case when the proper time squared vanishes it is possible to send a light signal directly from the origin to the event point and we therefore refer to this situation as a *lightlike* space-time vector difference.

There are other cases for which we will meet such lightlike vectors, e.g. when we want to describe massless particles such as the quanta of the electromagnetic field, *photons*. For them the energy (cf. Eq. (2.15)) is equal to the total momentum, i.e. $e = |\mathbf{k}| = |k|$. The corresponding rapidity y_ℓ as defined in Eq. (2.21) is directly expressible in terms of the angle, θ, between a given axis and the photon direction:

$$k_\ell = |k| \cos\theta$$
$$y_\ell = \frac{1}{2} \ln\left(\frac{1 + \cos\theta}{1 - \cos\theta}\right) = \ln\cot\left(\frac{\theta}{2}\right) \simeq -\ln\left(\frac{\theta}{2}\right) \qquad (2.30)$$

The last statement is an approximation valid for small angles.

Although Eq. (2.30) is strictly valid only for massless particles it is often a very good approximation (and then the variable is called the *pseudo-rapidity*) for other particles, those whose mass is small compared to their energy. In this way we obtain another intuitive way to look at the rapidity; *it is directly related to the angle with respect to the chosen longitudinal direction.*

While both the individual masses of two lightlike particles vanish, the sum of their energy-momenta is in general no longer lightlike but timelike:

$$k_j k_j \equiv k_j^2 = e_j^2 - (\mathbf{k}_j)^2 = 0$$

$$s_{12} \equiv (k_1 + k_2)^2 = 2k_1 k_2 = 2e_1 e_2 (1 - \cos\theta_{12}) = 4e_1 e_2 \sin^2\theta_{12}/2 > 0$$

$$(2.31)$$

unless the two lightlike vectors are parallel, which means that the angle between them $\theta_{12} = 0$.

It is always possible by means of a Lorentz boost to go to the *centre-of-mass system* (from now on the cms) of two lightlike or timelike vectors. This system is defined so that the total momentum vector vanishes. If the mass of the four-vector sum $\sqrt{s_{12}}$ from Eq. (2.31) is nonvanishing, the size of the velocity of the sum is less than c:

$$\mathbf{v}_{12} = \frac{\mathbf{k}_1 + \mathbf{k}_2}{|\mathbf{k}_1| + |\mathbf{k}_2|} \tag{2.32}$$

It is a useful exercise to prove to oneself that by a boost of $\mathbf{v}_{12}$ one reaches a Lorentz frame in which the two vectors in Eq. (2.31) have after the boost, the components

$$k'_{+1} = k'_{-2} = \sqrt{s_{12}}; \ k'_{-1} = k'_{+2} = 0; \ \mathbf{k}'_{t1} = \mathbf{k}'_{t2} = \mathbf{0} \tag{2.33}$$

Thus they have 'oppositely' directed lightcone components in the cms. Another way to formulate this is to note that *a timelike vector may be uniquely partitioned into two lightlike vectors* (oppositely directed in space in the restframe of the timelike vector).

2   Spacelike four-vectors

If the invariant length in Eq. (2.3) (generalised possibly by means of Eq. (2.6)) is negative then the four-vector is called *spacelike*. An example of a spacelike vector in space-time is the difference vector between two points in space measured at the same time.

Actually, it is always possible for a spacelike vector in space-time, to find a frame such that the time component vanishes. To see this let us assume that in the situation described above involving the two observers A and B event (1) has a spacelike difference vector with respect to the origin, e.g.

$$0 < t_{1A} < x_{1A} \quad \text{and} \quad y_{1A} = z_{1A} = 0 \tag{2.34}$$

(the sign choice of (t_{1A}, x_{1A}) being made for convenience). Then if the observer B moves at a velocity of size $v = t_{1A}/x_{1A}$ (although it appears to be a rather peculiar 'velocity' it is evidently smaller than $c = 1$) we obtain directly from (the first line of) Eq. (2.2) that event (1) will occur for B at the same time as he starts out from the origin.

For the observer B there is, however, a (space) distance between the origin and (1), that can be obtained from (the second line of) Eq. (2.2),

$$x_{1B} = \sqrt{x_{1A}^2 - t_{1A}^2} \qquad (2.35)$$

i.e. the invariant length, as expected.

When the difference vector between two space-time points is spacelike then it is impossible to send any kind of signal between them. Therefore, it is impossible for two physical events occurring at the two points to be *causally connected*. The occurrence of one of the events cannot affect the occurrence of the other. We will in the course of this book have many occasions to come back to such situations.

The typical spacelike vectors in energy-momentum space correspond to *momentum transfers*. If two particles with rest masses m_1 and m_2 are scattered elastically from each other then in general there is a momentum transfer between them. Elastic scattering means that the same kinds of particle occur in the initial state and in the final state.

The energy-momentum vectors in the initial state, p_{ji}, and in the final state, p_{jf}, of the particles indexed $j = 1, 2$ are, however, in general different. Energy-momentum conservation means that

$$\sum_{j=1}^{2} p_{ji} = \sum_{j=1}^{2} p_{jf} \qquad (2.36)$$

This implies that the difference vector, q, i.e. the momentum transfer between the two particles during the scattering, fulfils

$$q = p_{1f} - p_{1i} = -(p_{2f} - p_{2i}) \qquad (2.37)$$

If we analyse the situation in the cms, with the two particles approaching each other along the x-axis with $\mathbf{p}_{1i} = p\mathbf{n}_x = -\mathbf{p}_{2i}$ (see Fig. 2.1) we conclude that

I The absolute sizes of the momenta of the final-state particles are the same as for the initial-state particles. To see this we note that

1 The total momentum in the cms vanishes also in the final state. Therefore the two final-state particles must have oppositely directed momentum vectors of equal size also.

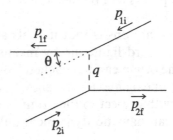

Fig. 2.1. Two particles experience elastic scattering against each other with notation described in the text.

2 Each of the particle energies is given by the momentum size, e.g. $e_{ji} = \sqrt{p^2 + m_j^2}$ and in order to conserve the total energy, cf. (2.36), the final-state momentum sizes therefore must be p, too.

II In the cms the momentum transfer four-vector, q, has no energy component, and we obtain for the invariant momentum transfer (conventionally called t or $-Q^2$)

$$-Q^2 = t \equiv q^2 = -4p^2 \sin^2(\theta/2) \simeq -p_t^2 \qquad (2.38)$$

in terms of the scattering angle θ (see Fig. 2.1) and in the small angle limit, $\sin(\theta/2) \simeq (\sin\theta)/2$, in the last line with the transverse momentum $p_t = p\sin(\theta)$.

3 Minkowski space

The vector space endowed with the *metric* defined by the Lorentz-invariant four-vector product in Eqs. (2.3), (2.6) is called Minkowski space. Although ordinary space-time contains three space dimensions, it frequently occurs that physical models are formulated in lower-dimensional regions, corresponding to one- or two-dimensional space. (It is, of course, sometimes useful to make use of larger dimensions both for time and space but we shall not need to do so in this book.)

Minkowski space can be subdivided into the three different parts, considered above, i.e. into timelike, lightlike and spacelike points with respect to the origin (or for that matter with respect to any other point).

The lightlike vectors form three-dimensional regions, called *lightcones*, in between the other two classes, which are both four-dimensional. It is possible to further classify a lightcone into a *positive (forward)* part and a *negative (backward)* part, according to the sign of the time component, i.e.

$$t = \pm\sqrt{\mathbf{r}^2} \qquad (2.39)$$

In the same way timelike points can be inside the *forward* or the *backward* lightcones.

The significance of these notions is that it is always possible to reach a point inside, or on, the forward lightcone by means of a signal from the origin. In a similar way the origin can be reached from all the points inside, or on, the backward lightcone by means of a signal. All the spacelike points are, however, *non-causal* with respect to the origin, i.e. as mentioned above, events in the two points can have no dynamical influence on each other.

2.4   The electromagnetic field equations and some of their consequences

We will start with the notion of gauge invariance and after that turn to the properties of dielectrics. The rationale for introducing dielectrics is the following. The vacuum in a quantum theory, which intuitively corresponds to the no-particle state, behaves owing to quantum fluctuations in a way effectively similar to a dielectric medium.

1   Gauge invariance

The two Maxwell equations corresponding to Faraday's induction law and the absence of magnetic charges connect the electric field $\mathscr{E}$ and the magnetic field $\mathscr{B}$ in the following ways:

$$\nabla \times \mathscr{E} + \frac{\partial \mathscr{B}}{\partial t} = 0, \quad \nabla \cdot \mathscr{B} = 0 \tag{2.40}$$

These equations can be solved by introducing the four-vector potential $A_\mu \equiv (A_0, \mathbf{A})$:

$$\mathscr{B} = \nabla \times \mathbf{A}, \quad \mathscr{E} = -\nabla A_0 - \frac{\partial \mathbf{A}}{\partial t} \tag{2.41}$$

It is well known that these relations do not completely determine A_μ from a knowledge of $\mathscr{E}, \mathscr{B}$. It is always possible to introduce the change

$$\mathbf{A} \rightarrow \mathbf{A} + \nabla \Lambda, \quad A_0 \rightarrow A_0 - \frac{\partial \Lambda}{\partial t} \tag{2.42}$$

and still obtain the same electric and magnetic fields.

The transformation in Eq. (2.42) is a *local gauge transformation*. The word *local* means that it is possible to choose the function Λ so that it varies from point to point in space and time.

In somewhat loose language this means that the vector field A_μ contains redundant, non-observable, degrees of freedom and that one must by convention fix these degrees of freedom in order to be able to discuss its quantum properties.

Such gauge-fixing conventions of a more or less 'physical' kind have been suggested and used but it is essential to understand that one convention is, from a dynamical point of view, just as good as another. *Any observable result of a calculation must be gauge-independent.*

One should always remember when considering the emission of the quanta of A_μ that, with a certain gauge-fixing condition, the quanta may seem to be emitted from some particular part of the emitting current. It may well be the case, however, that the same observable quanta would seem to be emitted from a completely different part of the current if one were to use a different gauge condition (or as a matter of fact the same gauge condition but a different Lorentz frame). We will discuss these matters in more detail when we come to matter fields in Chapter 11 and to gluon radiation in Chapter 16.

If we introduce the energy-momentum-space quantities (we use the notation $\tilde{A}(q)$ or $\mathscr{A}(q)$ for the Fourier transform of a space-time quantity $A(x)$, with q the Fourier transform variable) a gauge transformation is

$$\tilde{A}(q) \to \tilde{A}(q) + iq\tilde{\Lambda}(q) \tag{2.43}$$

This means that, for a radiation field, when the vector potential $A = \epsilon \exp(ikx)$ describes a photonic quantum with energy-momentum vector k ($k^2 = 0$ for real photons) and polarisation vector ϵ, the physics results should be independent of the change

$$\epsilon \to \epsilon + ik\tilde{\Lambda}(k) \tag{2.44}$$

for any Λ.

In order to understand the relation in Eq. (2.44) we consider a boost along the direction of motion of the quantum, i.e. along the direction of **k**. In the new frame the size of the momentum $|\mathbf{k}|$ and therefore also the energy are changed. For the polarisation vector ϵ this change can be compensated by a gauge transformation according to Eq. (2.44). Therefore in a charge-free region only the polarisation-vector components transverse to the direction of motion (that are invariant with respect to such boosts, i.e. those with $k\epsilon = 0$) are physically important (cf. the (brief) discussion of *helicity* in Chapter 5).

2 *The notion of dielectrics*

Besides the two equations mentioned above there are in Maxwell's treatment also Coulomb's and Ampère's laws, which tell us how to construct the fields from a knowledge of the charges and currents. They are expected to be precise in the microscopic sense (we use small letters to denote the microscopic fields and large letters for the corresponding macroscopic ones):

$$\nabla \cdot \mathbf{e} = \eta, \quad \nabla \times \mathbf{b} - \frac{\partial \mathbf{e}}{\partial t} = \iota \tag{2.45}$$

Here η and ι are the 'local' charge and vector current densities, stemming from e.g. individual atomic charges. A quantum field does not really make sense as an operator acting as a single point (although with suitable care it is often possible to write quantum field operators in that way) because it is *distribution valued*. It should be smoothed out over a region by means of a 'test-function' f, [31]:

$$\mathbf{e}(f) \equiv \mathcal{E}(f) = \int dx f(x) \mathbf{e}(x) \tag{2.46}$$

We have here assumed that the test function f is nonvanishing (mathematically 'has support in') a region of suitable size around the point x. The typical atomic dimension is of the order of 10^{-8} cm (about twice the Bohr radius for hydrogen), and depending upon the system under consideration we may need this or other length units when we consider this averaging procedure. Jackson gives a lucid description, to which we refer the interested reader.

The result of the averaging procedure is, however, that not only 'the true' charges will affect the fields; there are also induced dipole moments, $\mathcal{P}$ and $\mathcal{M}$, stemming from the polarisation and magnetisation of the medium. The effective values of charge and current vector densities are thus changed; it is necessary to take into account also the polarisation charge, the polarisation current and the magnetic moment current. We then arrive at the macroscopic equations containing the free charge (ρ) and current ($\mathbf{j}$) densities (the difference from the rapidly changing local η and ι densities in Eq. (2.45), which describe individual atomic charges in motion, is that these microscopic fluctuations are averaged out, giving relatively smooth and slowly varying macroscopic quantities):

$$\nabla \cdot \mathbf{D} = \rho, \quad \nabla \times \mathbf{H} - \frac{\partial \mathbf{D}}{\partial t} = \mathbf{j}$$
$$\mathbf{D} = \mathcal{E} + \mathcal{P} = \epsilon \star \mathcal{E}, \quad \mathbf{H} = \mathcal{B} - \mathcal{M} = \frac{1}{\mu} \star \mathcal{B} \tag{2.47}$$

Here $\mathbf{D}$ is the *electric displacement vector* and $\mathbf{H}$ is the *magnetic field*; ϵ and μ are the *dielectricity* and the *magnetic permeability*, of the material under investigation. The symbol $\star$ is used in order to indicate the possibility that, e.g.

$$\mathbf{D}(x) \equiv \epsilon \star \mathcal{E} = \int dx' \epsilon(x - x') \mathcal{E}(x') \tag{2.48}$$

This would correspond to an energy-momentum-dependent displacement

$$\mathscr{D}(q) = [1 + \tilde{\xi}(q)]\mathscr{E}(q) \tag{2.49}$$

where we have introduced the *index of refraction* $\tilde{\xi} = \tilde{\varepsilon} - 1$.

If we consider plane-wave solutions to the electromagnetic equations, (2.47), in a (true) charge- and current-free medium we may write (with the convention in classical physics that we are supposed to take the real part of all complex quantities)

$$\mathscr{E} = \mathscr{E}_0 \exp i(k\mathbf{n} \cdot \mathbf{x} - \omega t), \quad \mathscr{B} = \mathscr{B}_0 \exp i(k\mathbf{n} \cdot \mathbf{x} - \omega t) \tag{2.50}$$

We then obtain the following requirements:

$$k^2\mathbf{n}^2 - \tilde{\mu}\tilde{\varepsilon}\omega^2 = 0, \quad \mathbf{n} \cdot \mathscr{E}_0 = \mathbf{n} \cdot \mathscr{B}_0 = 0, \quad \mathscr{B}_0 = \sqrt{\tilde{\mu}\tilde{\varepsilon}}\mathbf{n} \times \mathscr{E}_0 \tag{2.51}$$

At this point we may consider a few limiting situations. Suppose firstly that $\tilde{\varepsilon}$ is a constant and (for simplicity) $\tilde{\mu} = 1$. This means that **D** and **H** = **B** are completely local fields. We may in particular consider the vector **n** to be a unit vector. Then we will according to the last two equations of (2.51) have transverse waves in the medium. According to the first equation in (2.51) there is also a relation, usually referred to as a *dispersion relation*, between the wavenumber $k \simeq 1/\lambda$, with λ the wavelength, and the frequency ω.

To see what this relation implies we note that the transport velocity of the field energy-momentum is given by the ratio of the (space-time averaged) Poynting vector **S** ($|\mathbf{S}| \equiv S$) and the (space-time averaged) energy density u:

$$S = \frac{1}{2}|\mathscr{E} \times \mathscr{H}| = \frac{\sqrt{\tilde{\varepsilon}}}{2}|\mathscr{E}|^2, \quad u = \frac{1}{4}(\tilde{\varepsilon}\mathscr{E} \cdot \mathscr{E}^* + \mathscr{B} \cdot \mathscr{B}^*) = \frac{\tilde{\varepsilon}}{2}|\mathscr{E}|^2 \tag{2.52}$$

The factor $\frac{1}{2}$ results from averaging the squared harmonic waves and we find in this way that the velocity has changed from $c = 1$ to $v = 1/\sqrt{\tilde{\varepsilon}}$. Thus we require $\tilde{\varepsilon} > 1$ in order that the transport velocity of the energy should not exceed the velocity of light in the vacuum. We note that the *phase velocity* of the waves, which is ω/k, then coincides with v.

Another case of interest is an electron plasma in the limit $\omega \gg \omega_p$, where ω_p is the plasma frequency. Then (cf. Jackson) $\tilde{\varepsilon} = 1 - (\omega_p/\omega)^2$ and we obtain the same relation between k and ω as for a particle with mass ω_p (this is the only true Higgs-phenomenon we know of at present, i.e. the velocity of the electromagnetic waves in a medium is smaller than the vacuum velocity; this is tantamount to give a mass, corresponding to the plasma frequency, to the field quanta):

$$\omega^2 = k^2 + \omega_p^2 \tag{2.53}$$

In this case the phase velocity of the waves, ω/k, is greater than the

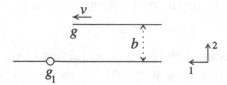

Fig. 2.2.   A charged particle, g approaches a charged observer at the origin with velocity v along a direction with impact parameter b.

velocity of light. The true velocity, called the *group velocity*, is then instead the variation of ω with respect to k, $d\omega/dk = k/\omega < 1$, as we find by the well-known construction of local wave-packets from the waves in Eq. (2.50), cf. Jackson and Merzbacher. Consequently the index of refraction in Eq. (2.49) may be both positive and negative in real life situations.

We finally note that the index of refraction, $\tilde{\xi}$, may have an imaginary part. This corresponds to an absorption of the waves, i.e. to an interaction between the medium and the waves. There is a general set of relations, the Kramers-Kronig relations, [89], [88], between the real and the imaginary parts of the index of refraction. They stem from the causality requirement that there can be no effect until the waves have reached the medium. This leads to analyticity properties for $\tilde{\xi}$. We will meet the same properties in connection with the vacuum polarisation functions in quantum field theory in Chapter 4.

2.5   The method of virtual quanta

In this section we consider the electromagnetic field of a fast-moving charge and show how to express it in terms of its field quanta. The problem will be phrased as follows:

- Describe the field of an electric charge (size g), moving with velocity v along a direction (the 1-direction) having impact parameter b (for definiteness in the 12-plane) with respect to an observation point at the origin $x_1 = x_2 = 0$.

We assume that there is an observer, i.e. a detector carrying charge g_1, at the origin (Fig. 2.2). We expect that the approach of g will be noticeable as a pulse of radiation energy for this charged observer. This pulse will now be described in a semi-classical framework.

The Lorentz rest frames of the charges g and g_1 will be assumed to coincide at time $t_1 = t = 0$. Then we may calculate the Coulomb force field in the rest system of the charge g (where it is the usual spherically symmetric field falling off with distance R as $\propto 1/R^2$).

After that we may use the rules of special relativity and translate this field by a boost (with velocity $-v$) along the 1-axis to obtain the field components in the rest system of the charge g_1 (Jackson does it for us so we will not dwell upon the details):

$$\mathscr{E}_1 = -\frac{g\,v\,t\,\gamma}{r^3}, \quad \mathscr{E}_2 = -\frac{g\,b\,\gamma}{r^3}, \quad \mathscr{B}_3 = v\,\mathscr{E}_2 \tag{2.54}$$

(with r defined below in Eq. (2.55)). Note that the components in the 2- and 3-directions basically constitute a 'radiation field', i.e. $\mathscr{B} = \mathbf{v} \times \mathscr{E}$, when $v \sim c = 1$. We are now going to investigate that field.

The γ-factor is as usual $1/\sqrt{1-v^2}$ and the space extensions of the field components are Lorentz-contracted. Therefore, apart from the times $t \sim 0$, when the charges are close to each other, the distance r is a large number:

$$r = \sqrt{b^2 + (vt\gamma)^2} \tag{2.55}$$

The field components in Eq. (2.54) provide two Poynting-vector pulses, one along the 1-axis and one along the 2-axis. The latter is small and we will neglect it from now on. The main 1-axis radiation pulse is strongly Lorentz-contracted and looks like a bell-shaped curve in the time variable with a width (noticeable from Eq. (2.55)) around $t = 0$ of δt, where

$$\delta t = \frac{b}{v\gamma} \tag{2.56}$$

Note that this *typical passage time*, δt, can be written as

$$\delta t = \frac{mb}{p} \tag{2.57}$$

where m is the rest mass and $p \simeq e$ (for large $v \simeq c = 1$) are the momentum and energy of the charge g.

We can describe these results in terms of frequency (Jackson provides the exact formulas but we do not need the details). The differential intensity of the 1-axis pulse, $dI(\omega)$, where ω is the frequency will be essentially constant from a low-frequency value ω_{min} (where the wavelength becomes so long that there is nothing to observe) up to a maximum (determined by Eqs. (2.56), (2.57)):

$$\omega_{max} \simeq \frac{1}{\delta t} = \frac{p}{mb}. \tag{2.58}$$

This follows from the properties of the Fourier transform and also comes out of Jackson's formulas in terms of combinations of Bessel functions. We obtain approximately (note that the Poynting vector corresponds to the surface density of the field momentum)

$$dI(\omega, b) \simeq d\omega dA_t \frac{\hbar \alpha}{\pi^2 b^2} \tag{2.59}$$

For values of $\omega > 1/\delta t$ the distribution contains an exponential tail, with fast falloff. Here $dA_t = 2\pi b db$, i.e. the increase in the transverse area per unit impact parameter b. We have also defined the fine structure constant $\alpha = g^2/(4\pi\hbar c)$ under the assumption that g is a unit, i.e. electron, charge. We have been careful to keep Planck's constant in the expression (although we usually put $\hbar = 1$ according to the conventions in the Introduction) because up to now there has been no reference to quantum mechanics.

We may, however, now make the time-honoured transition to quantum mechanics by noting that for a fixed frequency ω the number of quanta, dn (in this case photons) in the pulse dI is given by

$$dI = \hbar \omega dn \tag{2.60}$$

This means that the whole field energy is carried by individual field quanta, each with an energy proportional to its frequency according to Einstein's proposal.

Therefore we have found an (approximate) expression for the number of field quanta which will be available for an interaction with the charge g_1 at the origin:

$$dn = \left(\frac{\alpha}{\pi}\right) \left(\frac{dA_t}{\pi b^2}\right) \frac{d\omega}{\omega} \tag{2.61}$$

This is basically a classical formula (but with quantum mechanics sneaked in through Eq. (2.60)). It describes the *flux factor* in connection with the interaction of the charged-particle field quanta. If the scattering cross section for the individual quanta is known then we simply multiply by this flux in order to get the cross section for the whole charged field.

Before the flux factor can be used we note, however, that it is singular in two different ways. The first way corresponds to the singularity for large wavelengths, $\omega \to 0$, to which reference already has been made. (The Lund model is everywhere infrared stable and we will therefore not consider the problems corresponding to infrared singularities. The main point is that when the number of quanta increases indefinitely at small frequencies then the dynamical behaviour is not given by their number but instead by their 'combined action', which corresponds to the action of a classical field.)

The second singularity is the logarithmic divergence for small values of b. This is a typical problem in all situations involving a charged particle. *It is necessary to define what is meant by the energy of the particle itself and what should be attributed to the field. This is called mass renormalisation,* i.e. it is necessary to provide the particle with a given rest energy equal to its mass, independently of the field surrounding it.

Classically the field energy from a point particle is always infinite and therefore after the discovery of the electron it was described not as a 'point' but as a small charged sphere with a radius $r_0 > 0$ such that its (Coulomb) field energy was exactly equal to the mass, m_e:

$$\frac{e^2}{4\pi r_0} = m_e \qquad (2.62)$$

This quantity r_0, the classical electron radius, is approximately 3×10^{-15} m (using the conventions of $c = \hbar = 1$ to convert to metres) and occurs in the cross section for the interaction between an electron and low-frequency radiation, $\omega \to 0$:

$$\frac{d\sigma}{d\Omega} = \frac{r_0^2}{2}(1 + \cos^2\theta) \qquad (2.63)$$

This is the Thompson cross section in the solid angle $d\Omega = \sin\theta\, d\theta d\phi$, where θ is the scattering angle and ϕ the azimuthal angle around the beam direction. It should, however, be understood that as far as we know (and this is at least down to 10^{-17} m because of the results of the LEP experiments at CERN) there is no extended space structure of the electron. The Thompson cross section therefore corresponds to the size of the Coulomb field around the particle rather than to some 'solid-sphere' behaviour.

The necessary cutoff in impact parameter depends upon the problem one is considering. It is either the Compton wavelength of the particle that is used or the characteristic size of the quantity that is probed by the field (but it is always the largest of the parameters). The Compton wavelength is $\lambda_C = \hbar/m$ and this b-cutoff therefore means that ω_{max} as defined in Eq. (2.58) will be given by

$$\omega < \omega_{max} = \frac{p}{m\lambda_C} = p \simeq e \qquad (2.64)$$

This is not an unreasonable requirement. After all you cannot radiate away more energy than you have got!

The above representation is not normally used in connection with quantum field theory, where one usually describes the field not in terms of the energy and the impact parameter of the field quanta but instead in terms of their energy and transverse momentum.

The impact parameter vector **b** is, as we will see later in Chapter 10, the canonically conjugate variable to the transverse momentum $\mathbf{k}_t$ in a high-energy scattering event. Therefore one obtains the distribution of one from the other by means of a Fourier transform of the transition amplitude.

We note that the formulas above contain (as always for observables in quantum mechanics) the square of the amplitude (in this case $|\mathscr{E}|^2$) but from the scaling behaviour (no dimensional constants) we may guess that the relation between the distribution in impact parameter and the transverse momentum will be

$$\frac{2\pi b\,db}{\pi b^2} = \frac{db^2}{b^2} \rightarrow \frac{dk_t^2}{k_t^2} \tag{2.65}$$

and this turns out to be the right answer.

It is also conventional to rearrange the ω-dependence into a dependence upon the scaled variable $x = \omega/e$, e being the moving charged particle's energy. In that way we may write

$$dn = \left(\frac{\alpha}{\pi}\right)\left(\frac{dk_t^2}{k_t^2}\right)\frac{dx}{x} \tag{2.66}$$

which we will later meet as the spectrum for dipole bremsstrahlung radiation. The scaled variable x evidently has a range $x < 1$ according to Eq. (2.64).

Thus the method of virtual quanta (MVQ) redefines the interaction ability of a charged particle in terms of a flux of available (but virtual) field quanta, with precise properties with respect to interactions. Note that the word 'virtual' is appropriate: the field quanta are available but do not do anything until they find something to interact with.

3

The harmonic oscillator
and the quantum field

3.1 Introduction

In this and the next chapter we will consider some properties of quantum fields. The examples taken will be mostly scalar fields and only when necessary will we invoke the complexities stemming from the vector nature of the interactions in QED and QCD; there are many good text-books devoted to a detailed treatment of the subject.

We need only intuition and a set of understood formulas for the investigations contained in this book. We start with a discussion of the quantum mechanical harmonic oscillator coupled to an external force. There are several reasons to dwell on this particular system. Firstly its sine and cosine behaviour in time is matched by the corresponding harmonic behaviour of the plane wave solutions for the quanta in a field theory.

It was noted even in the first papers on quantum field theory that a free or weakly interacting quantum field is in a rather precise way a superposition of an infinite, although enumerable, set of harmonic oscillators, one for each degree of freedom.

A real interacting-field theory does not behave in this way with respect to its excitations. There is always, however, at the basis of any experiment in high-energy particle physics the idea of a three-act scenario in time.

1 In the first phase, a long time before the interaction, the initial states are prepared with production setups in general arranged so that each state is isolated.

2 After that there will be a more or less violent encounter in the second phase.

3 In the final state the produced quanta are observed by means of detectors placed far apart, a long time after the interaction.

27

Therefore the descriptions of the initial and final states are expected
to correspond to the states of free non-interacting fields. For a confining
theory like QCD this particular asymptotic before-and-after scenario does
not hold but there is instead another asymptotics, the *asymptotic freedom*
of the theory in which the free-field theories are expected to be relevant.

The second reason for considering both the free and the interacting
harmonic oscillator is that from a mathematical point of view they correspond to very well-behaved systems. This is not the case in general for
interacting quantum fields, which contain many different mathematical
complications. But it turns out that almost all the things which can be
done in a simple and precise way for the single harmonic oscillator can
also, albeit after a large amount of cumbersome mathematics, be done
for infinite-dimensional quantum fields. It is therefore easier to present
the methods in a well-behaved manner for those who are not particularly
interested in the mathematical complexities but nevertheless would like to
understand what they are doing inside a computable framework.

After we have rehearsed the properties of interacting harmonic oscillators from an elementary quantum mechanical point of view we will exhibit
the corresponding properties for a scalar quantum field. We will in particular consider quantum states which correspond as closely as possible to
classical fields (coherent states). At the same time we will introduce the *S*-
operator, which connects the initial- and final-state free fields, mentioned
above as phases 1 and 3 in the interaction.

After that we consider interacting fields. It is then necessary to provide
a more precise definition of the *S*-operator. We introduce the Feynman-
Dyson prescription of time-ordering and, for simple cases, show how to
make calculations in this framework. We consider the Feynman propagator and show its significance with regard to Heisenberg's indeterminacy
requirements. We also calculate the scattering cross section in a simple
situation. Finally we exhibit some features of the lightcone formulation of
a field theory, often referred to as 'a field theory in the infinite-momentum
frame'.

3.2   The quantum field as a sum of harmonic oscillators

This section will firstly contain a few reminders of the properties of
the one-dimensional harmonic oscillator. After we have shown how the
harmonic oscillator reacts to a time-dependent external force we discuss
the corresponding properties of a scalar quantum field coupled to an
external current.

In both cases we obtain a set of states called *coherent states*. They are
the closest correspondence to classical behaviour which can be found for

simple quantum systems. Therefore they are often used as models for more complex situations. When we go from the single harmonic oscillator to quantum fields it will be necessary to introduce some cutoff procedures, which are used repeatedly in connection with the calculation of observables such as cross sections later in the book.

1 The one-dimensional harmonic oscillator

I *The equation of motion.* The (classical) equation of motion of a one-dimensional harmonic oscillator in an external field, $j(t)$ is

$$m\ddot{x} + m\omega^2 x = j(t) \tag{3.1}$$

Here the dot(s) correspond to time derivative(s) and the harmonic oscillator frequency ω has been explicitly introduced.

Equation (3.1) can be derived from Hamilton's equations:

$$\dot{x} = \frac{\partial H}{\partial p}, \quad \dot{p} = -\frac{\partial H}{\partial x} \tag{3.2}$$

where

$$H = \frac{p^2}{2m} + \frac{m\omega^2 x^2}{2} - xj \tag{3.3}$$

II *The commutation relations.* Quantum considerations are introduced by means of the Heisenberg commutator relations

$$[p, x] = -i, \quad [p, p] = [x, x] = 0 \tag{3.4}$$

For the harmonic oscillator it is useful to introduce two adjoint operators a^* and a, usually referred to as the *raising* and *lowering* (or in more colorful language *creation* and *annihilation*) operators:

$$x = \frac{a + a^*}{\sqrt{2m\omega}}, \quad p = \frac{i\sqrt{m\omega}(a^* - a)}{\sqrt{2}} \tag{3.5}$$

Their commutation relations are obtained from Eq. (3.4):

$$[a, a^*] = 1, \quad [a, a] = [a^*, a^*] = 0 \tag{3.6}$$

III *The case of no disturbance.* For the case when $j = 0$ the hamiltonian $H \equiv H_0$ can be written as

$$H_0 = \omega\left(a^*a + \frac{1}{2}\right), \quad \langle\alpha|H_0|\alpha\rangle \geq \frac{\omega}{2} \equiv E_0 \tag{3.7}$$

for any state $|\alpha\rangle$. There is a lowest energy eigenstate $|0\rangle$ with an x-space representation, $\psi_0(x)$, obtained from the requirement $a\psi_0\ x) = 0$, i.e.

$$\left(\sqrt{\frac{m\omega}{2}}x + \frac{1}{\sqrt{2m\omega}}\frac{d}{dx}\right)\psi_0(x) = 0, \quad \psi_0 = \sqrt{\frac{m\omega}{2\pi}}\exp\left(-\frac{m\omega x^2}{2}\right) \tag{3.8}$$

with ψ_0 normalised to 1: $\int dx |\psi_0(x)|^2 = 1$. It obviously fulfils $H|0\rangle = E_0|0\rangle$.

IV *The excited states.* All other eigenstates of the hamiltonian are given by $N_n(a^*)^n |0\rangle \equiv |n\rangle$; in an x-space representation these are polynomials of nth degree in x multiplying ψ_0. Using

$$[a, (a^*)^n] = n(a^*)^{(n-1)}, \quad [a^*a, (a^*)^n] = n(a^*)^n \qquad (3.9)$$

the normalisation constant N_n can by iteration be shown to be

$$N_n = \frac{1}{\sqrt{n!}} \quad \text{so that} \quad |n\rangle = \frac{(a^*)^n}{\sqrt{n!}} |0\rangle \qquad (3.10)$$

The corresponding eigenvalue is $E_n = (n + 1/2)\omega$.

V *Normal-ordering.* It is useful to introduce the notion of *normal-ordering.* This means that in an operator expression O containing both a and a^* operators the normal-ordered O, denoted $:O:$, contains all the a-operators to the right of the a^*-operators. In particular this means that $\langle 0|:O:|0\rangle = 0$ if O contains a nonzero number of operators.

VI *The time dependence.* The time dependence of the operators a and a^* is found, in the Heisenberg picture (for $j = 0$), as follows:

$$\begin{aligned} \frac{da}{dt} &= i[H_0, a] = -i\omega a \;\Rightarrow\; a(t) = a\exp(-i\omega t) \\ \frac{da^*}{dt} &= i[H_0, a^*] = i\omega a^* \;\Rightarrow\; a^*(t) = a^*\exp(i\omega t) \end{aligned} \qquad (3.11)$$

We also note that the identification of the canonical momentum p with $m\dot{x}$ is consistent with the time development:

$$p = mi[H_0, x] \qquad (3.12)$$

VII *Time-independent disturbance.* When j is nonvanishing but *independent of time* the hamiltonian can be rewritten as

$$H = \frac{p^2}{2m} + \frac{m\omega^2(x - x_0)^2}{2} - \frac{m\omega^2 x_0^2}{2}, \quad x_0 = \frac{j}{m\omega^2} \qquad (3.13)$$

We can then choose to re-express everything using a new coordinate $x' = x - x_0$ and a new hamiltonian $H' = H + j^2/(2m\omega^2)$ (where we have introduced the expression for x_0 in the energy change).

The new ground-state wave function, ψ_{0j}, obviously corresponds to a translation of the old one:

$$\psi_{0j}(x) = \psi_0(x - x_0) = \left\{\exp[-j^2/(2m\omega^3)]\right\} \{\exp(xj/\omega)\} \psi_0(x) \quad (3.14)$$

and can therefore by a suitable expansion be expressed in terms of the the old set $|n\rangle$ (this applies, of course, to any other state as well).

The exponential of an operator should be interpreted in terms of a power series expansion and can be handled in almost the same way as an ordinary exponential.

- We will use two simple properties of the operator $T = \exp(jx/\omega)$ expressed in terms of the original operators a and a^*. In general if A and B are operators and if $[A, B] = c$, c being an operator-independent constant (conventionally called a *c-number*),

$$(\exp A)B \exp(-A) = B + [A, B]$$
$$(\exp A)(\exp B)\exp(-[A, B]/2) = \exp(A + B)$$

(3.15)

The first relation in Eq. (3.15) can be obtained from a Taylor series expansion of the function $f(\lambda)$ around $\lambda = 0$, where

$$f(\lambda) = \exp(\lambda A)B \exp(-\lambda A)$$

(3.16)

Consider the derivatives of f (note the careful ordering!)

$$\frac{df}{d\lambda} = Af(\lambda) - f(\lambda)A = [A, f(\lambda)], \quad \frac{d^n f}{d\lambda^n} = [A, [A, \cdots, [A, f(\lambda)] \cdots]]$$

(3.17)

As $f(\lambda = 0) = B$ we obtain that all but the first of the derivatives of f vanish at $\lambda = 0$:

$$f(\lambda) = B + \lambda[A, B]$$

(3.18)

The result in the first line in Eq. (3.15) then corresponds to $\lambda = 1$.

For the second equation in (3.15) consider the function $g(\lambda)$, where

$$g(\lambda) = \exp(\lambda A)\exp(\lambda B)\exp\{-\lambda(A + B)\}$$

(3.19)

Using the first equation in (3.15), we obtain for the derivative of g:

$$\frac{dg}{d\lambda} = \lambda[A, B]g(\lambda)$$

(3.20)

This is a differential equation with a plain number λc in front of g on the right-hand side. We conclude that g, which is equal to 1 for $\lambda = 0$ from its definition, is the following simple function:

$$g(\lambda) = \exp(c\lambda^2/2)$$

(3.21)

which again provides the expected result for $\lambda = 1$. Note that we have extensively used that the commutator of A and B is a plain number. Setting $A = ja^*/(\sqrt{2m\omega^3})$ and $B = ja/(\sqrt{2m\omega^3})$ we obtain

$$[A, B] = -\frac{j^2}{2m\omega^3}$$

(3.22)

so that already expressed in a normal-ordered form the operator T becomes

$$T = \exp\left[j^2/(4m\omega^3)\right]\exp(ja^*/\sqrt{2m\omega^3})\exp(ja/\sqrt{2m\omega^3})$$

(3.23)

From Eq. (3.14) this means that

$$\psi_{0j} = \exp\left[-j^2/(4m\omega^3)\right] \exp(ja^*/\sqrt{2m\omega^3})\psi_0 \tag{3.24}$$

or

$$|0j\rangle = \exp\left[-j^2/(4m\omega^3)\right] \sum_{n=0}^{\infty} \left(\frac{j}{\sqrt{2m\omega^3}}\right)^n \frac{1}{\sqrt{n!}} |n\rangle \tag{3.25}$$

Therefore the application of a constant force j to the harmonic oscillator will bring it into a new ground state with the property that the transition amplitudes will fulfil

$$|\langle n|0j\rangle|^2 = \frac{\bar{n}^n}{n!} \exp(-\bar{n}) \tag{3.26}$$

This corresponds to a Poisson distribution with the mean excitation $\bar{n}$ given by

$$\bar{n} = \frac{j^2}{2m\omega^3} \tag{3.27}$$

This is, however, dynamically incorrect: there is no way to change the system unless we use a time-dependent scenario so that there is energy pumped in or out of the system.

VIII *A time-dependent scenario.* In order to describe an actual dynamical situation we assume that the force j introduced above is nonvanishing and changes in time, t, during a finite period $t_1 < t < t_2$ so that we can talk about the situation 'before', $t \leq t_1$, and 'after', $t \geq t_2$ (the 'three-way-scenario' mentioned before!). Then the hamiltonian will be

$$\begin{aligned} H &= \omega(a^*a + 1/2) - g(t)a - g^*(t)a^* \equiv H_0 + H_1 \\ H_1 &= -j(t)x = -g(t)a - g^*(t)a^* \end{aligned} \tag{3.28}$$

where we have written $j \to g(t) = g^* = j(t)/\sqrt{2m\omega}$ in anticipation of a more general situation, when g is a complex function.

The equations of motion become

$$\frac{da}{dt} = i[H,a] = -i\omega a + ig^*(t), \quad \frac{da^*}{dt} = i[H,a^*] = i\omega a^* - ig(t) \tag{3.29}$$

We will assume that there are *initial-state* operators $a_i(t)$, $a_i^*(t)$, which, like the operators in Eq. (3.11), describe the undisturbed system before $t = t_1$ (when $g(t) = g^*(t) = 0$ so that the equations of motion coincide) and likewise *final-state* operators $a_f(t)$, $a_f^*(t)$, which describe the system after $t = t_2$.

Then the equations (3.29) can be solved in a general way by means of the Green's function method. We define the functions $G_R(t)$ and $G_A(t)$ as the solutions of the equation

$$\frac{dG}{dt} + i\omega G = \delta(t) \qquad (3.30)$$

with boundary conditions

$$
\begin{aligned}
G_R(t) &= 0 \quad \text{if} \quad t < 0 \\
G_A(t) &= 0 \quad \text{if} \quad t > 0
\end{aligned}
\qquad (3.31)
$$

They are called the retarded and the advanced *Green's function*, respectively, and are in this case rather easily constructed:

$$G_R(t) = \Theta(t)\exp(-i\omega t), \quad G_A(t) = -\Theta(-t)\exp(-i\omega t) \qquad (3.32)$$

where Θ is the Heaviside distribution, which is equal to 1 for a positive argument and vanishes elsewhere.

The fact that the solutions of Eq. (3.30) should correspond to step-functions at $t = 0$ can be understood from an integration of the equation from $t = -\epsilon$ to $t = +\epsilon$ when $\epsilon \to +0$:

$$\lim_{\epsilon \to 0}\left[G(\epsilon) - G(-\epsilon) + i\omega \int_{-\epsilon}^{\epsilon} dt\, G(t)\right] = 1 \qquad (3.33)$$

Here we have used the following property of the δ-distribution: $\int dt\, \delta(t) = 1$, if the integration region includes $t = 0$. The fact that the contribution from the integral in Eq. (3.33) vanishes as ϵ is left for the reader to prove.

In this way we obtain the following solutions for $a(t)$:

$$
\begin{aligned}
a(t) &= a_i(t) + i \int_{-\infty}^{t} dt'\, g^*(t')\exp\left[-i\omega(t - t')\right] \\
a(t) &= a_f(t) - i \int_{t}^{\infty} dt'\, g^*(t')\exp\left[-i\omega(t - t')\right]
\end{aligned}
\qquad (3.34)
$$

Therefore the final-state operators can be expressed in terms of the initial-state ones by a translation (noting that they all have the trivial time dependence $\exp(\pm i\omega t)$, which can be divided away):

$$a_f = a_i + i \int_{-\infty}^{\infty} dt'\, g^*(t')\exp(i\omega t') \equiv a_i + i\tilde{g}^*(\omega) \qquad (3.35)$$

Consequently the final-state operators, a_f, a_f^*, in a similar way to VII above have been translated with respect to the initial ones, a_i, a_i^*, this time, however, by the Fourier transform of the force!

IX *The S-operator.* It is possible to construct a *unitary operator S*, which transforms the initial states into the final states in this simple situation:

$$
\begin{aligned}
S^*S &= 1 \Leftrightarrow S^{-1} = S^*, \quad a_f = S^* a_i S = a_i + i\tilde{g}^*(\omega) \\
S^* |0i\rangle &= |0f\rangle \quad \Leftrightarrow \quad S |0f\rangle = |0i\rangle
\end{aligned}
\qquad (3.36)
$$

(note that this also fixes the relation between a_f^* and a_i^* !). The operator S provides a complete mapping of the eigenstates of the final system onto the initial eigenstates:

$$|nf\rangle = S^* |ni\rangle \tag{3.37}$$

It is easy to find by means of the results we have obtained in VII and VIII:

$$S = \exp\{i[\tilde{g}^*(\omega)a_i^* + \tilde{g}(\omega)a_i]\} \tag{3.38}$$

The expression in the exponent can be neatly reformulated by noting that

$$
\begin{aligned}
\int_{-\infty}^{\infty} dt\, j(t)x(t) &= \int_{-\infty}^{\infty} dt[g(t)a_i(t) + g^*(t)a_i^*(t)] \\
&= \int_{-\infty}^{\infty} dt[a_i g(t)\exp(-i\omega t) + a_i^* g^*(t)\exp(i\omega t)] \\
&= \tilde{g}(\omega)a_i + \tilde{g}^*(\omega)a_i^* \tag{3.39}
\end{aligned}
$$

Then the S-operator can be expressed as

$$S = \exp\left\{i \int_{-\infty}^{\infty} dt[g(t)a_i(t) + g^*(t)a_i^*(t)]\right\} = \exp i \int_{-\infty}^{\infty} dt\, j(t)x(t) \tag{3.40}$$

This is a general result in the perturbative treatments of quantum field theory, which holds also when j is an operator-valued function. We obtain the (negative) difference between the operator H in Eq. (3.28) and the 'free' harmonic oscillator hamiltonian H_0 in Eq. (3.2), integrated over time, as the exponent in the expression for the S-operator.

In this more general case the exponential must be treated with care because operators for different times have complicated commutation relations. One cannot without a prescription for *ordering* use the ordinary exponential property that the exponent of a sum is equal to the product of the exponents of the terms in the sum.

X *The transition probabilities.* For the case when j is an external 'nice' function 'real' transitions are possible. An original state such as the initial ground state, $|0i\rangle$, will afterwards become some outgoing, possibly excited, state:

$$\langle nf|0i\rangle = \langle ni|S|0i\rangle = \left[\exp\left(-\frac{|\tilde{g}|^2}{2}\right)\right]\frac{(\tilde{g}^*)^n}{\sqrt{n!}} \tag{3.41}$$

In VII we presented the transition probabilities $|\langle nf|0i\rangle|^2$ as a Poisson distribution in the free harmonic oscillator states. This is evidently still true and the mean excitation level, $\bar{n}$, for the Poissonian will be for the general case:

$$\bar{n} = \frac{1}{2m\omega}\left|\int dt\, j(t)\exp(iwt)\right|^2 \tag{3.42}$$

The result in the case (3.27) is characteristic for a single sudden change in the force. A suitable force (corresponding to a limiting situation when $\epsilon > 0$ approaches 0 after the integral has been performed) would be

$$j(t) = j \exp(-\epsilon t)\Theta(t). \tag{3.43}$$

Before we go over to quantum fields we note another property of the states. The state $|0i\rangle$ is actually an *eigenstate of the operator* a_f:

$$a_f |0i\rangle = a_f S |0f\rangle = a_f \sum_{n=1}^{\infty} \left[\exp\left(-\frac{|\tilde{g}|^2}{2} \right) \right] \frac{(\tilde{g}^*)^n}{\sqrt{n!}} |nf\rangle$$
$$= \tilde{g}^* S |0f\rangle = \tilde{g}^* |0i\rangle \tag{3.44}$$

This also implies that the expectation value in the initial ground state of the final-state operator $x_f(t) = [a_f \exp(-i\omega t) + a_f^* \exp(i\omega t)]/(\sqrt{2m\omega})$ is

$$\hat{x}(t) = \langle 0i|x_f(t)|0i\rangle = \frac{\tilde{g}^* \exp(-i\omega t) + \tilde{g} \exp(i\omega t)}{\sqrt{2m\omega}}$$
$$= \int_{-\infty}^{\infty} dt' \frac{1}{m\omega} j(t') \cos[\omega(t' - t)] \tag{3.45}$$

This is the final-state harmonic motion in a classical mechanics situation when one starts out with a harmonic oscillator at rest and then applies the external force $j(t)$ over a finite time interval $t_1 < t' < t_2$. Evidently the integrand in Eq. (3.45) is only nonvanishing over this time region and we consider $t > t_2$.

In order to prove (3.45) it should be noted that the equations of motion in Eqs. (3.2) and (3.29) also work classically for the quantities a, a^* defined in Eqs. (3.5). The whole formalism involving Green's functions that relate the initial-state and final-state quantities a_i, a_i^* and a_f, a_f^* is just as valid when the a's and a^*'s are classical c-numbers!

2   A scalar quantum field coupled to an external current

We will now consider the corresponding situation for a scalar quantum field $\phi(x)$. We will firstly show that it has the same behaviour as a superposition of an infinite number of independent harmonic oscillators. It will then follow that we can take over everything we have done in I to X when we treat $\phi(x)$. Every time one introduces an infinity, however, it is necessary to worry a little about convergence problems. We will soon find that there are plenty of such things to worry about when we go to interacting quantum fields!

XI *The Klein–Gordon equation.* A scalar field, $\phi(\mathbf{x}, t)$, which fulfils the Klein–Gordon equation

$$(\Box + M^2)\phi \equiv \ddot{\phi} - \Delta\phi + M^2\phi = j \tag{3.46}$$

where stated earlier the Laplacian $\Delta = \nabla^2$ is given by

$$\Delta = \frac{\partial^2}{\partial x_1^2} + \frac{\partial^2}{\partial x_2^2} + \frac{\partial^2}{\partial x_3^2} \tag{3.47}$$

will, in momentum space, $\phi \to \tilde{\phi}(t)\exp(i\mathbf{x}\cdot\mathbf{k})$, fulfil the equation

$$\ddot{\tilde{\phi}} + \omega^2(\mathbf{k})\tilde{\phi} = \tilde{j} \tag{3.48}$$

This essentially coincides with Eq. (3.1) for the single harmonic oscillator with frequency $\omega \to \omega(\mathbf{k}) = \sqrt{\mathbf{k}^2 + M^2}$.

In order to facilitate this transfer to momentum space we assume that the whole system is enclosed in a large box with three space dimensions and volume V, and that only those waves that fit into the box with periodic boundary conditions are included. This means that instead of a field ϕ defined at every space point we obtain an *enumerable set of amplitude fields for the momentum-space waves*.

The allowed momenta, e.g. in the 1-direction with a large box-length L_1, are, for any integer n_1,

$$k_{1,n_1} = \frac{n_1 2\pi}{L_1} \tag{3.49}$$

A sum over n_1 can be made into an integral over dk_1 by the formal exchange (which is valid when we sum and integrate over 'nice' functions)

$$\sum_{n_1} \to \int dn_1 = \frac{L_1}{2\pi}\int dk_1 \Rightarrow \sum_{n_1,n_2,n_3} \to \frac{V}{(2\pi)^3}\int d^3k \tag{3.50}$$

With this construction we have the following identities

$$\int_V d^3x \exp\left[i(\mathbf{k}-\mathbf{k}')\cdot\mathbf{x}\right] = V\delta_{\mathbf{k},\mathbf{k}'}$$
$$\sum_{\mathbf{k}} \exp(i\mathbf{k}\cdot\mathbf{x}) = V\delta(\mathbf{x}) \tag{3.51}$$

In the first equation the symbol on the right-hand side is equal to 1 when the two arguments coincide and vanishes elsewhere. The second equation contains the usual δ-distribution in three dimensions.

The results in Eq. (3.51) stem directly from Fourier analysis and correspond to the orthonormality and completeness relations of Fourier waves. We will later see that in all formulas describing physical observables the volume V will disappear.

XII The *hamiltonian formulation*. The field equation can also be described by a variation of the hamiltonian H in which $\phi(x), \Pi(x)$ are the canonical coordinates at every space point $\mathbf{x}$:

$$H_0 = \tfrac{1}{2} \int_V d^3x [\Pi^2 + (\nabla\phi)^2 + M^2\phi^2] \tag{3.52}$$

$$H = H_0 + H_1$$

with

$$H_1 = -\int_V d^3x\, j(\mathbf{x}, t)\phi(\mathbf{x}) \equiv \int_V d^3x\, \mathscr{H}_1 \tag{3.53}$$

The fields Π and ϕ can be decomposed as sums over the different momentum components similar to the single harmonic oscillator in Eq. (3.5):

$$\phi = \sum_{\mathbf{k}} \frac{1}{\sqrt{2V\omega(\mathbf{k})}} (a(\mathbf{k})\exp[i\mathbf{k}\cdot\mathbf{x}] + a^*(\mathbf{k})\exp(-i\mathbf{k}\cdot\mathbf{x}))$$

$$\tag{3.54}$$

$$\Pi = \sum_{\mathbf{k}} \frac{i\sqrt{\omega(\mathbf{k})}}{\sqrt{2V}} [-a(\mathbf{k})\exp(i\mathbf{k}\cdot\mathbf{x}) + a^*(\mathbf{k})\exp(-i\mathbf{k}\cdot\mathbf{x})]$$

We note that the field ϕ in this way is written as a set of harmonic oscillators (cf. Eq. (3.5)) $\mathbf{x} = \sum_j (1/\sqrt{2m\omega})(a_j + a_j^*)\mathbf{e}_j$, although this time the (euclidean) vectors $\mathbf{e}_j$ (with $\mathbf{e}_j\mathbf{e}_m = \delta_{km}$) are exchanged for the normalised eigenfunctions $\exp(\pm i\mathbf{k}\cdot\mathbf{x})/\sqrt{V}$, which are vectors in a Hilbert space, i.e. an infinite-dimensional generalisation of a euclidean space. This also implies that the field ϕ has energy dimension $\dim\phi = 1$ (corresponding to a negative length dimension -1). We will use similar dimensional arguments many times later in the book.

This dimensional assignment for ϕ is necessary in order that the hamiltonian H_0 in Eq. (3.52) should also have energy dimension 1 ($\dim d^3x = -3$, $\dim M^2 = 2$ and $\dim\nabla = 1$). In the same way we conclude that for H_1 to have energy dimension 1 the current j must have $\dim j = 3$.

It is straightforward to prove that the commutation relations

$$[a(\mathbf{k}), a^*(\mathbf{k}')] = \delta_{\mathbf{k},\mathbf{k}'}, \quad [a(\mathbf{k}), a(\mathbf{k}')] = [a^*(\mathbf{k}), a^*(\mathbf{k}')] = 0 \tag{3.55}$$

imply

$$[\Pi(\mathbf{x}), \phi(\mathbf{x}')] = -i\delta(\mathbf{x} - \mathbf{x}'), \quad [\phi(\mathbf{x}), \phi(\mathbf{x}')] = [\Pi(\mathbf{x}), \Pi(\mathbf{x}')] = 0 \tag{3.56}$$

if we use Eqs. (3.55), (3.50) and (3.51). The sets of commutation relations in Eqs. (3.55) and (3.56) are thus equivalent and are obvious generalisations of the harmonic oscillator relations in Eqs. (3.6) and (3.4).

XIII *The ground-state energy.* For an undisturbed set of harmonic oscillators the hamiltonian in terms of operators is

$$H_0 = \sum_{\mathbf{k}} \omega(\mathbf{k})[a^*(\mathbf{k})a(\mathbf{k})] + C \tag{3.57}$$

The constant C corresponds to the sum of the energies of all the zero-point modes of the oscillators, i.e. $C = \sum_{\mathbf{k}} \omega(\mathbf{k})/2$. In that way it is simply the energy of the vacuum and is consequently not an observable quantity.

There are, however, situations when *the difference in energy between the 'total' vacuum fluctuations in C and those from a particular boundary configuration can be measured*, [41]. This effect is outside the scope of this book. It is, nevertheless, of great interest because it exhibits *experimentally* the existence of quantum field fluctuations in the vacuum state.

XIV *The time dependence.* To obtain the time dependence we use the same relations as in Eqs. (3.11) and (3.12):

$$\frac{da_{\mathbf{k}}}{dt} = i[H_0, a_{\mathbf{k}}] = -i\omega(\mathbf{k})a_{\mathbf{k}} \quad \Rightarrow \quad a_{\mathbf{k}}(t) = a_{\mathbf{k}} \exp[-i\omega(\mathbf{k})t]$$

$$\Pi(\mathbf{x}) = \dot{\phi} = i[H_0, \phi(\mathbf{x})]$$

(3.58)

In this way $\phi(\mathbf{x}) \to \phi(\mathbf{x}, t)$ by including the time dependence of the a- and a^*-operators. We note in passing that this will result in Lorentz-invariant exponential factors $\exp \pm (i\mathbf{k} \cdot \mathbf{x} - \omega t) = \exp \mp (ik_\mu x^\mu) \equiv \exp(\mp ikx)$ multiplying the a- and a^*-operators.

When the current j is nonvanishing the time dependences will take on the form of Eqs. (3.29):

$$\frac{da_{\mathbf{k}}}{dt} = i[H, a_{\mathbf{k}}] = -i\omega(\mathbf{k})a_{\mathbf{k}} + ig^*(\mathbf{k}, t)$$

$$\frac{da_{\mathbf{k}}^*}{dt} = i[H, a_{\mathbf{k}}^*] = i\omega(\mathbf{k})a_{\mathbf{k}}^* - ig(\mathbf{k}, t)$$

(3.59)

$$g(\mathbf{k}, t) = \int_V d^3x \frac{1}{\sqrt{2V\omega(\mathbf{k})}} j(\mathbf{x}, t) \exp(i\mathbf{k} \cdot \mathbf{x})$$

(3.60)

Thus here $g(t) \to g(\mathbf{k}, t)$, the Fourier transform of the external current. This means that the numbers $g(\mathbf{k}, t)$ are in general complex but for a real-valued current $j(x)$ they fulfil $g^*(\mathbf{k}, t) = g(-\mathbf{k}, t)$.

All these steps from the definition of the Green's functions to the resulting equation for the S-operator in Eqs. (3.30) to (3.40) can then be performed separately for each wavenumber $\mathbf{k}$. The final S-operator is a product over all components and can be written as

$$S = \exp\left[-i \int_{-\infty}^{\infty} dt H_{1i}(t)\right] = \exp\left[i \int d^4x \phi_i(x) j(x)\right]$$

(3.61)

The index i is introduced in order to stress that we are using the initial-state fields, i.e. those that describe the state a long time before the interaction is turned on. The time dependence in $H_{1i}(t)$ contains also the free-field time dependence of the oscillators so that $a_i(\mathbf{k})$ is changed into $a_i(\mathbf{k}) \exp(-i\omega(\mathbf{k})t)$. The integration symbol $\int d^4x = \int_V d^3x \int_{-\infty}^{\infty} dt$.

An interesting observable is the probability that the vacuum before the interaction is turned on (the no-quanta state) is still the vacuum after the interaction, i.e. the probability that there has been no excitation due to the onset of the current j

$$| \langle 0f|0i \rangle |^2 = | \langle 0i|S|0i \rangle |^2 = \exp(-U)$$

$$U = \sum_{\mathbf{k}} \frac{1}{2\omega(\mathbf{k})V} \left| \int_V d^3x dt j(\mathbf{x}, t) \exp[i\omega(\mathbf{k})t - \mathbf{k}x] \right|^2 \tag{3.62}$$

The quantity U is the sum over all the mean excitations for the Poisson-distributed oscillators (cf. Eq. (3.42)). It can be rearranged by changing the sum over $\mathbf{k}$ to an integral, see Eq. (3.50); we then arrive at (with the vector $\delta x = (t - t', \mathbf{x} - \mathbf{x}')$)

$$U = \int \frac{d^3k}{2(2\pi)^3\omega(\mathbf{k})} \int d^4x d^4x' j(x)j(x') \exp\{i[\omega(\mathbf{k})(\delta t) - \mathbf{k}(\delta \mathbf{x})]\}$$

$$= \int dx dx' j(x)\Delta_+(\delta x)j(x') \tag{3.63}$$

$$\Delta_\pm(x) = \frac{1}{(2\pi)^3} \int \frac{d^3k}{2\omega(\mathbf{k})} \exp\{i[\pm\omega(\mathbf{k})t - \mathbf{k} \cdot \mathbf{x}]\} \tag{3.64}$$

We firstly note that the volume V has vanished from these expressions (when we have taken the limit $V \to \infty$ we use the symbol dx instead of d^4x). Secondly we note that the functions $\Delta_\pm$ defined in the last line of Eq. (3.63) are Lorentz-invariant. In order to show that we use the following property of the δ-distribution:

$$\int dadb\Theta(\pm a)\delta(a^2 - b^2)f(a, b)$$

$$= \int dadb\Theta(\pm a)\frac{[\delta(a - |b|) + \delta(a + |b|)]f(a, b)}{2|b|}$$

$$= \int \frac{dbf(\pm|b|, b)}{2|b|} \tag{3.65}$$

For Eq. (3.64) we have

$$\int \frac{d^3k}{2\omega(\mathbf{k})} f(\mathbf{k}, \pm\omega(\mathbf{k})) = \int dk\delta^\pm(k^2 - M^2)f(\mathbf{k}, k_0) \tag{3.66}$$

where the symbols $dk \equiv d^3k dk_0$ and $\Theta(\pm k_0)\delta(k_0^2 - \mathbf{k}^2 - M^2) \equiv \delta^\pm(k^2 - M^2)$ will be used from now on. (Note that the prescription $k_0 > 0$ is Lorentz-invariant together with the δ-distribution!)

Thus the functions $\Delta_\pm$ become (changing $\mathbf{k}$ to $-\mathbf{k}$ for Δ_-):

$$\Delta_\pm(x) = \frac{1}{(2\pi)^3} \int dk\delta^\pm(k^2 - M^2) \exp(ikx) \tag{3.67}$$

The distribution $\Delta_+(x)$ actually corresponds to the matrix element

$$\langle 0i|\phi_i(x_1)\phi_i(x_2)|0i\rangle = \langle 0i|\phi_i^-(x_1)\phi_i^+(x_2)|0i\rangle$$
$$= \sum_{\mathbf{k}} \frac{\exp[ik(x_2 - x_1)]}{2V\omega(k)} = \Delta_+(x_2 - x_1) \quad (3.68)$$

We have here introduced the notation $\phi = \phi^- + \phi^+$ where we include the sum of all the a-operators (a^*-operators) in ϕ^- (ϕ^+). The second line stems from the fact that the only (nonvanishing) intermediate state is a single quantum, which can be created by ϕ_i^+ and annihilated by ϕ_i^-. For the third and fourth lines we have used Eqs. (3.63) and (3.64).

We also note that the (in-)vacuum expectation value of the field $\phi_f(x)$ is

$$\langle 0i|\phi_f(x)|0i\rangle = \int dx \left[\Delta^+(x - x') + \Delta^-(x - x')\right] j(x') \quad (3.69)$$

which in the same way as for Eq. (3.45) is the classical solution to the field equation in Eq. (3.46) after the interaction.

In conclusion we have shown the following:

- quantum fields, including that of the single harmonic oscillator, which are coupled to an external current contain excitations of a Poissonian nature, the mean number of quanta being determined from the Fourier components of the current;

- they also have vacuum expectation values that coincide with the classical c-number solutions for the interaction;

- the phases of the states, called *coherent states*, are well defined by the Fourier components of the external current.

3.3   Feynman's time-ordering prescription

In this section we will generalise the expression we have derived for the S-operator in Eqs. (3.40) and (3.61) from the simple case when the current j is an external c-number function to the general case when j is operator-valued. This will lead us to ways to calculate high-energy multiparticle production amplitudes in perturbation theory.

It is necessary to provide an ordering prescription for the S-operator in Eq. (3.61) when the current j is operator-valued. The right prescription (first introduced by Feynman and Dyson) is that *all expressions should be time-ordered*. If we would like to express the S-operator solely in the initial-state fields then

$$S = \mathscr{T}\left\{\exp\left(i \int d^4x \mathscr{H}_{1i}\right)\right\} \equiv 1 + \sum_{n=1}^{\infty} \frac{(-i)^n}{n!} \int \mathscr{T}\left\{\prod_{j=1}^{n} dt_j H_{1i}(t_j)\right\} \quad (3.70)$$

with the *time-ordering* symbol $\mathscr{T}$ implying that all operators should be written so that those with a later time are to the *left* of those with an earlier time.

Intuitively the prescription is rather easy to understand. The free initial quantum fields get distorted as time goes by. Each new distortion evidently follows the earlier ones and must therefore be applied after one has applied the previous interactions. (If we would like for some reason to write everything in terms of the final-state fields then we must *anti-time-order* everything, i.e. all operators should be arranged so that those with a later time are to the *right* of the others.)

As an example of the time-ordering procedure consider the second-order term in Eq. (3.70):

$$\mathscr{T}\left\{H_{1i}(t_1)H_{1i}(t_2)\right\} = \Theta(t_1 - t_2)H_{1i}(t_1)H_{1i}(t_2) + \Theta(t_2 - t_1)H_{1i}(t_2)H_{1i}(t_1)$$
$$(3.71)$$

We have now defined two different ordering prescriptions, *normal-ordering* where all annihilation operators a are to the right of all creation operators a^* and *time-ordering* where all earlier-time operators are to the right of all the later-time operators. There is a mathematical manipulation theorem, Wick's theorem, which provides a connection between these orderings; you will find it described in great detail in many text-books.

1 Time-ordered products and the Feynman propagator, causality and locality

In order to understand some features of quantum fields we will show how Wick's theorem works in connection with the time-ordered product of a free field ϕ at two different space-time points. Again using the notation $\phi^{\pm}$ from Eq. (3.68) we obtain

$$\mathscr{T}\left\{\phi_i(x_1)\phi_i(x_2)\right\} = \phi_i^+(x_1)\phi_i^+(x_2) + \phi_i^-(x_1)\phi_i^-(x_2)$$
$$+\Theta(t_1 - t_2)\left\{\left([\phi_i^-(x_1), \phi_i^+(x_2)]\right.\right.$$
$$\left.+\phi_i^+(x_2)\phi_i^-(x_1) + \phi_i^+(x_1)\phi_i^-(x_2)\right\}$$
$$+\Theta(t_2 - t_1)\left\{\left([\phi_i^-(x_2), \phi_i^+(x_1)]\right.\right.$$
$$\left.+\phi_i^+(x_1)\phi_i^-(x_2) + \phi_i^+(x_2)\phi_i^-(x_1)\right\} \quad (3.72)$$

We have thus item by item brought the time-ordered operators into normalordering. The result is evidently

$$
\begin{aligned}
\mathscr{T}\{\phi_i(x_1)\phi_i(x_2)\} &= :\phi_i(x_1)\phi_i(x_2): +\Theta(t_1 - t_2)\left[\phi_i^-(x_1), \phi_i^+(x_2)\right] \\
&\quad +\Theta(t_2 - t_1)\left[\phi_i^-(x_2), \phi_i^+(x_1)\right] \\
&\equiv :\phi_i(x_1)\phi_i(x_2): +\Delta_F(x_2 - x_1, M)
\end{aligned}
\tag{3.73}
$$

The function Δ_F (F stands for Feynman) could have been constructed directly from the fact that the normal-ordered product $:\phi_i(x_1)\phi_i(x_2):$ has a vanishing vacuum expectation value. We then obtain

$$
\Delta_F(x_2 - x_1, M) = \langle 0i|\mathscr{T}\{\phi_i(x_1)\phi_i(x_2)\}|0i\rangle
\tag{3.74}
$$

Using the result from Eq. (3.68) in Eq. (3.74) we may write the following expression for Δ_F :

$$
\Delta_F(x_2 - x_1) = \Theta(x_1 - x_2)\Delta_+(x_2 - x_1) + \Theta(x_2 - x_1)\Delta_+(x_1 - x_2)
\tag{3.75}
$$

(Note the order of the arguments in the Δ_+ distributions. For each this is related to the time dependence of the creation and annihilation operators.)

Before we construct an expression for Δ_F we note from the result in Eq. (3.68) the following result for the general commutator:

$$
[\phi_i(x_1), \phi_i(x_2)] = \Delta_+(x_2 - x_1) - \Delta_-(x_2 - x_1) \equiv -i\Delta(x_2 - x_1)
\tag{3.76}
$$

The notation is conventional and the factor i introduced to make Δ real.

The general commutator Δ, just like the $\Delta_\pm$-distributions, can be computed by straightforward means. We will give Δ in detail because it has two properties of direct interest for what follows:

$$
\Delta(x) = -\frac{\epsilon(x)}{2\pi}\left[\delta(x^2) - \frac{M}{2\sqrt{x^2}}J_1(M\sqrt{x^2})\Theta(x^2)\right].
\tag{3.77}
$$

We have used the conventional sign-distribution $\epsilon(x) \equiv \epsilon(x_0) = \Theta(x_0) - \Theta(-x_0)$ and the Bessel function of the first rank J_1 in Eq. (3.77).

Firstly note that the *commutator distribution* Δ *vanishes for spacelike vectors x.* This is our first encounter with practical *causality.* There is no possible signal connecting two space-time points with a spacelike difference. Therefore two *local* field operators taken at two such points commute. *They are independent and a measurement of the observable correponding to one of the operators at one point cannot influence a measurement of the observable corresponding to the other operator at another point separated from the first by a spacelike difference.*

The word *local* is essential, however. All the field operators are singular from a strict function-definition point of view (note the occurrence of the δ- and ϵ-distributions in Eq. (3.77)). Mathematically such expressions should be defined by means of a *test function* f, [31]:

$$\phi(f) = \int dx \phi(x) f(x) \tag{3.78}$$

A local operator is such that if we choose the test function f to be strongly localised around a point x (i.e. vanishing outside a suitably small region around x) *then also all the matrix elements of the operator $\phi(f)$ should have this property.*

If we consider the definition of Δ_F from Eq. (3.73) we find that this function can also be defined by means of commutators. But these are commutators of field operators which are not local. None of the $\phi^\pm$ is local because they contain only positive or negative frequencies, respectively. There is no way to localise anything in time by means of a function containing only frequencies of a definite sign.

The distribution Δ_F can instead, according to the result in Eq. (3.75), be written e.g. as

$$\Delta_F(x) = -i\Theta(-x)\Delta(x) + \Delta_+(x) \tag{3.79}$$

and only the first term on the right-hand side is local in the sense used above.

Secondly we note from Eq. (3.77) that the commutator is highly singular along the lightcones. Although the quanta have mass M and therefore always move with a velocity below $c = 1$ the corresponding quantum fields can influence each other in principle at infinite distances along the lightcones. It is also worthwhile to note that the *principal singularity* (the second term inside the large parentheses of Eq. (3.77) approaches a constant for $x^2 \to 0$) *is independent of the mass-value M.*

2 *The formula for the Feynman propagator, the lightcone singularities*

We will next provide a formula for Δ_F using a distribution-valued integral we have referred to in Eq. (3.43):

$$\Theta(x) \equiv \Theta(x_0) = \lim_{\epsilon \to 0} \int \frac{-i dk_0'}{2\pi(k_0' - i\epsilon)} \exp(ik_0' x_0) \tag{3.80}$$

From Eq. (3.75) we may then use the result in Eq. (3.80) to obtain an integral representation for Δ_F. We will subsequently not write out the limit sign but we will keep ϵ as a small but arbitrary number.

$$\Delta_F(x) = \frac{i}{(2\pi)^4} \int \frac{d^3 k}{2\omega(\mathbf{k})} \left(\frac{dk_0'}{k_0' + i\epsilon} \exp(i(\omega + k_0')x_0 - \mathbf{k} \cdot \mathbf{x}) \right.$$
$$\left. - \frac{dk_0'}{k_0' - i\epsilon} \exp[i(-\omega + k_0')x_0 + \mathbf{k} \cdot \mathbf{x}] \right)$$

$$= \frac{i}{(2\pi)^4} \int \frac{d^3k}{2\omega} dk_0 \exp(ikx) \left(\frac{1}{k_0 - \omega + i\epsilon} - \frac{1}{k_0 + \omega - i\epsilon} \right)$$

$$= \frac{i}{(2\pi)^4} \int \frac{dk \exp(ikx)}{k^2 - M^2 + i\epsilon} \tag{3.81}$$

Here we have introduced the result of Eq. (3.80) together with the corresponding result for $\Theta(-x)$ and then changed the integration variable k_0' to $k_0 = k_0' \pm \omega$ (as well as replacing $\mathbf{k}$ by $-\mathbf{k}$ in the second term). In the last line we have gathered the two denominators into one.

The final result corresponds to the limiting situation when the number ϵ approaches 0. This means that Δ_F is actually singular for all values of the vector $\mathbf{k}$ which correspond to a 'real' particle with mass M.

When we want to consider a physical observable that is sensitive to the limit then it is necessary to be more precise in the definition of the size of ϵ. An example of this is provided in Chapter 14.

From a mathematical point of view Δ_F is a *distribution*, which must be defined by means of integration over suitable test functions, as mentioned above. It is also the Fourier transform of the boundary value $\epsilon \to 0$ of an *analytic function* defined on complex-valued vectors k with $\mathrm{Im}\, k^2 > 0$. In that case it can be described as analytic and Lorentz-invariant with poles whenever $k^2 = M^2$.

In Chapter 6 we will provide a formula for the behaviour of the Feynman propagator for spacelike arguments. That formula will be based upon the property that Δ_F satisfies the Klein-Gordon equation

$$(\Box + M^2)\Delta_F(x, M^2) = 0 \tag{3.82}$$

everywhere outside the origin, $x = 0$.

For the investigations in Chapter 19 it is also of interest to know the space-time singularities of both the Feynman propagator Δ_F and the function Δ_+. We will not give the formulas for the general case but only for the case when the mass $M = 0$ because *just as for the function Δ in Eq. (3.77) the main singularities of all the functions are independent of the mass.*

The following formal development may be used in such a derivation. We firstly note that

$$\frac{i}{k^2 + i\epsilon} = \int_0^\infty d\alpha \exp(i\alpha k^2) \tag{3.83}$$

(the integral on the right-hand side converges when we add a small positive imaginary part to k^2). If we introduce this result into the formula for the Feynman propagator given in Eq. (3.81) we obtain gaussian integrals (which due to the imaginary parts are called Fresnel integrals):

$$\Delta_F(x, M^2 = 0) = \frac{1}{(2\pi)^4} \int d\alpha \int dk \exp(i\alpha k^2 + ikx)$$

$$= \frac{i}{4(2\pi)^2} \int \frac{d\alpha}{\alpha^2} \exp\left(\frac{-ix^2}{4\alpha}\right) = \frac{1}{(2\pi)^2(x^2 - i\epsilon)} \quad (3.84)$$

In the second line we have made the change of integration variable $1/\alpha \to \alpha$; performing the integral shows that x^2 must contain a small *negative* imaginary part, which ensures convergence.

If we perform the integrals for the function $\Delta_+(x)$ with the mass $M = 0$ (which is straightforward) we obtain the same result as in Eq. (3.84) but with the boundary value $x^2 \to x^2 + i\epsilon x_0$. This means that the imaginary part depends upon the sign of the time-component of the vector x.

At this point we will consider a particular distribution-valued boundary value. Suppose that we have a (test)function, $f(x)$, of a single real variable x and that we consider the result of integrating it together with the boundary value $1/(x - i\epsilon)$. We may then start by using the following formal manipulation:

$$\frac{1}{x - i\epsilon} = \frac{x + i\epsilon}{x^2 + \epsilon^2} = R + iI \quad (3.85)$$

If we start with the imaginary part then we obtain the result for I:

$$\int dx f(x) \frac{\epsilon}{x^2 + \epsilon^2} = \int dy f(\epsilon y) \frac{1}{y^2 + 1} \quad \to \quad f(0)\pi \equiv \int dx f(x) \pi \delta(x) \quad (3.86)$$

We have assumed that the function f vanishes sufficiently fast that we may take the limit $f(y\epsilon) \to f(0)$ outside the integral; then as is well known, $\int dy/(y^2 + 1) = \pi$.

We have in this way obtained a representation of the δ-distribution which is very useful. It is the difference between the boundary values:

$$\frac{1}{x - i\epsilon} - \frac{1}{x + i\epsilon} = 2i\pi\delta(x) \quad (3.87)$$

For the real part, R, in Eq. (3.85) we may use the trick of adding and subtracting the quantity

$$\int_{-\alpha}^{\alpha} f(0) \frac{x \, dx}{x^2 + \epsilon^2} = 0 \quad (3.88)$$

This result is obviously valid for any (finite) positive number α because the integrand is an odd function. For values outside $-\alpha < x < \alpha$ we now have no problem in taking the limit $\epsilon \to 0$ for R in Eq. (3.88) for a well-behaved function f (we again use the Heaviside function Θ):

$$R(f) = \int \left\{ \Theta(x^2 - \alpha^2)f(x) + \Theta(\alpha^2 - x^2)[f(x) - f(0)] \right\} \frac{dx}{x} \quad (3.89)$$

If afterwards we let $\alpha \to 0$ we find that we always have a well-defined integral, called the *principal part of f* and defined so that in the neighborhood of the singular point $x = 0$ we make the change $f(x) \to f(x) - f(0)$.

As a simple example for this limiting situation consider the relationship between the commutator distribution Δ and Δ_+. If we take the indicated difference in Eq. (3.76) we obtain just the lightcone δ-distribution in Eq. (3.77) from the result in Eq. (3.87) and the limiting behaviour of Δ_+ we mentioned above.

We have in this section stressed the following facts:

- a local quantum field must contain both positive and negative frequencies;

- the S-operator must be defined by means of time-ordering.

These are the origins of the Feynman propagator distribution.

It is, of course, possible to interpret the two parts of the time-ordering process in Eq. (3.75) as respectively 'forwards' and 'backwards' transmission in time for the quanta involved (the former would be 'particles' and the latter 'antiparticles'). There is, however, no reason to inflict nonsense upon one's physical intuition and we prefer to consider the propagator as a unity.

In the last section of this chapter we will show that in a lightcone dynamical scenario it makes sense to talk about the propagator in terms of old-fashioned energy denominators.

In the next subsection we will discuss the Fierz [61] interpretation of the Feynman propagator, which is how the physicists working with Stückelberg thought about it. This is done in order to convince the reader that the way in which it works is not only in accordance with the Heisenberg indeterminacy principle. *The Feynman propagator is actually as causal as it can be when the principle is fulfilled.*

3   An interpretation of the Feynman propagator

For a simple and intuitively useful example we will consider the case when $H_1 = g\phi(x){:}\psi^2(x){:}$ (with ϕ and ψ free independent scalar quantum fields), an interaction which we will discuss later in the book. This is meant to be a simplified version of the current-vector-potential interaction in a gauge theory.

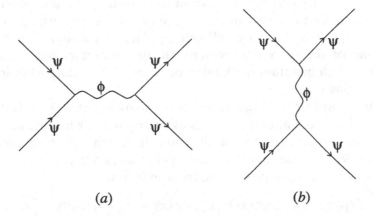

(a) (b)

Fig. 3.1. Two simple examples of Feynman graphs. The situation in (a) corresponds to the annihilation of two ψ-particles; the state then propagates as a virtual ϕ and finally two outgoing particles appear. In (b) there is scattering with the exchange of energy-momentum. The straight lines symbolise ψ-quanta and the wavy lines the ϕ-propagator.

For this case we will need the fact that $R_2 = (1/2)\mathcal{T}\{\mathcal{H}_{1i}(x_1)\mathcal{H}_{1i}(x_2)\}$ (cf. Eq. (3.71)) contains among many others the term

$$R_2' = \frac{g^2}{2} : \psi_i^2(x_1)\psi_i^2(x_2) : \Delta_F(x_2 - x_1, M_\phi) \tag{3.90}$$

The result in Eq. (3.90) corresponds to the scattering of two ψ-particles which come in, interact at the point x_1 and are either annihilated into a virtual ϕ (Fig. 3.1(a)) and afterwards reappear as outgoing ψ-particles at x_2 or exchange energy-momentum between points x_1 and x_2 through a virtual ϕ (Fig. 3.1(b)).

In this subsection we will simplify the working by assuming that there are two kinds of ψ-particle, which we call p- and e-flavored, which may interact via the common ϕ-field. This assumption does not change the argument in the least but makes it easier to discuss.

Any kind of interpretation of a physical quantity is always defined by means of a measurement that is at least theoretically possible. We will show that a measurement made in accordance with quantum mechanical requirements will preserve all causality and energy-momentum conservation properties and that this is due to the properties of the Feynman propagator.

In order to further simplify the problem we will assume that there are regions of space-time R_j within which we can measure what is going on in connection with the scattering. As always in a measurement process we expect these regions to be determined by some some size parameters. We will solely be interested in the *time slices* of the regions, i.e. the time intervals they span; these we will call T_j. Thus we assume that there is in any one of the space-time regions an ideal detector (but working in accordance with quantum mechanics, of course!) recording what is going on as time passes.

We then consider the case when an e_1-particle scatters against a p_2-particle and goes out after the process as an e_3-particle while the p_4-particle recoils. This corresponds to the situation described diagrammatically in Fig. 3.1(b). We assume that their energy-momenta are k_j $j = 1,\ldots,4$ and we will now write the transition matrix element as

$$\mathcal{M} = \int dx_1 dx_2 g^2 \left\langle k_4 \middle| : \psi_p^2(x_1) : \middle| k_2 \right\rangle \Delta_F(x_2 - x_1) \left\langle k_3 \middle| : \psi_e^2(x_2) \middle| k_1 \right\rangle. \quad (3.91)$$

We then change the integral over all space-time into an integral over the regions where we have the detectors:

$$\int dx_1 dx_2 = \sum_{j,k} \int_{R_j} dx_1 \int_{R_k} dx_2 \quad (3.92)$$

The only argument of which we are going to make use is related to the energies so it is not necessary that we expand the Δ_F-function in plane waves; energy harmonics $\exp(\pm i\omega x_0)$ are sufficient. The next thing is to go back to the definition of Δ_F, Eq. (3.75), and rewrite $\mathcal{M}$ in Eq. (3.91) as (note that we must include both time-orderings!)

$$g^2 \sum_{j,k} \int_{R_j} dx_1 \int_{R_k} dx_2 \left\langle k_4 \middle| : \psi_p^2(x_1) : \middle| k_2 \right\rangle \left\langle k_3 \middle| : \psi_e^2(x_2) \middle| k_1 \right\rangle$$

$$\times \left[\Theta(x_1 - x_2)\Delta_+(x_2 - x_1) + \Theta(x_2 - x_1)\Delta_+(x_1 - x_2) \right] \quad (3.93)$$

If we write out the time dependence of the first term we will find for regions R_1 and R_2 (spreading over the times $T_j, j = 1, 2$; note that k_0 must be positive as it corresponds to the argument in the Δ_+-distribution)

$$\exp[-i(\omega_2 - \omega_4)x_{01} - i(\omega_1 - \omega_3)x_{02}]\Theta(x_{01} - x_{02})dk_0 \exp[ik_0(x_{02} - x_{01}]$$
$$(3.94)$$

Now we gather the terms containing x_{01} and x_{02}, respectively, and assume that the time slices T_j for the detector configuration are such that

$$T_1|\omega_2 - \omega_4| \gg 1 \quad \text{and} \quad T_2|\omega_1 - \omega_3| \gg 1 \quad (3.95)$$

This is what Heisenberg would require in order that we should be able to measure the energies in each of the detectors so precisely that we can distinguish between the energies of p_2 and p_4 and between those of p_1 and p_3. It is necessary to have sufficiently long times available for such measurements, at least several frequency periods. But we note that there is then little to work on if we are to obtain a nonvanishing value for the integrals. The only possibility is to choose the value of k_0 such that

$$k_0 \simeq \omega_4 - \omega_2 \simeq \omega_1 - \omega_3 \tag{3.96}$$

This requirement is a direct result of the properties of the Fourier integrals, for which it is necessary not to have strongly fluctuating integrands if we want nonvanishing results.

We conclude that, as the time in region R_2 is earlier than the time in region R_1, according to the Θ-distribution, and as k_0 is positive:

- the energy of the *e*-flavor particle decreases from ω_1 to ω_3 by emission of the (virtual) ϕ-quantum in the region R_2;

- then the *p*-flavor particle absorbs the ϕ-quantum in the region R_1 and so increases its energy from ω_2 to ω_4;

- in both cases it is necessary to have time slices T_j large enough to measure the energy loss and energy increase, respectively, with sufficient precision.

In the other term in Eq. (3.93) the region R_1 is before the region R_2 in time; this correponds to the opposite process. *The basic point is that the Feynman propagator describes emission and absorption (within the requirements of Heisenberg) in a causal way.*

3.4 The method for calculating the scattering cross sections

Here we consider the steps that are necessary to get from the transition amplitude to the scattering cross section for a multiparticle interaction. The reasons for doing this are two-fold. On the one hand we have introduced a cutoff procedure with the box V and we want to show why it does not appear in our final formulas. On the other hand, in the last section, at Eq. (3.90) and Fig. 3.1, we considered a particular scattering process. To understand the physics of that process we will calculate its properties in some detail. The result will serve as an example of other formulas that we will meet later on.

We will consider the matrix element $\mathcal{M}$ between two incoming ψ-particles (energy-momentum k_1, k_2) and two outgoing ψ-particles (k_3 and k_4) interacting via the field ϕ according to the interaction term

$$\int H_1(t)dt = \int dxg : \psi^2(x) : \phi(x) \tag{3.97}$$

From Eq. (3.90) we know the term responsible for the transition and so we obtain for the matrix element $\mathcal{M}(k_3, k_4; k_1, k_2)$

$$\int d^4x d^4x' \left\langle k_3, k_4 \left| \tfrac{1}{2}g^2 : \psi^2(x) :: \psi^2(x') : \Delta_F(x'-x) \right| k_1, k_2 \right\rangle$$

$$= \int d^4x d^4x' \frac{2g^2}{4V^2\sqrt{\omega_1\omega_2\omega_3\omega_4}} \times$$

$$\{\exp[ix(k_3-k_1)+ix'(k_4-k_2)] + \exp[ix(k_4-k_1)+ix'(k_3-k_2)]$$

$$+ \exp[-ix(k_1+k_2)+ix'(k_3+k_4)]\}\,\Delta_F(x'-x)$$

$$= \frac{2g^2}{4V^2\sqrt{\omega_1\omega_2\omega_3\omega_4}}(2\pi)^8\delta(k_1+k_2-k_3-k_4)\frac{i}{(2\pi)^4}$$

$$\times \left[\frac{1}{(k_1-k_3)^2-M_\phi^2} + \frac{1}{(k_1-k_4)^2-M_\phi^2} + \frac{1}{(k_1+k_2)^2-M_\phi^2} \right]$$

$$\equiv AB \tag{3.98}$$

We have here introduced in the second line of the equation the wave functions for the incoming and outgoing particles, i.e. the factors multiplying the necessary annihilation and creation operators in the representation of the operators ψ. In the third line we have, after the introduction of the Fourier representation of the Feynman propagator, performed the space-time integrals. In the last line we re-express the three terms inside the square bracket as B and the remaining factors as A. We note in particular that the energy-momentum conserving δ-distribution appears in A.

The cross section, according to Fermi's Golden Rule, is obtained by multiplying the transition rate per unit time by the inverse of the incoming particle flux and by the final-state density. We are going to introduce and discuss these factors in turn.

The *transition rate* is obtained from the square of the matrix element $\mathcal{M}$ and we immediately encounter the difficulty of squaring a δ-distribution in the factor A. If we go back to Eq. (3.51) we note that the distribution for a finite box V is, for the momentum part,

$$(2\pi)^3\delta(\mathbf{k}-\mathbf{k}') \to V\delta_{\mathbf{k},\mathbf{k}'} \tag{3.99}$$

Consequently the square of the space-momentum part is, formally,

$$[\delta(\mathbf{k}-\mathbf{k}')]^2 \to \frac{V}{(2\pi)^3}\delta(\mathbf{k}-\mathbf{k}') \tag{3.100}$$

For the energy part we note that the δ-distribution stems from an integral

$$\delta(\Delta E) = \frac{1}{2\pi}\left[\lim_{t_0\to\infty}\int_{-t_0}^{t_0} dt\,\exp(it\Delta E)\right] = \lim_{t_0\to\infty}\left[\frac{\sin(t_0\Delta E)}{\pi\Delta E}\right] \tag{3.101}$$

The last expression is a well-known representation of the δ-distribution. We always have in mind the physical picture that there should be a finite time overlap for the interaction; this corresponds to a finite 'effective' value of t_0. Therefore this representation is in accordance with our intuition. If we formally square the last line and note the well-known relation

$$\lim_{x \to 0} \left[\frac{\sin(xy)}{x} \right] = y \tag{3.102}$$

we find the following formal definition of the square of the energy part of the δ-distribution (with $\Delta t = 2t_0$ the 'interaction time'):

$$[\delta(\Delta E)]^2 \to \frac{\Delta t}{2\pi} \delta(\Delta E) \tag{3.103}$$

Thus the transition rate per unit time is

$$\frac{w}{\Delta t} = \frac{(2g^2)^2}{(4V^2)^2 \omega_1 \omega_2 \omega_3 \omega_4} (2\pi)^8 \frac{V}{(2\pi)^4} \delta(k_1 + k_2 - k_3 - k_4)|B|^2 \tag{3.104}$$

The *incoming flux*, i.e. the number of states interacting per unit time and unit transverse area, is v_r/V, where v_r is the relative velocity of the particles. If we divide the formula in Eq. (3.104) by this flux factor we notice that we obtain two factors V in the numerator, one from the (space-momentum) δ-distribution and one from the flux. These two compensate the two factors V in the denominator stemming from the *two incoming particle wave functions*.

The remaining factors from the incoming wave functions, $4\omega_1\omega_2$, combine in the denominator with the velocity v_r so that we have

$$4\omega_1\omega_2 v_r = 4\omega_1\omega_2 |v_1 - v_2| = 4 \, ||\mathbf{k}_1|\omega_2 - |\mathbf{k}_2|\omega_1|$$

$$= 4M_1 M_2 |\sinh(y_1 - y_2)| = 4M_1 M_2 \sqrt{\cosh^2(y_1 - y_2) - 1} \tag{3.105}$$

$$2\sqrt{(s - M_1^2 - M_2^2)^2 - 4M_1^2 M_2^2} \equiv 2\sqrt{\lambda(s, M_\psi^2, M_\psi^2)} \to 2s$$

with s the squared cms energy $s = (k_1 + k_2)^2$. Here we have first introduced the relative velocity and used that each particle velocity is $v_j = |\mathbf{k}_j|/\omega_j$ and that energies and momenta can be written in terms of rapidities $\omega_j = M_j \cosh y_j$, $|\mathbf{k}_j| = M_j \sinh y_j$. The rest is simple manipulation and we note that the function $\lambda(a, b, c)$ is totally symmetric:

$$\lambda(a, b, c) = a^2 + b^2 + c^2 - 2ab - 2ac - 2bc \tag{3.106}$$

The quantity λ is very useful for quick calculations of Lorentz boosts. Thus the cms momenta of two particles (indexed 1 and 2) with a common cms energy $\sqrt{s}$ has the common cms momentum

$$|\mathbf{k}_{j,cms}| = \frac{\sqrt{\lambda(s, M_1^2, M_2^2)}}{2\sqrt{s}} \tag{3.107}$$

while in the rest frame of particle 1, particle 2 has momentum

$$|\mathbf{k}_{2,lab}| = \frac{\sqrt{\lambda(s, M_1^2, M_2^2)}}{2M_1} \tag{3.108}$$

In the rest frame of 2 we simply exchange exchange the indices.

The third factor in the cross section, the *final-state density* is the number of momentum states available and is given by Eq. (3.50). We note that it will contain in the numerator as many V-factors as particles. This will compensate the corresponding denominator V-factors from the *final state particle wave functions*. All in all this final-state density therefore combines with the wave function factors into

$$\prod_{j_f} \frac{d^3 k_{j_f}}{2\omega_{j_f}(2\pi)^3} = \prod_{j_f} \frac{dk_{j_f}}{(2\pi)^3} \delta^+(k_{j_f}^2 - M_{j_f}^2) \tag{3.109}$$

where we have used Eq. (3.66).

The full cross section then will appear as (for n_f final-state particles)

$$d\sigma = \frac{2g^4}{(2\pi)^{(3n_f - 4)}\sqrt{\lambda(s, M_\psi^2, M_\psi^2)}} |B|^2$$
$$\times \prod_{j_f} dk_{j_f} \delta^+(k_{j_f}^2 - M_{j_f}^2)\delta(k_1 + k_2 - \sum_{j_f} k_{j_f}) \tag{3.110}$$

The general phase-space factors in Eq. (3.110) will always occur in two-body to many-body processes but the factor $2g^4|B|^2$ (with *the matrix element B* defined in Eq. (3.98)) is specific to the particular process we have considered. We will meet the result repeatedly later in the book and we note that it is manifestly Lorenz-invariant.

3.5   The propagators in lightcone physics in the infinite-momentum frame

1   The formalism

We will in this section provide a different picture of the the Feynman rules by exhibiting the properties of perturbation theory when lightcone coordinates are used. The propagator in energy-momentum space will then have strong similarities to the old-fashioned energy denominators occurring in time-dependent perturbation theory in nonrelativistic dynamics.

Basically the scenario describes a two-dimensional field theory in transverse dimensions with a varying mass parameter which corresponds to one of the lightcone components. The whole idea stems from early investigations by Weinberg, [111], into the possibility of simplifying the

Feynman rules by performing all the integrals in a frame moving very fast in some direction. This has been called the 'infinite-momentum frame'. The discussion is based upon the development in [87].

The formalism is useful to understand intuitively some of the features of the parton model which is discussed in Chapter 5. We will use some of the results in connection with heavy quark fragmentation in Chapter 13.

We begin by defining the *lightcone components* η, H and τ, ζ of the energy-momentum and space-time operators:

$$\eta = \frac{P_0 + P_3}{\sqrt{2}}, \quad \tau = \frac{t + x_3}{\sqrt{2}}$$
$$H = \frac{P_0 - P_3}{\sqrt{2}}, \quad \zeta = \frac{t - x_3}{\sqrt{2}} \tag{3.111}$$

We will call the 1- and 2-components the *transverse components* of the corresponding four-vector and denote these by $\mathbf{p}_\perp$ and $\mathbf{x}_\perp$.

According to the ordinary commutation relations we have

$$[\eta, \tau] = [\eta, H] = [H, \zeta] = [\tau, \zeta] = 0$$
$$[\eta, \zeta] = [H, \tau] = i \tag{3.112}$$

and all these components commute with the transverse ones.

The mass-shell condition for a free particle means that

$$m^2 = P_0^2 - P_3^2 - \mathbf{p}_\perp^2 \quad \Rightarrow \quad H = \frac{\mathbf{p}_\perp^2}{2\eta} + V_0 \tag{3.113}$$

where $V_0 = m^2/2\eta$ is similar to a potential term. This is evidently a reduction of the problem to the two transverse dimensions using the variable 'mass'-parameter η.

We next consider the Feynman propagator and rewrite it in terms of the variables given above:

$$\Delta_F(x) = \frac{i}{(2\pi)^4} \int \frac{dk \exp(ikx)}{k^2 - M^2 + i\epsilon}$$
$$= \frac{i}{(2\pi)^4} \int d^2 p_\perp \int d\eta \exp i(\eta\zeta - \mathbf{p}_\perp \cdot \mathbf{x}_\perp)$$
$$\times \int dH \exp(iH\tau)(2\eta H - \mathbf{p}_\perp^2 - M^2 + i\epsilon)^{-1} \tag{3.114}$$

We note that by use of the results in Eq. (3.80) we may now write the following formula for the Feynman propagator:

$$\Delta_F(x) = \frac{1}{2(2\pi)^3} \int d^2 p_\perp \int_0^\infty \frac{d\eta}{\eta} [\Theta(\tau) \exp(-ipx) + \Theta(-\tau) \exp(ipx)] \tag{3.115}$$

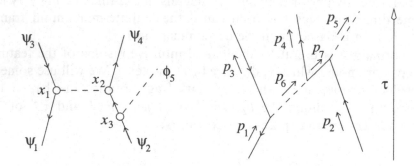

Fig. 3.2. One of the possible Feynman diagrams in the process $\psi_1 + \psi_2 \rightarrow \psi_3 + \psi_4 + \phi_5$ and the same diagram ordered according to one of the possible orderings along the lightcone.

where $px = H\tau + \eta\zeta - \mathbf{p}_\perp \cdot \mathbf{x}_\perp$ and H is defined by the mass-shell condition above.

In order to obtain the result in Eq. (3.115) we have divided the integration region of η into positive and negative parts to obtain the sign of the limiting imaginary part and then changed sign for the negative part. This provides the signs in the complex exponents.

We have thus come back to expressions with the properties described before. The 'effective' energy H is like a nonrelativistic kinetic energy term related to the generalised 'time', i.e. the lightcone coordinate τ.

2   *An example*

We will next provide an example of how the Feynman rules work when lightcone coordinates are used; we consider the Feynman diagram in Fig. 3.2(*a*). This corresponds to the $g\phi : \psi^2$:-theory we have discussed before and contains the scattering of two ψ-particles together with the emission of a ϕ-particle in a bremsstrahlung process. We note that there are several more diagrams which will contribute to the process.

In Fig. 3.2(*b*) we have drawn a version of the diagram in which there is a particular ordering of the τ-variables. A little thought will convince us that if we have n vertices in the primary Feynman diagram then there are $n!$ such ordered diagrams possible. That means six in this case and we have considered the one corresponding to the ordering $\tau_1 \leq \tau_2 \leq \tau_3$.

In the ordered diagram we must perform the τ-integrals with this ordering requirement, *which means that only one of the Θ-terms in the representation of Eq. (3.115) survives the requirement.*

There are two propagator terms and three τ-integrals. Note that all the transverse integrals and the ζ-integrals can easily be performed to give at each vertex a δ-distribution contribution

$$(2\pi)^3 \delta(\mathbf{p}_{\perp i} - \mathbf{p}_{\perp f})\delta(\eta_i - \eta_f) \tag{3.116}$$

where the indices i, f correspond to the 'in'- and 'out'-contributions at that vertex. Note that we have directed the vectors in Fig. 3.2(b).

The τ-integrals are given by

$$I = \int d\tau_1 d\tau_2 d\tau_3 \Theta(\tau_3 - \tau_2)\Theta(\tau_2 - \tau_1) \exp\{-i[(H_1 - H_3 - H_6)\tau_1$$
$$+ (H_6 - H_4 - H_7)\tau_2 + (H_7 + H_2 - H_5)\tau_3]\} \tag{3.117}$$

If we introduce the natural variables $T_0 = \tau_1, T_1 = \tau_2 - \tau_1, T_2 = \tau_3 - \tau_2$ then the integrals are transformed to give

$$I = \int dT_0 \exp\left[-i(\mathcal{H}_i - \mathcal{H}_f)T_0\right] \int_0^\infty dT_1 \exp\left[-i(\mathcal{H}_1 - \mathcal{H}_f)T_1\right]$$
$$\times \int_0^\infty dT_2 \exp\left[-i(\mathcal{H}_2 - \mathcal{H}_f)T_2\right] \tag{3.118}$$

where we have introduced the notation

$$\mathcal{H}_i = H_1 + H_2, \quad \mathcal{H}_f = H_3 + H_4 + H_5$$
$$\mathcal{H}_1 = H_3 + H_6 + H_2, \quad \mathcal{H}_2 = H_2 + H_7 + H_4 \tag{3.119}$$

Again the indices i, f correspond to the energies of the incoming and outgoing states (this time for the whole diagram, with signs) and the two indices 1 and 2 correspond to the *intermediate states*. If we consider Fig. 3.2(b) it is obvious what is meant by the intermediate states. They refer to those particles which exist at a particular τ-slice, for the index 1 the slice between τ_1 and τ_2, for the index 2 the slice between τ_2 and τ_3.

The T_0-integral, which is taken over the whole lightcone time, provides a δ-distribution for overall energy conservation. The T_1- and T_2-integrals only cover the positive regions and each give

$$\int_0^\infty dT \exp(i\mathcal{H}T) = \frac{i}{\mathcal{H} + i\epsilon} \tag{3.120}$$

This means that the total result will contain, besides an overall energy-momentum-conserving δ-distribution, 'mass'-conserving , i.e. η-conserving, and transverse-momentum-conserving δ-distributions at each vertex, something very similar to old-fashioned energy denominators:

$$(\mathcal{H}_1 - \mathcal{H}_f)^{-1}(\mathcal{H}_2 - \mathcal{H}_f)^{-1} \tag{3.121}$$

one for each intermediate state. It is not difficult to see that this structure survives for all the different contributions. Further, as one may guess, it is possible to do the same for any kind of field theory, although there are often more singular parts of the propagators (for QED cf. [87]) than those we encountered in the simple scalar theory.

It is worthwhile to note that that the η-terms we find everywhere are nothing other than the quantities

$$\frac{dp_l}{2e} = \frac{dy}{2} = \frac{d\eta}{2\eta} \qquad (3.122)$$

which we met before in the method of virtual quanta in Chapter 2, and also will meet later as Feynman's 'wee spectrum' of partons.

In this way each n-vertex Feynman graph can be reduced to $n!$ old-fashioned energy denominator integrals. This might not seem to be a major achievement. But this formalism often makes it easier to perform reasonable approximations among the many diagrammatic contributions to a particular scattering situation or bound-state configuration. It also provides an intuitively appealing picture of the difference between the longitudinal and the transverse dynamics.

4

The vacuum as a dielectric medium; renormalisation

4.1   Introduction

In this chapter we will consider some major problems in quantum field theory. They are related to the understanding of *polarisation effects in the vacuum state*. Although this state in the mean is empty it nevertheless embraces the continuous production and annihilation of virtual particle-antiparticle pairs due to quantum fluctuations. All the real charges and currents then behave as if they were moving in a dielectric medium. In connection with QED this effect is small (although readily observable). For QCD, on the other hand, it plays a major role.

The first kind of problem is mathematical, related to ill-defined series expansions in perturbation theory and also to undefined integrals. The second is general in physics: it is necessary to isolate the effective dependence on the theoretical parameters in all the calculated expressions for the observables (note that this dependence is in general complicated when one deals with non-linear equations). This is the *renormalisation procedure*, which always must be performed in order to relate the parameters in a theoretical expression to the observables in an experiment.

It is true that physicists are, compared to most other scientists, privileged because the components of many systems in physics can be isolated. In this situation the properties of each component can be determined. Afterwards the whole system can be brought back into interaction, with well-defined values of the parameters which govern the behaviour of each subsystem. For an interacting quantum field it is, however, not straight-forward to isolate the 'real' quanta from the surrounding fields and the quantum fluctuations. This was found for an electron in connection with the method of virtual quanta in Chapter 2: the electron energy can only be isolated from the surrounding field energy by means of an impact parameter cutoff. Similarly the properties of a field quantum in an interacting-field theory cannot be described in terms of the corresponding free-field behaviour without some limiting procedures and the introduction of cutoff parameters.

It is a surprising and gratifying result that we are able to solve both the above-mentioned problems at the same time. It turns out that all the 'bad' mathematical expressions occur, for a wide class of field theories, just where we would anyhow have had to redefine these expressions in order that the coupling constants and the masses should have the values observed for the free initial- and and final-state quanta.

We will consider two different examples with some similarities. The first one corresponds to the scalar field theory we exhibited before in Chapter 3, with two ψ-operators coupled to a single ϕ-field. We will choose the quanta of the ϕ-field to be massless, $M_\phi = 0$, in order to connect with the QED and QCD field theories. These will provide our second example, with massless vector particles, photons and gluons, coupled to spin $1/2$ fermions corresponding to the scalar ψ-fields in the first example.

We will find that, apart from *mass renormalisation*, the scalar field theory is a finite field theory, called *super-renormalisable*. This feature is related to the dimensions of the coupling constant. For a super-renormalisable theory the coupling constant dimension is positive in terms of energy dimensions. Then the theory contains at most a finite number of undefined diagrammatic contributions in perturbation theory and this can be understood in the following way. Undefined, divergent, integrals in perturbation theory stem from the fact that there are too many energy-momentum integration variables as compared with the energy denominators (from the propagators). Then the integrals are not sufficiently damped for large values of the energy-momentum variables (and are therefore called *ultra-violet divergent*). If we consider this phenomenon in space-time then the divergences in the energy-momentum integrals correspond to singular behaviour of the space-time integrals for small values of the relative coordinates of the field operators. The singularities stem from the *distribution properties* of the field operators, which we have already encountered in Chapter 3. In general one is not allowed to multiply field operators at the same space-time point (which we would like to do when we consider local interactions between fields).

Let us consider a physical quantity $\mathscr{I}$ which by a suitable rescaling is dimensionless. When it is defined in perturbation theory at the nth order in the coupling constant g one obtains $g^n I_n$ with I_n some integral. The integral I_n must then have the (energy) dimension $\dim I_n = -n \times \dim g$. Therefore in a super-renormalisable theory the (energy) dimension will become more and more negative with n. This means that the number of energy denominators must be increasing faster than the number of integration variables, which means that we obtain integrals which are more and more damped for larger energies.

QED and QCD have dimensionless coupling constants and in these cases the argument above does not work. They are nevertheless *renormalisable* in the following sense. One finds that in each order of perturbation theory there will be logarithmically divergent integrals (which in practice stem from non-allowed changes of integration order and undefined limits). It is then necessary to introduce in each order of perturbation theory a method to make the results finite. For renormalisable theories it turns out that *all the undefined quantities can be incorporated as multiplicative constants in the field operators and in the coupling constants* after mass renormalisation has been performed. This means that the 'new' *renormalised field theory* contains just as many parameters as the original one. When these parameters have been fixed by the observed values then *all the remaining observable quantities are finite and predicted by the theory.*

To be more precise we may imagine that we have a fixed external electric charge (size g_0) and that we make use of it in order to measure the properties of a quantum field coupled to the charge via QED. As a thought experiment we will consider the scattering of a field quantum with momentum transfer q^2 from this external source.

Now let us take into account the influence of the quantum fluctuations in the state, i.e. what we have earlier referred to as the dielectric properties of the surrounding vacuum state. All these properties can be calculated in perturbation theory but (unless one is particularly careful about the distribution properties of the fields) the expressions will correspond to ill-defined integrals and series. The calculations can nevertheless be performed with different degrees of sophistication. We will then obtain results which can be expressed in terms of the original (unrenormalised) parameters of the theory together with some suitable cutoff parameters to make the mathematical expressions well defined.

In this way we will obtain an expression for the scattering amplitude (cf. Eq. (3.98)) which should be of the generic shape

$$\frac{g_0 g(q^2)}{q^2} \times \text{k.f.} \tag{4.1}$$

where k.f. corresponds to the necessary kinematical factors and $g(q^2)$ corresponds to the coupling constant at the 'test frequency', q^2, at which we perform the calculation (the theoretically evaluated quantity $g(q^2)$ also depends upon the the cutoff parameters, of course). We may also calculate other quantities, such as the value for which there will be a pole in the field propagator. This obviously corresponds to the squared mass of the corresponding quanta (as seen by a probe with the frequency q^2) and from the size of the pole term we may calculate the number of quanta which are available at the scale q^2 in the field (the 'field strength').

These calculations provide us with a value for the effective coupling constant, $g(q^2)$, as well as values for the mass(es) and the normalisation(s) of the field(s) at the 'frequency' q^2, and they are all expressed in terms of the unrenormalised parameters and the cutoff parameters. We may then choose these numbers to coincide with our expectations (giving *coupling constant, mass* and *wave function renormalisation*, respectively). But note that this free choice can in general only be made for a single value of the frequency! For other frequencies there will be changes but *in a renormalisable theory all such changes are computable and finite* although all quantities will seemingly depend upon the value of q^2 for which the original definition is made. There is, however, no reason to prefer one value of q^2 to another and we may then freely move between different 'normalisation points'. But the values of our parameters at these different points are all related, i.e. for any given value q^2 and our choices of the parameters at that value we may compute the result for any other frequency value. And for any particular value q_p^2 we will obtain the same observable results, independent of the normalisation point!

This is the content of the renormalisation group theory. After we have sketched the general behaviour of any renormalisable field theory we will derive the *Callan–Symanzik equations*, [108], which relate the behaviour of the matrix elements and the effective coupling constant $g(q^2)$ at different values of the momentum transfer by means of differential equations. We will use these equations again in Chapter 19 to derive the QCD predictions for the scale breaking in the parton structure functions, which governs the behaviour of the inelastic lepto-production cross sections.

We start by introducing the Källén–Lehmann representation as a convenient tool to perform the renormalisation procedure. This will also provide an opportunity to show the occurrence of some of the phase space factors we will meet further on. We also show how to calculate the polarisation correlations which occur when one couples spin 1/2 (Dirac) particles to spin 1 particles (photons or gluons) as is done in QED and QCD. The particular polarisation properties of the QCD field theory are treated in some detail and we will then also consider the relationship between the weight function and the full polarisation function in a Källén–Lehmann representation, i.e. we will introduce the notion of 'cut diagrams'. We will finally show how to calculate the color factors which occur in QCD.

4.2 The Källén-Lehmann representation, the *n*-particle phase space

We start out with the following general expression for a propagator

$$\Delta_{FA}(x) = \langle 0| \mathscr{T} \{A(0)A(x)\} |0\rangle$$
$$= \Theta(x) \langle 0| A(x)A(0) |0\rangle + \Theta(-x) \langle 0| A(0)A(x) |0\rangle \quad (4.2)$$

where A is a local (for simplicity also self-adjoint, i.e. real) operator of any kind expressed in terms of the in-fields (we omit all i-indices from now on). We may introduce a complete set of states $\sum_n |n\rangle \langle n| = 1$ in between the operators. Further we note that (due to translation invariance)

$$\langle 0| A(x) |n\rangle = \exp(-ik_n x) \langle 0| A(0) |n\rangle$$
$$\langle n| A(x) |0\rangle = \exp(ik_n x) \langle n| A(0) |0\rangle \quad (4.3)$$

with k_n the total energy-momentum of the state n. We then rearrange the expression for Δ_{FA} into

$$\Delta_{FA}(x) = \sum_n [\Theta(x) \exp(-ik_n x) + \Theta(-x) \exp(ik_n x)] |A_{0n}|^2$$

$$= \int \frac{dq}{(2\pi)^3} [\Theta(x) \exp(-iqx) + \Theta(-x) \exp(iqx)] \, da\delta^+(q^2 - a)G_A(a)$$

$$G_A(q^2) = (2\pi)^3 \sum_n \delta(q - k_n)|A_{0n}|^2$$

$$\quad (4.4)$$

where we have used the shortened version $A_{0n} = \langle 0| A(0) |n\rangle$.

The fact that G_A is a Lorentz invariant will be exhibited below. Then the resulting expression for Δ_{FA} is

$$\Delta_{FA} = \int da\Delta_F(x, a)G_A(a) = \frac{i}{(2\pi)^4} \int dq \frac{\exp(iqx)}{q^2 - a + i\epsilon} da G_A(a) \quad (4.5)$$

which is the Källén-Lehmann representation for the general propagator. The structure is a sum of ordinary Feynman propagators with contributions from the squared masses of all the possible intermediate states which can be reached by A.

We note that the weight function G_A, if we use the distribution described by Eq. (3.87), is essentially the real part of (the Fourier transform, i.e. the energy-momentum space version, of) Δ_{FA}. This general feature is in Chapter 2 referred to as the Kramers-Kronig relations: the imaginary part of the dielectricity is determined by the real part. From Eq. (4.5) we find the content of this statement, i.e. the total energy-momentum space propagator is determined by its real part. We will elaborate this result further on in this chapter.

In order to investigate the weight function G_A we start by considering the case $A(x) = :\psi^2(x):$. Then there is only a single intermediate state, a two-particle ψ-quantum state, and we obtain for this situation

$$G_A = (2\pi)^3 \sum_n \delta(q - k_n)|<0|:\psi^2:|n>|^2$$

$$= \int \frac{V^2 d^3k_1 d^3k_2}{(2\pi)^3} \delta(q - k_1 - k_2)\frac{1}{4V^2\omega_1\omega_2}$$

$$\to \frac{1}{(2\pi)^{3n_f-3}} \int \prod_{j_f=1}^{n_f} dk_{j_f}\delta^+(k_{j_f}^2 - M_{j_f}^2)\delta\left(q - \sum_{j_f=1}^{n_f} k_{j_f}\right) \quad (4.6)$$

where in the last line we have gone over to the result for A, $:\prod_{j_f=1}^{n_f} \psi_{j_f}:$, in order to show the general structure of any G_A-expression containing normal-ordered local-field operators. *The main point is the occurrence of the manifestly Lorentz-invariant n_f-particle phase space* .

For the scalar field theory case the probability of producing real states with the mass square a is given simply by this phase space factor. We will later find a difference when we have spin $1/2$ particles coupled to a vector field; then there is also a spin-correlation term.

We will now calculate the phase space integrals, I_{n_f}, for the cases when $n_f = 2, 3$ because we will need them later. We start with I_2:

$$I_2(q^2, a_1, a_2) = \int \frac{dk_1 dk_2}{(2\pi)^3} \delta^+(k_1^2 - a_1)\delta^+(k_2^2 - a_2)\delta(q - k_1 - k_2) \quad (4.7)$$

Evidently q must be a timelike vector with $\sqrt{q^2} \geq \sqrt{a_1} + \sqrt{a_2}$. In order to simplify our formulas, we will make use of the Lorentz invariance to choose the particular system where q is at rest (the cms of particles 1 and 2). Then $q = (W, \mathbf{0})$. Performing the k_2-integral by means of the energy-momentum-conserving δ-distribution we obtain in this frame

$$I_2 = \frac{1}{(2\pi)^3} \int dk_1 \delta^+(k_1^2 - a_1)\delta^+((q - k_1)^2 - a_2)$$

$$= \int \frac{k^2 dk d\Omega}{2\omega(2\pi)^3} \delta^+(W^2 - 2\omega W + a_1 - a_2) \quad (4.8)$$

with the notation $k_1 = (\omega = \sqrt{k^2 + a_1}, \mathbf{k})$. We have chosen a spherical coordinate system with $d^3k = k^2 dk d\Omega$. We may then transform to the integration variable ω and obtain

$$I_2 = \frac{1}{(2\pi)^2} \int k d\omega \delta(2\omega W - (W^2 + a_1 - a_2)) = \frac{\sqrt{\lambda(W^2, a_1, a_2)}}{(4\pi)^2 W^2} \quad (4.9)$$

where λ is again the symmetrical function defined in Eq. (3.105). In particular the expression can be written as

$$I_2 = \frac{2p_{cms}}{(4\pi)^2 W} \qquad (4.10)$$

Thus the two-particle phase space integral vanishes linearly when the relative velocity vanishes and approaches the constant $1/(4\pi)^2$ for large W^2-values. We note that *the phase space for two particles is dimensionless* (with our conventions, i.e. when $c = \hbar = 1$).

If we look back we notice that for $n = 2$ there are $n \times 4$ integration variables with dimension mass. But there is a four-dimensional δ-distribution, with dimensions $4 \times (-1)$, and $n = 2$ (mass-shell) δ-distributions with dimensions -2. This means that the n-particle phase space has the energy dimension $\dim I_n = 4n - 4 - 2n = 2n - 4$.

We note also that the number of degrees of freedom is $4n-4-n$ because the mass-shell δ-distributions fix only one of the four energy-momentum variables describing each particle. There is, however, also the question of orienting the event. It takes three Euler angles (cf. Goldstein) to fix the coordinate system. If there is no outside direction to relate to, these angles will always be integrated out. Thus for the internal dynamics of the n-particle state there are effectively $3n - 7$ degrees of freedom.

For the three-particle phase space we find an energy dimension 2 and also that there are two internal degrees of freedom. This is a sign that it is a density in two energy variables. We note that if we again go to the cms, i.e. choose the vector $q = (W, \mathbf{0})$, then the energy δ-distribution requires the three cms energies to satisfy

$$\sum_{j=1}^{3} \omega_j = W \qquad (4.11)$$

We may then choose two of these to be independent variables, e.g. the pair ω_1, ω_2. We will only calculate in detail the result when all the three particles are massless; we then obtain

$$
\begin{aligned}
\frac{d^2 I_3}{d\omega_1 d\omega_2} &= \frac{1}{(2\pi)^6} \int d^3k_1 d^3k_2 dk_3 \prod_{j=1}^{3} \delta(k_j^2)\delta\left(q - \sum_{j=1}^{3} k_j\right) \\
&= \frac{1}{(2\pi)^6} \int d^3k_1 d^3k_2 \prod_{j=1}^{2} \delta(k_j^2)\delta\left((W - \omega_1 - \omega_2)^2 - (\mathbf{k}_1 + \mathbf{k}_2)^2\right) \\
&= \frac{\pi}{(2\pi)^5} \omega_1 \omega_2 \int \sin\theta d\theta \delta\left(W^2 - 2W(\omega_1 + \omega_2) + 2\omega_1\omega_2(1 - \cos\theta)\right)
\end{aligned}
$$

$$(4.12)$$

In the second line we have introduced $k_3 = (W - \omega_1 - \omega_2, -(\mathbf{k}_1 + \mathbf{k}_2))$ and then performed the integrals over everything besides the relative angle θ

between the vectors $\mathbf{k}_1$ and $\mathbf{k}_2$. The final step leads to the result

$$\frac{d^2 I_3}{d\omega_1 d\omega_2} = \frac{\pi^2}{(2\pi)^6} \Theta(W - \omega_1 - \omega_2)$$
$$\times \Theta \left(2(\omega_1 + \omega_2) - W\right)(W - 2\omega_1)(W - 2\omega_2)) \quad (4.13)$$

One way to make the whole thing symmetric is to introduce the new dimensionless quantities $x_j = 2\omega_j/W$ for $j = 1, 2, 3$ and to rewrite the distribution as

$$d^3 I_3 = \frac{W^2}{(4\pi)^4} \delta \left(\sum_{j=1}^3 x_j - 2 \right) \prod_{j=1}^3 dx_j \Theta(x_j) \Theta(1 - x_j) \quad (4.14)$$

The expressions for the higher-order phase space factors become more and more complicated to handle. Van Hove [81] devised the idea of 'longitudinal phase space', which means that one projects the total n-particle phase space onto a single direction. He was in that way rather successful in obtaining low-energy dynamical information from the experimental distributions. But even in this simplified case one cannot make do with fewer than n coordinates for n particles so this method fails to give information as soon as we go away from the resonance region.

4.3   A scalar-field-theory propagator in the Källén-Lehmann representation

In this section we will make use of the Källén-Lehmann representation together with the structure of the perturbative expansion as given in Dyson's equation to study some very general properties of the propagator.

We will as an example consider the time-ordered product

$$T_{prop} = \mathscr{T} \{\phi_f(x_1)\phi_f(x_2)\} = \mathscr{T} \{S^*\phi_i(x_1)\phi_i(x_2)S\} \quad (4.15)$$

for the simple $g : \phi\psi^2$:-theory. To second order in the coupling constant the (in-)vacuum expectation value of the operator T_{prop} contains two terms:

$$\langle 0i| T_{prop}|0i\rangle = \Delta_F(x_2 - x_1, M_\phi) + 4g^2 \int dx_3 dx_4 \Delta_F(x_2 - x_3, M_\phi)$$
$$\times \Delta_F^2(x_3 - x_4, M_\psi)\Delta_F(x_4 - x_1, M_\phi) \quad (4.16)$$

The result is presented in Feynman graph language in Fig. 4.1.

It is not too difficult to continue towards higher-order approximations (although there are some problems with respect to counting the number of contributions to each particular diagram in accordance with combinatorics). In Fig. 4.2 we show the relevant contributions in the next order; it is then possible to deduce the general structure.

Fig. 4.1. The first two orders in the expansion of the ϕ-propagator described by Feynman diagrams in the simple $g\phi:\psi^2:$-theory. Solid (broken) lines correspond to ψ- (ϕ-)propagators.

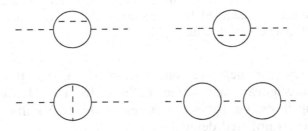

Fig. 4.2. The next-order contributions to the propagator in the $g\phi:\psi^2:$-theory.

There is a 'master' part, ρ, which is called the *polarisation function*. It is the sum of all the contributions from diagrams (with one ϕ-line in and one out) with the following connection structure:

- they are everywhere two-line (at least) connected, i.e. all parts are connected to the rest by at least two lines (this is called *one-particle irreducible*). It means that you cannot disconnect one part from the others by cutting a single line (whether it is broken, corresponding to a ϕ-propagator, or solid, a ψ-propagator).

The contributions in Fig. 4.2 are one-particle irreducible for the first three cases shown but the fourth contribution can be divided easily by cutting the line in between the 'blobs'.

We will assume that it is possible to sum up the contributions to ρ. Unfortunately it can be proved that in a scalar field theory the contributions are, at the $2n$th approximation level, positive and the number of contributions increases more than $n!$ [82]. Therefore the power series in the coupling constant g^2 cannot converge in the usual sense.

This behaviour can be described in very sophisticated mathematical ways but the major physical reason is that the interaction term is not well-behaved, in this case the interaction term $\propto \phi:\psi^2:$ is not *positive definite*. Therefore it is possible to find state configurations with a positive energy in the original free-field case (we may e.g. chose large negative ϕ-field contributions). For the total energy operator $\mathscr{H}$ such configurations

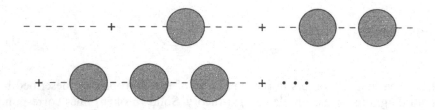

Fig. 4.3. The result of summing all one-line irreducible diagrams into the polarisation function ρ (denoted by a shaded oval) and then adding all these one-particle reducible contributions.

will provide very large negative contributions. Then the Hilbert space of the free-field configurations becomes different from the Hilbert space of the interacting fields, i.e. for some states of the free Hilbert space the interaction term is not well defined.

(You can find a similar behaviour if you introduce e.g. a seemingly small but singular perturbation $\propto \epsilon|x|^{-1-\delta}$ into the one-dimensional Schrödinger equation with a binding potential at the origin; for any $\epsilon, \delta > 0$ there is at least one state, the ground state ψ_0, which is not allowed in the Hilbert space of states of the total hamiltonian because the perturbation term is not defined on the state ψ_0.)

In Fig. 4.2 we note in the fourth contribution the appearance of a repeated part from Fig. 4.1. After a little thought we may conclude that to all orders in the expansion the result can be described as a single (free) ϕ-propagator connecting ρ's in accordance with Fig. 4.3. This means that if we introduce the Fourier transform $\tilde{\rho}$ of ρ by

$$\rho(x) = \frac{i}{(2\pi)^4} \int dq \, \tilde{\rho}(q) \exp(iqx) \tag{4.17}$$

then we obtain the total result after Fourier transformation (using $\tilde{\Delta}$ for the full Feynman ϕ-propagator in energy-momentum space and $\tilde{\Delta}_0^{-1}(q) = q^2 + i\epsilon$ for the corresponding free ϕ-propagator):

$$\tilde{\Delta} = \tilde{\Delta}_0(q) - \tilde{\Delta}_0(q)\tilde{\rho}(q)\tilde{\Delta}_0(q) + \tilde{\Delta}_0(q)\tilde{\rho}(q)\tilde{\Delta}_0(q)\tilde{\rho}(q)\tilde{\Delta}_0(q) + \cdots$$

This can be expressed as an algebraic equation:

$$\tilde{\Delta} = \tilde{\Delta}_0(q) - \tilde{\Delta}_0(q)\tilde{\rho}(q)\tilde{\Delta}(q) \tag{4.18}$$

with solution

$$\tilde{\Delta} = \frac{\tilde{\Delta}_0(q)}{1 + \tilde{\Delta}_0(q)\tilde{\rho}} = \frac{1}{q^2 + \tilde{\rho}(q) + i\epsilon} \tag{4.19}$$

We have then in effect summed a geometrical series without worrying about convergence problems. This has at least a formal meaning in connection with a perturbative expansion. Equation (4.18) was first derived by Dyson and corresponds to his propagator equation. We conclude that *in order to learn about the general propagator it is enough to know the polarisation function $\tilde{\rho}$ in energy-momentum space.*

Actually we have in the simple $\phi : \psi^2 :$-theory already calculated the lowest-order contribution to ρ, i.e. the contribution $\rho^{(1)}$ corresponding to the second term in Fig. 4.1. We note that this has exactly the structure of the vacuum expectation value of a time-ordered product (cf. Eq. (4.16)):

$$4\Delta_F^2(x_4 - x_3, M_{\psi^2}) = \left\langle 0 \left| \mathcal{T} \left\{ :\psi^2(x_3) :: \psi^2(x_4): \right\} \right| 0 \right\rangle \qquad (4.20)$$

We may then use the Källén-Lehmann representation for such an expression and pick up the result directly from Eqs. (4.5), (4.6), (4.9):

$$\tilde{\rho}^{(1)}(q^2) = \frac{4g^2}{(4\pi)^2} \int_{4M^2} \sqrt{1 - \frac{4M^2}{a}} \left[\frac{da}{a - q^2 - i\epsilon} \right] \qquad (4.21)$$

with $M \equiv M_\psi$. The weight function in the integral is just the size of the intermediate two-particle phase space. The integral does not vanish for $q^2 = 0$; as a matter of fact it does not even converge! This is due to a too-cavalier treatment of limits in the calculations. But even if the integral were finite we would have to (re)define it so that $\tilde{\rho}$ vanishes for $q^2 = 0$. *This is called mass renormalisation and corresponds to the requirement that our physical ϕ-field also should have massless quanta.*

It can be done easily in this case:

$$\tilde{\rho}^{(1)}(q^2) \to \tilde{\rho}^{(1)}(q^2) - \tilde{\rho}^{(1)}(0)$$

$$= \frac{4g^2}{(4\pi)^2} \int_{4M^2} \sqrt{1 - \frac{4M^2}{a}} \left[\frac{da}{a - q^2 - i\epsilon} - \frac{da}{a} \right]$$

$$= q^2 \frac{4g^2}{(4\pi)^2} \int_{4M^2} \sqrt{1 - \frac{4M^2}{a}} \left[\frac{da}{a(a - q^2 - i\epsilon)} \right] \equiv q^2 \iota_\phi^{(1)}(q^2) \qquad (4.22)$$

Then we obtain for the total propagator the expression

$$\Delta_F(x) = \frac{i}{(2\pi)^4} \int dq \frac{\exp(iqx)}{q^2 + i\epsilon} \left[\frac{1}{1 + \iota_\phi(q^2)} \right] \qquad (4.23)$$

where the correction term $[1 + \iota_\phi(q^2)]^{-1} \simeq 1 - \iota_\phi(q^2)$ is to lowest order equal to a convergent integral:

$$1 - \iota_\phi^{(1)}(q^2) = 1 - \frac{4g^2}{(4\pi)^2} \int_{4M^2} \sqrt{1 - \frac{4M^2}{a}} \left[\frac{da}{a(a - q^2 - i\epsilon)} \right] \qquad (4.24)$$

A very similar calculation can be done for the full ψ-propagator and in this case we obtain as the lowest-order correction term the Källén-Lehmann contribution, corresponding to Eq. (4.20):

$$\langle 0| \mathscr{T} \{ {:}\phi\psi(3){:}\,{:}\phi\psi(4){:} \} |0\rangle = \Delta_F(x_4 - x_3, M_\psi^2)\Delta_0(x_4 - x_3) \quad (4.25)$$

which again leads to the necessity of defining the mass pole for the ψ-propagator. We end up with an expression for the propagator similar to the one we obtained for the ϕ-propagator in Eq. (4.23) and with a denominator in the integral containing an (inverse) correction term $1 + \iota_\psi$.

The quantities in Eqs. (4.25) and (4.20) are usually referred to as 'self-energy contributions', indicating that the $\psi(\phi)$-particle may fluctuate into a $\phi\psi(\psi\psi)$-state and back again, i.e. interact with 'its own field'. Just as for the electron in the method of virtual quanta, cf. Chapter 2, it is then necessary to distinguish between the $\psi(\phi)$-quantum itself and the surrounding quantum field, i.e. it is necessary to define the mass of the quantum.

It turns out that in this field theory there are now no other undefined Feynman diagram integrals. After mass renormalisation it contains in each order of perturbation theory only well-defined expressions. As mentioned above, the number of contributions increases very fast with perturbation order and therefore the theory as a whole is not definable by means of our present formulation of perturbation theory.

There is, however, one particular feature which is valid both for ι_ϕ and ι_ψ: *they are both positive-definite functions for spacelike values of $q^2 < 0$. This can be traced back to the properties of the weight functions.* It has the evident consequence that there is a dielectricity function, $\tilde{\varepsilon} \simeq 1 - \iota_\phi^{(1)}(q^2)$ to the lowest order, *which must be always smaller than* 1 (to all orders if it can be defined at all).

This is the most general feature we can prove for any renormalisable or super-renormalisable field theory in which the Källén–Lehmann representation is valid in the form Eq. (4.5). The main point is that the *weight function G in the integral is positive-definite because we are in reality calculating the phase space size of the real intermediate states.*

Actually the weight function generally has the meaning of a *production rate*, i.e. the probability of emitting a ψ-quantum pair from an external (unit) ϕ-source, ϕ_e, carrying energy-momentum P with $P^2 \geq 4M_\psi^2$. To see this we note that the matrix element $\mathscr{M}$ and the transition rate ω will be

$$\mathscr{M} = \frac{g}{2V\sqrt{E_2 E_3}} \int dx \phi_e(x) \exp\left[i(k_2 + k_3)x\right]$$

$$\omega = \frac{g^2}{(2\pi)^3} |\tilde{\phi}_e(P)|^2 dP I_2(P^2, M_\psi^2, M_\psi^2)$$

$$(4.26)$$

where $\tilde{\phi}_e$ is the Fourier transform of the external source $\phi_e(x)$ and I_2 is the two-particle phase space in Eq. (4.8). With normalisation such that $\int |\tilde{\phi}_e(P)|^2 \delta^+(P^2 - a)dP/(2\pi)^4 = 1$ we obtain directly from the distribution-valued limit in Eqs. (3.85), (3.87) that ω agrees with the (negative) real part of the first-order polarisation contribution in Eq. (4.21).

This is just the Kramers-Kronig result for this case: the absorption cross section for the ϕ-field, i.e. the rate of producing ψ-pairs, determines the real part of the dielectricity function while the imaginary part stems from an integral over that quantity, cf. Eq. (4.21).

4.4   The photon propagator in QED and the gluon propagator in QCD

1   Introduction

Before we consider the renormalisation process further we will discuss the results for the propagators in QED and QCD corresponding to those in the previous section. We will start with the properties of the polarisation function and methods for calculating the spin-averaged current matrix elements in QED and QCD.

We will use the results from this calculation repeatedly in the book. It is possible to understand the simple structure without ever entering into the complexities of the Dirac spinors if we use

1 helicity conservation,

2 Lorentz covariance,

3 common sense and simple algebra.

Of these only the first item has not been used before. It is a general property, valid for all massless particles with spin, that the spin must always be directed either along the direction of motion of the particle (*positive helicity*) or in the opposite direction (*negative helicity*). This feature was noticed by Wigner, [112], in his fundamental classification of the Lorentz group. Actually we already know from Chapter 2 that a real (massless) photon, which is a quantum of an electromagnetic radiation field $(\mathcal{E}, \mathcal{B})$ with its motion along the Poynting vector $\mathbf{P} = \mathcal{E} \times \mathcal{B}$, has its polarisation plane in a direction transverse to $\mathbf{P}$ (conventionally along $\mathcal{E}$). Its spin component is then either $+1$ or -1 along the Poynting vector direction (remember how the spherical harmonics Y_1^m look for $m = 0, \pm 1$).

The same goes for a massless spin $1/2$ particle and it is also a good approximation when the particle's rest mass can be neglected compared to its energy ($m \ll e$). For a particle with energy of order its restmass it is always possible to go to its restframe and prepare the spin in any suitable direction and then (although some care is needed in the Lorentz transformations of spins, cf. Chapter 14), it will have a definite direction in any other Lorentz frame. In particular a massive spin 1 particle will have three possible values of its spin, $\pm 1, 0$, along any direction.

There is a precise statement that the electromagnetic current matrix element between an incoming electron and an outgoing electron vanishes (if we neglect the electron's mass) unless they have the same helicity. This is evidently also true for the massless q- and $\bar{q}$-particles in QCD. The implication is that *QED and QCD interactions conserve the helicity of massless charged particles* or in other words the current only couples to the transverse degrees of freedom of the vector potential.

2   *The vector nature of the field theories QED and QCD*

The two major differences between QED (QCD) and the simple scalar version we discussed in section 4.3 are that *QED and QCD are vector theories*, which means that all the operators carry Lorentz vector or tensor indices and that *they have different dimensional properties.*

The fact that the currents are conserved also means restrictions on the different operator matrix elements. In particular the polarisation distribution will in this connection be a tensor, $\rho^{\mu\nu} \equiv \rho^{\mu\nu}(x)$, where

$$\rho^{\mu\nu}(x) = \langle 0| \mathscr{T} \{j^\mu(x)j^\nu(0)\} |0\rangle \tag{4.27}$$

which in space-time and energy-momentum space must fulfil

$$\partial_\mu \rho^{\mu\nu} = \partial_\nu \rho^{\mu\nu} = 0 \quad \Rightarrow \quad q_\mu \tilde{\rho}^{\mu\nu} = q_\nu \tilde{\rho}^{\mu\nu} = 0 \tag{4.28}$$

because it is constructed from conserved currents.

There is only one Lorentz-covariant tensor fulfilling Eq. (4.28) that can be built from a *single* vector q; its Fourier transform has the shape

$$\tilde{\rho}_{\mu\nu} = (q^2 g_{\mu\nu} - q_\mu q_\nu)\tilde{\rho}(q^2), \quad \tilde{\rho}(q^2) = \alpha\tilde{\rho}'(q^2) \tag{4.29}$$

In this way we have defined the *polarisation function* $\tilde{\rho}$ and in the second equation indicated that it is proportional to the fine structure constant, i.e. the squared electric coupling constant $\alpha = e^2/4\pi$. As well as having tensor indices $\tilde{\rho}_{\mu\nu}$ must be expressible in a Källén-Lehmann representation because it fulfils all the requirements needed to derive Eq. (4.4) (note in particular that the current is a real operator). Therefore it should be possible to write for the polarisation function

$$\tilde{\rho}(q^2) = \int \frac{da\sigma(a)}{a - q^2 - i\epsilon} \tag{4.30}$$

where the *polarisation weight function* $\sigma(a)$ stems from the sum over intermediate states with squared mass a. (We note that it is in this case also necessary to be able to sum over the spin of the quanta in these states and we will devise methods for that in the next subsection.)

Further the free photon propagator is

$$\langle 0| \mathscr{T} \{A_\mu(0)A_\nu(x)\} |0\rangle_0 = \frac{i}{(2\pi)^4} \int dq \exp(iqx)\tilde{D}_0(q^2)(g_{\mu\nu} + \text{g.t.}) \quad (4.31)$$

$$\tilde{D}_0(q^2) = \frac{1}{q^2 + i\epsilon}$$

The notation g.t. stands for gauge terms and we have used the conventional notation D_0 for the photon propagator in QED. We have already, in Chapter 2, pointed out that owing to gauge invariance it is possible to make the change $A_\mu \to A_\mu + \partial_\mu \Lambda$ without changing the physical results in any calculation. This is due to the fact that the interaction term can be expressed as follows:

$$\int d^4x g j^\mu(x)A_\mu(x) \to \int d^4x g j^\mu(x)A_\mu(x) - \int d^4x g \Lambda(x)\partial_\mu j^\mu(x) \quad (4.32)$$

On the right-hand side we have performed a partial integration and we find that the added gauge term vanishes owing to current conservation. Evidently gauge invariance and current conservation are intimately connected! Depending upon the gauge choice there are different tensor-indexed contributions to the gauge term g.t. in Eq. (4.31) but *when the field and its propagator are coupled to a conserved current we can ignore these terms*.

The second difference between the simple scalar version and the full QED is the dimensions of the currents. For a scalar field we have already noted that the field operator formally has (positive) energy dimension 1. Therefore the term $:\psi^2:$, which in the last subsection corresponds to the current, has energy dimension 2. In order to obtain the right dimensions for the interaction term it is necessary that the coupling constant, g, multiplying $\phi:\psi^2:$ in the interaction term, also has energy dimension 1. The theory is then super-renormalisable, according to the introduction to this chapter.

For QED and QCD (fermion) currents, which are constructed from Dirac operators, we have instead an energy dimension 3. This means that the coupling constant in Eq. (4.32) is dimensionless and also that *the polarisation tensor has energy dimension* 2 in this case. It corresponds to the matrix element in Eq. (4.27). Comparing to Eq. (4.17) we note that the (positive) energy dimension 6 of the coordinate space $\rho^{\mu\nu}$ is after Fourier transform changed to 2, for $\tilde{\rho}^{\mu\nu}$.

This means that the quantity $\tilde{\rho}$ in Eq. (4.29) is dimensionless and it is also obviously a Lorentz invariant and has a Källén-Lehmann representation. We will now provide a more detailed expression for this quantity.

3   The current matrix elements

In order to obtain the correspondence to Eq. (4.21) for the quantity $\tilde{p}$ in Eq. (4.29) we need a method to sum over the spins in the intermediate states. We start with the contribution to the polarisation tensor from the lowest-mass state. We need the matrix element between the vacuum state and any state containing an electron-positron pair, $\langle k_1, k_2 | j_\mu | 0 \rangle$. Then we may define the sum over the spin states of the tensor γ (we will only write out tensor indices when it is necessary to avoid confusion):

$$\gamma_{\nu\mu} = \sum_{spin} \langle 0 | j_\nu | k_1, k_2 \rangle \langle k_1, k_2 | j_\mu | 0 \rangle \tag{4.33}$$

It is useful to introduce the *reduced matrix element*, denoted by $\{\}$:

$$\langle k_1, k_2 | j_\mu | 0 \rangle \equiv \frac{1}{2V \sqrt{k_{10} k_{20}}} \{ k_1, k_2 | j_\mu | 0 \} \tag{4.34}$$

i.e. we take out the 'ordinary' volume and energy factors from the matrix element. This means that the energy dimension of the reduced matrix element is 1. We obtain the corresponding tensor γ^r (which is Lorentz-invariant due to our conventions in the definition of the weight function in Eq. (4.4) and has energy dimension 2) in terms of these reduced matrix elements:

$$\gamma = \frac{1}{4V^2 k_{10} k_{20}} \gamma^r \tag{4.35}$$

We note that, in order to keep the current conservation condition, γ and therefore also γ^r must fulfil

$$q^\mu \gamma^r_{\nu\mu} = q^\nu \gamma^r_{\nu\mu} = 0 \tag{4.36}$$

with $q = k_1 + k_2$. Further, due to the fact that electromagnetic interactions are parity conserving it must be constructed directly from the vectors k_1, k_2 or from the $g_{\mu\nu}$. This means that γ^r must be constructed from the two tensors T_j, $j = 1, 2$ because these are the only independent combinations that fulfil Eq. (4.36):

$$T_{1\mu\nu} = g_{\mu\nu} q^2 - q_\mu q_\nu, \quad T_{2\mu\nu} = (k_1 - k_2)_\mu (k_1 - k_2)_\nu \tag{4.37}$$

In order to have the right energy dimension, γ^r must then be a linear combination of the T's with coefficients which are dimensionless:

$$\gamma^r = u T_1 + w T_2 \tag{4.38}$$

If the coefficients u, w are to be Lorentz-invariant they can only depend upon the available Lorentz invariants $k_1^2, k_2^2, k_1 k_2$ and if they are to be dimensionless then the dependence must be upon the ratios of these three quantities. For massless particles they must then be plain numbers and,

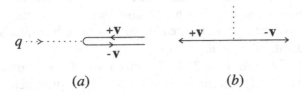

Fig. 4.4. The Breit frame and the cms description of an electron coming in and bouncing back due to a momentum transfer q and an electron-positron pair going apart, respectively.

unless the theory is very singular when the mass approaches 0 (which it is not in this connection), then u, w *must be plain numbers in the general case, too.*

Before we continue we also consider the matrix element $\langle k_1| j_\mu |k_2\rangle$, i.e. the current matrix element between the electron energy-momentum states k_1 and k_2. This will be of interest in connection with lepton scattering, cf. Chapter 5. In that case, in order to calculate the cross section we will need the spin-summed matrix element combination

$$\hat{\gamma}_{\nu\mu} = \sum_{spin} \langle k_2| j_\nu |k_1\rangle \langle k_1| j_\mu |k_2\rangle \tag{4.39}$$

We may again introduce the corresponding reduced matrix element, defined in an obvious way, and the corresponding Lorentz-covariant tensor $\hat{\gamma}^r$. Current conservation again must hold *but this time we must change the definition of q to $q = k_1 - k_2$.*

The fact that with the reduced matrix elements and tensors we obtain the same result for γ^r and $\hat{\gamma}^r$ with the exchange $k_2 \to -k_2$ is obvious for the scalar field theory we discussed in the earlier section. It is called *crossing symmetry.* It takes a little effort to prove that it also works for the vector theories QED and QCD but it is nevertheless true and *it is one of the few very general properties which is valid in any field theory.*

To see that u and w must be equal we take recourse to helicity conservation for a vanishing lepton mass. This means that the current matrix elements only couple to the transverse degrees of freedom of the electromagnetic (four)-potential A. The transverse directions are well defined when the electron and positron go out in opposite directions e.g. along the 3-axis, as they do in the cms, for the tensor γ^r. For the tensor $\hat{\gamma}^r$ the same is true in the so-called Breit frame (see Fig. 4.4). This is sometimes called the 'brick-wall frame' for easily understood reasons, i.e. the electron comes in and after the interaction bounces out again with the same energy backwards along the 3-axis.

Thus helicity conservation means that if $k_j^2 = 0$ then the 00- and 33-components of the tensors γ^r (in the cms) and $\hat{\gamma}^r$ (in the Breit frame) should vanish. We leave it to the reader to convince him-/herself that this happens if and only if $w = u$.

The fact that the sum $T_1 + T_2 \equiv T_\gamma$ only has tensor components in the directions transverse to the momentum transfer (lepton scattering in the Breit frame) or the production axis ($e^+ e^-$ annihilation in the cms) means that all its time components vanish and its space part is proportional to the tensor $t(\gamma)$ (using $\mathbf{k}$ as a vector along one of these directions):

$$t(\gamma)_{jl} = \delta_{jl} - \frac{k_j k_l}{\mathbf{k}^2} \tag{4.40}$$

This space tensor occurs when we use transverse wave solutions to describe a photon (gluon in QCD) with energy-momentum $k = (k_0, \mathbf{k})$, i.e. $\mathbf{A} = \epsilon \exp(ikx)$, and would like to sum over the polarisation directions of the square of the wave function:

$$\sum_{polarisation} \epsilon_k \epsilon_l \equiv t(\gamma)_{kl} \tag{4.41}$$

(note that 'transverse' means that $\mathbf{k} \cdot \epsilon = 0$ and the normalisation comes from the fact that there are two transverse directions). The tensor T_γ is a continuation of $t(\gamma)$ to values of k outside the mass-shell $k^2 = 0$ for a real photon (gluon). The result is a consequence of the relationship between current conservation and gauge invariance, cf. Eq. (4.32).

We may now calculate the polarisation weight function σ, occurring in Eq. (4.30), to lowest order:

$$(q^2 g_{\mu\nu} - q_\mu q_\nu)\sigma^{(1)}(q^2) = (2\pi)^3 \sum \delta(k_1 + k_2 - q)\gamma_{\mu\nu}$$
$$= \frac{e^2}{(2\pi)^3} \int dk_1 dk_2 \delta(k_1^2 - M^2)\delta(k_2^2 - M^2)$$
$$\times \delta(k_1 + k_2 - q)\gamma_{\mu\nu}^r \tag{4.42}$$

The simplest way to obtain an expression for $\sigma^{(1)}(q^2)$ is to take the trace of the tensors on both sides of the equation. We note that $\operatorname{tr}(q^2 g_\mu^\nu - q_\mu q^\nu) = 3q^2$ and that $\operatorname{tr}\gamma^r = u(2q^2 + 4M^2)$ (prove that with $u = w$!) and therefore we obtain immediately

$$\sigma^{(1)}(q^2) = \frac{2e^2 u}{3(2\pi)^3}\left(1 + \frac{2M^2}{q^2}\right)$$
$$\times \int dk_1 dk_2 \delta(k_1^2 - M^2)\delta(k_2^2 - M^2)\delta(k_1 + k_2 - q) \tag{4.43}$$

We recognize in the integrand the expression for the polarisation function of the simpler case in Eq. (4.21) (*the two-particle phase space*). It is multiplied by a factor $1 + 2M^2/q^2$ from the tensor structure (*the spin-correlation factor*) and a different factor $2e^2 u/[3(2\pi)^3]$ in front.

The (squared) scalar coupling constant (which due to combinatorics is multiplied by the factor 2 in Eq. (4.21)) is exchanged for $2ue^2/3$. The factor $2/3$ stem from the fact that massless fermions only couple to two (the transverse ones) of the three vector degrees of freedom (cf. the discussion of the tensor $t(\gamma)$ in Eq. (4.40)). Therefore the unknown quantity u should equal unity, which is confirmed in more elaborate calculations with the full Dirac formalism.

It is worthwhile to note that the spin-correlation factor, within the large parentheses, contains a term proportional to M^2/q^2 which corresponds to a correction for massive particles. Such terms occur frequently but evidently vanish in the limit of large squared momentum transfer (or cms energy) q^2. They are known as 'higher-twist corrections'.

In this way we obtain the result for the first-order perturbative correction to $\tilde{p}$:

$$\tilde{p}_{QED}^{(1)} = \frac{\alpha}{3\pi} \int_{4M^2} \sqrt{1 - \frac{4M^2}{a}} \left(1 + \frac{2M^2}{a}\right) \frac{da}{a - q^2 - i\epsilon} \qquad (4.44)$$

Before we end this subsection we note that the tensor T_2 defined above can be written solely in terms of the initial electron energy-momentum k ($= k_1$) and the momentum transfer $q = k_2 - k_1$. Thus

$$k_1 + k_2 = 2k + q = 2(k - (kq/q^2)q) \equiv 2\hat{k} \qquad (4.45)$$

which is true for elastic scattering because of the identity

$$k_2^2 = (q + k)^2 = k^2 + 2qk + q^2 \Rightarrow q^2 = -2kq \qquad (4.46)$$

when the lepton is on the mass shell before and after the interaction. Note, however, that the vector $\hat{k}$ fulfils $\hat{k}q = 0$ independently of the mass-shell condition. We will meet this vector later in connection with inelastic scattering situations.

4 Dyson's equation for QED

Dyson's equation, Eq. (4.18), is for the full photon propagator $\tilde{D}$

$$\tilde{D}_{\mu\nu} \equiv \tilde{D}g_{\mu\nu} + \text{g.t.} = \tilde{D}_0(q^2)g_{\mu\nu} - \tilde{D}_0(q^2)\tilde{p}_{\mu\lambda}\tilde{D}_\nu^\lambda + \text{g.t.} \qquad (4.47)$$

From this expression we obtain, using the results of the earlier subsections, the solution

$$\alpha_u \tilde{D} = \frac{\alpha_u}{(q^2 + i\epsilon)[1 + \alpha_u \tilde{p}'(q^2)]} \qquad (4.48)$$

The gauge terms, which do not contribute if the propagator is coupled to a conserved current, may be neglected. We have explicitly exhibited the dependence on the (unrenormalised) coupling α_u according to Eq. (4.44).

There are two features of this result worth pondering:

- Owing to the tensor character and the (energy) dimensions of the polarisation tensor we have in Eq. (4.48) obtained the mass-renormalised photon propagator without the subtraction necessary in Eq. (4.23). *The photon must always be massless and this can be traced back to gauge invariance and current conservation.*

- The function $\tilde{\rho}'$ is defined by a non-convergent integral. This is noticeable for the lowest-order term in Eq. (4.44). A few further terms are known in the perturbation-theoretical expansion of $\tilde{\rho}'$. They exhibit the same sign and scaling behaviour as the one written out in Eq. (4.44). The *sign of the correction term can again be traced back to the positive-definiteness of the corresponding weight function G in Eq. (4.5)*, i.e. to the fact that we obtain positive contributions from the real intermediate states in the weight function.

Before we perform the necessary renormalisations for QED we will consider the differences for the equations derived above in QCD. In this case the current coupling to the gluon propagator contains contributions both from the quark-antiquark currents and from the field self-interaction, the *three-gluon vertex coupling* (there is also a 'local' four-gluon vertex necessary to keep to the symmetries of the theory but it does not change the conclusions). This field self-interaction is different because it corresponds to a coupling between three *vector* particles. We will find that this contribution means a large difference between the polarisation function in QCD and that in QED, where there is no such interaction possible between the chargeless photons (although they also are vector particles).

The fermion contribution is the same as we have met before. Thus the $q\bar{q}$ intermediate state will give a contribution per flavor (evidently each flavor provides an independent contribution) equal to the result in Eq. (4.44) with the exchange $\alpha_{QED} \rightarrow \alpha_s/2$. The factor 2 is due to an unfortunate convention in the normalisation of the QCD coupling constant and we will meet it further on also.

For the gluonic contributions to the weight function we find the surprising result that the total contribution is no longer positive, [68]. This is very disturbing because we have repeatedly pointed out that the definite sign in the Källén-Lehmann description of the polarisation function stems from the fact that we sum over positive contributions from the intermediate physical states. Depending upon the gauge choice there are different ways to obtain the result but the gauge-independent result is a function with the same properties as in Eq. (4.44) although with the opposite sign.

4.5 Two reasons why in QCD the polarisation tensor behaves differently; the introduction of cut diagrams

In this subsection we will provide two ways of getting an intuitive understanding for the negative contributions to the polarisation weight function for the gluons (in subsection 2 of the next section we will present a third way to see the difference between QED and QCD bremsstrahlung emission). At the same time it will provide us with the possibility of introducing higher-order corrections, such as the *vertex corrections* (usually termed 'virtual corrections'), in a natural way. In order to clarify the relationship between a Feynman diagram and the weight function of its Källén-Lehmann representation we will define the notion of *cut diagrams*.

The first argument for the behaviour of σ_{QCD} in Eq. (4.29) is that the negative contributions stem from a lack of phase space for the real emitted gluons in the intermediate states. The second reason we provide is that there is a difference between the states containing transversely polarised gluons and those containing Coulomb interaction gluons.

According to the first argument, when we calculate to a certain order of perturbation theory and two gluons are emitted too close in phase space (i.e. too close in angle or rapidity) then they will be reabsorbed into a single gluon again, at the next order. This is at the specified order noticeable as an available phase space for real gluon emission and as a *larger* phase space for the absorption, i.e. for the virtual corrections to this emission process. *This will result in a negative contribution* to the polarisation weight function σ in Eq. (4.30). (The implication is that the theory should be formulated in terms of 'effective gluons', which are not reabsorbed; we will do that in sections 18.5 and 18.6, where we introduce an approximation method called discrete QCD.)

For the second argument we note that the Coulomb gluons are not real degrees of freedom to be quantised in the QCD field (there is always a Coulomb field around any gauge theory charge). If, nevertheless, the interactions with the Coulomb fields are incorporated into the Feynman diagrammatical description then the occurrence of Coulomb gluons in a state provides negative contributions to the state sum (they have a negative metric in the Hilbert space of the states, cf. the Gupta-Bleuler formalism in e.g. [30]). Therefore the weight function σ in the Källén-Lehmann representation does not need to provide positive contributions from the states containing Coulomb gluons (needless to say the two descriptions of the phenomena are equivalent!).

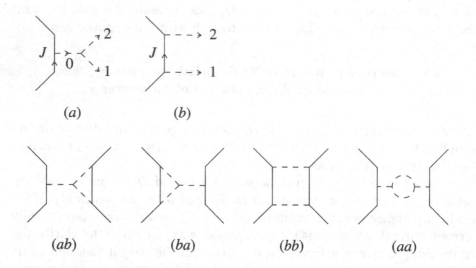

Fig. 4.5. The diagrammatic description of the matrix elements between the current and the two gluons: the contributions $\mathscr{J}_k, k = a, b$, and the three contributions to the polarisation function obtained by squaring the matrix element $\mathscr{J}$ (note that there is a second symmetric vertex contribution in C_{ab}).

In order to relate to the QED calculations above we consider a current emitting two gluons g_1, g_2. (Gluon bremsstrahlung emission is treated in great detail in Chapters 16, 17 and 18.) We consider the process in a transverse gauge, i.e. with the $g_j, j = 1, 2$ polarised e.g. transverse to the current direction. This process can occur according to perturbative QCD diagrams in two different ways. Either there is a 'first' emission of a gluon: $J \to Jg_0$, with J the current. Afterwards the gluon decays via the three-gluon vertex as $g_0 \to g_1 g_2$, see Fig. 4.5(a). This is similar to the decay into a (fermionic) e^+e^--pair of a massive (i.e. off-shell) photon stemming from the emission of a QED current. *But this time we are dealing with two vector particles in the final state and this makes a major difference.*

There is, to the same order in perturbation theory, a second way to emit the gluons, i.e. sequentially as $J \to Jg_1 \to Jg_1 \oplus g_2$, see Fig. 4.5(b). The total matrix element for the bremsstrahlung emission is therefore a sum of two contributions, $\mathscr{J} \equiv \mathscr{J}_a + \mathscr{J}_b$ (in easily understood notation). As the contribution to the polarisation weight function contains the absolute square $|\mathscr{J}|^2$ there will be in principle three contributions, two from real gluon emission, $C_{aa} = |\mathscr{J}_a|^2$, $C_{bb} = |\mathscr{J}_b|^2$, and a correction from the interference term $C_{ab} = 2\,\mathrm{Re}(\mathscr{J}_a^* \mathscr{J}_b)$. A closer examination tells us, however, that (see Figs. 4.5(aa), (bb), (ab) and (ba)) only C_{aa} and C_{ab} correspond to corrections to the polarisation function of second order in the coupling constant.

To clarify this statement we note the simple relationship between the weight function and the polarisation function in the Källén-Lehmann representation. The weight function will contain the square of the matrix elements (obtained in a certain perturbative order) between the initial state and a state containing some particular on-the-mass-shell configuration, e.g. the particles $p_1, \ldots, p_n$. For the case discussed above the initial state is a current in the vacuum and the intermediate state contains also a two-particle state, which may be emitted from the current.

If the matrix element contains several terms, each leading to this state, then we must consider the overlap of all the terms. To obtain the sum over the intermediate state it is necessary to consider the product of one term, say $\mathscr{J}_a$, and the complex conjugate of another term, say $\mathscr{J}_b^\star$, etc. All these overlap integral terms can be considered as diagrammatic contributions as exhibited in Fig. 4.5. But we note that C_{bb} in this way corresponds to two-gluon exchange for the current, i.e. it is not part of the corrections to single-gluon emission.

The difference between the weight function and the polarisation function is that the weight function is obtained by putting the intermediate state on the mass shell, i.e. each line corresponds to $\delta^+(p^2 - m^2)$, while the polarisation function corresponds to using the corresponding Feynman propagator $(p^2 - m^2 - i\epsilon)^{-1}$. Actually we are again invoking the distribution-valued relationship obtained in Eqs. (3.85), (3.87). The operation of introducing δ-distribution(s) instead of propagator(s) is called *cutting the diagrams* and we will meet this notion later on in the book.

We will now consider the contributions in more detail, using the transverse gauge. We assume that the two gluons g_1 and g_2 are emitted with compensating transverse momenta $\pm \mathbf{k}_\perp$ with respect to the polarisation direction. Further we assume that their combined squared mass a (corresponding to the 'virtuality' of g_0 and to the a-variable in Eq. (4.44)) is very large, $a \gg \mathbf{k}_\perp^2$. Then the available rapidity region for the emission in the contribution C_{aa} is $\Delta y = \log(a/\mathbf{k}_\perp^2) - 11/6$. The result (including the peculiar number $11/6$) is further clarified in section 18.5.

There are two comments on the result. The first is that this is evidently a large rapidity region, growing logarithmically with $a/\mathbf{k}_\perp^2$, and secondly it is a result typical of vector emission. If we consider the emission of massless fermions, i.e. the contribution $g_0 \to q\bar{q}$, then *there is no such logarithmic contribution to the available rapidity region.*

The difference is that if we emit two spin $1/2$ particles from a vector then helicity conservation (cf. section 4.4 above) implies that they would like to be close together in phase space (to make $1/2 + 1/2 = 1$ with respect to the helicity states). Then the contribution to the weight function is constant for large values of a, as seen in Eq. (4.44) (the result $\alpha/3\pi$ means that the effective rapidity difference will be $2/3$ as we will see in detail in section 18.5). But for the vector emissions the final-state vector gluons must go in different directions to conserve the helicity. Therefore *vectors will tend to spread apart in rapidity space*. A more precise mathematical statement is that the (relative) rapidity (y) dependence for a given $k_\perp$ is proportional to dy for the vector (gg) emissions and to $dy\exp(-y)$ for the $q\bar{q}$ emission.

The vector emission contribution will therefore provide a factor proportional to the available rapidity region, i.e. it grows logarithmically with the integration variable a in the Källén-Lehmann representation. It is not difficult to see that for states containing more gluons there will be logarithmic factors with a power growing with the number of gluons in the intermediate state.

It is nevertheless a fact that QCD is renormalisable (although t'Hooft, who was first to provide the proof, had to work very hard!). The reason is that the logarithmic rapidity-difference term from C_{aa} is cancelled by the C_{ab} corrections, the 'vertex corrections'. If we calculate the interference term C_{ab} in the transverse gauge we find that, just as for the gluon emission in C_{aa}, it depends upon the rapidity difference $\delta y = \log(a/\mathbf{k}_\perp^2)$. *It will provide a contribution δy with the opposite sign to the contribution Δy of the emission term C_{aa}.* Therefore to this order in the coupling constant (and it can be shown to all orders, too, which actually is necessary for the renormalisability property) there is no δy-dependence in the weight function of the polarisation tensor in QCD.

There is, however, the term $-11/6$ left over from combining the vector emission and vertex correction terms and this really has the meaning, according to section 18.5, that there is a depletion of gluon emission close to an already emitted gluon. Therefore the gluon contribution to the polarisation weight function in QCD will for large a-values go to a constant, just as do the fermionic contributions $(N_c\alpha/2\pi)(-11/6)$, with $N_c = 3$ the number of colors, cf. section 18.6, subsection 1.

Another way to understand this result is to note that every charged particle is surrounded by a Coulomb field and this also goes for the gluonic (octet) charges. As soon as we produce a 'physical transverse' gluon then it is necessary to handle the interaction between this gluon and its Coulomb field. Therefore gluons in QCD do not behave like the photons described by the method of virtual quanta (MVQ) (cf. section 2.5). The gluons are not independent of the fields, i.e. they will reinteract on the way out. Actually such Coulomb vector particle interactions do not provide positive-definite contributions to the Källén-Lehmann weight function because the wave functions are not positive-definite in the state space. We may intuitively say that in order to be able to have room for the vector Coulomb fields the two vectors must have an effective rapidity

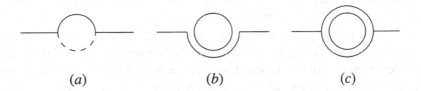

Fig. 4.6. (*a*) The diagrammatic description of a self-energy contribution, i.e. in QCD the (color-3) *q*-field propagator (full line) turns into a *qg*-state firstly emitting and afterwards absorbing the color-8 *g*-quantum (broken line); (*b*) the color flow in diagram (*a*); (*c*) the corresponding color flow in a $g \to gg$ intermediate state.

difference $-11/6$. In the last subsection of this chapter we provide one further intuitive picture of the result, this time related to one of its major implications, asymptotic freedom.

1 The color factors of QCD

In the last subsection we considered the emission of gluon states from a QCD current but we did not specify the current in any detail. Suppose, however, that the current is a quark current so that we consider the emission of color-8 gluons from a color-3 current. Then there is a subtle but necessary color factor correction in the sum over colors in the squared matrix element. To see this we consider Fig. 4.6(*a*), which is a self-energy correction corresponding to Eq. (4.25), i.e. a quark *q* (propagator) fluctuates into a *qg*-state and back again.

We may compare that to the situation when a gluon decays into two gluons and afterwards rearranges into a single gluon as in Fig. 4.5(*aa*). In both cases we find that there is principally a new color produced, i.e. we may draw the color lines as in Figs. 4.6(*b*) and (*c*) with a closed color ring in the middle.

For the *q*-state we note that we start out and end in a coherent color-3 state, containing *r*, *g* and *b*. For the sake of argument we may project e.g. onto the initial state color *r*. In the intermediate state this color-3 can then turn into a *g* or *b* by the emission of a $r\bar{g}$ or a $r\bar{b}$, which are both true color-8 states, i.e. they correspond to the gluon in the intermediate state.

But if the gluon emission corresponds to $r\bar{r}$ then there is a color-coherence suppression factor because only two out of the three possible states are really color octets. The third color combination is a color singlet, i.e. not a gluon. Therefore only $2 + 2/3$ of the possible 3 choices are really gluons. We obtain a factor $3 - 1/3 \to N_c - 1/N_c$ multiplying the kinematical matrix element instead of the expected factor N_c from the closed color ring, with N_c the number of colors.

For the gluon propagator we are, however, reassured that due to the coupling we always obtain a true gluon and therefore the relative color weight between the two states is $1 - 1/N_c^2 = 8/9$. This is also the relative coupling between the emission of gluons from q- or $\bar{q}$-currents and the emission from a g-current.

2   *The operations in multiplicative renormalisation*

We will in this subsection exhibit the way one can rearrange the propagator equations by means of a *multiplicative renormalisation scheme*. One basic assumption for what we are going to do is that we already have performed mass renormalisation for the photon (gluon) propagator. We have seen that within QED this is trivially possible by making explicit use of gauge invariance and current conservation for the photon propagator in Eq. (4.48), and there is a correspondence in QCD.

We will start with the result in Eq. (4.48) and note that we may rearrange it in the following way:

$$\alpha_u \tilde{D} = \frac{1}{(q^2 + i\epsilon)} \frac{1}{\left\{[1/\alpha_u + \tilde{\rho}'(-\mu^2)] + [\tilde{\rho}'(q^2) - \tilde{\rho}'(-\mu^2)]\right\}} \Rightarrow$$

$$\alpha_\mu \hat{D}_\mu = \frac{1}{(q^2 + i\epsilon)} \frac{1}{\left\{1/\alpha_\mu + \hat{\rho}'(q^2, \mu^2)\right\}} \Rightarrow$$

$$\hat{D}_\mu = \frac{1}{(q^2 + i\epsilon)} \frac{1}{\left\{1 + \alpha_\mu \hat{\rho}'(q^2, \mu^2)\right\}} \tag{4.49}$$

with

$$\frac{1}{\alpha_\mu} = \frac{1}{\alpha_u} + \tilde{\rho}'(-\mu^2)$$

$$\hat{D}_\mu = Z_{3,\mu^2}^{-1} \tilde{D}$$

$$Z_{3,\mu^2} \alpha_u = \alpha_\mu \tag{4.50}$$

$$\hat{\rho}'(q^2, \mu^2) = (q^2 + \mu^2) \int_{4M^2} \frac{da\sigma(a)}{(a + \mu^2)(a - q^2 - i\epsilon)}$$

$$\sigma^{(1)}(a) = \frac{\alpha}{3\pi} \sqrt{1 - \frac{4M^2}{a}} \left(1 + \frac{2m^2}{a}\right)$$

In Eq. (4.49) we have item by item rearranged the unrenormalised quantities $\alpha_u, \tilde{D}, \tilde{\rho}'(q^2)$ so that only the renormalised correspondences (defined at the effective frequency $q^2 = -\mu^2$) $\alpha_\mu, \tilde{D}_\mu, \tilde{\rho}'(q^2, \mu^2)$ occur. In the last line of Eq. (4.50) we have written out the lowest-order approximation to the weight function σ occurring in the Källén-Lehmann representation for the polarisation function (cf. Eq. (4.30)).

What we have achieved by these operations can be formulated in the following way:

R1 We have introduced a *coupling constant* α_μ, *renormalised at the scale* μ^2, by multiplying the unrenormalised coupling constant α_u by the quantity Z_{3,μ^2}, which is formally defined by

$$Z_{3,\mu^2} = 1 - \alpha_\mu \tilde{\rho}'(-\mu^2) \tag{4.51}$$

R2 We have introduced a *propagator* $\hat{D}_\mu$ *renormalised at the scale* μ^2, by multiplying the (unrenormalised) photon propagator $\tilde{D}$ by the inverse quantity Z_{3,μ^2}^{-1}. This is equivalent to exchanging the unrenormalised photon-field operator for a new scaled operator, $A_\nu^{(\mu^2)}$, the *renormalised photon operator*:

$$A_\nu^{(u)} \rightarrow A_\nu^{(u)} = A_\nu^{(\mu^2)} Z_{3,\mu^2}^{1/2} \tag{4.52}$$

Note that the 'size' of a field operator is not observable. The only requirement is that the propagator, i.e. the expectation value of the square of the field operator in the vacuum state, should correspond to the contribution from a single massless quantum at the renormalisation scale $q^2 = -\mu^2$.

R3 We have defined all the new renormalised quantities by a subtraction at the arbitrary (negative) value $q^2 = -\mu^2$. We could, of course, also have done it at $\mu^2 = 0$ or any other value $q^2 < 4M^2$ such that our integrals converge. If we make the exchange $\mu^2 \rightarrow \mu_1^2$ we can again do all the changes in the same way and obtain a new set, $\alpha_{\mu_1}, A_\nu^{(\mu_1^2)}, Z_{3,\mu_1^2}$, which is related to the old one by the same equations. The fact that we may do repeated changes of scale $\mu_1^2 \rightarrow \mu_2^2 \rightarrow \cdots \rightarrow \mu_n^2$ and still end up with the same μ_n^2-dependent renormalised quantities means that there is a group character to the procedure, that of the renormalisation group.

R4 In particular, if we chose to define α at the point $\mu^2 = 0$ we would find for the fine structure constant the well-known value $\alpha \equiv \alpha(0) \sim 1/137$, which is observed for static interactions. It is of interest to note that at LEP with a (cms-)energy value of the annihilating e^+e^--pair ~ 90 GeV one obtains an effective coupling $\alpha(90) \simeq 1/128$. This is in accordance with this finite renormalisation group prediction of a change in α of $0 \rightarrow 90$ GeV.

In this way we have exhibited in some detail the procedure for redefining the photon field, the photon propagator and the coupling constant in QED. Both within QED and QCD there are other undefined quantities and also other integrals which need a redefinition in the same way as the photon operator in Eq. (4.52). In particular the fermion operators will need a renormalisation such that $\psi^{(u)} \to Z_{2,\mu^2}^{-1/2} \psi^{(\mu^2)}$, with the conventional wave function renormalisation constant Z_{2,μ^2}. The quantity $Z_{2,\mu^2}^{-1/2}$ is defined at the same effective frequency $q^2 = -\mu^2$ as that of the photon, Z_{3,μ^2}, and in this way the renormalised operator $\psi^{(\mu^2)}$ will describe a single quantum at this frequency. There is finally the vertex renormalisation constant, conventionally called Z_{1,μ^2}, which corresponds to a renormalisation $\Gamma^{(u)} \to Z_{1,\mu^2} \Gamma^{\mu^2}$ of every Feynman graph vertex, so that $\Gamma^{\mu^2} = 1$ for the particular momentum transfer $q^2 = -\mu^2$ at the vertex.

One essential result in QED, which also has a correspondence in QCD, is called the Ward identity: $Z_{1,\mu^2} = Z_{2,\mu^2}$. This relation stems from the current conservation and gauge invariance properties of the theory. The consequence is that for every vertex in QED for which two fermion operators and one photon operator are connected, one obtains the rescaling factor $Z_3^{-1/2}(Z_2^{-1/2})^2$. At the same time the vertex itself provides the rescaling Z_1 and the coupling constant the rescaling $Z_3^{1/2}$ according to the third line in Eq. (4.50). In this way we obtain the result that at every vertex there is a renormalised contribution $Z_3^{-1/2}(Z_2^{-1/2})^2 Z_1 Z_3^{1/2} \equiv 1$. *The result is that all the Feynman diagrams in the theory will have the same form of expression as before but now in terms of renormalised propagators and charges.* And now everything is finite (although μ^2-dependent)!

Thus the result is that if we consider a scattering situation or a multiparticle production diagram in QED containing n_γ in- or out-going photons and n_ψ in- or out-going Dirac particles, which in the unrenormalised form looks like

$$F_{n_\gamma, n_\psi}^{(u)}(k_1, \ldots, k_{n_\gamma}; p_1, \ldots, p_{n_\psi}; \alpha_u) \tag{4.53}$$

it will after renormalisation look like (note that the 'external' renormalisations are not cancelled!)

$$Z_1^{-n_\psi/2} Z_3^{-n_\gamma/2} F_{n_\gamma, n_\psi}(k_1, \ldots, k_{n_\gamma}; p_1, \ldots, p_{n_\psi}; \alpha) \tag{4.54}$$

While the quantity in Eq. (4.53) is independent of the renormalisation point, the one in Eq. (4.54) will contain a μ-dependence both in the scaled out Z-factors, in the renormalised coupling constant and in all the renormalised propagators. We will not in this book go into further details of the renormalisation process because we do not need it here. The formalism can be found in any field theory text-book and does not provide much more physical insight than the results we have already encountered.

4.6 The Callan-Symanzik equations for the renormalisation group

1 The equations and what they imply for QCD

There is evidently nothing sacred about the particular value μ we have chosen in connection with the renormalisation procedure described in the last section. The quantity μ can be varied at will (within the region of no singularities). Therefore we can formulate the dependence upon μ easily. The *unrenormalised function F in Eq. (4.53) is independent of μ*:

$$\mu \frac{dF^{(u)}}{d\mu} = 0 \tag{4.55}$$

If we rewrite this in terms of the renormalised function we obtain immediately a partial differential equation:

$$\left(\mu \frac{\partial}{\partial \mu} + \beta \frac{\partial}{\partial \alpha} - \frac{n_\gamma \gamma_\gamma}{2} - \frac{n_\psi \gamma_\psi}{2} \right) F = 0 \tag{4.56}$$

with

$$\beta = \mu \frac{\partial \alpha}{\partial \mu}, \quad \gamma_\gamma = \mu \frac{\partial \log Z_3}{\partial \mu}, \quad \gamma_\psi = \mu \frac{\partial \log Z_1}{\partial \mu} \tag{4.57}$$

In the partial differentiations of Eq. (4.57) the unrenormalised coupling constant α_u and any cutoff parameters used in order to make the integrals finite should be kept fixed. Equation (4.56) is the *Callan-Symanzik equation*, [108] and it connects different possible renormalisation points (note that as it may contain several related functions it may be of a matrix character). The β-function in Eq. (4.57) for QED is thus, in the lowest order of perturbation theory, given by differentiating the following expression (cf. the first line of Eq. (4.50)),

$$\alpha = \frac{\alpha_u}{1 + \alpha_u \tilde{\rho}'(-\mu^2)} \tag{4.58}$$

with

$$\tilde{\rho}'(-\mu^2) = \frac{1}{3\pi} \int_{4M^2}^{\Lambda^2} \sqrt{1 - \frac{4M^2}{a}} \left(1 - \frac{2M^2}{a} \right) \frac{da}{a + \mu^2}$$

$$\simeq \frac{1}{3\pi} \log \left(\frac{\Lambda^2}{\mu^2} \right) \tag{4.59}$$

where we have assumed that $4M^2 \ll \mu^2 \ll \Lambda^2$ so that we may neglect all the dependence upon the finite-mass parameters and only keep the logarithmic singularity of the integral.

It is then easy to see that

$$\beta_{QED}^{(1)} = \frac{2\alpha^2}{3\pi} \qquad (4.60)$$

In general it is evident that in the limit in which we only keep the (logarithmically) divergent terms then *all the functions β and γ, defined in Eq. (4.55), are solely functions of the renormalised coupling constant.*

The β-function in QCD can also be calculated and one obtains to the same order as in Eq. (4.60) the result, [68],

$$\beta_{QCD}^{(1)} = -\alpha^2 \left(\frac{11}{4\pi} - \frac{n_f}{6\pi} \right) \qquad (4.61)$$

We note the different signs in front of the squared coupling constants for QED and for QCD (at least as long as there are less than 16 flavors!).

We will end this subsection by solving the Callan-Symanzik equations for the two cases of QED and QCD. We will use the following notation for the β- and γ-functions:

$$\beta_{QED} = b_e \alpha^2, \quad \beta_{QCD} = -b_c \alpha^2$$
$$\gamma_e(\alpha) = \ d_e \alpha, \quad \gamma_c(\alpha) = d_c \alpha \qquad (4.62)$$

where b_e, b_c are positive numbers. The choice for γ, that it is linear in α, is the case we are going to use in Chapter 19 when we encounter the following Callan-Symanzik equation:

$$\left(\mu \frac{\partial}{\partial \mu} + \beta \frac{\partial}{\partial \alpha} - \gamma \right) F(\log(Q^2/\mu^2), \alpha) = 0 \qquad (4.63)$$

We here assume that the distribution F depends (logarithmically) upon a single scaled Lorentz invariant variable Q^2/μ^2 and upon the coupling constant α and we neglect all other dimensional scales in the problem, such as e.g. mass thresholds etc. (cf. [102] where possible observables stemming from the contributions from the mass thresholds are given).

The variable q^2 we used before for the propagators is related to the variable $Q^2 = -q^2$, i.e. we assume that the (Lorentz-invariant) function $F = F(q)$ is taken for large spacelike energy-momentum vectors.

The Callan-Symanzik equations are linear partial differential equations of a kind which occurs very often both in physics and in other disciplines. They are usually called *gain-loss equations*. They correspond to situations when a distribution changes in 'time', which here corresponds to

$$t = \log \mu, \qquad (4.64)$$

by a gain term, in this case γ times the value of the distribution, and by a loss term, in this case β times the derivative of the distribution with respect to some variable, here the coupling constant α.

As an example, taken from Coleman's Erice Lectures, [45], assume that the distribution F corresponds to the density of a population of bacteria moving with a fluid along a pipe. The fluid has velocity $\beta(\alpha)$ with α a position coordinate along the pipe. As they move along there is a changing illumination γ, which determines their rate of reproduction.

We will later consider another example corresponding to the increase in gluon multiplicity and phase space size due to the change in the resolution scale of a parton cascade, cf. Chapter 18. There is a very simple way, called the method of rays, by means of which we can solve this kind of equation.

We start by noting that if we define the *effective coupling constant*, α_{eff}, by means of the equation

$$\frac{d\alpha_{eff}}{dt} = \beta(\alpha_{eff}) \tag{4.65}$$

in terms of the variable t in Eq. (4.64) then Eq. (4.63) becomes

$$\left(\frac{d}{dt} - \gamma(\alpha_{eff})\right) F(L_Q - 2t, \alpha_{eff}(t)) = 0 \tag{4.66}$$

We have then used the notation $L_Q = \log Q^2$ and rearranged the dependence upon α into a dependence upon the effective coupling constant. The earlier partial differential equation is in this way changed into an ordinary differential equation with a t-dependent $\alpha \equiv \alpha_{eff}$.

This means that the quantity $\mathscr{F}$ is a constant

$$\mathscr{F} = F(L_Q - 2t, \alpha_{eff}) \exp\left[-\int^t dt' \gamma(\alpha_{eff}(t'))\right] \tag{4.67}$$

along all 'rays'; those correspond to the solutions for Eq. (4.65).

For QCD and QED we may construct these rays from Eq. (4.62):

$$\alpha_{eff,QED} = \frac{1}{c_e - b_e t}, \quad \alpha_{eff,QCD} = \frac{1}{c_c + b_c t} \tag{4.68}$$

where c_e, c_c are constants. The main property is that independently of the value of the constant c_c if we choose the scale $\mu = \exp t$ sufficiently large *then for QCD the effective coupling constant will vanish but for QED the effective coupling will instead increase with μ* (cf. the result in connection with the LEP experiments in remark R4 above).

This means that the exponential factor in Eq. (4.67) is for QCD given by (introducing the expression for $\gamma(\alpha_{eff}(t))$ from Eq. (4.62))

$$\exp\left[-\int^t dt' \gamma(\alpha_{eff}(t'))\right] = (c_c + b_c t)^{-d_c/b_c} \equiv [\alpha_{effQCD}(t)]^{d_c/b_c} \tag{4.69}$$

and for QED with obvious changes there is a corresponding result.

Now, let us assume that we would like to know the function F for some scale corresponding to μ_1, where the coupling constant is α_1. From Eq.

(4.67) we may then immediately write for the QCD case

$$F(\log(Q^2/\mu_1^2, \alpha_1)(\alpha_1)^{d_c/b_c}$$
$$= F(\log(Q^2/\mu^2), \alpha_{eff,QCD}(t)) \left[\alpha_{eff,QCD}(t)\right]^{d_c/b_c} \tag{4.70}$$

In particular there is nothing to stop us from choosing the scale $t = L_Q/2$.
From this we conclude that

$$\lim_{Q^2 \equiv \mu^2 \to \infty} \left\{ F(\log(Q^2/\mu_1^2, \alpha_1)) \right\}$$

$$= \lim_{Q^2 \to \infty} \left\{ F(0, \alpha_{eff\,QCD}(L_Q/2)) \left[\frac{\alpha_{eff,QCD}(L_Q/2)}{\alpha_1}\right]^{d_c/b_c} \right\} \tag{4.71}$$

As the effective coupling constant for QCD vanishes in this limit we may
write in the second line $F(0, \alpha_{eff,QCD}(L_Q/2)) \simeq F(0,0)$. Thus we have found
a simple and powerful way to calculate the limiting behaviour of F as just
a power in the coupling constant times a number $F(0,0)$ corresponding
to the behaviour of the function F for a free-field theory, for which the
coupling constant is 0!

2   *The running coupling constant of QCD*

The above procedure does not work at all for QED, nor as a matter
of fact for any other kind of theory known to date besides nonabelian
gauge theories. The positive-definiteness of the weight function in the
Källén-Lehmann representation of the polarisation function results for
other theories in a positive value of the β-function, which means that the
effective coupling increases with the scale.

The β-function may evidently turn over to negative values again for
larger-order terms in the perturbation series (although this would mean
that the theory contains states which effectively provide a negative phase
space contribution according to the Källén-Lehmann representation!).
Such a behaviour would lead to an *attractive fixed point for the coupling*
at the value α^* for which $\beta(\alpha^*) = 0$. This means that when the energy
increases the effective coupling constant will be attached to this value. We
will, however, not pursue this discussion any further because there is for
the cases of interest in this book no known example of such behaviour.

The very fact that the β-function goes from 0 for $\alpha = 0$ to negative
values for a nonabelian gauge theory like QCD (and it is known to have
the same behaviour also for the next order in perturbation theory) means
that there is an attractive fixed point for a vanishing coupling constant.
And a vanishing coupling constant in principle means a free-field theory.

In reality, though, we find that the theory is not completely free. There
are evidently some logarithmic power corrections and we will see in
Chapter 19 that this means scale-breaking corrections to the parton model.

The QCD effective coupling, usually referred to as the *running coupling of QCD*, can be written (with the number of colors $N_c = 3$)

$$\frac{\alpha_s(Q^2)}{4\pi} = \frac{3}{(11N_c - 2n_f)\log(Q^2/\Lambda^2_{QCD})} \tag{4.72}$$

by a suitable redefinition of the constant c_c in Eq. (4.69) and the introduction of the value for b_c given above.

We will end with a simple picture of why the coupling constants in QED and QCD behave so differently. We consider an ordinary electric charge in the vacuum and note that this will imply that the vacuum will be polarised in the way described above. In particular there will be some screening of the bare charge, because all the time it will be surrounded by a (virtual) cloud of charged particle-antiparticle pairs. These pairs will arrange themselves in a dipole-like manner so that viewed from afar we will see a diminished charge.

Now suppose that we send a set of probes towards this (pointlike) charge, corresponding to shorter and shorter wavelengths, i.e. we will observe the results from larger and larger values of the momentum transfer Q^2. The probes will evidently come closer and closer to the original bare charge and therefore 'see' more and more of it without the charge screening. Thus the effective charge will become larger with increasing Q^2. The main point in this argument is that the virtual pairs can in effect move and spread freely around the original charge. But note that the field quanta, i.e. the photons, are uncharged so that the charge is pointlike inside the virtual cloud of dipole pairs.

Let us now consider the corresponding situation in QCD. In this case, the field itself also contains charge, because the gluons are color-8's. This means that any original color charge will be smeared out over the region where the field is. A long-wavelength probe will then not be affected, i.e. it will see the whole, bare, charge. On the other hand, of course, as always in quantum mechanics short-wavelength probes will either 'see' the whole charge or nothing. But there will be a decreasing probability of finding the charge the smaller the region that is probed. In this way *the effective QCD charge actually corresponds to a charge multiplied by a 'form factor'*. We will show in Chapter 18 that the size and the behaviour of the β-function in QCD do in fact correspond to an interval in rapidity space within which we can expect modifications of the field.

5
Deep inelastic scattering and the parton model

In this chapter we will consider the notion of *partons*, in the way Feynman introduced them. The parton model (PM) corresponds to a very clever application of the concepts behind the method of virtual quanta, which we described in Chapter 2. The theoretical reasons why the PM provides a relevant description of the hadronic constituents are, however, very complicated and this chapter only contains a first introduction.

The road to the PM goes through experiment. Over many years physicists have performed in various contexts a type of experiment which can be traced back to Rutherford. They have used a charged particle to extract information on the charge and mass structure of smaller and smaller constituents of matter. Rutherford made use of α-radiation on nuclear targets and very quickly made two essential observations.

He and his assistant were able to detect the scattering of the α-particles by direct observation of the flashes that they produced on a screen. They found, firstly, that *most of the beam particles simply continued through the target as if it was empty of matter*. But, secondly, *every now and then they found quite an appreciable deviation*.

It was Rutherford's genius that not only he did take his observations seriously but also used them to provide a description of the atom. We are going to consider his result, together with the necessary corrections due to relativity, spin and the internal structure of the target.

He explained the source of the α-particle deviations by a classical mechanics calculation of the orbits of charged particles in a Coulomb field and he attributed this Coulomb field to a precise charge value placed inside a very tiny region indeed, i.e. an atomic *nucleus*. He was pretty lucky, however, that his classical mechanics calculation agreed with the quantum mechanical results.

This is by no means trivial. In principle Nature could have chosen to use something other than an inverse square law for the force between electrically charged particles (although this would have been difficult to accommodate with many other phenomena, among them ourselves!). Then Rutherford would have obtained a result which subsequent quantum mechanical corrections would have made obsolete: he did not at that time know anything about quantum mechanics and his beautiful atomic model would have been irrelevant.

The Rutherford scattering cross section is also at the basis of high $p_\perp$-scattering among hadronic constituents. Therefore the results will occur again in connection with deep inelastic scattering in the linked dipole chain model, in section 20.7, when we consider the hadronic wave function in a Feynman diagrammatic description of perturbative QCD. The (color-)charged constituents (the 'partons') will be sensitive to the strong Coulomb fields between them (such fields are inherent properties of any gauge field theory). In particular, when we use small wavelength probes, Heisenberg's indeterminacy principle implies that the observable partons must have large energy-momenta, i.e. their interactions will correspond to large momentum transfers.

After Rutherford, when more energetic beams of charged particles became available, experiments were performed on nuclear targets directly. A great amount of information was extracted about the charges inside (or actually mostly on the surface of) the nucleus. Still later, people were able to study scattering from the simplest nucleon, i.e. the proton itself and for a long time there was a general understanding that the proton was a complex charged object but that the charge seemed to be smeared out in a continuous way. It was necessary, in order to describe the reaction of a proton to an electromagnetic field pulse, to introduce a *form factor*. Such a form factor corresponds classically to an extended charge distribution.

When I was a young student, my teacher Källén referred to the next possible observational tool, the Stanford linear accelerator (SLAC), as the 'Monster'. It was understood from the beginning that the Monster might provide beams sufficiently high in energy to smash the proton but there were few people around who believed that this would lead to a new concept of constituents. The young Bjorken was around, however, and based upon theoretical investigations in current algebra he predicted that *one should find a 'scaling' cross section*.

Physicists have always used dimensional analysis to derive results of the kind usually referred to as 'back-of-an-envelope' calculations. Thus when one considers a particular dynamical situation there are always dimensional parameters. The typical space size may in a quantum mechanical description of a particle either be the Compton wavelength $1/m$, the Bohr radius $1/m\alpha$ or the 'classical charge radius' α/m (which occurs in the Thompson cross section for long-wavelength radiation scattering on a charged particle) with m the particle mass and $\alpha \simeq 1/137$ the fine structure constant. Based upon such quantities it is in general easy to find the possible size of an effect, besides some plain (usually combinatorial) numbers such as $3! = 6$ and factors like $2/3$ (from spin) or (multiple) 2π's. (Note that π is almost a dimensional number because high-energy physicists generally obtain it either from the conversion of Planck's constant $h \to \hbar = h/2\pi$ – the π's in the conversion of the volume factors to cross sections are generally of that kind – or from integrals over the azimuthal angle.) We will use such considerations repeatedly in this book.

For the proton it was already known that there was a scale involved in connection with the form factor. This length scale corresponds to the extension of the proton charge distribution and it is of the same order as the inverse proton mass. Bjorken's statement can be rephrased to mean that *there should be no new length scales deeper inside the proton.*

The process, which is called deep inelastic scattering (DIS), will be discussed further within the Lund model in Chapter 20 and within the conventional QCD scenario in Chapter 19. It contains three dimensional numbers: the squared momentum transfer to the proton from the impinging electron, conventionally called $-Q^2$; the squared mass of the final-state (smashed) system, conventionally called W^2; and then the squared mass of the original system, i.e. the squared proton mass m_p^2.

The reason why Källén and his contemporaries called the machine the Monster was the fact that it would produce beams such that $m_p^2 \ll Q^2$ and/or W^2. Bjorken's suggestion was that the cross section should depend (besides a trivial Q^2-dependence) only on the ratio Q^2/W^2 of the two larger dimensional numbers. This turned out to be essentially correct.

According to Dick Taylor, who was present at the time, Feynman used to come over to SLAC to learn about the experimental results. One day he presented the experimentalists with the PM as an explanation for the scaling phenomena. Since Feynman's proposal there have been few high-energy theorists who have not produced some kind of work on the PM at some time in their career. We who have worked on the Lund model were very late arrivals on the scene.

In order to exhibit the PM we will provide a brief description of Rutherford's classical mechanics calculation and then show how to obtain the same result in a potential scattering model in quantum mechanics. This discussion is relevant to lepton-hadron scattering when the hadron can be considered as very heavy, i.e. its mass is much larger than any parameter with energy dimension in the problem. We will after that turn to the question of scattering on a composite system and introduce the idea of a *form factor*. This will lead to the Rosenbluth formula, which describes elastic scattering within the most general framework possible in a Lorentz-covariant and parity-invariant setting.

We will finally consider inelastic scattering, in which the incident lepton

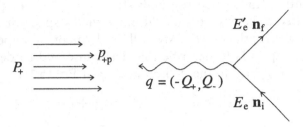

Fig. 5.1. The inelastic scattering of an electron from the field quanta of a hadron with notation described in the text.

produces field pulses, i.e. momentum transfers, which are so large that the initial hadron disintegrates. We will start with an excursion into lightcone physics and in particular indicate some of the steps that led Bjorken to suggest scaling cross sections.

Finally, we will use the results to exhibit the PM. We will show how parton flux factors arise and, in particular, the importance of spin and the other quantum numbers of the quark-partons for the resulting description.

5.1 The parton model: Feynman's proposal

Feynman used the results of the method of virtual quanta (MVQ), cf. Chapter 2, in an ingenious way. He assumed that the interaction ability of a hadron with respect to an electromagnetic field pulse is defined by a set of quanta which he called partons. Partons are at this stage operationally defined by the single property that *they are able to scatter elastically with an electron by absorbing a radiation quantum.*

In order to give a precise description we will assume that an accelerator provides us with electrons, of high energy E_i, coming in along a well-defined direction $\mathbf{n}_i$. We also assume that such an electron is scattered in the field of the hadron so that afterwards we observe it to have energy $E_f < E_i$ moving outwards in a direction $\mathbf{n}_f$ described by the angle θ (i.e. $\mathbf{n}_i \cdot \mathbf{n}_f = \cos \theta$, see Fig. 5.1).

From this situation we conclude that the electron has been exposed to a four-momentum transfer, conventionally called q:

$$q \equiv (q_0, -l\mathbf{n}) = (E_i - E_f, \ p_i\mathbf{n}_i - p_f\mathbf{n}_f) \tag{5.1}$$

As we have seen in Chapter 2 this four-vector must be spacelike, i.e. q^2 must be negative, $q^2 = -Q^2$, in order that the incoming and outgoing electrons stay on the mass shell $E_i^2 - p_i^2 = E_f^2 - p_f^2 = m_e^2$.

The momentum transfer corresponds (for large values of Q^2) to a very highly collimated electromagnetic field pulse with a space-time size of the order of the wavelength, $1/\sqrt{Q^2}$. We will use lightcone components along the vector **n** in Eq. (5.1) to describe this field pulse and so define positive $Q_\pm$ with $Q_+ Q_- = Q^2$ (note the definition of q in Eq. (5.1))

$$-Q_+ = q_0 - l, \quad Q_- = q_0 + l \qquad (5.2)$$

In Fig. 5.1 the hadron comes in as a cloud of (massless) partons together having a large positive-lightcone component P_+. The interaction between the radiative pulse described by q and one of the partons with a positive-lightcone component p_{+p} corresponds to an absorption of this radiation quantum. In order to stay on the mass shell the parton will have to reverse direction so that after the collision it will have a negative-lightcone component p_{-p}. Note that, as the parton is massless and is assumed to move along the direction $\pm$**n**, it will before and after have a single nonvanishing lightcone component in this picture.

From energy-momentum conservation we conclude that all the kinematical properties of the interaction are fixed by

$$p_{\pm p} = Q_\pm \qquad (5.3)$$

There are two observable (large) Lorentz invariants, i.e. $Q^2 = Q_+ Q_-$ and $2Pq \simeq Q_- P_+$. We have neglected the hadronic mass and we note that in this approximation the final-state mass square of the smashed hadron has increased to $W^2 = (P + q)^2 \simeq 2Pq - Q^2$. Because the cross section depends only upon the ratio of these Lorentz invariants it must therefore depend only upon the fraction of the energy-momentum of the hadron, which is carried by the scattered parton (the index refers to Bjorken)

$$x_B = \frac{-q^2}{2Pq} = \frac{Q_+}{P_+} = \frac{p_{+p}}{P_+} \qquad (5.4)$$

This sole dependence upon x_B can be understood as follows: *the interaction depends only upon the number of partons with that particular value of the fractional energy-momentum.* Thus the hadron has been reduced to a flux of partons with respect to the interaction, just as in the MVQ a charged particle is described by the flux of photons.

This assumption of Feynman about the interaction between the field pulse and the constituents implies the possibility of an experimental study of the flux of the partons, i.e. to decide upon the detailed structure of the hadron under study. It is then only necessary to consider the electron before and after the interaction. The probability of finding a large momentum transfer is directly related to the amount of suitable absorbers, i.e. partons, in the hadron.

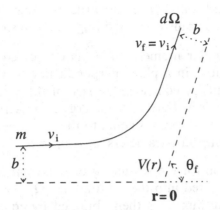

Fig. 5.2. Particles moving in a central force field are deflected in a definite direction characterised by the solid angle $d\Omega$.

Large values of the fraction x_B correspond to the partons which carry a large part of the total energy-momentum of the hadron. Therefore they should be major constituents of the hadronic wave function. For smaller values of x_B Feynman suggested that there should be a bremsstrahlung spectrum like the one we found for the photons in a moving Coulomb field according to the MVQ,

$$\sim dx_B/x_B \tag{5.5}$$

This is usually referred to as 'Feynman's wee parton spectrum'.

5.2 Rutherford's formula from classical mechanics

A detailed derivation of the Rutherford formula is given in Goldstein's book and we will only provide a brief description. In classical mechanics everything is completely determined by the force law and the initial conditions on the particle(s) involved. Consequently there is always a definite orbit along which every particle moves in space-time and a corresponding trajectory in phase space.

We assume that a particle with mass m is approaching the force centre in a field described by a potential $V(r)$, see Fig. 5.2, which vanishes as $r \to \infty$. Thus the force is spherically symmetric, $\mathbf{F} = - [dV(r)/dr]\,\mathbf{e}_r$ where $\mathbf{e}_r$ is a unit vector pointing radially outwards. We also assume that the particle has velocity v_i far from the centre, impact parameter b_i and orientation along some azimuthal angle ϕ_i.

This means that we can define an incident flux $I\,d\phi_i b_i db_i$ of such particles. All these particles will move along the same orbit and after the encounter will end up moving outwards in a definite direction, which we will characterise in terms of a solid angle $d\Omega_f = \sin\theta_f d\theta_f d\phi_f \equiv d\Omega$.

RI The orbital angular momentum, $\mathbf{L}$, is conserved and therefore the particles will move in a plane perpendicular to $\mathbf{L}$. This means that the angles ϕ_i and ϕ_f coincide. The size of $|\mathbf{L}| = L$ is from the initial conditions $L = mv_i b_i$. Further the energy is conserved, cf. RIV below. Therefore the initial speed is equal to the final one and thus the same is true for the impact parameters, $b_i = b_f \equiv b$.

RII The cross section for the scattering of these particles is the fraction of particles scattered into the solid angle $d\Omega_f$ per unit time, divided by the incoming flux. It is then obtained by equating the outgoing and the ingoing fluxes:

$$\frac{d\sigma}{d\Omega_f} I \, d\Omega_f = -I b_i db_i d\phi_i \qquad (5.6)$$

The minus sign is introduced because the larger the value of b the smaller the force and therefore the smaller the scattering. From this equation we conclude that

$$\frac{d\sigma}{d\Omega} = \frac{-b_i}{\sin\theta_f} \frac{db_i}{d\theta_f} \qquad (5.7)$$

Therefore we must calculate the relationship between the impact parameter and the scattering angle.

RIII In order to calculate this orbit relation we use cylindrical coordinates $r(t)$, $\theta(t)$, so that the velocity is $\mathbf{v} = \dot{r}\mathbf{e}_r + r\dot{\theta}\mathbf{e}_\theta$ (with dots indicating time derivatives). We obtain for the Lagrangian

$$\mathcal{L} = T - V(r) \text{ with } T = m\mathbf{v}^2/2 = m(\dot{r}^2 + (r\dot{\theta})^2)/2 \qquad (5.8)$$

As $\mathcal{L}$ is independent of the angle θ the corresponding angular momentum component is conserved:

$$p_\theta = \frac{d\mathcal{L}}{d\dot{\theta}} = mr^2\dot{\theta} \equiv L \qquad (5.9)$$

This can be used to reorganise the time dependence of $r(t)$ and $\theta(t)$ and from this we obtain an equation for the orbit $r = r(\theta)$:

$$\dot{r} = \frac{dr}{dt} = \frac{dr}{d\theta}\frac{d\theta}{dt} = \frac{L}{mr^2}\frac{dr}{d\theta} = -\frac{L}{m}\frac{d}{d\theta}\left(\frac{1}{r}\right) \qquad (5.10)$$

Using $u = 1/r$ and $u' = du/d\theta$ we can then write the kinetic energy term T in Eq. (5.8) as

$$T = \frac{L^2}{2m}(u')^2 + \frac{L^2 u^2}{2m} \qquad (5.11)$$

For an attractive Coulomb force between a charge $-e$ and a charge Ze we obtain for the potential term:

$$V = V(r) = -\frac{Ze^2}{4\pi r} = -Z\alpha u \tag{5.12}$$

(where we have introduced the fine structure constant α).

RIV As the total energy is conserved and expressible in terms of T and V we obtain

$$E \equiv \frac{mv_i^2}{2} = T + V$$
$$= \frac{L^2}{2m}[(u')^2 + u^2] - Z\alpha u = \frac{L^2}{2m}[(u')^2 + (u - u_0)^2 - u_0^2] \tag{5.13}$$

where u_0, the displacement of u, is given by

$$u_0 = \frac{Z\alpha m}{L^2} \tag{5.14}$$

Equation (5.13) is equivalent to the harmonic oscillator relationship discussed in Chapter 3 and we can immediately write down the solution:

$$u \equiv \frac{1}{r} = u_0(1 + \epsilon \cos \theta) \tag{5.15}$$

This is the equation for a hyperbola since ϵ, the eccentricity, is larger than 1:

$$\epsilon = \sqrt{1 + \left(\frac{2Eb}{Z\alpha}\right)^2} \tag{5.16}$$

RV There are then two values of θ for which $r \to \infty$; these are given by $\cos \theta = -1/\epsilon$ and the angle between these directions is evidently $\pi - \theta_f$ (see Fig. 5.2). A little algebra then leads to the result that

$$b_i \equiv b = \frac{Z\alpha}{2E} \cot\left(\frac{\theta_f}{2}\right) \tag{5.17}$$

The final result for the Rutherford cross section is from Eq. (5.6)

$$\frac{d\sigma}{d\Omega} = \left(\frac{Z^2\alpha^2}{16E^2}\right)\frac{1}{\sin^4(\theta_f/2)} \tag{5.18}$$

We will meet the same expression when we do the calculations using quantum mechanics. The energy E in Eq. (5.18) is given by the nonrelativistic kinetic energy $(mv_i^2)/2$.

The formula is singular for small scattering angles because the small-angle region corresponds to large impact parameters $b = b_i$ according to Eq. (5.17). The particles come in far from the force centre and are consequently deflected very little. The Coulomb force *per se* has infinite range but it is evident that any charge centre in real life will be screened by other charges (e.g. by its own electrons if it is an atomic nucleus).

In order to get an estimate of the cross section for a screened situation we will assume that the impact parameter is equal to w times the corresponding Bohr radius, i.e.

$$b_i = \frac{w}{Z m \alpha} \equiv w r_B \qquad (5.19)$$

Then from Eq. (5.16) the parameter $\epsilon = \sqrt{1 + (wE/E_0)^2}$ with E_0 equal to the corresponding Rydberg energy:

$$E_0 = \frac{m(Z\alpha)^2}{2} \qquad (5.20)$$

We now consider a fixed energy E much larger than E_0. This means that the velocity v_i will be much greater than $Z\alpha \simeq Z/(137)$, where we have introduced the well-known value for the fine structure constant in QED. This leaves, at least for small Z-values, a region where we may neglect relativistic corrections and still fulfil the requirement. We then obtain $\theta_f \sim 2E_0/(wE)$.

If we exchange the angular variation for one with respect to the parameter w we obtain a smooth behaviour,

$$d\sigma \simeq 2w dw \pi r_B^2 \qquad (5.21)$$

and the cross section is independent of the energy E as long as $wE \gg E_0$.

Note that the cross section only depends upon the square of the charge combination $Z\alpha$. Therefore we obtain the same formula if the two charges have the same sign, i.e. if the attractive Coulomb potential in Eq. (5.12) is exchanged for a repulsive one: $-Z\alpha \to Z\alpha$. The displacement u_0 will in that case change sign, however. This means that the force centre will no longer be the internal focus of the hyperbola but instead the external one. Or, in other words, while the particle will go around the force centre for an attractive force it will go in an outside hyperbola if the force is repulsive. But the scattering angles are the same!

5.3   Rutherford's formula in relativistic quantum mechanics

1   *The calculation of the cross section*

We will in this section again consider the scattering of a charged particle from a Coulomb potential. This is a preliminary for treating the scattering of two charged particles. We will again meet Rutherford's result although this time in terms of the square of the Fourier transform of the potential. We use the transition operator $\mathcal{T} = \int dx j_\mu A^\mu$ and assume that the external potential A^μ depends only upon the space coordinates. At the end we shall specialise to the ordinary Coulomb shape $A_\mu = -\delta_{\mu,0} Ze/(4\pi r)$, which was used in the previous section.

The transition matrix element between an incoming electron (energy-momentum k) and an outgoing one (energy-momentum k') is

$$\langle k' | \mathcal{T} | k \rangle = \frac{e}{2V\sqrt{k_0 k_0'}} \int dx \{ \mathbf{k}' | jA(\mathbf{x}) | \mathbf{k} \} \exp\left[-ix(k-k')\right] \quad (5.22)$$

We have here introduced the reduced matrix elements of the current operator, which we discussed in Chapter 4.

Time integration produces an energy-conserving δ-distribution and space integration leads to the Fourier transform of the vector potential:

$$\langle k' | \mathcal{T} | k \rangle = \frac{2\pi e}{2V\sqrt{k_0 k_0'}} \{ \mathbf{k}' | j\mathscr{A}(\mathbf{q}) | \mathbf{k} \} \delta(k_0 - k_0') \quad (5.23)$$

with $q = k' - k$. Momentum is not conserved in this case, because the infinitely heavy potential takes up the recoil. To calculate the cross section we use the techniques described in Chapter 3:

$$
\begin{aligned}
d\sigma &= \left(\frac{w}{\delta t}\right)\left(\frac{V}{v}\right)\left(\frac{V d^3 k'}{(2\pi)^3}\right) \\
&= \frac{e^2}{(2\pi)^2}\left(\frac{1}{4|\mathbf{k}|k_0'}\right) \int d^3 k' \delta(k_0 - k_0') |\{ \mathbf{k}' | j\mathscr{A}(\mathbf{q}) | \mathbf{k} \}|^2 \\
&= d\Omega' \frac{\alpha}{4\pi} |\{ \mathbf{k}' | j\mathscr{A}(\mathbf{q}) | \mathbf{k}) \}|^2 \quad (5.24)
\end{aligned}
$$

The first factor in the first line is the transition probability per unit time, the second the (inverse) flux of incoming particles with $v = |\mathbf{k}|/k_0$ and the third the number of final states. In the second line we have rewritten the whole expression and in the third gone over from the integration variable $|\mathbf{k}'|$ to k_0' and performed the integral by means of the δ-distribution.

We may now make use of the analysis presented in Chapter 4 for the reduced matrix element combination summed over the spin states:

$$\sum_{spins} \{ \mathbf{k}' | j_\mu | \mathbf{k} \} \{ \mathbf{k} | j_\nu | \mathbf{k}') \} = (T_1 + T_2)_{\mu\nu},$$

$$T_{1\mu\nu} = g_{\mu\nu} q^2 - q_\mu q_\nu, \quad T_{2\mu\nu} = (k+k')_\mu (k+k')_\nu \quad (5.25)$$

2   *The Mott cross section and the form factors*

Gathering the different factors and assuming that the four-vector potential only has a time component, $A_0 = V(r)$, we obtain the cross section

$$\frac{d\sigma}{d\Omega} = \frac{\alpha}{2\pi} \left[\mathbf{k}^2(1 + \cos\theta) + 2m^2 \right] |\mathscr{V}(\mathbf{q})|^2 \tag{5.26}$$

where θ is the scattering angle. (We have neglected a few steps, leaving it to the reader to obtain this result.) There are two terms multiplying the squared Fourier transform of the potential. Depending upon whether the lepton rest mass m or the momentum $|\mathbf{k}|$ dominates we obtain a nonrelativistic or an extreme relativistic approximation.

For the Coulomb potential of a point particle with charge Ze we obtain

$$\mathscr{V}(\mathbf{q}) = -\frac{Ze}{4\pi} \int \frac{d^3x}{|\mathbf{x}|} \exp(-i\mathbf{x} \cdot \mathbf{q}) = \frac{Ze}{q^2} \tag{5.27}$$

The simplest way to see this is to use the coordinate-space differential equation for the Coulomb potential,

$$\Delta V(\mathbf{x}) = Ze\delta(\mathbf{x}) \tag{5.28}$$

and perform the Fourier transform, thereby changing the Laplacian Δ to $-\mathbf{q}^2 \equiv q^2$ (Note that $\Delta \exp[i\mathbf{q} \cdot \mathbf{x}] = -\mathbf{q}^2 \exp[i\mathbf{q} \cdot \mathbf{x}]$).

It is at this point that Rutherford was lucky in his classical mechanics approach. The squared Fourier transform of the Coulomb potential evidently contains an inverse power of the squared momentum transfer $(q^2)^2 = (-|\mathbf{q}|^2)^2 = 4k^2(1 - \cos\theta)^2 = 16k^4 \sin^4(\theta/2)$ (where k is the cms conserved momentum of the particles), which is just what Rutherford obtained from his calculation of the variation of the impact parameter with angle. This relation between the Fourier transform of the potential and the variation of the impact parameter is only true for a Coulomb potential.

This leads to the so-called Mott cross section in the limit where we may neglect the electron mass:

$$\frac{d\sigma}{d\Omega}_{Mott} = \left(\frac{Z^2\alpha^2}{4E^2} \right) \frac{\cos^2(\theta/2)}{\sin^4(\theta/2)} \tag{5.29}$$

There is a factor $4\cos^2(\theta/2)$ as compared to the Rutherford formula. If we go back to Rutherford's derivation we find that it is based upon nonrelativistic kinematics. The projectile mass is assumed to be much larger than its kinetic energy. This means according to Eq. (5.26) that $\mathbf{k}^2 \ll m^2$ and we obtain in this limit

$$\frac{d\sigma}{d\Omega} = \frac{\alpha}{\pi}(m^2)|\mathscr{V}(\mathbf{q})|^2 = \left(\frac{Z^2\alpha^2}{16(\mathbf{k}^2/2m)^2} \right) \frac{1}{\sin^4(\theta/2)} \tag{5.30}$$

which is Rutherford's result (with $E = E_{kin} = |\mathbf{k}|^2/2m$).

If the electron encounters not a point charge but a charge distribution $Zef(\mathbf{x})$ then on the right-hand side of Eq. (5.28) the exchange $Ze\delta(\mathbf{x}) \to Zef(\mathbf{x})$ should be made; this evidently means that in place of Eq. (5.27) we will have

$$\mathscr{V}(\mathbf{q}) = -\frac{Ze}{\mathbf{q}^2}\tilde{f}(\mathbf{q}), \quad \tilde{f}(\mathbf{q}) = \int d^3x f(\mathbf{x})\exp(-i\mathbf{x}\cdot\mathbf{q}) \qquad (5.31)$$

The normalisation condition $\int d^3x f = 1$ corresponds to $\tilde{f}(|\mathbf{q}|^2 = 0) = 1$. We conclude that with the introduction of the charge distribution f the Mott (or Rutherford) cross section is changed as follows:

$$\frac{d\sigma}{d\Omega}_{Mott} \to \frac{d\sigma}{d\Omega}_{Mott}|\tilde{f}(\mathbf{q})|^2 \qquad (5.32)$$

Provided that the momentum transfer $\sqrt{|q^2|}$ is smaller than the inverse of any length scale in the charge distribution, or in other words provided that the wavelength of the electromagnetic pulse cannot resolve the target structures, then we have the same pointlike cross section. For larger momentum transfers the scattering experiment can be used to measure (the Fourier transform of) the charge distribution. The function $\tilde{f}$ is known as a form factor.

5.4 The target recoil and the general elastic cross section for the scattering of spin $1/2$ particles

The form factor introduced at the end of the last section is too simple to describe scattering from a baryon target. Firstly, one cannot consider baryons as merely charge distributions. They also have magnetic moments and an electromagnetic pulse will influence that aspect of the baryon structure, too. Secondly, they are not infinitely heavy and so we must include also the recoil of the target, i.e. we must introduce not only energy but also momentum conservation in the scattering.

We have already, in Chapter 3 on field theory, considered a simplified model for this scattering situation, the scalar $g:\psi^2:\phi$-model. From the results in Eqs. (3.104)–(3.110) we now generalise the situation to two different ψ-particles, ψ_e indexed $1,3$, and ψ_B, indexed $2,4$, with $1,2$ the incoming pair (Fig. 5.3). We have in mind particles such as electrons and baryons and as they are both spin $1/2$ particles the interaction is governed by the four-vector currents $j_e \propto :\psi^*\psi:$ and j_B likewise expressed in terms of Dirac spinors.

This means that the coupling constant factor $4g^2$ in Eq. (3.110) should be replaced by e^2 (this is plain combinatorics). Further the factor B in Eq. (3.110) contains three pole terms. Due to the fact that the lepton

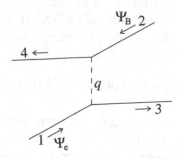

Fig. 5.3. The elastic scattering of an incoming electron (index 1) from a baryon (2) to the final state $(3,4)$ with the exchange of a virtual photon.

and baryon cannot mutually annihilate or be exchanged there is in the present situation only one of the factors left, the momentum-transfer pole $1/(k_1 - k_3)^2 = 1/(k_2 - k_4)^2$ (with $M_\phi = 0$ for the photon).

With these modifications we can use the result in Eq. (3.110):

$$d\sigma = \frac{e^2}{2(2\pi)^2 \sqrt{\lambda(s, M_e^2, M_B^2)}} |B|^2$$

$$\times \prod_{j_f=3}^{4} dk_{j_f} \delta^+(k_{j_f}^2 - M_{j_f}^2) \delta(k_1 + k_2 - k_3 - k_4) \qquad (5.33)$$

this time with B expressed in terms of the reduced matrix elements

$$B = \sum_{spins} \left(\frac{\{k_3 e | j^\mu | k_1 e\}\{k_4 p | j^\mu | k_2 p\}}{(k_2 - k_4)^2} \right)$$

Comparing with the result in Eq. (5.24) we find that the Fourier transform $\mathscr{A}^\mu$ of the four-vector potential A^μ has been replaced as follows:

$$\mathscr{A}^\mu \rightarrow \frac{ie}{2V\sqrt{E_2 E_4}} \frac{\{k_4 p | j^\mu | k_2 p\}}{(k_2 - k_4)^2} \qquad (5.34)$$

This is exactly in accordance with our physical intuition that we should now obtain the four-vector potential A^μ from the baryon current, j_B^μ:

$$A^\mu(x) \rightarrow \int dx' D_0(x - x') \langle k_4 | j_B^\mu(x') | k_2 \rangle + \text{g.t.} \qquad (5.35)$$

where g.t. again stands for gauge-dependent terms of no interest because of the coupling to the conserved (electron) current. The result in Eq. (5.35) is, as easily seen, equal to the energy-momentum space result in Eq. (5.34) and corresponds to a solution of Maxwell's equations for the vector potential in terms of the (baryon) current j_B^μ (note that the Green's function $\Box D_0(x) \propto \delta(x)$ is chosen as the Feynman propagator).

If we sum over the final-state spins and average over the initial ones for both the baryon and the electron we obtain two tensors (cf. Eq. (4.36)), one for the electron (which we have already written out in Eq. (5.25)) and one for the baryon, similarly with the two parts:

$$T_B = T_{1B} + T_{2B},$$
$$T_{1B\mu\nu} = g_{\mu\nu}q^2 - q_\mu q_\nu, \tag{5.36}$$
$$T_{2B\mu\nu} = 4\left[k_2 - q(qk_2)/q^2\right]_\mu \left[k_2 - q(qk_2)/q^2\right]_\nu$$

For T_{2B} we have used the form explained in connection with Eq. (4.45).

Multiplying the electron and baryon tensors together we obtain the cross section. It is at this point useful to write it in an invariant form because we will need this later. To that end we introduce the two invariants corresponding to the energy and the scattering angle, the cms squared energy s and the squared momentum transfer $q^2 = -Q^2$:

$$\hat{s} \equiv s - M^2 \simeq 2k_1k_2 \simeq 2k_3k_4,$$
$$Q^2 = -q^2 = 2k_1k_3 = 2(k_2k_4 - M^2) \tag{5.37}$$

Here we shall neglect the lepton mass and write $M \equiv M_B$. Note that in this case $\sqrt{\lambda} \simeq \hat{s}$. We obtain (note the factor $(1/2)^2$ from the initial spin-averaging)

$$\frac{d\sigma}{dQ^2} = \frac{4\alpha^2 De}{\hat{s}Q^4}I, \quad De = Q^4 - 2(\hat{s} + M^2)Q^2 + 2\hat{s}^2$$
$$I = \int dk_3 dk_4 \delta(k_3^2)\delta(k_4^2)\delta(k_1 + k_2 - k_3 - k_4)\delta((k_1 - k_3)^2 + Q^2) \tag{5.38}$$

We have here used the same trick as before, introducing a derivative of a function by means of a δ-distribution, this time in Q^2.

The integral I is more complicated (because of the δ-distribution in Q^2) than the phase-space integrals we have encountered before. To calculate it we introduce the vector $P = k_1 + k_2$, the total energy-momentum in the cms where $P = (W, \mathbf{0})$ and we place the vector k_1 along the 3-axis. For simplicity we shall calculate the integral in detail for the case when we can neglect the mass M, although we will at the end introduce it into the result. We obtain

$$I = \int dk_3 \delta(\mathbf{k}_3^2 - E_3^2)\delta(W^2 - 2WE_3)\delta(-2E_1E_3(1 - \cos\theta) + Q^2)$$
$$= \frac{\pi}{2W^2} \rightarrow \frac{\pi}{2\hat{s}} \tag{5.39}$$

We have here performed the E_3-integral by means of the second δ, the $|\mathbf{k}_3|$-integral by the first δ and then the $d\Omega(= d\theta \sin\theta d\phi)$-integrals by means of the last δ. The last line contains the generalisation to the case when $M \neq 0$.

In this way we arrive at the following result for the invariant Rutherford scattering cross section:

$$\frac{d\sigma}{dQ^2} = \frac{2\pi\alpha^2[Q^4 - 2(\hat{s} + M^2)Q^2 + 2\hat{s}^2]}{(Q^2)^2\hat{s}^2} \tag{5.40}$$

(the factor 2π corresponds to the fact that in a spin-averaged cross section there is no dependence upon the azimuthal angle).

We have obtained the cross section for the process $e_1 + p_2 \rightarrow e_3 + p_4$ by the use of the spin sums over the current matrix element in (5.33). According to crossing symmetry (mentioned after Eq. (4.37)) we may from this result easily obtain the result for the process $e_1 + \bar{e}_3 \rightarrow \bar{p}_2 + p_4$, i.e. the annihilation of the pair $e_1\bar{e}_3$ into $\bar{p}_2, p_4$ by the exchanges $p_3 \rightarrow -p_3$ and $p_2 \rightarrow -p_2$ in the matrix element. At the same time we note that the (squared) cms energy is in this situation $(p_1 + p_3)^2 \simeq 2p_1p_3$ while the momentum transfer variable is $Q^2 = -(p_1 - p_2)^2 \simeq 2p_1p_2$, i.e. we obtain the relevant cross section with $s \simeq \hat{s} \leftrightarrow Q^2$ (neglecting the masses). We will later only need the result for the case when all the particles are massless and we obtain after some straightforward calculations the (spin- and azimuthal angle-averaged) annihilation cross section

$$\frac{d\sigma_A}{dQ^2} = \frac{2\pi\alpha^2(s^2 - 2sQ^2 + 2Q^4)}{s^4} \tag{5.41}$$

5.5 The extension to non-pointlike baryons, form factors

Written in this form it is easy to evaluate the above cross section in any Lorentz frame. Conventionally we use the laboratory (lab) frame, in which the baryon is initially at rest.

In the lab frame the electron energies before and after the interaction, E and E' respectively, are different and in particular fulfil the relations

$$\frac{E'}{E} = \frac{1}{1 + (E/M)(1 - \cos\theta)}$$
$$s = M^2 + 2ME \tag{5.42}$$
$$Q^2 = 2EE'(1 - \cos\theta)$$

We shall leave the reader to prove these and also to show that

$$\left|\frac{dQ^2}{\sin\theta \, d\theta}\right| = 2E'^2 \tag{5.43}$$

Using these relations we obtain by straightforward means the cross section in the lab frame from Eq. (5.40):

$$\frac{d\sigma}{d\Omega}\bigg|_{lab} = \frac{d\sigma}{d\Omega}\bigg|_{Mott} \frac{E'}{E} \left[1 + \frac{Q^2}{2M_B^2} \tan^2\left(\frac{\theta}{2}\right) \right] \tag{5.44}$$

There are two new factors: the electron energy is not the same before and afterwards in the lab system; as a Dirac particle, the baryon also has a magnetic moment.

We will not go into detail with respect to the electric and magnetic interaction properties of a Dirac particle. Just as there are different electric and magnetic fields in different Lorentz frames, these properties are also frame dependent. It is useful to remember, however, that if we multiply in the factor $\cos^2(\theta/2)$ from the numerator in the Mott cross section then the factor inside the brackets in Eq. (5.44) becomes

$$\cos^2\left(\frac{\theta}{2}\right) + \frac{Q^2}{2M_B^2} \sin^2\left(\frac{\theta}{2}\right) \tag{5.45}$$

Here we have two obviously independent terms stemming from the parts $2\hat{s}^2 - 2(\hat{s} + M^2)Q^2$ and Q^4 of the factor De in Eq. (5.40).

We have up to now treated the baryon as a point Dirac particle; however, according to experiment it is not. It turns out that there are two independent form factors, just as we saw in Eq. (5.36) that the squared current leads to two independent tensors, T_{1B} and T_{2B}. These form factors can be introduced in different ways. The most symmetrical version involves the so-called electric and magnetic form factors G_E and G_M.

These play the roles of electric and magnetic couplings in the Breit frame, [84]. But their main importance is that they can be shown to be invariants, i.e. to depend only upon Q^2, and that they occur in a simple way. The bracketted terms in Eq. (5.44) are then exchanged as follows:

$$1 + \frac{Q^2}{2M_B^2} \tan^2\left(\frac{\theta}{2}\right) \rightarrow \frac{G_E^2 + (Q^2/4M_B^2)G_M^2}{1 + (Q^2/4M_B^2)} + \frac{Q^2}{2M_B^2} \tan^2\left(\frac{\theta}{2}\right) G_M^2$$

$$\tag{5.46}$$

With this exchange in Eq. (5.44) we obtain the general elastic cross section formula for lepton-baryon scattering when parity is conserved. It is called the *Rosenbluth formula* and has been thoroughly investigated experimentally. One finds that both the electric and the magnetic form factors behave in the same way:

$$G_E \propto G_M \propto [1 + Q^2/(M_0)^2]^{-1}, \quad M_0 \simeq 0.71 \, \text{GeV} \tag{5.47}$$

In the early days of investigation of the proton and neutron this result lead to many speculations. Actually, the finding that the form factors were pole-dominated even led to the prediction that there should be

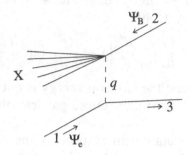

Fig. 5.4. The inelastic scattering of an electron from a baryon with one-photon exchange (a single electromagnetic pulse q) leading to a final state in which the baryon has fragmented into a complex system X.

(resonance) particles, [97], with the quantum numbers needed for the form factors, i.e. spin 1 particles. The fact that the ρ- and ω-particles fulfil these requirements and also have masses close to M_0 created particular attention. It is, however, not possible to prove from first principles that the elastic form factors should be analytic functions of $q^2 = -Q^2$ in the same way as we proved via the Källén-Lehman representation that the propagator should be analytic; within the Källén-Lehman formalism developed in Chapter 4 it would be natural to obtain a pole from an intermediate state.

Depending upon temperament and taste one may consider Eq. (5.47) as either a surprising finding or a reason for building a model. Such a model, *the vector dominance model* for the evaluation of matrix elements containing operators with the quantum numbers of the electromagnetic currents, [60], has been extensively used but is outside the scope of this book.

5.6 The inelastic scattering of electrons on baryons; lightcone physics

We will now consider the seemingly much more complex situation when the electromagnetic pulse q from the electron towards the baryon is such that the baryon breaks up into many final-state fragments (see Fig. 5.4).

The way in which we have introduced the elastic cross section makes it, however, rather easy to extend the formalism to the inelastic case, at least if we are only going to observe the electron before and after the interaction. According to Eqs. (5.34) and (5.35) the baryon is observable only through its current. For the case at hand, with a final state $\langle X|$ for the baryon containing all kinds of fragment particles, we need only to make the exchange

$$\langle k_4 | j_B^\mu(0) | k_2 \rangle \to \langle X | j_B^\mu(0) | k_2 \rangle \tag{5.48}$$

We then obtain the same cross section but the baryon current parts are then described by (after averaging over the initial baryon spin)

$$\begin{aligned} W^{\mu\nu} &= \frac{V E_2}{4\pi M} \sum_X (2\pi)^4 \delta(q + k_2 - k_X) \langle k_2 | j_B^\mu(0) | X \rangle \langle X | j_B^\nu(0) | k_2 \rangle \\ &= \frac{V E_2}{4\pi M} \int dx \exp(iqx) \sum_X \langle k_2 | j_B^\mu(x) | X \rangle \langle X | j_B^\nu(0) | k_2 \rangle \\ &= \frac{V E_p}{4\pi M} \int dx \exp(iqx) \langle p | j_B^\mu(x) j_B^\nu(0) | p \rangle \end{aligned} \tag{5.49}$$

where in the last line, we have gone over to the conventional notation $p = (E_p, \mathbf{p})$ instead of the earlier k_2. In the second line we have re-defined the δ-distribution as a Fourier transform using $\langle p | j_B^\mu(x) | X \rangle = \langle p | j_B^\mu(0) | X \rangle \exp[ix(p - k_X)]$. In the third line we have used the completeness relation $\sum_X |X\rangle \langle X| = 1$ to arrange the result into a two-current matrix element in the initial (spin-averaged) baryon state $|p\rangle$.

We evidently need the the factor $2V E_p$ to cancel a volume factor and to obtain the invariant combination $E_p |\mathbf{k}_1| \to s - M^2$. The same factors are also needed in Eq. (5.49) to make the tensor W into an invariant according to our conventions. The momentum transfer four-vector q is defined in terms of the initial- and final-state (observable) lepton energy-momenta: $q = k_1 - k_3$. Finally, the factor $2M$ is introduced for conventional reasons.

It is useful at this point to note that

$$\begin{aligned} &\int dx \exp(iqx) \langle p | j_B^\mu(0) j_B^\nu(x) | p \rangle \\ &= \sum_{X'} (2\pi)^4 \delta(q + k_{X'} - p) \langle p | j_B^\mu(0) | X' \rangle \langle X' | j_B^\nu(0) | p \rangle = 0 \end{aligned} \tag{5.50}$$

because in this case the masses of the states X' must be smaller than the baryon mass and there are no such states containing a baryon (the electromagnetic interactions conserve baryon number). To see this we note that the mass M_X of a state X occurring in Eq. (5.49) must fulfil

$$M^2 \le M_X^2 = (p + q)^2 = M^2 - Q^2 + \nu \Rightarrow \nu \ge Q^2 \tag{5.51}$$

where $\nu = 2pq$ (note that different authors use somewhat different definitions of ν). Therefore the mass of X' in Eq. (5.50) must fulfil $M_{X'}^2 = (p - q)^2 \le M^2 - 2Q^2 < M^2$.

We may use this fact to rewrite the tensor W in terms of a commutator matrix element:

$$W^{\mu\nu} = \frac{V E_p}{4\pi M} \int dx \exp(iqx) \langle p | [j_B^\mu(x), j_B^\nu(0)] | p \rangle \tag{5.52}$$

The parton model

In Chapter 3 we argued that due to causality the commutator of a local-field operator at two different points vanishes if the points are spacelike with respect to each other. This means that the integral in Eq. (5.52) is actually not over all space-time but only over the lightcones and their interior, i.e. $x^2 \geq 0$. It turns out to be reasonable to make the case that only the lightcone itself plays a role in the limit $v \to \infty$ with $x_B = Q^2/v$ nonvanishing. (Note that Eq. (5.51) implies the limit $x_B \leq 1$).

We will present a few steps in connection with such an argument (which is basically the scaling argument presented by Bjorken). We firstly choose to make use of the baryon rest frame in which $q = (q_0, \mathbf{0}_\perp, -|\mathbf{q}|)$ and note that in this frame

$$v = 2Mq_0 \quad \Rightarrow \quad |\mathbf{q}| = \sqrt{\frac{v^2}{4M^2} + Q^2} \simeq \frac{v}{2M} + Mx_B \qquad (5.53)$$

so that the lightcone components of q along the 3-axis are approximately

$$q_- = q_0 + |\mathbf{q}| \simeq \frac{v}{M}, \quad q_+ = q_0 - |\mathbf{q}| \simeq -Mx_B \qquad (5.54)$$

Then we consider a simplified model of the causal tensor function $W^{\mu\nu}$ in Eq. (5.52):

$$W(v, x_B) = \int dx \exp(iqx) F(x^2, px) \qquad (5.55)$$

where $F = 0$ if $x^2 < 0$. (Note that there are only three possible invariants that the integrand F for a scalar W can depend upon, x^2, px, p^2, and that the third of these is a constant, $p^2 = M^2$.)

The argument in the oscillating exponent is then $iqx = i(q_-x_+ + q_+x_-)/2 \simeq i(x_+v/M - x_-Mx_B)/2$. According to the theory of the Fourier transform the function W can then only obtain significant contributions from the integration regions $x_+ \leq M/v$ and $x_- \leq 1/Mx_B$. For the limit $v \to \infty$ this evidently means the region $0 \leq x^2 = x_+x_- - \mathbf{x}_\perp^2 \leq 1/(x_Bv) - \mathbf{x}_\perp^2$. Therefore the inverse of $Q^2 = x_Bv$ limits the transverse area inside which the integral obtains significant contributions and we are then led towards the lightcone itself when $Q^2 \to \infty$.

There are several pitfalls in this argument and it only works for sufficiently well-behaved functions F in the integrand. If F is of that kind we may continue the argument a little further and assume that the main contribution to such an F constructed from scalar currents,

$$F = \frac{VE_p}{2\pi} \langle p| [j(x), j(0)] |p \rangle \qquad (5.56)$$

will be a singularity along the lightcone, similar to the one obtained in

Eq. (3.77) for the ordinary commutator, multiplied by a function $f(px)$:

$$F = \frac{i\epsilon(x)}{2\pi}\delta(x^2)\left[\int da\exp(iapx)\tilde{f}(a) + \cdots\right] \qquad (5.57)$$

The dots indicate less singular terms and we have written f in terms of its Fourier transform, $\tilde{f}$. If this is introduced into Eq. (5.55) we obtain, neglecting the terms indicated by ellipses and using the Fourier transform occurring in Eq. (3.77),

$$W = 2\pi \int da\tilde{f}(a)\epsilon(q + ap)\delta((q + ap)^2)$$

$$\simeq \frac{2\pi}{v}\int da\tilde{f}(a)\delta(a - x_B) = \frac{2\pi}{v}\tilde{f}(x_B) \qquad (5.58)$$

This is apart from the dimensional factor v^{-1} a result which only depends upon the Bjorken scaling variable x_B through the Fourier transform of f.

It is of particular interest to note that the scaling variable x_B in this way occurs as the inverse Fourier transform variable (the 'canonical coordinate') of the quantity px, which intuitively describes the variations of the matrix elements along the lightcone $x^2 = 0$. The result stems from the assumption that the (scalar) current commutator behaves as the free-field commutator in Eq. (3.77). The argument can, however, be generalised to include more complex situations where the lightcone singularity contains derivatives of δ-distributions. The main point throughout is that no new scale is involved. The lightcone *per se* is evidently the same everywhere.

5.7 **The parton model revisited**

We have seen in the previous section how to make use of some simple causality arguments, and some perhaps optimistic limits, to obtain the scaling behaviour of the inelastic cross section. In this section we will arrive at the same result by an analysis of the cross section we obtain from the inelastic scattering situation. We obtain, by introducing the tensor W into Eq. (5.33),

$$d\sigma = \frac{e^2}{2(2\pi)^2(s - M^2)t^2}dk_3\delta^+(k_3^2)2MT_{\mu\nu}W^{\mu\nu} \qquad (5.59)$$

The tensor W can be constructed from the two T-tensors we have used before. In conventional notation we write (with two scalar form factors W_j):

$$W_{\mu\nu} = e^2\left[W_1\left(-g_{\mu\nu} + \frac{q_\mu q_\mu}{q^2}\right) \right.$$

$$\left. + W_2\frac{1}{M^2}\left(p_\mu - \frac{pq}{q^2}q_\mu\right)\left(p_\nu - \frac{pq}{q^2}q_\nu\right)\right] \qquad (5.60)$$

which leads to

$$T_{\mu\nu}W^{\mu\nu} = e^2 \left\{ 2Q^2 W_1 + W_2 \left[\frac{\hat{s}(\hat{s}-v)}{M^2} - Q^2 \right] \right\} \tag{5.61}$$

using $v = 2pq$. Next, we introduce instead of the vector k_3 the two invariants Q^2, v by means of the usual trick:

$$dQ^2 dv \int dk_3 \delta^+(k_3^2) \delta(Q^2 - 2k_1k_3)\delta(v - 2p(k_1 - k_3)) = \frac{\pi}{2\hat{s}} dQ^2 dv \tag{5.62}$$

The cross section is then given by

$$d\sigma = \frac{2\pi\alpha^2 dQ^2 dv}{\hat{s}^2 Q^4} \left[2MQ^2 W_1 + \frac{W_2}{M}(\hat{s}^2 - \hat{s}v - M^2 Q^2) \right] \tag{5.63}$$

According to Feynman's suggestion this cross section should be expressible in terms of a flux factor $\sum_j e_j^2 f_j(x)dx$ of partons, all massless and scattering like point particles (with squared charge $e_j^2 e^2$) from the electron. Their cross section should then be given by the invariant cross section in Eq. (5.40), so that

$$d\sigma = dQ^2 \frac{2\pi\alpha^2(Q^4 - 2\hat{s}Q^2 + 2\hat{s}^2)}{Q^4 \hat{s}^2} dv\delta(v - Q^2) \tag{5.64}$$

If everywhere we replace the parton energy-momentum p by xp, this implies the following changes:

$$Q^2 = -q^2 \rightarrow Q^2, \quad \hat{s} = 2pk_1 \rightarrow x\hat{s}, \quad v = 2pq \rightarrow xv \tag{5.65}$$

We then obtain, by comparing coeffients,

$$\frac{W_2}{M} = \int 2 \sum_j e_j^2 f_j(x) x dx \delta(xv - Q^2) \quad \Leftrightarrow \quad \frac{vW_2}{2M} = \sum_j e_j^2 f_j(x_B) x_B \tag{5.66}$$

$$2MW_1 = \int \sum_j e_j^2 f_j(x) Q^2 dx/x \delta(xv - Q^2) \quad \Leftrightarrow \quad 2MW_1 = \sum_j e_j^2 f_j(x_B)$$

where $x_B = Q^2/v$ is the Bjorken scaling variable. In this way we have been able to give a precise relationship between Feynman's parton flux factors and the inelastic form factors W_1 and W_2.

We note that the fact that we have assumed the partons to be spin 1/2 particles provides a very precise relationship between the two structure functions W_1 and W_2, i.e.

$$\frac{vW_2}{2M} = x2MW_1 \equiv xf_B(x) \tag{5.67}$$

using the subscript B to denote a baryon target. There will be different parton flux factors (or, as they are called, *structure functions*) for a proton (p) and a neutron (n) as we will see below.

If we had assumed that the partons were spin 0 particles the corresponding analysis would have led to the result that W_2 still has the same scaling shape but now $W_1 = 0$. Needless to say the original SLAC experiments proved conclusively that W_j fulfils the scaling laws in Eq. (5.66) and therefore that the partons involved were, just as the quarks should be, spin $1/2$ particles.

It is instructive to compare these results with the case of elastic scattering and the corresponding form factors from the Rosenbluth formula (Eqs. (5.44), (5.46). If we put $G_E = G_M = G$ then we obtain the correspondences

$$W_1 \rightarrow \frac{Q^2}{4M^2} G^2(Q^2)\delta(v - Q^2), \quad W_2 \rightarrow G^2(Q^2)\delta(v - Q^2) \qquad (5.68)$$

In this case there is another scale, $M_0 \simeq 0.71$ GeV, from the dipole formula for the baryon elastic form factors. Therefore it is impossible to rearrange these expressions into a scaling form.

We note, however that for $x_B = 1$ we go from the inelastic to the elastic contribution. In real-life experiments it is not actually a δ-peak, although it does stand out by several orders of magnitude (depending upon the way one plots it) from the inelastic background. In the neighborhood $x_B \simeq 1$ there are also contributions from several nucleon resonances and it is interesting that the inelastic cross section as described above takes over in a very smooth way. If we take an average over these resonances then we smoothly go over to the general inelastic cross section (the Drell-Yan-West relations, [54]). This means that the nucleon splits up into partons as smoothly as possible.

5.8   The partons as quarks

We will mention, just for completeness, a few properties of the structure functions for baryons when the partons are identified as quarks, in accordance with Gell-Mann's and Neeman's suggestion. For more extensive discussions we refer to [77].

With the wild proliferation of new particles, found in high-energy interactions at the end of the 1950s and in the 1960s, it quickly became clear that all these quantities could not be fundamental quanta. Therefore several different classification schemes were suggested, all of them building upon some idea of a basic symmetry in the interactions. The one which was successful, the SU(3)-group classification, contains besides the singlet, octet and higher representations a triplet also (corresponding to the spin $1/2$ representation in SU(2)). This triplet (which we from now on will call 3_f, f for *flavor*) contains three 'building blocks', the u, d and s quarks (q-particles, or q's). Together with the corresponding antitriplet,

$\overline{3}_f$, containing the antiquarks, the $\bar{u}$, $\bar{d}$ and $\bar{s}$ ($\bar{q}$-particles, or $\bar{q}$'s), they can be used to build up all known higher representations of the hadrons without charm and bottom flavors. The quarks must have a set of internal quantum numbers in order to be useful.

Q1 *The quark electric charges* are $+2e/3$ for the u and $-e/3$ for the d and s, with e the fundamental electric charge. As it is the square of the charges which occurs in the cross sections (the square of the matrix elements) the u will couple four times as strongly as the d and s to electromagnetic interactions.

This means that the effective flux factors for electromagnetic interactions contain a different weighting between the quark species so that the observed flux must be proportional to

$$f(x) = \tfrac{4}{9}\left[u(x) + \bar{u}(x)\right] + \tfrac{1}{9}\left[d(x) + \bar{d}(x) + s(x) + \bar{s}(x)\right] \qquad (5.69)$$

when electromagnetic probes are used. We use the notation $x_B = x$ and the quark names for the distributions.

Q2 *The pairs u, d and $\bar{u}, \bar{d}$ each form an isospin $1/2$ doublet.* The s and $\bar{s}$ do not carry isospin but instead *strangeness* and *antistrangeness*). This means that the SU(3) flavor-group contains fundamental building blocks both in abstract isospin space (in both directions, u 'up' and d 'down') and along the strangeness direction.

The strong interaction conserves these quantum numbers so that the total isospin I and the strangeness content is conserved; further they do not care about the directions in isospin space. This means that states with the same I but different I_3 (i.e. different steps in the u- or d-directions) react in the same way to the strong interaction.

In particular the proton, p, and the neutron, n, form an isospin doublet with $I = 1/2$; they contain uud and ddu respectively. Therefore a knowledge of the u-content (u_p) of the p is equivalent to a knowledge of the d-content (d_n) of the n. The same goes for $d_p = u_n$.

Q3 *SU(3)-symmetry of the 'ocean'.* One usually assumes (for lack of evidence to the contrary) that there are two particular kinds of parton distributions, for *valence constituents* and for *'ocean' q- and $\bar{q}$-particles*. Thus $u_p = u_{pv} + u_{po}$, i.e. the sum of the valence and the ocean contributions and a similar relation holds for d_p.

Further one often assumes that *all the ocean parts are equal*, i.e. $u_{po} = \bar{u}_p = d_{po} = \bar{d}_p = s_p = \bar{s}_p \equiv o$ (note that for a baryon all the antiquarks then belong to the ocean).

Then we can rewrite the relations for the effective structure functions of the p, n, f_p and f_n, and their difference, as

$$f_p(x) = \tfrac{1}{9}(4u_{pV} + d_{pV}) + \tfrac{4}{3}O(x)$$
$$f_n(x) = \tfrac{1}{9}(4d_{pV} + u_{pV}) + \tfrac{4}{3}O(x) \qquad (5.70)$$
$$f_p(x) - f_n(x) = \tfrac{1}{3}(u_{pV} - d_{pV})$$

Q4 *The q- and q̄-partons carry spin* 1/2, as we have shown above.

Taken together this means that (if property Q3 is fulfilled) that there are three different structure functions for the quarks in the baryons. There is also the gluon structure function $g(x)$, which is often taken as closely related to the ocean quark properties.

The experimental results provide both a direct measurement of some combinations of the structure functions and also constraints on all of them. We will end by pointing out that the original SLAC experiments had already given constraints on the behaviour of $g(x)$. It is evident that the following integral will contain all the momentum carried by q and $\bar{q}$:

$$\int_0^1 x\, dx(u + \bar{u} + d + \bar{d} + s + \bar{s}) = I \qquad (5.71)$$

From their measurements on protons and neutrons the experimentalists were able to determine that

$$\int_0^1 x\, dx f_p(x) \simeq \tfrac{4}{9}I_u + \tfrac{1}{9}I_d \simeq 0.18$$
$$\qquad (5.72)$$
$$\int_0^1 x\, dx f_n(x) \simeq \tfrac{1}{9}I_u + \tfrac{4}{9}I_d \simeq 0.12$$

with the approximation that one neglects the strange and antistrange contributions.

From here we conclude that the fraction of the proton's energy-momentum carried by the u and $\bar{u}$, I_u, and the fraction carried by the d and $\bar{d}$, I_d, are approximately 0.36 and 0.18, respectively. Therefore in this approximation I for the proton is 0.54. This means that about 50% of the proton momentum is carried by the field or, as we will in general say, the field quanta, the *gluons*.

6

The classical motion of the massless relativistic string

6.1 Introduction

In this chapter we start to consider the properties of the *massless relativistic string* (the MRS). We will begin with a simple situation in which the MRS plays the role of a constant force field, acting upon a 'charge' and an 'anticharge' placed at the endpoints of an open MRS. This means that the motion will be in one space dimension along the force direction. We will refer to it as the *yoyo-mode* for reasons that will become clear when it is exhibited.

In later chapters we will come back to more complex modes involving several dimensions. All these modes are used in the Lund model as semi-classical models for different high-energy interactions between hadrons. The yoyo-mode is used both to describe an e^+e^- annihilation event and as a simple model for stable hadrons. In the last section of this chapter we provide a possible dynamical analogy between the QCD vacuum and superconductivity as a justification for using string dynamics to describe hadronic states and interactions.

In the yoyo-mode the two charges at the endpoints of the string move like point particles, i.e. the momentum of the state is localised in these endpoint particles of the MRS force field. At any moment the total energy of the state can be decomposed into the energy in the force field, corresponding to a linearly rising potential, and the kinetic energies of the particles at the endpoints. We will use the situation to exhibit in detail the *causality* and the *relativistic covariance* properties of the MRS.

In the Lund model the endpoints of an open MRS are always identified with triplet, 3, or antitriplet, $\bar{3}$, color charges, i.e. with quark, q, or antiquark, $\bar{q}$, properties. In connection with the description of baryonic particles, cf. Chapter 13, we will consider more complex charge configurations.

114

6.2 The MRS as a constant force field

1 The equations of motion

The equations of motion in relativistic particle dynamics are, in general, complex in a consistent theory. The finiteness of the maximal velocity, i.e. that of light, implies a *causality requirement*. A message about changes in the system, such as e.g. the change in the state of motion of a charge somewhere, takes a finite time to be transmitted to any other part of the system. Consequently, the reaction of the system to the change, i.e. the ensuing force action, is of a retarded character.

More precisely, some cause at the origin at time t_0 will affect what happens at a point $\mathbf{R}$ only after a message has been able to reach that far. If this moves with the velocity of light, $c = 1$, in a straight line, it will cause an effect at time t with $t = t_0 + |\mathbf{R}|$. The calculations including the retarded times then become rather complicated.

There is one particular situation, that of a *constant force*, that is easy to work with (because then the retardation effects are not noticeable). The historical start of what is now known as the Lund string model was based upon the consideration of such a force, [14]. We only later learned that the ensuing motion is a simple variety of the modes of the MRS [24].

If we consider the motion of a relativistic particle in space-time (t, x), with rest mass m, energy E and momentum p, under the influence of a constant force $-\kappa$, we have the force equation

$$\frac{dp}{dt} = -\kappa. \tag{6.1}$$

The solution is evidently

$$p = p(t) = p_0 - \kappa t \equiv \kappa(t_0 - t) \tag{6.2}$$

The velocity of the particle is

$$\frac{dx}{dt} = \frac{p}{E} = \frac{dE}{dp}, \quad E = \sqrt{p^2 + m^2} \tag{6.3}$$

(The first equation of (6.3) corresponds to one of Hamilton's equations, the hamiltonian being given by the relativistic particle energy.)

From Eqs. (6.1) and (6.3) it is possible to obtain an equation for the variation of the energy with respect to the space coordinate, if we use the chain rule for differentiation:

$$\frac{dE}{dx} = \left(\frac{dE}{dp}\right)\left(\frac{dp}{dt}\right)\frac{dt}{dx} = \frac{dp}{dt} = -\kappa \tag{6.4}$$

This equation has, similarly, a simple solution:

$$E = E(x(t)) = E_0 - \kappa x \equiv \kappa(x_0 - x) \tag{6.5}$$

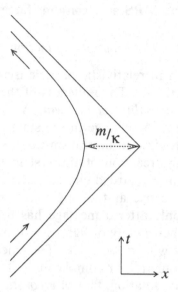

Fig. 6.1. The motion in space and time of a particle with mass m under the influence of a constant force $-\kappa$. The distance between the hyperbola and the intersection (t_0, x_0) between the asymptotes is m/κ.

From the relationship between energy, momentum and mass we conclude that the orbit of the motion is

$$m^2 = E^2 - p^2 = \kappa^2 \left[(x_0 - x)^2 - (t_0 - t)^2 \right] \tag{6.6}$$

i.e. a hyperbola in space-time, centred at (t_0, x_0) and with a size parameter m/κ (see Fig. 6.1).

At large negative times the particle comes in from the region of large negative space coordinates with its momentum pointing along the positive coordinate axis. The momentum decreases and the particle is, at time $t = t_0$, momentarily at rest at the *classical turning-point* $x - x_0 = -m/\kappa$. Afterwards it moves with increasingly negative momentum back to large negative space coordinates.

We note that if the mass vanishes then the particle will move along the lightcones $|t - t_0| = x_0 - x$ throughout and it will look as though it 'bounces' back (changing from velocity $+c$ to $-c$ with vanishing energy and momentum at the origin (t_0, x_0)).

We will use massless particles from now on because of the simplifications in the ensuing pictures of the motion. We would like to stress, however, that the dynamics we are going to consider is basically independent of this assumption (cf. the considerations in Chapter 12).

2   The Schwinger model and confinement

A particularly interesting dynamical situation arises when there is a constant force and a linearly rising potential; this occurs in one-spacedimensional electrodynamics. There are also in three space dimensions situations that can be approximated by one-dimensional dynamics, e.g. the field between two condensor plates.

Then the usual four-vector potential $A_\mu = (A_0, \mathbf{A})$ can by a gauge choice be arranged so that only the scalar potential $A_0 \equiv V$ is nonvanishing. The single component of the electric field $\mathscr{E} = -dV/dx$ will in a charge-free region fulfil Gauss's law, i.e.

$$\frac{d^2 V}{dx^2} = 0 \tag{6.7}$$

which means exactly a linear potential. This constant force is approximately realised in a capacitor.

A quantised version of one-dimensional electrodynamics was investigated by Schwinger, [101]. He was able to show that an electric field coupled to massless fermion particles is (essentially, i.e. leaving aside some peculiar modes) *equivalent to a free, non-interacting, but massive, quantum field theory.*

The quanta of this field are massive and electrically neutral. Their mass is a function of the electric charge, $m^2 = g^2/\pi$. Note that the charge g, as defined by a gaussian 'integral' (in a one-dimensional world there are no transverse dimensions to integrate over)

$$g = \mathscr{E} \tag{6.8}$$

does not have the same dimensions as in the usual three-dimensional case. The dimensions of the electric field $\mathscr{E}$ can be read out from the usual energy density requirement, that half the square of the field strength is equal to the energy density, $dE/dx = \mathscr{E}^2/2$. This means that the electric field has (energy) dimensions dim$\mathscr{E}$ equal to 1. Therefore g^2 has the dimensions of a squared mass in this case.

The fact that the quanta are electrically neutral is very surprising because it seems as if the original electric charges have vanished. It turns out, however, that *the resulting free-quantum field, ϕ, corresponds to a dipole density.* The original massless fermions are arranged two by two with a positive and a negative charge bound together as a dipole.

This is a realisation of confinement, i.e. the original massless fermions are not observable by themselves but only in particular combinations. In the Schwinger model the original fermions and antifermions can only occur in pairs as bound states with one of each kind.

In this one-dimensional setting this means that one of the charges must be to the left of the other, thereby producing a dipole moment.

We may compare this with with the case of colored quanta, where the hadronic states are built from color combinations corresponding to no-color singlets. In the Lund fragmentation model the hadrons are modelled by the massless relativistic string, corresponding to a color field spanned between two endpoints associated with quark (color-3) and antiquark (color-$\bar{3}$) charges (the 'ultimate dipoles' in Chapters 7–14).

We will also introduce this dipole character in the description of multi-gluon bremsstrahlung in the dipole cascade model (Chapters 16–18). In this case the emitting current has only a direction and a very small space extension. Similarly in the linked dipole chain model, which describes the properties of deep inelastic scattering (Chapter 20) we will again find the same dipole structures, describing the (squared) wave functions of the hadrons (the structure functions).

In the Schwinger-model case confinement is related to the infinitely rising field energy necessary in order that a charge should be moved away from all the other charges. In our calculations in subsection 1 we found a constant energy density along the whole negative axis beyond where the particle reaches its classical turning point.

We will carry the model on a little further to a simulated particle-production situation, like the one described in [39]. These authors investigated the situation where an external current is composed of a $\pm g$ charged pair. The charges set out at the time $t = 0$ in opposite directions along the single space dimension, the 1-axis. We assume that they move with velocity $v = c = 1$. This means that there is a current (j_0^{ext}, j_1^{ext}), where

$$j_0^{ext} = g\epsilon(x_1)\delta(\epsilon(x_1)x_1 - t), \quad j_1^{ext} = g\delta(\epsilon(x_1)x_1 - t) \qquad (6.9)$$

(note the appearance of the sign function $\epsilon = \pm 1$, depending upon the sign of its argument, which describes the way the charges $\pm g$ move). This current corresponds to an external dipole density

$$\phi^{ext} = \frac{g}{m}\Theta(t + x_1)\Theta(t - x_1) \qquad (6.10)$$

where the fields are normalised somewhat differently from that in [39]. Our choice is in accordance with the one-dimensional equivalent to the fields introduced in Chapter 3; thus the quantum field ϕ is, using $\omega \equiv \omega(k) = \sqrt{k^2 + m^2}$ and L for the length of the one-dimensional 'quantisation box',

$$\phi(x_1, t) = \sum_k \frac{1}{\sqrt{2\omega L}} \{a \exp i(kx_1 - \omega t) + a^* \exp[-i(kx_1 - \omega t)]\} \quad (6.11)$$

Then we may write out the equations of motion for the fields, the Klein-Gordon equation

$$(\Box + m^2)\phi = m^2 \phi^{ext} \qquad (6.12)$$

perform all the operations for quantisation and solve the equations to obtain as solution a coherent-state field like those of Chapter 3. Then the quanta in every state will be distributed in a Poissonian manner with an excitation probability described by the mean occupation number $\bar{n}(k)$ (cf. Eqs. (3.25), (3.41)):

$$h = \frac{1}{\sqrt{2\omega L}} \int_0^\infty dt \int dx m^2 \phi^{ext} \exp[i(kx - \omega t)] = \frac{1}{\sqrt{2\omega L}} \left(\frac{2g}{m}\right), \quad (6.13)$$

$$\bar{n} = |h|^2 = \frac{4g^2}{2\omega L m^2} = \frac{2\pi}{\omega L}$$

We have performed the integral in the first line by adding a small negative imaginary part to ω (remember the three-act scenario described in the first section of Chapter 3) and used the relationship between the mass and the coupling constant in the second line.

This means that when we go to the limit $L \to \infty$ we obtain for $\bar{n}$

$$\bar{n}\Delta n \to \bar{n}dk \frac{L}{2\pi} = \frac{dk}{\omega} \equiv dy \qquad (6.14)$$

in terms of the rapidity variable y. This is nothing other than the wee parton spectrum of Feynman or, if you like, the distribution of photons in the method of virtual quanta in Chapter 2.

Consequently, an external excitation in the Schwinger model tends to spread as a Poissonian fluctuating production of dipole quanta of average size one quantum per unit rapidity!

3   *The yoyo-mode at rest*

As a classical model corresponding to Schwinger's dipole quanta we consider the motion of a system of two massless particles, a q- and a $\bar{q}$-particle, which are acting upon each other with an attractive constant force.

In Fig. 6.2 we consider the situation when the q and $\bar{q}$ go apart with the same energy E_0 from a common origin but in opposite directions. Such a system evidently has a total energy $E_{tot} = 2E_0$. This coincides with the system mass m as the total momentum vanishes.

According to the results in subsection 1 the particles will move along the two different lightcones and each will lose energy-momentum κ per length and time unit. The starting situation corresponds to the q and $\bar{q}$ each having lightcone energy-momentum $2E_0$.

The ensuing motion can most easily be described in terms of a series of fixed-time snapshots (the lines on the right-hand side of the figure, although the space-time picture given on the left of Fig. 6.2 provides a total view of the system):

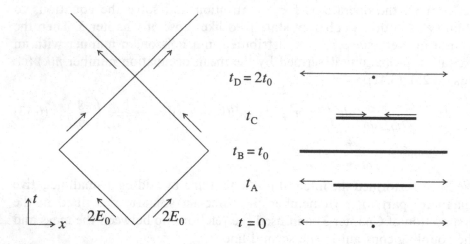

Fig. 6.2. On the left, a space-time diagram for the motion of a $q\bar{q}$-state, in which the particles always have the same energy, i.e. the yoyo-mode at rest. The different times mentioned in the text are shown, with the length of the arrowed lines corresponding to the energy of the particles and the length of the thicker lines corresponding to their separation, i.e. the field in between.

A After a time t_A $(< E_0/\kappa \equiv t_0)$ they will be a distance $2t_A$ apart, each with energy $E_0 - \kappa t_A$. The 'lost' energy has gone into the force field, which now has energy κ times its length, i.e. $2\kappa t_A$.

B At time $t_B = t_0 = E_0/\kappa$ they have lost all their energy and they will then turn back and move towards each other.

C At the time $(2t_0 >) t_C > t_0$ they will be at a separation $4t_0 - 2t_C$, each with energy $\kappa t_C - E_0$. This energy has been obtained from the force field, which now is dragging them towards each other.

D At time $t_D = 2t_0$ they will meet again but this time they have exchanged their modes of motion compared to the starting point. The q moves along the original $\bar{q}$-direction and vice versa.

As can be seen by a straightforward extrapolation of the argument, after the time $4t_0 = 4E_0/\kappa \equiv 2E_{tot}/\kappa$ the q- and $\bar{q}$-particles will come back exactly to the starting position. Actually a little thought will tell us that the system is always in the same mode of motion at the times t and $t + 2E_{tot}/\kappa \equiv t + t_{per}$. This fact that *the period of motion is equal to* $t_{per} = 2E_{tot}/\kappa$ *is true for all modes of the* MRS, as we will see later.

Another general property of the MRS is that *the total area $\tilde{A}$ spanned by the force field in space-time during one period is related to the squared*

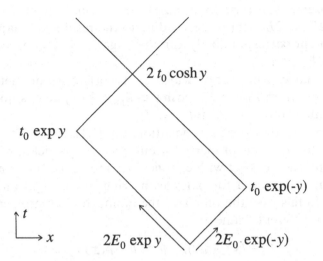

Fig. 6.3. The yoyo-mode after a Lorentz boost along the positive direction; the times and the lightcone energy-momenta from Fig. 6.2 are shown in the new system.

mass of the system. It is easy to see that the relationship is

$$\kappa^2 \tilde{A} = \kappa^2 8 \frac{E_0^2}{2\kappa^2} = m^2 \qquad (6.15)$$

in our case; there are exactly eight identical triangles with side and height lengths $t_0 = E_0/\kappa$.

In this particular mode the q and the $\bar{q}$ will continue to move in and out along the lightcones and the name 'yoyo-mode' has a self-evident meaning. The energy and momentum are at different times divided in different ways between the endpoint particles and the force field. We note for future reference that, averaged over a period, half of the energy is in the endpoint particles and half of it is in the field. This is the same result for energy sharing between the quarks and the gluons in a hadron that we quoted in Chapter 5 from the experimental results.

4   Lorentz covariance and causality properties

The model is Lorentz-covariant; we will now demonstrate this by an explicit calculation.

We will consider the situation after we have boosted the system (see Fig. 6.3) longitudinally, i.e. along its axis, with the rapidity y. Then the q-particle, which moves along the positive direction, will by the corresponding Lorentz transformation change its original (positive) lightcone component $2E_0$ to $2E_0 \exp(-y)$ according to the results in Chapter 2. For the $\bar{q}$ we obtain correspondingly for the negative lightcone component that $2E_0 \to 2E_0 \exp y$.

Thus the total system energy, which at the origin is completely in the $q\bar{q}$-pair, changes from E_{tot} to $E_{tot} \cosh y \equiv E'_{tot}$. The system is now moving with a total momentum $-E_{tot} \sinh y \equiv P'_{tot}$.

It is not obvious that the force equation, Eq. (6.1), is Lorentz-invariant. But it is easy to show this property for our massless particles, which move along the lightcones $x = \pm t$ with energies and momenta $E = \pm p$; the plus and minus signs are valid for particles moving to the right and the left, respectively. In this case the time and the momentum component of such a particle in a different frame are

$$t' = t \exp(\pm y), \quad p' = p \exp(\pm y) \tag{6.16}$$

and we immediately obtain that

$$\frac{dp'}{dt'} = \frac{dp}{dt} \tag{6.17}$$

A more general but also more complex argument could be based upon the properties of the electromagnetic field and its interactions with particles; then all dynamical variables evidently have simple covariance properties. The constant force will occur in one-dimensional QED as mentioned in connection with the discussion of the Schwinger model.

Thus, in the new frame the particles will also be acted upon by a constant force of the same size. The main difference is that the q now has a diminished, and the $\bar{q}$ an increased, original energy. Therefore, in this case they will not stop at the same time. Again using the equal-time snapshot technique we have, from Fig. 6.3,

A   The q will stop and turn around at time $t'_A = t_0 \exp(-y)$ (at the space point $t_0 \exp(-y)$) and after that move behind the $\bar{q}$ at a distance $2t_0 \exp(-y)$.

During the ensuing motion the $\bar{q}$ is losing its energy to the field and the q will be increasing its energy from the field, both of them at the same rate. In somewhat vivid language the q 'eats', and the $\bar{q}$ 'spits out', the field as they move along.

B   At time $t'_B = t_0 \exp y$ (at the spot $x' = -t_0 \exp y$) the $\bar{q}$ has used up its energy and turns around towards the q.

From Fig. 6.3 we also deduce the following three properties:

C1 The two particles will meet again at time $t_0 \exp y + t_0 \exp(-y) = 2t_0 \cosh y$.

C2 The meeting point has x'-coordinate given by $-2t_0 \sinh y$.

C3 By the time they arrive at the meeting point the two particles have exchanged their energies and momenta, i.e. the q has gained exactly as much energy as the $\bar{q}$ has lost, and vice versa (although the gain and loss have not occurred at the same times but rather through the field).

After a second such yoyo 'round' the q- and $\bar{q}$-particles will be back at their original energy-momentum conditions.

The time it has taken is, however, longer than in the rest system, i.e. instead of $4t_0$ it is $4t_0 \cosh y$. But we note that the period is again given by twice the total energy divided by κ: $2E'_{tot}/\kappa = 2E_{tot}/\kappa \cosh y$. This is the MRS version of the time-dilation effect, described in Chapter 2.

The Lorentz-contraction phenomenon implies that the field sizes are correspondingly always shorter. We note, however, that *the Lorentz-contraction and the time-dilation effects combine in such a way that the space-time size spanned by the field during the period will again satisfy Eq. (6.15)*. We leave the proof of this statement to the reader.

Finally, we note from the above exercise that during such a full period the system has moved a distance $\delta x'$ from the origin to the meeting point:

$$\delta x' = 2[t_0 \exp(-y) - t_0 \exp y] = -4t_0 \sinh y \equiv 2P'_{tot}/\kappa. \qquad (6.18)$$

This is another general property of the MRS: *during a period $t_{per} = 2E_{tot}/\kappa$ the system will be translated by the vector $\mathbf{x}_{per} = 2\mathbf{P}_{tot}/\kappa$.*

There are two comments to add to this result:

- when the system is at rest as in the previous subsection then $\mathbf{P}_{tot} = 0$ by definition of 'at rest';

- the system will move during a period as if it had a mean velocity $\mathbf{x}_{per}/t_{per} = \mathbf{P}_{tot}/E_{tot}$, which is just the usual velocity for a particle with energy-momentum $(E_{tot}, \mathbf{P}_{tot})$.

This moving extended system contains three parts and behaves in a surprising manner. The two particles are moving with the velocity of light in the same or opposite directions and therefore contain both energy and momentum. There is, further, the field, which throughout seems to be longitudinally at rest, i.e. it contains only energy and no momentum. But the field nevertheless does change its position because it only exists in the region between the charges!

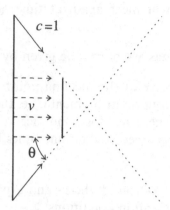

Fig. 6.4. The yoyo-mode after a Lorentz boost transverse to the field direction. The field is shown by the thick solid lines, the endpoints move with the velocity of light $c = 1$ along the thin solid lines and the field moves with velocity v along the direction of the broken lines. The dotted lines are the continuations of the motion of the endpoints.

With respect to causality we note that the two particles meet every half period, but meanwhile are often at spacelike distances with respect to each other. From Figs. 6.2 and 6.3 we note that each particle can while in motion in principle send away lightlike and even timelike messages 'via the field'; these can be received by its partner during the second part of the half-period. Thus the typical communication time can be short (when the particles move together in a strongly Lorentz-contracted string field) or long (when they move apart). It is necessary to introduce some kind of measurement procedure to define the notion of 'communication' and we will not speculate further on the subject at this point.

The result is, however, that there is always a finite delay time for any message travelling through the system. If one of the endpoint particles were acted upon by some outside agent then it would take some time before the other one would 'know'. This feature will be more noticeable when we consider the reaction of the yoyo system to an external momentum transfer, in Chapter 20.

5   A transverse boost of the yoyo-mode

It is instructive to consider the yoyo-state in a frame that is boosted transversely to the directions of motion of the two endpoints. We will then find that the field this time actually must contain also momentum. The situation is shown in Fig. 6.4 for two different times.

We are now going to analyse the situation using the following two rules.

1 The two endpoints always move with the velocity of light.

2 The energy-momentum is conserved throughout and we will see that it is even *locally* conserved owing to causality.

The left-hand vertical line in Fig. 6.4 corresponds to a time when in its rest frame the system is stretched out as far as possible, i.e. at a turning time for the two endpoint charges. Then, in a frame where the string is moving with velocity v with respect to the rest frame the field contains both energy and momentum. If the field length is $2l$ then its rest frame energy is $2\kappa l$. In the moving frame that means (cf. Chapter 2):

$$E = 2\kappa l\gamma(v), \quad P = 2\kappa l v\gamma(v) \tag{6.19}$$

where $\gamma(v) = 1/\sqrt{1-v^2}$. Note that the force field does not change its shape or size as it is boosted transversely. Equation (6.19) evidently gives the total energy-momentum of the system.

After a time δt (measured in the new frame) the endpoints have moved the distance δt and a point in the middle of the field has moved $v\delta t$. From Fig. 6.4 we conclude that the velocity v is related to the angle θ by

$$v = \cos\theta \tag{6.20}$$

The length of the force field is now $2(l - \delta t\sin\theta)$ and therefore the energy-momentum of the field is proportionally smaller.

In particular the field energy has decreased by an amount

$$\delta E = 2\kappa\delta t\sin\theta\gamma(v) \tag{6.21}$$

Using Eq. (6.20) we obtain

$$\gamma(v) = \frac{1}{\sin\theta} \tag{6.22}$$

which implies that

$$\delta E = 2\kappa\delta t \tag{6.23}$$

This field-energy loss is easy to understand from what we already know. It means that each endpoint particle will obtain (from the field) an increased energy $\delta e = \kappa\delta t$ while it moves the distance δt.

Further, we note that the momentum of the field along the boost direction has decreased by

$$\delta P = 2\kappa\delta t\sin\theta v\gamma(v) = 2\kappa\delta t\cos\theta \tag{6.24}$$

(again using Eqs. (6.20), (6.22)). This is the amount of momentum $\delta p\cos\theta$ which the q- and $\bar{q}$-particles have picked up along that direction.

They have also (in order to fulfil the masslessness condition $\delta e = |\delta p|$) acquired two compensating components $\pm \delta p \sin \theta$ along the field direction. In this way one can describe the force from the field on the particles as $\pm \kappa \sin \theta = \pm \kappa \sqrt{(1 - v_\perp^2)}$ in the moving frame. We have seen before that the force is not changed for Lorentz boosts along the string but, owing to the time dilation effect, it is in this way affected by transverse boosts.

Consequently the energy-momentum is redistributed between the end-point particles and the field in a *local* way. Once again we can talk of them as 'eating' or 'spitting out' the field in their neighborhoods.

From the two rules given above it is possible to trace any complicated motion of the force field, as we will see in later sections. A useful exercise at this point is to consider the necessary Lorentz transformations and the ensuing motion if one were to boost the 'flat' yoyo-mode in a direction between the longitudinal (exhibited in the previous subsection) and the transverse as discussed here.

You will then notice that it is only the transverse part of the field velocity (transverse, of course, with respect to the field direction) that plays a role for the field momentum. This means that the field only contains momentum with respect to its transverse motion, i.e. longitudinal momentum-carrying modes of the field do not exist for the MRS field (but they do occur for the endpoints). This is once again quite in accordance with good old classical string motion, where only transverse degrees of freedom play a role.

6.3   The QCD vacuum as a color superconductor

Both the Schwinger model and QCD are confining in the sense that the real charges (respectively electromagnetic and color) cannot be isolated from each other and only occur in particular singlet combinations. Confinement is, however, also expected to lead to restrictions on the spatial extension of the force fields between the charges. Calculations in the lattice approximation of QCD tend to confirm this behaviour.

The MRS, as a model of a confining force field in which the charges are identified as the endpoints, evidently has both these properties. In this section we will provide a motivation for the use of the MRS in hadron dynamics. We introduce a color superconductor as a simple model for the QCD vacuum state. We will also briefly mention another well-known model, the *bag model* for hadrons, and point out its relation to the MRS.

1 The London equations and types I and II superconductors

Electromagnetic superconductors have many wonderful properties and we mention only a few here:

- According to condensed-matter physics there is a tiny attractive interaction between two electrons close to the fermi surface, owing to the exchange of phonons associated with the crystal lattice of the material. Therefore there exists a (very) loosely bound state of two electrons, a *Cooper pair*, with spin 0. The spatial extent of the state, called ξ, is often in the μm range, i.e. it may be of macroscopic size. Due to this bosonic nature many such states may overlap in space and behave as a degenerate (although charged) Bose gas. The pairs move freely through the material and there is no resistance.

- According to Lenz's law an applied magnetic field will produce a (super) current of Cooper-pair states that will expel the applied field. Thus a magnetic field will only have an exponentially falling penetration depth (called λ) in a superconductor. If the temperature or the field is increased beyond a critical size, the states will be excited and break up and there is thus a phase transition from the superconducting to the normal state.

Due to the relative sizes of ξ and λ, such ordinary superconductors have one of two rather different behaviours at the critical point. We will now consider the two cases, called types I and II superconductors. *The shape of the normal-state field regions depends upon the superconductor type.*

If $\xi \gg \lambda$ the boundary regions between the superconducting state and the rest will be empty because neither the magnetic field nor the Cooper pairs can spread there. These regions are then inactive from a dynamical viewpoint. Nature will according to the gospel of thermodynamics then try to *minimise the boundaries* of a type I superconductor.

At the opposite extreme, $\lambda \gg \xi$, both the Cooper-pair density and the magnetic fields can populate the boundary region and Nature will consequently *maximise the boundaries* between the superconducting and the normal state in a type II superconductor.

It is known, [98], that there are in QCD possible color magnetic field configurations with energy below the no-particle state. In these states gluon combinations take the place of the Cooper pairs in an electromagnetic superconductor and the color electric field is in this case neutralised by the vacuum fields. The sizes of the corresponding lengths ξ and λ are not known from first principles. If the QCD vacuum corresponds to such a state then the appearance of color charges and fields in between them will correspond to regions with normal-state properties. Such regions will then be surrounded by such a vacuum color superconductor. In particular the boundary regions between the superconducting and normal states are interesting.

For the type I superconductor, the region where the (color) field expands (the normal-state region) will have boundaries that are as small as possible. For a localised excitation, the field will arrange itself as an (isolated) 'resonant cavity field', cf. Jackson, with standing waves inside this, in general, spherical region. The total field energy is proportional to the volume and we note that a sphere has the smallest boundary-to-volume ratio possible.

If the field has a longitudinal extension then the whole field will stay inside a cylindrical 'wave guide'. Once again the field energy will be proportional to the volume and if the longitudinal size is given then a connected cylinder shape will have minimal surface area.

There are, in QCD, analogy models for the two cases. The first corresponds to an isolated hadronic state, containing valence-quark color charges and color field energy organised into a spherical *bag*. The second corresponds to the production of an outward-moving $q\bar{q}$-state with its field energy organised into a *flux tube*. We will not go into details here but the basic idea involves introducing a 'bag-pressure' from the vacuum. This is neutralised at the boundary by the pressure from the fields inside so that there is a stable boundary.

To explain the different behaviour of a type II superconductor we consider a slab of matter (width L) in an (electromagnetic) superconducting state. Both for types I and II there is a minimal critical field, $\mathscr{B}_{c1}$, for which the superconducting state breaks down. We assume the field exists inside a region of total area A. Outside A there is still a superconducting state. For a type I superconductor the region will be homogeneous and the boundary region will have area $R_I = 2\sqrt{\pi A L}$. The whole field passes through A and so the total energy deposited in the slab is $E = \mathscr{B}_{c1}^2 A L$ and the total flux is $\Phi = \mathscr{B}_{c1}A$. For the type II case there is also a second critical field strength, $\mathscr{B}_{c2} > \mathscr{B}_{c1}$. For a field strength in between $\mathscr{B}_{c1}$ and $\mathscr{B}_{c2}$, the region will be penetrated by many thin *vortex-line* fields *each of a quantised size*. The core size is typically ξ and there is a weak repulsive interaction which keeps the vortex lines apart so that the field strength will vary inside A.

We may for simplicity consider the area A as divided into n circular non-connected regions. You will then find the same flux and the same energy deposit but the boundary region now has area $R_{II} \sim \sqrt{n}R_I$. Thus to maximise the boundary it is profitable to subdivide the region. When the field strength is greater than $\mathscr{B}_{c2}$ the whole region becomes filled with vortex lines and it will behave as for the type I case.

We shall exhibit a few steps in the London theory of superconductivity, [91], and in particular show the quantisation of the flux lines.

We consider a constant Cooper-pair density $n(\mathbf{x}, t)$ and a corresponding current $\mathbf{j}(\mathbf{x}, t) = -2en\mathbf{v}$, with $\mathbf{v}$ the velocity field. The continuity equation as well as the Lorentz force law will give (with a Cooper-pair mass m and charge $-2e$) for the stable state

$$\nabla \mathbf{j} = \nabla \mathbf{v} = 0, \quad \frac{d\mathbf{v}}{dt} = -\frac{2e}{m}(\mathscr{E} + \mathbf{v} \times \mathscr{B}) \qquad (6.25)$$

The total change in time of the velocity field should be regarded as the change in time for a fixed coordinate plus the change in the coordinate for a fixed time; thus

$$\frac{d\mathbf{v}}{dt} = \frac{\partial \mathbf{v}}{\partial t} + \nabla\left(\frac{\mathbf{v}^2}{2}\right) - \mathbf{v} \times (\nabla \times \mathbf{v}) \qquad (6.26)$$

Then the Lorentz force law is equivalent to

$$\frac{\partial \mathbf{v}}{\partial t} + \frac{2e}{m}\mathscr{E} + \nabla\left(\frac{\mathbf{v}^2}{2}\right) = \mathbf{v} \times \left(\nabla \times \mathbf{v} - \frac{2e}{m}\mathscr{B}\right) \qquad (6.27)$$

We may now apply the differential vector operator 'curl' ($\nabla \times$) on both sides of this equation and note that, according to Faraday's induction law (cf. Chapter 2), $\nabla \times \mathscr{E} = -\partial \mathscr{B}/\partial t$ and also that $\nabla \times \nabla a = 0$ for any function a.

Then one obtains the resulting equation for the vector $\mathscr{L}$:

$$\frac{\partial \mathscr{L}}{\partial t} = \nabla \times (\mathbf{v} \times \mathscr{L}) \quad \text{where} \quad \mathscr{L} = \nabla \times \mathbf{v} - \frac{2e\mathscr{B}}{m} \qquad (6.28)$$

When both fields and current vanish $\mathscr{L} = \mathbf{0}$. The Londons, [91], made the fundamental assumption that $\mathscr{L}$ *should always vanish inside a superconductor*. This implies immediately an equation for the magnetic field because a vanishing $\mathscr{L}$ means that

$$\mathscr{B} = -\frac{m}{4ne^2}(\nabla \times \mathbf{j}) = -\frac{m}{4ne^2}(\nabla \times (\nabla \times \mathscr{B})) = \frac{m}{4ne^2}\triangle\mathscr{B} \qquad (6.29)$$

Equations (6.28), (6.29) are known as the *London equations*. To exhibit the result in (6.29) we have used Ampère's law (assuming a static situation, $\partial \mathscr{E}/\partial t = 0$) and also the absence of magnetic charges (cf. Chapter 2). The solutions to Eq. (6.29) correspond to magnetic fields which are exponentially falling with a rate equal to the parameter λ mentioned above, which fulfils:

$$\lambda = \sqrt{\frac{m}{4ne^2}} \qquad (6.30)$$

The inverse of this λ is identical to the plasma frequency we met in the discussion of the behaviour of the dielectricity in Chapter 2 (although here, for the Cooper pairs, the charge is $-2e$).

2   Solutions of the differential equation

We will need a particular solution of Eq. (6.29), i.e. the one corresponding to cylindrical symmetry around the 3-axis, with no variation along that axis, $\mathscr{B} = \mathscr{B}e_3$ with $\partial\mathscr{B}/\partial x_3 = 0$. We will solve that equation at the same time as we also exhibit the behaviour of the Feynman propagator in spacelike regions (as promised in Chapter 3).

Let us consider symmetrical solutions $f \equiv f(x^2)$ to the equation

$$\tfrac{1}{4}(\triangle_{2d} - A_{2d}^2)f = 0 \qquad (6.31)$$

(for $x^2 > 0$) where $2d$ is the dimension of the space and A_{2d} is a positive number. It is instructive to note that in both of the following cases,

$$\triangle_{2d} = \sum_{j=1}^{2d} \frac{\partial^2}{\partial x_j^2}, \quad x^2 = \sum_{j=1}^{2d} x_j^2 \qquad (6.32)$$

$$\triangle_{2d} = \sum_{j=1}^{2d-1} \frac{\partial^2}{\partial x_j^2} - \frac{\partial^2}{\partial t^2}, \quad x^2 = \sum_{j=1}^{2d-1} x_j^2 - t^2$$

we obtain directly the following equation in $z = x^2$:

$$4\left(z\frac{d^2f}{dz^2} + d\frac{df}{dz}\right) - A_{2d}^2 f = 0 \qquad (6.33)$$

Assuming that the solution is of the kind

$$f(z) = (\zeta)^{2\alpha}g(\zeta) \text{ where } \zeta = \sqrt{z} > 0 \qquad (6.34)$$

the equation can be brought into the form (dots mean derivatives with respect to ζ)

$$\zeta^2\ddot{g} + (2d + 4\alpha - 1)\zeta\dot{g} + [4\alpha(d + \alpha - 1) - A_{2d}^2\zeta^2]g = 0 \qquad (6.35)$$

Then if we choose $\alpha = (1 - d)/2$ we obtain a modified Bessel differential equation,

$$\zeta^2\ddot{g} + \zeta\dot{g} - [(1 - d)^2 + A_{2d}^2\zeta^2]g = 0 \qquad (6.36)$$

For the case we started with, i.e. $d = 1$ with $A_{2d} = \lambda$ in Eq. (6.30), we have

$$f \equiv \mathscr{B} = CK_0\left(\frac{\sqrt{x^2}}{\lambda}\right) \tag{6.37}$$

where K_0 is the modified Bessel function of rank 0, which is exponentially falling and behaves for large values of its argument as follows:

$$K_0(x) \simeq \sqrt{\frac{\pi}{2x}}\exp(-x) \tag{6.38}$$

In order that $\mathscr{B}$ should be a proper magnetic field the normalisation constant C must have (energy) dimension 2.

For the Feynman propagator *for spacelike values of x^2* we obtain ($d = 2$ and $A_{2d} = m$) the same exponential falloff as in Eq. (6.38) but a power in front:

$$\Delta_F(x, m) \propto \frac{m}{\sqrt{x^2}}K_1(m\sqrt{x^2}) \tag{6.39}$$

3   *The quantisation of the magnetic flux*

The result in Eq. (6.37) has a logarithmic singularity for $x^2 = 0$:

$$\mathscr{B} \simeq C\log(\lambda/\sqrt{x^2}) \tag{6.40}$$

The corresponding magnetic flux, Φ, through the 12-plane is

$$\Phi = \int dx_1 dx_2 \mathscr{B} = 2\pi C\int_0^\infty x\,dx K_0\left(\frac{x}{\lambda}\right) = 2\pi C\lambda^2 = \left(\frac{2\pi}{2e}\right)\frac{Cm}{n} \tag{6.41}$$

We note that the quantity Cm/n is a dimensionless number (n, being a three-dimensional space density, then has energy dimension 3 using our ordinary convention with $c = \hbar = 1$).

We also note that the Cooper-pair (super)current $\mathbf{j}$ is given by

$$\mathbf{j} = \nabla \times \mathscr{B} = -\mathbf{e}_\phi\frac{d\mathscr{B}}{d\sqrt{x^2}} \tag{6.42}$$

where the derivative can be expressed in terms of the modified Bessel function K_1 and therefore again falls off exponentially in directions normal to the 3-axis. *It is, however, singular, like $1/\sqrt{x^2}$, along the 3-axis.*

We also note that *the current flows around the origin, i.e. the 3-axis.* (The unit vector $\mathbf{e}_\phi$ circulates around this axis in the direction of increasing azimuthal angle ϕ.) Thus the Cooper pairs circulate, thereby producing a magnetic field similar to that in a solenoid. This is the reason for the nonvanishing magnetic flux through the 12-plane and the singularity along the 3-axis.

In order to understand what is going on we go back to the London condition for a superconducting state, $\mathscr{L} = \mathbf{0}$, which we write as

$$m\mathscr{L} = \nabla \times m\mathbf{v} - 2e\mathscr{B} = \nabla \times (m\mathbf{v} - 2e\mathbf{A}) = 0 \qquad (6.43)$$

where we have introduced the vector potential **A**. In Chapter 11 we will study this expression further and show that the canonical momentum of a particle with kinetic momentum $m\mathbf{v}$ and charge $-2e$ is, in an electromagnetic field,

$$\mathbf{p} = m\mathbf{v} - 2e\mathbf{A} \qquad (6.44)$$

We further note that the flux Φ in Eq. (6.41) is given by

$$-2e\Phi = \int dx_1 dx_2 m\mathscr{L} = \oint d\mathbf{s} \cdot \mathbf{p} \qquad (6.45)$$

Here we have used Stoke's theorem. This result was noted by F. London and he interpreted it correctly, along the lines of a Bohr-Sommerfeld quantisation condition: the integral should be equal to an integer times Planck's constant h. In this way we obtain that the combination Cm/n in Eq. (6.41) is an integer, N, and that the flux $\Phi = -N/2e$.

The result may at first sight seem like witchcraft. The vector $\mathscr{L}$ was assumed to vanish, according to the London prescription, *inside the superconductor*. The fact that its surface integral is nonvanishing and in particular equal to an integer times a flux unit must then mean that the whole surface is *not inside the superconductor*. We have already pointed out that there is a singularity for the solution along the 3-axis. In other words there is a thin 'hole' along the axis and we may conclude that it should be of the order $\xi \ll \lambda$ and correspond to a lack of Cooper pairs. This is a *vortex line*.

F. London suggested on the basis of these results that it should be possible to produce a *magnetic flux trap*. Suppose that we have a ring of matter in a normal state inside a magnetic field and that we then bring the ring into a superconducting state. This will produce a supercurrent of Cooper pairs in the ring. Further suppose that after this we remove the magnetic field and investigate the magnetic flux *through the hole in the ring*, caused by the supercurrent (which must continue inside the superconductor because there is no 'stopping force'!). A set of clever experiments, [49], were later performed, which justified both the flux trapping and, in particular, the quantisation of the flux.

We conclude that the solution we have obtained, which corresponds to a *vortex line*, penetrates the superconductor to a small depth and contains a definite flux corresponding to an integer times the inverse charge of a Cooper pair. This corresponds to the typical type II superconductor breakdown. The superconductor is penetrated by as many isolated vortex lines as the field flux permits and we now understand the subdivision of the slab discussed above.

A dynamical vortex line, i.e. one connected to moving charges must have a dynamics very similar to the MRS and therefore if the QCD vacuum state has the properties of a superconductor type II our use of the MRS as a model for the color electric force field is natural. We will later consider the question of the width of the Lund string field, cf. Chapter 11, and will find that its radius is typically 0.3–0.4 fm.

7

The decay kinematics of the massless relativistic string

7.1 Introduction

In this chapter we consider the situation when a $q\bar{q}$-state is produced with a large amount of energy at a single space-time point. It will be called the original pair and we assume that q and $\bar{q}$ interact through a constant attractive force, κ. The pair will then form a yoyo-hadron state as described in the previous chapter and immediately start to separate.

The state composed of the two particles and the force field, if it contains a larger mass than that of the stable hadrons, will decay into smaller-mass particles. Such a decay process is of course of a quantum mechanical nature.

Although we will at this point use semi-classical arguments, we will later show that the resulting formulas fit into both a quantum mechanical tunnelling process and a statistical mechanics scenario.

The major assumption will be that *a string state may decay by the production of new pairs of $q\bar{q}$-particles along the force field*. Using the earlier interpretation that a q or $\bar{q}$ corresponds to the endpoint of a string, the production process corresponds to creating new endpoints, i.e. to breaking up the original string into smaller pieces.

The q- and $\bar{q}$-particles will be treated as massless during the discussion. This assumption is necessary in a semi-classical framework for the conservation of energy-momentum. A massless pair produced at a single space-time point does not take any energy from the field. A massive pair (mass μ) will, however, in classical physics need a field region $\delta x = 2\mu/\kappa$. We will later consider the quantum mechanical modifications which are necessary in order to treat the production of massive pairs.

The production point of a new pair is called a *vertex*. Figure 7.1 shows the development in space-time of parts of a $q\bar{q}$-state, with some of the vertices produced.

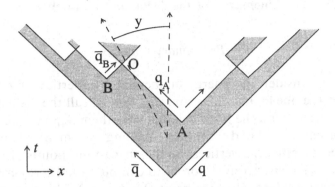

Fig. 7.1. Space-time development in a breakup situation showing some of the vertices produced together with the state S_{AB} discussed in the text. The rapidity y of the state S_{AB} is the hyperbolical angle between the broken-line directions.

We note that due to *causality* the two original endpoint particles will know nothing about the breakup vertices 'behind' them, at least not for some considerable time. As they are massless and move with the velocity of light there is no possibility of reaching them with a signal until they have turned around.

We further note that a produced pair will immediately start to separate owing to the forces exerted by the two adjoining string pieces. The new particles in that way use up the field energy between them, i.e. the string field in between them vanishes. Their parting situation is actually irrevocable – they will never meet again.

In this way the notion of *confinement* is smuggled in. A string force field is always confining in the sense that the force field vanishes at the endpoint 'charges'. This is in contrast to the situation in electrodynamics, where a newly produced electron-positron pair will continue to interact even if pulled apart by external forces.

In our case, at every vertex there will be two independent string pieces with endpoint particles moving away in opposite directions. There may be several vertices along the string, as shown in Fig. 7.1. In this way every vertex actually partitions the set of all vertices into two parts, those belonging to the string piece moving to the left and those belonging to the string piece moving to the right. This observation will later on provide us with a convenient way to order the vertices.

7.2   The kinematics of the decay and its implications

1   *Preliminary remarks*

We will now consider the energy-momentum properties of one of the string pieces, the one ending in q_A and $\bar{q}_B$. We will call the state consisting of the two particles and the force field between them S_{AB} and we note that it is after formation isolated from the remaining system. The two particles are produced at adjacent vertices, at the space-time points $A = (x_A, t_A)$ and $B = (x_B, t_B)$, respectively. In order to compute the energy-momentum of S_{AB} we consider the space-time point $O = (x_O, t_O)$. This is, according to Fig. 7.1, the first meeting point of q_A and $\bar{q}_B$ and there is no field between them when they are at this point.

According to the equations of motion given in Chapter 6 the energies E_j and momenta p_j $(j = A, B)$ at this point (note that momentum is counted positive along the positive x-axis) are given by

$$E_A = \kappa(x_A - x_O), \quad E_B = \kappa(x_O - x_B)$$
$$p_A = \kappa(t_A - t_O), \quad p_B = \kappa(t_O - t_B)$$

(7.1)

Therefore the state S_{AB} will have a *total energy-momentum depending only upon the space-time difference between the production vertices A and B*:

$$E = E_A + E_B = \kappa(x_A - x_B), \quad p = p_A + p_B = \kappa(t_A - t_B)$$

(7.2)

For reference we note that there is a relationship between some of the quantities in Eq. (7.1) because the positive (negative) lightcone component of the point labelled O is equal to the corresponding component for the vertex A (B):

$$t_O + x_O = t_A + x_A, \quad t_O - x_O = t_B - x_B$$

(7.3)

If the state S_{AB} corresponds to a meson state with mass m then the vertices A and B must lie on the two branches of the hyperbola

$$\frac{E^2 - p^2}{\kappa^2} = \frac{m^2}{\kappa^2} = (x_A - x_B)^2 - (t_A - t_B)^2$$

(7.4)

Therefore there is a strong correlation between two vertices corresponding to the production of a definite mass in between. One can, assuming that one knows one of the vertices (e.g. A) draw the hyperbola branch along which B must be found (see Fig. 7.2) and vice versa.

It is also useful to note that the velocity of the 'particle' produced

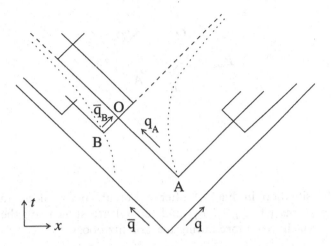

Fig. 7.2. Two neighboring vertices A and B with the requirement that they should each lie on a hyperbola. The hyperbolas are indicated for A and B.

between the vertices A and B is given by

$$v_{AB} = \frac{p}{E} = \frac{\Delta t}{\Delta x} \tag{7.5}$$

where Δ indicates the differences between the A and B coordinates. We remember from Chapter 2 that this result is to be expected in connection with spacelike vectors. The system is evidently at rest when q_A and $\bar{q}_B$ are produced at the same time. The rapidity of the system is given by the hyperbolic angle, y, shown in Fig. 7.1 and we note that the faster the system, the more tilted towards the lightcone is its velocity direction:

$$y_{AB} = \frac{1}{2} \log \left(\frac{1 + v_{AB}}{1 - v_{AB}} \right) = \frac{1}{2} \log \left(\frac{\Delta x + \Delta t}{\Delta x - \Delta t} \right) \tag{7.6}$$

2   The consequences

The distance between the vertices A and B must be spacelike in order that the mass should be real, according to Eq. (7.4). Thus the two production points are not causally related and no signal can be sent between the vertices. This has some interesting consequences, which we will now consider. According to Fig. 7.1 vertex A appears earlier than vertex B in the ordinary time sense. This is, however, a statement which depends upon the Lorentz system if A and B are spacelike with respect to each other, since then we can always, according to Chapter 2, find a Lorentz boost to another frame such that the vertex B (in its new position B_y) will seem to appear earlier than vertex A (A_y, see Fig. 7.3).

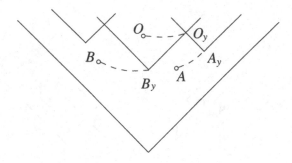

Fig. 7.3. The situation in Fig. 7.1 after a Lorentz boost along the negative direction. The points (A, A_y), (B, B_y) and (O, O_y) are shown together with the hyperbolas on which they move during the Lorentz boost.

The same considerations also apply to every other pair of adjacent vertices. We conclude that *all the vertices must be spacelike with respect to each other* for the produced states to have positive masses. Therefore no statement about (ordinary) time-ordering in the breakup process is Lorentz-invariant. There is consequently no 'first' vertex in this sense; the vertices all occur, in a relativistic setting, at the same time. We will later see that there are other possible ways to order the process and also other ways to define a useful time variable.

Thus, for the description of the decay process to be Lorentz-invariant then *there can be no vertex that is more significant than any other*. Each vertex has the same property, i.e. it divides the system into two parts, the vertices to its left and those to its right. Evidently these parts can also be described as two independent groups of particles moving apart. One often uses the term 'jet' for such a connected group. (It may then happen that a jet will contain only a single particle, *viz.* if we consider the outermost vertex on one end.)

It is an important constraint, when we provide a probabilistic description of the process, that all the vertices must be treated in the same way. This is what causality and Lorentz invariance imply.

The fact that all the vertices occur at spacelike distances with respect to each other also seems to be necessary from the point of view of ordinary common sense. It seems evident that the field cannot break up at a space time point if such a breakup has already occurred earlier, i.e. in the backward lightcone with respect to the point. In accordance with what has been said above there is then no longer any field left, and therefore there is no energy left, and so on.

7.3 Ordering of the decay process along the lightcones

Another property that we may deduce from the mass-shell condition (7.4) is that for every yoyo meson there is only a single degree of freedom. We may prescribe either the energy-momentum component $p_+ = E + p$ or the energy-momentum component $p_- = E - p$ (the positive and negative lightcone components) of the system S_{AB}. They are linked by the mass-shell condition

$$p_+ p_- = m^2 \qquad (7.7)$$

(Let the reader be warned, as Carter Dickson or any other honest mystery writer would say. A very sophisticated reader might note that we are at this point introducing a slight mismatch between the ordinary space-time coordinates and the lightcone coordinates. We have already shown that the squared mass is given by the area spanned by the string during a complete period and not by a half period as Eq. (7.7) implies. The difference corresponds to using, instead of the normal metric $dxdt$, the lightcone metric $dx_+ dx_- = 2dxdt$. We will go on employing this mismatch in order to avoid writing several factors of 2 or $\sqrt{2}$ in our formulas.)

From the calculations in connection with Eq. (7.1) we note that for the state S_{AB} the positive lightcone component is actually carried by the $\bar{q}_B$-particle and the negative one by the q_A-particle at the time of their first meeting to form the final-state yoyo-hadron. (It is necessary to make use of Eq. (7.3) to prove this statement.) This property is in the same sense valid for all the yoyo-hadrons, i.e. that the positive (negative) lightcone energy-momentum is, at the meeting points, carried by the corresponding $\bar{q}(q)$-particle. The assignment to the particles of positive and negative lightcones is of course related to the choice of directions of motion for the original pair.

This observation provides a useful way of ordering the process. Consider Fig. 7.4, which exhibits the decay of a whole string system stemming from an original pair $q_0, \bar{q}_0$ with lightcone energy-momenta p_{+0}, p_{-0} into many yoyo-hadrons, which go off in different directions (i.e. with different velocities). From the remarks above we conclude that *the production process is easily ordered along one of the lightcones.* Then the corresponding lightcone energy-momentum of the yoyo-meson indexed j (composed of $q_j, \bar{q}_j$ from adjacent vertices) is given by the lightcone component of either the q_j (the p_{-j} if we use the negative lightcone ordering) or the $\bar{q}_j$ (the p_{+j} for the positive lightcone ordering). The other component can be computed from Eq. (7.7). We will normally choose to number the yoyo-hadrons along the positive lightcone.

The sum of these components will, of course, add up to the light-cone components of the original pair; this corresponds to total energy-

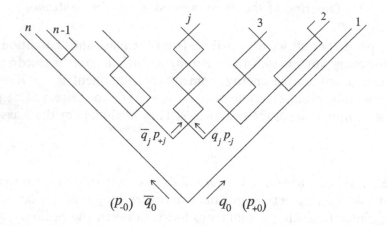

Fig. 7.4. A high-energy string breakup of a pair $q_0, \bar{q}_0$ having lightcone energy-momenta p_{+0}, p_{-0}.

momentum conservation:

$$\sum_{j=1}^{n} p_{\pm j} = p_{\pm 0} \qquad (7.8)$$

Thus *the production process can be characterised as a set of choices for the lightcone components of one set of constituents of the yoyo-hadrons, i.e. of either the q_j or the $\bar{q}_j$.*

These lightcone components are evidently obtained from the field (remember that all the pairs are produced 'at rest'). Therefore *another way to describe the energy-momentum of the final-state yoyo-hadrons is to state the size of the space-time region within which the constituents have been acted upon by the string force field.* In order to state the energy-momentum of the system S_{AB} in Fig. 7.1 we may therefore prescribe a *lightcone distance*, either $\Delta t + \Delta x = \Delta x_+$ or $\Delta t - \Delta x = \Delta x_-$ ($\Delta t = t_A - t_B$, $\Delta x = x_A - x_B$). The other of these is then given by Eq. (7.7) rewritten as

$$\Delta x_+ \Delta x_- = -\frac{m^2}{\kappa^2} \qquad (7.9)$$

In this way *the production process can be considered as a series of 'steps' along the positive (negative) lightcone. Each step corresponds to the lightcone distance between two adjacent vertices.* Then energy-momentum conservation according to Eq. (7.8) corresponds to stepping all the way from the turning point of the original q_0 ($\bar{q}_0$) back to the origin.

After each step it is necessary to go along the opposite lightcone a distance Δx_{-j} (Δx_{+j}) in order to keep the yoyo-meson on the mass shell. In that way the string decay process corresponds to a Markovian stochastic process, where each vertex in the process is determined solely by the previous starting point, i.e. the vertex already reached, and by the probability of taking a particular step along the lightcone.

It is convenient to define the scaled lightcone components z_+ and z_- by means of the equations

$$z_{\pm} = \frac{p_{\pm}}{p_{0\pm}} \tag{7.10}$$

where $p_{0\pm}$ are the corresponding lightcone components for the original q- and $\bar{q}$-particles. The quantitities $z_{\pm}$ are Lorentz invariants, being the ratio between two quantities which transform with the same factors $\exp(\pm y)$ under a Lorentz boost along the x-axis.

The total production process may then be looked upon as a set of steps $\{z_{+j}\}$ along the positive lightcone (or equivalently $\{z_{-j}\}$ along the negative lightcone). Energy-momentum conservation means that all the steps add up to unity. Each step corresponds to the production of a new meson containing a fraction of the original q- (or $\bar{q}$-) particle's energy-momentum that corresponds to the step size.

7.4 Iterative cascade fragmentation models

The above situation as viewed in a frame boosted along the positive x-axis with a large velocity is shown in Fig. 7.5. We note that, while in Fig. 7.4 the hadrons in the centre are the slowest and also the first to be produced in time in that system, in Fig. 7.5 it is instead the hadrons which are furthest out along the lightcone (usually the fastest in Fig. 7.4) that are the slowest and the first to be produced (cf. the discussion of velocities and rapidities in connection with Eqs. (7.5), (7.6)). This is again a very general property of all Lorentz-covariant production processes and we will return to this observation in the next section.

Up to now we have not been concerned with the conservation of internal quantum numbers, e.g. the flavor quantum numbers of the newly produced $q\bar{q}$-pairs. We will from now on assume that the pairs produced are actually a quark and its antiparticle, an antiquark with the opposite flavor, i.e. *the pairs will together have the quantum numbers of the vacuum*.

This means that it is possible to relate adjacently produced hadrons also by means of their flavor quantum numbers. We will introduce the notion of 'rank' in the following sense. The first-rank meson contains the quantum number of the original q-particle together with the antiflavor of the $\bar{q}$-particle produced at the first vertex along the lightcone.

In the same way we define a second-rank particle as the particle composed of the q-particle from the first vertex and the $\bar{q}$-particle from the

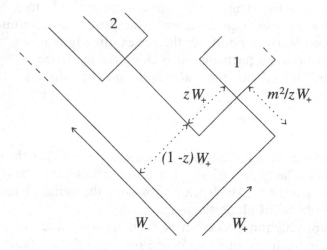

Fig. 7.5. The situation of Fig. 7.4 in a frame boosted along the positive lightcone direction in such a way that the first-rank particle is at rest. For simplicity we write $z \equiv z_{+1}$.

next, etc. It is evidently possible to introduce rank also by starting with the original $\bar{q}$-particle and the negative lightcone. Thus ordering by rank and flavor corresponds, in this kind of model, to an ordering along the lightcone(s).

From Fig. 7.5 we notice that the first vertex along the lightcone, V_1, actually divides the decay event into a single first-rank particle moving to the right and all the remaining ones as a combined jet moving to the left.

After the production of the first meson with lightcone fraction z_{+1} the remainder of the system will share the fraction $1 - z_{+1}$. This means that the remaining system will have a squared mass s_1 equal to (using for simplicity z for z_{+1})

$$s_1 = (1 - z)W_+ \left(W_- - \frac{m^2}{z W_+} \right) = (1 - z) \left(s - \frac{m^2}{z} \right) \qquad (7.11)$$

where we suppose the original system to have squared mass $s = W_+ W_- (= p_{+0} p_{-0}$, due to Lorentz invariance).

The different parts of this formula have simple geometrical interpretations. The first term, i.e. the scaled-down mass-square is immediately recognised. For the second term it is easy to convince oneself that the area of the region below the first vertex, $V_1 = \kappa(x_{+1}, x_{-1})$, and above the production point of the original pair, is

$$\Gamma_1 = \kappa x_{+1} \kappa x_{-1} \qquad (7.12)$$

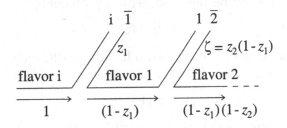

Fig. 7.6.   An iterative cascade chain.

This is (apart from the factor κ^2) the squared proper time $\tau_1^2 = t_1^2 - x_1^2$ of the vertex V_1. The positive lightcone component of the vertex V_1 (with respect to the origin) is what is left of the original q's energy-momentum, $\kappa x_{+1} = (1 - z)W_+$. The negative lightcone component is similarly what was taken by the first particle, i.e. $\kappa x_{-1} = m^2/zW_+$. Therefore the quantity Γ_1 is equal to minus the second term in Eq. (7.11):

$$\Gamma_1 = (1 - z)W_+ \frac{m^2}{z W_+} = (1 - z)\frac{m^2}{z}. \tag{7.13}$$

In the Lund model formulas both terms are taken into account and the model therefore exhibits complete energy-momentum conservation, i.e. every new particle takes away not only its forward lightcone energy-momentum zW_+ but also the negative fraction needed to put it on the mass shell.

We will later see that the proper times of the vertices are generally of a limited size. For large values of s we may then neglect Γ_1 and approximate the *remainder system* as being the same as the original one apart from a scaling down of the positive lightcone component by the factor $1 - z \equiv 1 - z_1$.

The basic idea of regarding particle production at high energies as a scaling process was conceived many years ago, [90], to describe the fragmentation regions in hadronic interactions. Later similar ideas were used in partonic scenarios as *iterative cascade fragmentation schemes*, [13]. Then one assumes that there is a certain probability

$$f_{i1}(z_1)dz_1 \tag{7.14}$$

of producing the first-rank hadron (indexed by the original q's flavor i and the produced $\bar{q}_1$'s antiflavor) with fractional energy-momentum z_1, leaving the system with a q_1-particle at the endpoint and with a scaled down energy-momentum $1 - z_1$ (see Fig. 7.6).

Then the process can be repeated, with a probability

$$f_{12}(z_2)dz_2 \tag{7.15}$$

of producing a second-rank meson with flavors 12 (the second flavor-index refers to a $\bar{q}$-particle) and with energy-momentum fraction

$$\zeta_2 = z_2(1 - z_1) \tag{7.16}$$

After that the system is left with a q_2-particle at the end and with a scaled down energy-momentum equal to

$$1 - z_1 - \zeta_2 = (1 - z_1)(1 - z_2) \tag{7.17}$$

Thus at each step a new flavor is produced, a certain probability distribution is applied to find the fraction z_j and the remainder system is scaled down by a further factor $1 - z_j$.

In this way the problem has been reduced to finding a set of probability distributions $f_{ij}(z)$ and then repeatedly applying them to the situation at hand. This is the basis of what is often referred to as the *iterative cascade jet* or *Feynman-Field model* in honor of two of the main contributors. We will consider some of their main features in section 9.4

In the next chapter we will see that *there is a unique form for the distribution(s) f in the Lund model*. To prove that we will require that the final-state meson production process should be statistically the same if we describe it in terms of steps along the positive or along the negative lightcones (*left-right symmetry*).

We will end this chapter with a few remarks on a possible problem, to my knowledge first raised by Bjorken for the iterative cascade models, in the well-known Landau-Pomeranchuk 'formation time' concept.

7.5   The formation time and iterative cascade jets

Landau and Pomeranchuk considered the notion of a formation time in the context of QED bremsstrahlung. In its simplest setting the problem is as follows:

- at what time can one distinguish between a state containing a single charged particle and a state containing the particle accompanied by a photon?

They pointed out that in a Lorentz frame where the particle moves along one axis and the photon is moving transversely to this axis then it it is necessary to wait at least a time corresponding to the photon's wavelength to make a measurement that can distinguish the photon. Since the wavelength is inversely proportional to the transverse momentum of the photon $k_\perp$, it is thus necessary to wait a time

$$\tau_0 \simeq k_\perp^{-1}. \tag{7.18}$$

In a frame where the photon has energy E there will be a time-dilation factor $\gamma(v) = E/k_\perp$ and one obtains

$$\tau = \tau_0 \frac{E}{k_\perp} \simeq \frac{E}{k_\perp^2} \tag{7.19}$$

With the wavelength exchanged for some rest frame typical production time, i.e. with $k_\perp$ exchanged for some 'virtuality' Q (e.g. the transverse mass of a hadron), this formation time should, in any relativistically covariant and causal setting, provide a time-ordering of the process. Therefore it is always the slowest particles which will be the first to be emitted while the higher-energy particles will take a time proportional to their energy.

In the iterative cascade models the first-rank particle, according to the considerations above, will take a fraction z_1 of its energy-momentum leaving a fraction $1 - z_1$ to the remaining ones. The second-rank particle then takes $z_2(1 - z_1)$, etc. The values z_j are assumed to be given stochastically by means of a distribution $f(z)dz$.

As we will later see, one basically obtains a geometrical series for the final-state particle energy-momentum fractions. Therefore the first-rank particle is generally faster than the rest, i.e. it will have a longer formation time. Bjorken's question was: 'how can it then be the first to be produced in the chain?'

In the Lund model there is evidently a simple answer to this problem. Rank-ordering, as we have seen, corresponds to an ordering along the lightcone of the production vertices. There is no contradiction to an ordinary time-ordering with respect to the original $q\bar{q}$ production point, which is in accordance with the Landau-Pomeranchuk prescription. In any frame it is always the slowest mesons which are the first to be produced, according to the Lund model.

8
A stochastic process for string decay

8.1 Introduction

In Chapter 7 we considered the kinematics of string decay. At the same time we found and formulated a set of constraints stemming from causality, confinement and Lorentz covariance which are necessary for a consistent description of the decay process.

The intention of this chapter is to show that *there is only one stochastical process for string decay which is consistent with the requirements derived in Chapter 7* and it contains essentially two parameters. The discussion is based upon results obtained in [19].

Once again only semi-classical physical arguments as well as probability concepts will be used during the discussion. We begin by listing the basic concepts which were derived in Chapter 7. They must all be incorporated into the stochastical process for which we are looking.

A   The process of string breakup corresponds to the production of a set of yoyo-states with given masses. Each yoyo-hadron is composed of a q-particle and a $\bar{q}$-particle stemming from adjacent vertices (i.e. string breakup points) together with the string piece between them.

B1   Each pair from a vertex is massless (*local energy-momentum conservation*); the particles start to move apart after their production, due to the force from the string field.

B2   There is no interaction between the q and $\bar{q}$ of such a vertex after their production, i.e. the string force field ends on the endpoint charges (this implies *confinement*).

C   The separation of the vertices is spacelike with adjacent vertices, in particular, on hyperbolas determined by the yoyo-hadron masses (this implies *causality conditions*).

146

D1 All vertices therefore are of the same dynamical status. There should not be a different treatment of any one of these decay situations.

D2 Each vertex corresponds to the partitioning of the final state into one left-moving and one right-moving set of final-state yoyo-hadrons.

E Each vertex pair contains the internal (flavor) quantum numbers of the vacuum (*local conservation of internal quantum numbers*).

With regard to the ordering and the variables we have found:

F A convenient ordering of the process is *rank-ordering*. Two hadrons of adjacent rank share a $q\bar{q}$-pair produced at a vertex and therefore (according to property E above) contain the corresponding internal quantum numbers (e.g. flavors and antiflavors). Rank-ordering corresponds to an ordering along either the positive or the negative lightcone. *The process should be independent of which lightcone we use.*

G Rank ordering also implies that the process can be described as a set of steps from one vertex to the next. The steps correspond to choosing a partitioning of the energy-momentum of the original $q\bar{q}$-pair p_{+0}, p_{-0} (which at the time of the breakup goes into field energy and is then given back to the produced particles). This implies *total energy-momentum conservation*.

H1 A convenient *Lorentz-invariant* set of variables are the scaled lightcone energy-momentum fractions $p_{\pm j}/p_{\pm 0}$, with $p_{\pm j}$ the positive or negative lightcone energy-momentum of the rank-j yoyo-hadron. The $p_{\pm j}$ are carried by the q- or $\bar{q}$-particle, respectively, at the time when they meet during the yoyo-cycle.

H2 The steps referred to under property G above correspond to the space-time interval during which the particles have obtained that energy-momentum, i.e. $\Delta x_{\pm j} = p_{\pm j}/\kappa$ where κ is the string tension.

It is necessary to introduce a further assumption, which later we will show to be consistent with the results.

J Even when the energy of the original pair becomes very large the proper times of the vertices stay finite.

At the end of the chapter we will bring up a different approach, the Artru-Menessier model, [26], which was further extended and improved by Bowler, [32] (it is therefore known as the AMB model). This model contains many similarities to the Lund model fragmentation formulas

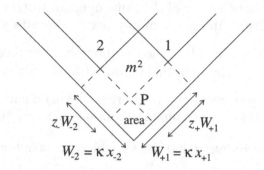

Fig. 8.1. The production of a hadron with mass m between two adjacent vertices 1 and 2 (the notation is explained in the text).

and it was conceived many years before we started our work. It was not until Artru, [24], pointed out to us that our considerations of hadrons produced in a linear potential (i.e. the yoyo string modes [14]) were similar to his results that we realised that these states actually correspond to some particular modes of the MRS. The two models, the Lund model and the AMB model, nevertheless contain major differences which we will briefly consider at the end of this chapter.

8.2 The unique breakup distribution for a single hadron

If the squared mass $s = p_{+0}p_{-0}$ of the original $q\bar{q}$-pair is very large then there will be many yoyo-hadrons produced, i.e. the process will contain many steps. A hadron produced at the centre will be little affected by the original pair and will be essentially independent of the many steps and production points that occur before its own production (or 'after'). We are introducing the idea that the process leads to a *steady-state fragmentation behaviour*. Property J, above, means that the density of hadrons will stay finite in the centre, as we will see further on.

1 The distributions H and f

We now consider two adjacent vertices at the space-time points 1 and 2, a hadron of mass m being produced in between (see Fig. 8.1).

We may describe this process as the result of taking many steps along the positive lightcone to reach vertex 1 and then one further step to reach vertex 2, thereby producing the hadron m. Another way would be to consider vertex 2 as the result of many steps along the negative lightcone, the production of m being one further step from 2 to 1.

In the first description the positive-lightcone energy-momentum remaining before the hadron m is produced is given by $W_{+1} = \kappa x_{+1}$. Similarly in the second description the negative-lightcone energy-momentum remaining is given by the corresponding 2-component $W_{-2} = \kappa x_{-2}$. We are now going to make use of assumption J above and conclude that there is a finite probability of arriving at vertex 1 after many steps:

$$H'(1)dx_{+1}dx_{-1} \equiv H(\Gamma_1)d\Gamma_1 dy_1 \tag{8.1}$$

In this expression we have introduced hyperbolic coordinates Γ_1, y_1 instead of the lightcone variables for the vertex 1:

$$\Gamma_1 = \kappa^2 x_{+1}x_{-1}, \quad y_1 = \frac{1}{2}\log\left(\frac{x_{+1}}{x_{-1}}\right) \tag{8.2}$$

Owing to Lorentz invariance the distribution H can depend only upon Γ_1, the only Lorentz invariant available. From its definition it is obvious that Γ_1 is essentially equal to the squared proper time of vertex 1, $\kappa^2 x_{+1}x_{-1} = \kappa^2(t_1^2 - x_1^2)$ (cf. Chapter 2).

There is, of course, a corresponding probability of reaching vertex 2 after many steps along the negative lightcone:

$$H(\Gamma_2)d\Gamma_2 dy_2 \tag{8.3}$$

Given that we have arrived at vertex 1 the production of the hadron corresponds to taking a step to 2, with probability

$$f(z_+)dz_+ \tag{8.4}$$

of taking a fraction z_+ of the remaining energy-momentum W_{+1} defined above. Note that z_+ is defined by a scaling with W_{+1} instead of with the original energy-momentum p_{+0}. This is a convenient quantity to use at this point, *its range $1 > z_+ > 0$ being independent of the other variables.*

The joint probability of being at vertex 1 and of producing the hadron is then given by the product of the two probabilities in Eqs. (8.1), (8.4). The hadron is the result of the last in a long row of steps along the positive axis. On the other hand it may also be considered as the result of the final step of many along the negative axis. Then the joint probability is

$$H(\Gamma_2)d\Gamma_2 dy_2 f(z_-)dz_- \tag{8.5}$$

where z_- is likewise scaled with respect to the energy-momentum remainder, in this case W_{-2}.

We are now going to equate these two probabilities. Surprisingly enough we will then be able to prove that there is a single (two-parameter) solution for H and f. (To be more precise there will, in principle, be $n_f + 1$ parameters if there are n_f different $q\bar{q}$-flavors).

2 *The derivation of the distributions*

We start by noting that the two quantities $dy_{1,2}$ can evidently be taken to be equal and that there is a set of relations between the remaining variables $\Gamma_{1,2}$ and $z_\pm$. From Fig. 8.1 we obtain the relations

$$\Gamma_1 = (1 - z_-)W_{-2}W_{+1}$$
$$\Gamma_2 = W_{-2}(1 - z_+)W_{+1} \qquad (8.6)$$
$$m^2 = (z_-W_{-2})(z_+W_{+1})$$

Thus there are only two independent variables in the problem (assuming m^2 as fixed), which we may take as e.g. $z_\pm$. We obtain immediately

$$\Gamma_1 = \frac{m^2(1 - z_-)}{z_+z_-}, \quad \Gamma_2 = \frac{m^2(1 - z_+)}{z_+z_-} \qquad (8.7)$$

$$d\Gamma_1 \frac{dz_+}{z_+} = d\Gamma_2 \frac{dz_-}{z_-}$$

Therefore the requirement of equality introduced at the end of the last subsection reduces to

$$H(\Gamma_1(z_+, z_-))z_+ f(z_+) = H(\Gamma_2(z_+, z_-))z_- f(z_-) \qquad (8.8)$$

where the $z_\pm$-dependence has been explicitly written out.

Taking the logarithm of this equation we obtain with $h(\Gamma) = \log H(\Gamma)$ and $g(z) = \log(zf(z))$

$$h(\Gamma_1) + g(z_+) = h(\Gamma_2) + g(z_-) \qquad (8.9)$$

If this expression is differentiated first with respect to z_+ and then with respect to z_- (keeping the other one fixed, i.e. using partial differentiation) then *all the g-dependence vanishes*. We will be left with only the variations in h. The result is

$$\frac{dh(\Gamma_1)}{d\Gamma_1} + \Gamma_1 \frac{d^2h(\Gamma_1)}{d\Gamma_1^2} = \frac{dh(\Gamma_2)}{d\Gamma_2} + \Gamma_2 \frac{d^2h(\Gamma_2)}{d\Gamma_2^2} \qquad (8.10)$$

To obtain this result a $z_\pm$-dependent expression has been divided out from both sides. Further the chain rule for differentiation has been used:

$$\frac{\partial h(\Gamma_1)}{\partial z_+} = \frac{dh(\Gamma_1)}{d\Gamma_1} \frac{\partial \Gamma_1}{\partial z_+} = \frac{dh(\Gamma_1)}{d\Gamma_1} \left[-\frac{m^2(1 - z_-)}{z_- z_+^2} \right] \qquad (8.11)$$

An important property of the differential equation in Eq. (8.10) is that *the left-hand side only depends on Γ_1 and the right-hand side only on Γ_2*. The two Γ-variables are just as independent of each other as the two $z_\pm$-variables. The $z_\pm$ can of course be expressed in terms of the Γ's by the equations above. Since the Γ's are taken as independently varying quantities then *the only way that the equation can be fulfilled is if both sides are equal to the same constant, to be called* $-b$.

Then the differential equation for h is

$$\frac{d}{d\Gamma}\left(\Gamma \frac{dh}{d\Gamma}\right) = -b \tag{8.12}$$

which implies

$$h(\Gamma) = -b\Gamma + a\log \Gamma + \log C \tag{8.13}$$

In this way we obtain for $H(\Gamma)$ (neglecting the indices 1 and 2 as the equation works equally well for both)

$$H(\Gamma) = C\Gamma^a \exp(-b\Gamma) \tag{8.14}$$

The parameters b, a and C are all constants of integration. While b (which has the dimension of an inverse squared mass) must be the same for all the vertices the (dimensionless) constants a and C may have different values. They may e.g. depend upon the flavor quantum numbers of the pair produced at a particular vertex. The constant C plays the role of a normalisation constant for the distribution H. We will later show the significance of a and b.

If we introduce the results for h into the original equation for h and g, Eq. (8.9), it is possible to derive an expression for the original distribution $f(z)$. This can be arranged so that all the dependence on z_+ is on one side of the equation and all the z_--dependence on the other:

$$g_{12}(z_+) + \frac{bm^2}{z_+} - a_1 \log\left(\frac{m^2}{z_+}\right) - a_2 \log\left(\frac{1-z_+}{z_+}\right) + \log C_1$$

$$= g_{21}(z_-) + \frac{bm^2}{z_-} - a_2 \log\left(\frac{m^2}{z_-}\right) - a_1 \log\left(\frac{1-z_-}{z_-}\right) + \log C_2 \tag{8.15}$$

Then we use the same argument based upon independence to deduce that both sides must be equal to the same constant. The result for f is

$$f(z) = N\frac{1}{z}(1-z)^a \exp\left(-\frac{bm^2}{z}\right) \tag{8.16}$$

if there is only a single value of the a-parameter for all vertices. The quantity N is again a normalisation constant. When there are different values a_α, a_β at two adjacent vertices then we obtain, with a labelling such that the produced hadron stems from a step from vertex α to vertex β,

$$f_{\alpha\beta} = N_{\alpha\beta}\frac{1}{z}z^{a_\alpha}\left(\frac{1-z}{z}\right)^{a_\beta} \exp\left(\frac{bm^2}{z}\right) \tag{8.17}$$

From Eq. (8.15) we conclude that the normalisation constants N_{12} and N_{21} are related to the normalisation of the distributions H_j, $j = 1, 2$, by a common factor N_c:

$$N_{12} = \frac{N_c}{C_1 m^{2a_1}}, \quad N_{21} = \frac{N_c}{C_2 m^{2a_2}} \tag{8.18}$$

The combined probability of being at vertex 1 and of taking the step z_+ towards vertex 2, thereby producing the hadron m, is (for equal values of a):

$$CN \left[\Gamma(1 - z_+) \right]^a \exp \left[-b \left(\Gamma + \frac{m^2}{z_+} \right) \right] d\Gamma \frac{dz_+}{z_+}$$

$$= CN \left[m^2 \frac{(1 - z_-)(1 - z_+)}{z_- z_+} \right]^a \exp \left(-\frac{bm^2}{z_+ z_-} \right) \cdot \frac{m^2 dz_+ dz_-}{(z_+ z_-)^2} \tag{8.19}$$

From the second line we find that the distribution is the same if we decide to go 'in the opposite direction', i.e. express the distribution in terms of the variables relevant for the negative lightcone description. We leave it to the reader to derive the corresponding relations for the case when a and C are different at neighboring vertices. In particular it is useful to note that the product CNm^{2a} becomes N_c as defined in Eq. (8.18).

Phenomenologically it has not up to now been necessary in the Lund model to use different a-values to describe the data from the experiments. We will present an idea of Bowler, [32], in connection with the discussion of heavy flavor fragmentation in Chapter 13 which fits very nicely into the Lund model scenario and would require a different a-value for the first-rank hadron in the fragmentation of a heavy quark jet.

If we should, nevertheless, require to use several a-values then it would be necessary to normalise the distributions H_j (j being an index corresponding to different flavor values) to the relative occurrence of the different flavors in phase space and to choose the normalisation(s) of the distributions f_{jk} in a similar way. We will come back to these normalisations in a later chapter.

Thus, using a remarkably simple assumption, we have obtained a very precise result for the string-breaking process. For the Lund model to work it is essential that the expressions we have obtained really do fit the experimental data.

It is, however, necessary, before we can compare with data, to extend the model. We need to remember that the hadronic momenta are measured in a three-dimensional world: therefore the model must be extended outside $1 + 1$ dimensions. We also need to prescribe a way of normalising our distributions in the case where we would like to describe several different flavors and different hadrons (and one should not forget that we should also be able to account for baryon-antibaryon production!).

Before doing all these things in later chapters we will provide an interpretation of the results we have obtained.

3   The interpretation of the distributions H and f

We will start with the combined expressions occurring in the exponentials of the distributions in Eq. (8.19). For the case when we have arrived at vertex 1 and take a step z_+ we obtain the negative exponential of

$$b \left(\Gamma_1 + \frac{m^2}{z_+} \right) \tag{8.20}$$

From Fig. 8.1 we find that the sum multiplying the parameter b is the area which is spanned below the first meeting point of the two constituents (the q_1-particle from vertex 1 and the $\bar{q}_2$-particle from vertex 2) of the hadron; it is evidently common to the two situations because it can just as well be described as follows (if we are at the vertex 2 and take step z_-):

$$\Gamma_2 + \frac{m^2}{z_-} \equiv \frac{m^2}{z_+ z_-} \equiv W_{+1} W_{-2} \tag{8.21}$$

We leave it to the reader to prove the equality of the expressions in Eqs. (8.20) and (8.21).

Thus *the exponential suppression is related to the size of an area characteristic of the production process.* We will come back to this property later on in Chapter 11 when we provide a quantum mechanical interpretation of the Lund fragmentation distributions.

For the remaining non-exponential factors obtained by multiplying f and H in the two cases we obtain (for different a_α, a_β)

$$\frac{dz_+ dz_-}{z_+^2 z_-^2} \left(\frac{1 - z_-}{z_-} \right)^{a_\alpha} \left(\frac{1 - z_+}{z_+} \right)^{a_\beta} \tag{8.22}$$

(besides some constant factors). This expression is evidently again symmetric between the two vertices and can also be interpreted as the size of certain areas. For the case when $a_\alpha = a_\beta$ we obtain the symmetrical area marked *area* in Fig. 8.1 as the common factor, i.e.

$$(area)^a \tag{8.23}$$

From this result we conclude (parameter a being positive) that there is a (power-)suppression if we take too large a step in the production process, i.e. when any one of the variables $z_\pm$ is chosen to be close to unity.

We will later see that the appearance of the parameter a stems from the requirement of not using up all the remaining energy-momentum. The reason is, of course, that we are implicitly assuming in all our considerations that we are far from the end or the beginning of the process. The distributions f and H are called *inclusive* distributions, i.e. they are characteristic of a single-production event independent of anything else that comes before or after. But there is, of course, a tacit assumption that there are other particles produced, over which we are summing.

8.3   The production of a finite-energy cluster of hadrons

We will in this section derive the distribution for a finite number of hadrons which are rank-ordered, for definiteness along the positive light-cone. From the resulting formulas all other possible situations can be deduced. Such a group of particles is often called a *cluster* or a *single jet*. Together they will have a finite mass, conventionally called $\sqrt{s}$.

The first-rank particle will then contain the flavor f_0 of the original q_0 together with the antiflavor $\bar{f}_1$ of the $\bar{q}_1$ produced at the first vertex. The second-rank particle will contain the flavor f_1 and antiflavor $\bar{f}_2$ of the q_1 from the first vertex and the $\bar{q}_2$ from the second, etc.

The probability of obtaining a first-rank meson with mass m_{01} and with a fractional lightcone component z_1 of the total energy-momentum p_{+0} of the original q_0 is according to Eq. (8.17)

$$f(z_1)dz_1 = N\frac{dz_1}{z_1}z_1^{a_0}\left(\frac{1-z_1}{z_1}\right)^{a_1}\exp\left(-\frac{bm_{01}^2}{z_1}\right) \qquad (8.24)$$

In order to simplify the formulas we will from now on consider the case when all the a-values as well as the masses are the same. At the end of the derivation we will provide the formulas for the general case. We will also use the convention of writing z_{oj} for the lightcone energy-momentum fraction of the hadron of rank j, scaled with respect to the original quark's energy-momentum p_{+0}; we call z_{oj} the 'observable' fraction.

Thus the variable z_1 in Eq. (8.24) equals z_{o1} while for the second-rank hadron, which takes a fraction z_2 of the remaining energy-momentum, $(1-z_{o1})p_{+0}$, we have

$$z_{o2} = z_2(1-z_{o1}) \qquad (8.25)$$

The variable z_2 is again distributed according to the function f (for equal a-values cf. Eq. (8.16)). Therefore the combined distribution for producing first- and second-rank hadrons with observable fractional lightcone components z_{o1} and z_{o2} is

$$f(z_1)dz_1 f(z_2)dz_2 = f(z_{o1})dz_{o1}f\left(\frac{z_{o2}}{1-z_{o1}}\right)\frac{dz_{o2}}{1-z_{o1}}$$

$$= \left(\frac{Ndz_{o1}}{z_{o1}}\right)\left(\frac{Ndz_{o2}}{z_{o2}}\right)(1-z_{o1})^a\left(1-\frac{z_{o2}}{1-z_{o1}}\right)^a\exp[-b(A_1+A_2)] \qquad (8.26)$$

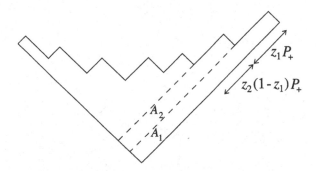

Fig. 8.2. The production of the first- and second-rank hadrons, with the areas in the exponent of Eq. (8.26) indicated.

The a-dependent factors obviously combine to give

$$(1 - z_{o1} - z_{o2})^a \tag{8.27}$$

and the fractional differentials can be reexpressed as follows:

$$\left(\frac{dz_{o1}}{z_{o1}}\right)\left(\frac{dz_{o2}}{z_{o2}}\right) = d^2 p_{o1} d^2 p_{o2} \delta^+(p_{o1}^2 - m^2)\delta^+(p_{o2}^2 - m^2) \tag{8.28}$$

Here $d^2 p = dp_+ dp_-$. We consequently introduce two new variables, in this case the negative-lightcone energy-momenta (note that $p_{+oj} = z_{oj} p_{+0}$). This is done by the introduction of two δ-distributions which fix their values. We have used the following properties of the δ-distribution, which was also used in Chapter 3 with the requirement that $C = D/B$:

$$dB dC \delta(BC - D) \rightarrow \frac{dB}{B} \tag{8.29}$$

The arrow implies that the left- and right-hand sides are equal if we actually perform the integral. We shall always use an equality sign even if we do not perform the integrals. The right-hand side of Eq. (8.28) explicitly exhibits the Lorentz invariance of the phase-space factors. The factor $A_1 + A_2$ in the exponential in Eq. (8.26) corresponds to the two regions indicated in Fig. 8.2 (the interpretation as an area size is given to the exponential factor in the fragmentation function in Eqs. (8.20), (8.21)).

From this result we may already guess what the result will be if we produce n particles with energy-momenta $\{p_{+oj}\} \equiv \{z_{oj} p_{+0}, p_{-oj}\}$:

$$dP(p_{o1}, \ldots, p_{on}) \equiv \left(1 - \sum_{j=1}^{n} z_{oj}\right)^a \prod_{j=1}^{n} N d^2 p_{oj} \delta^+(p_{oj}^2 - m^2) \exp(-bA_j). \tag{8.30}$$

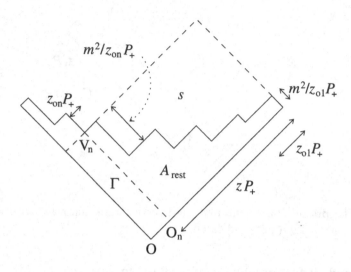

Fig. 8.3. An n-particle cluster with notation as explained in the text.

This formula is straightforward to prove; we will leave this to the reader.

The situation after n steps is depicted in Fig. 8.3. We note firstly that the total area in the exponent, $\sum_{j=1}^{n} A_j \equiv A_{tot}$, can be subdivided into two parts,

$$A_{tot} = A_{rest} + \Gamma \tag{8.31}$$

as shown in the figure. The quantity Γ then corresponds to the proper time (cf. Eq. (8.2)) of the 'last' vertex of the cluster.

Secondly we note that it seems as if the system of the n particles could have been produced just as well by the original q-particle and an antiparticle $\bar{q}_n$. This pair would then have started out at the point O_n in Fig. 8.3. We know in fact that the cluster is part of the system produced from the force field of the original $q\bar{q}$-pair which started at the point O and produced the pair $q_n\bar{q}_n$ at the vertex V_n. But we would not have that knowledge unless we had been able to observe some parts of the system outside the cluster!

The energy-momentum of the 'new' pair is then (W_{+n}, W_{-n}) where $W_{+n} = zp_{+0}$, $z \equiv \sum_{j=1}^{n} z_{oj}$ and

$$W_{-n} = \sum_{j=1}^{n} \frac{m_j^2}{z_{oj}p_{+0}} \tag{8.32}$$

The formulas for $W_{\pm n}$ are a somewhat complex way of writing the total energy-momenta of all the particles in the cluster: $W_{\pm n} = \sum_{j=1}^{n} p_{\pm oj}$. We conclude that the total squared mass of our n-particle system is

$$s = W_{+n}W_{-n} = \sum_{j=1}^{n} \frac{m^2 z}{z_{oj}} \qquad (8.33)$$

The variable s is also the size of an area according to Fig. 8.3. Using this fact it is easy to convince oneself that the area Γ defined in Eq. (8.31) is given by

$$\Gamma = \frac{s(1-z)}{z} \qquad (8.34)$$

which we again leave to the reader to prove.

Consequently all the interesting *external* properties of our n-particle system (i.e. its properties with regard to the original origin O) are given by the two Lorentz invariants s and z. It is useful to introduce these variables into the formulas and define (using Eq. (8.30))

$$dP(z,s;p_{o1},\ldots,p_{on}) \equiv dz\delta\left(z - \sum_{j=1}^{n} z_{oj}\right) ds\delta\left(s - \sum_{j=1}^{n} \frac{m^2 z}{z_{oj}}\right)$$
$$\times dP(p_{o1},\ldots,p_{on}). \qquad (8.35)$$

As $z > 0$ we may change the first δ-distribution as follows:

$$dz\delta\left(z - \sum_{j=1}^{n} z_{oj}\right) = \frac{dz}{z}\delta\left(1 - \sum_{j=1}^{n} \frac{z_{oj}}{z}\right) \qquad (8.36)$$

Then the two new δ-distributions (i.e. the s-definition and the above reorganised z-definition) only depend upon the *internal* variables

$$u_j \equiv \frac{z_{oj}}{z} = \frac{p_{+oj}}{W_{+n}} \qquad (8.37)$$

These would be the scaling variables if we consider the cluster as arising from the $q\bar{q}_n$-pair produced at the space-time point O_n in Fig. 8.3. We then obtain for the expression in Eq. (8.35)

$$dP(z,s;p_{o1},\ldots,p_{on}) = ds\frac{dz}{z}(1-z)^a \exp(-b\Gamma)\delta\left(1 - \sum_{j=1}^{n} u_j\right)\delta\left(s - \sum_{j=1}^{n} \frac{m^2}{u_j}\right)$$
$$\times \prod_{j=1}^{n} Nd^2 p_{oj}\delta^{+}(p_{oj}^2 - m^2)\exp(-bA_{rest}) \qquad (8.38)$$

By a further 'division trick' the two δ-distributions can be written as follows:

$$\delta\left(1 - \sum_{j=1}^{n} u_j\right)\delta\left(s - \sum_{j=1}^{n} \frac{m^2}{u_j}\right)$$

$$= \delta\left(W_{+n} - W_{+n}\sum_{j=1}^{n} u_j\right)\delta\left(W_{-n} - \sum_{j=1}^{n} \frac{m^2}{u_j W_{+n}}\right) \equiv \delta^2(P_{rest} - \sum_{j=1}^{n} p_{oj})$$

(8.39)

where the quantity P_{rest} is the energy-momentum of the n-particle system ($P_{rest} = (W_{+n}, W_{-n})$). The superscript 2 on the δ at the end of Eq. (8.39) indicates that here we use (the lightcone-component version of) the two-dimensional energy-momentum conservating δ-distribution.

In this way we have been able to partition the formula for the production probability of an n-particle cluster with a given endpoint (squared) proper time Γ, Eq. (8.34), and a given total energy-momentum $W_{\pm n}$ with a squared mass $s = W_{+n}W_{-n}$, into two parts. These will be called, according to the notions introduced above, the *external part*

$$dP_{ext} = ds\frac{dz}{z}(1 - z)^a \exp(-b\Gamma)$$

(8.40)

and the *internal part*:

$$dP_{int} = \prod_{j=1}^{n} Nd^2 p_{oj}\delta^+(p_{oj}^2 - m^2)\exp(-bA_{rest})\delta^2(\sum_{j=1}^{n} p_{oj} - P_{rest})$$

(8.41)

The external part corresponds to the (non-normalised) probability that the cluster as defined above will occur, while the internal part in the same way corresponds to the probability that the cluster will decay into the particular channel considered, containing the given n particles.

The general result for an n-particle cluster which starts at a vertex with the parameter a_0 and ends at a vertex with a_n is, for the external part,

$$dP_{ext} = ds\frac{dz}{z}z^{a_0}\left(\frac{1 - z}{z}\right)^{a_n}\exp(-b\Gamma)$$

(8.42)

The corresponding general formula for the internal part is

$$dP_{int} = \prod_{j=1}^{n} N_{j-1,j}\left(\frac{du_j}{u_j}\right)u_j^{a_{j-1}-a_j}$$

$$\times \exp(-bA_{rest})\delta\left(1 - \sum_{j=1}^{n} u_j\right)\delta\left(s - \sum_{j=1}^{n} \frac{m_{j-1,j}^2}{u_j}\right)$$

(8.43)

where we have kept the scaling variable description. We leave it to the reader to derive Eqs. (8.42), (8.43).

The result in Eq. (8.41) is evidently completely symmetric with respect to the different particles and therefore it has an obvious left-right symmetry with respect to the lightcones. This property is not so obvious in Eq. (8.43). We note, however, that the negative-lightcone variables v_j corresponding to the u_j obviously should fulfil

$$v_j u_j = \frac{m_{j-1,j}^2}{s} \tag{8.44}$$

(these are the mass-shell conditions). Therefore a change from the variables u_j to v_j can be carried through in a straightforward manner in Eq. (8.43). We obtain for the terms in Eq. (8.43):

$$\frac{du_j}{u_j} \to \frac{dv_j}{v_j}$$

$$N_{j-1,j} u_j^{a_{j-1}-a_j} \to N'_{j-1,j} v_j^{a_j-a_{j-1}} \tag{8.45}$$

$$\delta\left(1 - \sum_{j=1}^n u_j\right) \delta\left(s - \sum_{j=1}^n \frac{m_{j-1,j}^2}{u_j}\right) \to \delta\left(s - \sum_{j=1}^n \frac{m_{j-1,j}^2}{v_j}\right) \delta\left(1 - \sum_{j=1}^n v_j\right)$$

In the second line we have absorbed a $(j, j-1)$-dependent mass factor into the normalisation constant $N'_{j-1,j}$ and in the third line we have again made use of a 'division trick' for the two δ-distributions.

Obviously the result in Eq. (8.43) is, after these operations, the same in the u_j-language as in the v_j-language apart from the fact that we are now ordering the vertices as $j, j-1, \ldots$ along the negative lightcone.

In the following chapters we will investigate the internal- and external-part formulas in great detail and also exhibit several different interpretations from both quantum field theory and statistical mechanics.

8.4 The Artru-Menessier-Bowler model

We will now briefly consider a different approach to the decay of a high-mass string, the AMB model, [26], [32]. Here the idea is to take classical probability arguments, which also occur in the Lund model derivation as presented above, as far as they can go. There are two basic rules.

AMB1 There is a constant probability $\mathscr{P}$ per unit time and per unit length in the string's space-time history that it may break up by the production of $q\bar{q}$-pairs.

AMB2 The string cannot break up further in the forward lightcone with respect to an 'earlier' vertex.

The procedure can be visualised as the game of stochastic dart-throwing on a target corresponding to the original string's space-time history. The landing of each dart then produces a possible vertex and one accepts those vertices which have no other vertex in their prehistory.

A continuous mass spectrum will then be obtained for the produced particles. There is then a third rule to interpret the result.

AMB3 Using AMB1 and AMB2 one obtains a *first generation of breakups* producing a first generation of yoyo-hadrons. These states are then considered as 'resonances' and will be allowed to decay again, independently, according to the same rules.

If we go back to Fig. 8.1 then we conclude that one will obtain (just as for a radiative decay) that the probability for an allowed vertex at a point (Γ_1, y_1) is

$$dP_{AMB}(1) = bd\Gamma_1 dy_1 \exp(-b\Gamma_1) \qquad (8.46)$$

where $b = \mathscr{P}/\kappa^2$. We will consider this result in more detail below when we compare to the Lund model results.

Similarly there is a joint probability of having two primary neighboring AMB vertices at the two points 1 and 2 in Fig. 8.1. It is equal to

$$dP_{AMB}(12) = dP_{AMB}(1)dP_{AMB}(1 \to 2)$$
$$dP_{AMB}(1 \to 2) = b(W_{+1}dz_+)(W_{-2}dz_-) \exp[-b(W_{+1})(z_- W_{-2})] \quad (8.47)$$

with $dP_{AMB}(1 \to 2)$ *the conditional probability* that given 1 we may also obtain 2. We are using the notation of Fig. 8.1 and the Eqs. (8.6). This time there is no mass-shell condition to constrain the location of the two vertices 1 and 2. Therefore we need all four (independent) quantities Γ_1, y_1, z_+, z_-. (Note that due to Lorentz covariance there is no dependence on the rapidity variable y_1 in the formulas.)

The probability distribution $dP_{AMB}(1 \to 2)$ contains the negative exponential of the region (cf. Eq. (8.6))

$$W_{+1}W_{-2}z_- = \frac{m^2}{z_+} \qquad (8.48)$$

with m the mass produced between the adjacent vertices 1 and 2; together the exponentials of the two distributions $dP(1)dP(1 \to 2)$ contain the symmetrical surface $W_{+1}W_{-2}$ from Eqs. (8.20), (8.21). Therefore the joint distribution $dP(12)$ is symmetric with respect to vertices 1 and 2.

The distribution $dP_{AMB}(12)$ can be reformulated into a distribution in

z_+ and the mass m as

$$dP_{AMB}(12) = \frac{bdz_+dm^2}{z_+} \exp\left(-\frac{bm^2}{z_+}\right) \tag{8.49}$$

From this expression it is then possible to obtain the distribution in the mass m by means of an integral over z_+:

$$\frac{dP}{dm^2} = \int_0^1 \frac{bdz_+}{z_+} \exp\left(-\frac{bm^2}{z_+}\right) = bE_1(bm^2) \tag{8.50}$$

where E_1 is the exponential integral of the first rank. This function is singular when $m^2 \to 0$, which means that there is a large probability that the string in the AMB model breaks up into very tiny pieces. It is then necessary to introduce a lower cutoff in the mass spectrum. Such a cutoff is difficult to introduce in a consistent way if one wants to keep to the classical probability concepts which are at the basis of the model. It is nevertheless possible to interpret the resulting spectrum in a way similar to the resonance spectrum suggested by Hagedorn, [76] (although Hagedorn obtained a linear dependence upon the masses in the exponent).

The results of the AMB model are evidently (apart from the continuous mass spectrum) similar to the results of the Lund model. It contains an iterative structure based upon an area suppression law. It is, however, not possible to obtain the Lund model relations by the use of the probability concepts in the AMB model.

To see this, suppose that we specialise the AMB model to particular masses, e.g. a single mass with a width δm^2 around m^2. This would mean that a new vertex would only be allowed in a band along the mass hyperbola corresponding to m. If we are at vertex V and we are looking for the next vertex H in that band we may subdivide the band into many small boxes (see Fig. 8.4) and call them $1, 2, \ldots, n \ldots$. The boxes have areas $(\delta a)_j$ and the probability of finding a vertex in such a box is equal to $b(\delta a)_j$.

Then the probability of not finding a vertex in the first n boxes will be

$$\prod_{j=1}^n [1 - b(\delta a)_j] \to \exp\left[-\int bd(\delta a)\right] \tag{8.51}$$

Here the right-hand expression is the limit found when we subdivide the band indefinitely, i.e. when $n \to \infty$. The expression for $d(\delta a)$ is $\delta m^2 dz_+/z_+$ i.e. the width times an infinitesimal angular segment along the hyperbola.

Therefore the probability of finding a vertex at the value z_+ without having found it for any larger value of z_+ (i.e. for any 'earlier' vertex,

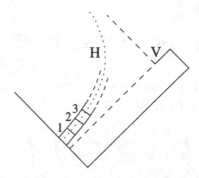

Fig. 8.4. The allowed region for finding the next vertex H, after the vertex produced at V, is a band around a hyperbola. This region can be subdivided into small boxes as discussed in the text.

closer to the origin, see Fig. 8.4) is

$$\frac{b\delta m^2 dz_+}{z_+} \exp\left(-b\delta m^2 \int_{z_+}^{1} \frac{dz'_+}{z'_+}\right) = b\delta m^2 dz_+ z_+^{b\delta m^2 - 1} \qquad (8.52)$$

This corresponds to a power law in z_+, owing to the fact that we no longer have a two-dimensional surface on which to apply the probability rule.

9

The properties of the Lund model fragmentation formulas; the external-part formulas

9.1 Introduction

In the previous chapter we derived a stochastical process for string fragmentation. The result is a unique process which is at the basis of the Lund model for the fragmentation of quark and gluon jets. We used only some general properties of a kinematical nature together with the necessary requirements of causality and relativistic covariance. The whole discussion is based on (semi-)classical arguments (quantum mechanics does of course enter into our assumptions on $q\bar{q}$-pair production).

In particular the process led to precise formulas for the production properties (we called these the external-part formulas) and the decay properties (the corresponding internal-part formulas, see chapter 10) of a finite-energy cluster of rank-connected hadrons.

The term 'external-part' is used to imply that the cluster is in general part of a larger-energy (possibly infinitely-large-energy) cluster. Two independent Lorentz invariants are necessary to specify the external properties of the cluster; these may be taken as the squared mass s and the lightcone fraction z used up by the cluster. They describe how the cluster starts and ends on some (space-time or energy-momentum-space) points that are inside (or on the border of) the larger external cluster.

In this chapter we will consider the external-part formulas in detail and in particular show the following.

E1 In the Lund model the cluster will be produced in accordance with the same formula as for a single particle (but with the squared hadronic mass $m^2 \to s$).

E2 The finite-energy version, H_s, of the space-time distribution of vertices, H in Eq. (8.14), approaches H very fast when s is larger than a few squared hadronic masses m^2.

163

We next consider the two functions H and f in the Lund fragmentation model in detail in order to understand some of their properties. After that we will exhibit some general properties of all iterative cascade fragmentation models of the Feynman-Field kind, [13]. We end the chapter with a discussion of an interesting analogy (first pointed out by Artru, [25]) between the proper time of a vertex in space-time and the momentum transfer between the group of particles produced to the left and those produced to the right of that vertex in energy-momentum space.

9.2   The production properties of a cluster

We start with the results in Eqs. (8.34), (8.40) and (8.42):

$$\Gamma = s\frac{1-z}{z}, \quad dP_{ext} = ds\frac{dz}{z}z^{a_0}\left(\frac{1-z}{z}\right)^{a_n}\exp(-b\Gamma) \tag{9.1}$$

Here Γ corresponds to the squared proper time of the last vertex, which has parameter a_n, and s is the squared mass of the particle cluster stretching between the vertex with parameter a_0 to the vertex with a_n. (Note that the expressions do not contain any relation to the decay of the cluster; in particular, the index n in this case does not indicate the multiplicity!)

These formulas can be rewritten in several different ways, each of which exhibits some particular feature of the Lund model fragmentation process.

1   *The vertex distribution in proper time for a finite energy*

If we use the first equation in (9.1) to solve for z in terms of Γ and then change the second equation into a distribution in s and Γ we obtain

$$dP_{ext} = dsd\Gamma\frac{\Gamma^{a_n}s^{a_0-a_n}}{(s+\Gamma)^{a_0+1}}\exp(-b\Gamma) \tag{9.2}$$

For a fixed and finite value of s we can read off an expression for the correspondence to the distribution $H(\Gamma)$ in Eq. (8.14):

$$H_{sn}(\Gamma) \sim \frac{s^{a_0}\Gamma^{a_n}}{(s+\Gamma)^{a_0+1}}\exp(-b\Gamma) = \frac{s^{a_0}H_{s\rightarrow\infty,n}}{C(s+\Gamma)^{a_0+1}} \tag{9.3}$$

In this way we have obtained the result we expected but multiplied by a factor in s and $\Gamma + s$, the power depending upon the starting vertex. (The indices on H in the final expression are meant to show that it is s-independent and has the correct power a_n.)

For any fixed value of s the function H_{sn} in Eq. (9.3) approaches 0 fast for large values of Γ owing to the exponential decrease. This feature is independent of s. A simple estimate implies for $\Gamma > \Gamma_0 \simeq (a_n + 1)/b$ that the exponential falloff dominates the distribution H_{sn}. Therefore for

$s \gg \Gamma_0$ (from phenomenological investigations Γ_0 correponds to a few GeV^2) a proper normalisation of H_{sn} will lead to an s-independent result. Then it is a good approximation, when $s \gg \Gamma_0$, that

$$dP_{ext} \simeq ds s^{-(a_n+1)} d\Gamma \frac{H_n(\Gamma)}{C_n} \qquad (9.4)$$

The constant $C \equiv C_n$ in Eq. (9.3) is, of course, the normalisation constant for H_n. In this way dP_{ext} depends only upon the flavor n of the final vertex. Actually this is just what we started with when we derived the distributions H_α and $f_{\alpha\beta}$: after many steps along a lightcone there is a certain probability of finding a vertex of a particular kind independently of where we started. We will come back to this *saturation property* later when we consider the internal-part formulas for the decay of a cluster.

This serves as a confirmation for the consistency of the assumption J, at the beginning of the last chapter, that there is, even in the limit $s \to \infty$, a finite number of vertices at the centre of phase space.

2 *The energy-momentum distribution of a finite-mass cluster*

Another obvious way to use the formula (9.1) is to exhibit the probability of obtaining a cluster with a given mass $\sqrt{s}$, thereby taking a fraction z of the positive lightcone component of the original system:

$$dP_{ext} = ds \exp(bs) \frac{dz}{z} z^{a_0} \left(\frac{1-z}{z}\right)^{a_n} \exp\left(\frac{-bs}{z}\right) \qquad (9.5)$$

We have then divided the expression for Γ from Eq. (9.1) into one z-dependent and one z-independent part in the exponential.

The remarkable feature of this result is that (besides the purely s-dependent parts and the normalisation) we evidently recover the 'old' formula, which was derived for a single particle, with the mass m exchanged for the mass of the cluster $\sqrt{s}$. Consequently, whether a single particle or a large-mass cluster arises in going between two vertices with a-parameters a_0 and a_n the (mass-dependent) probability distribution for picking a particular fraction of the energy-momentum is the same.

9.3 **The properties of the distributions H and f**

At this point it is worthwhile to consider the shape and the properties of the unique Lund model distributions in more detail.

1 *The properties of the proper time distribution H*

The distribution in proper time H is the mathematically well-known Γ-distribution (this is not a misguided pun!) which occurs e.g. in connection with radiative processes. Depending upon the values of the parameters it has a maximum at $\Gamma = a/b$, a mean value $\langle\Gamma\rangle = (a+1)/b$ and a variation around the mean $\langle(\Gamma - \langle\Gamma\rangle)^2\rangle = (a+1)/b^2$.

Typical phenomenological parametrisations for longitudinal jets (note the dependence of a and b upon the gluon radiation to be discussed in Chapter 17) would be $a \sim 0.5, b \sim 0.75$ GeV^{-2}. We conclude that for these values the typical proper time 'before' the string will break is somewhat more than 1 fm$/c$ but that the fluctuations around this value is of the same order of magnitude.

2 The properties of the fragmentation distribution f

The distribution f is a more complex kind of function. We note that it vanishes exponentially fast close to the origin (it has an essential singularity there, considered as an analytical function) and that it vanishes according to a power law for $z \to 1$. In between there is evidently a maximum.

In order to investigate this maximum in more detail we rewrite the distribution f as an exponential (considering only the case when all the a-parameters are equal):

$$f \sim \exp\Phi \quad \text{with} \quad \Phi = -\frac{bm^2}{z} - \ln z + a\ln(1-z) \qquad (9.6)$$

It is easy to prove that for $a = 0.5$ the quantity Φ has a maximum for

$$z = 1 + bm^2 - \sqrt{1 + (bm^2)^2} \simeq bm^2 - (bm^2)^2/2 \qquad (9.7)$$

We conclude that the typical z-values will increase with bm^2 and that the maximum of f will occur for a z-value around 0.3 using the value of b quoted above and a mass-value close to the centre of the mesonic mass spectrum, the ρ-mass $m \simeq 0.77$ GeV$/c^2$.

3 The typical hyperbola breakup

A useful exercise is to consider the relationship between the Γ-parameters of two adjacent vertices in the case where a hadron of mass m, taking a fraction z of the remaining lightcone energy-momentum, is produced in going from vertex 1 to vertex 2. It is left to the reader to prove that

$$\Gamma_2 = (1-z)\left(\Gamma_1 + \frac{m^2}{z}\right) \qquad (9.8)$$

From Eq. (9.8) we deduce that if there is a fluctuation in the value of z taken by the hadron the result will be a value of Γ_2 that is much larger (for $z \ll z_t$, where z_t is a typical value of z) or much smaller (for

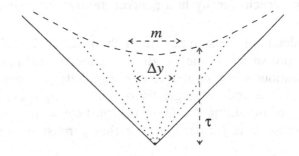

Fig. 9.1. The typical breakup hyperbola divided into particle-mass pieces.

$z \gg z_t$) than Γ_1. The first possibility is suppressed due to the area law (the exponential area suppression) while the second one is power suppressed.

The final result is that the Lund model fragmentation functions tend to produce vertices around a hyperbola (i.e. the locus of the points with a fixed value of $\Gamma \equiv \tau_t^2$), albeit with some fluctuations. The distance from the origin to the hyperbola, τ_t, is related to the typical mass of the hadrons in the cascade decay.

If we place all the vertices along this hyperbola the energy-momentum fractions taken by the hadrons form a geometrical series:

$$z_t, z_t(1 - z_t), \ldots, z_t(1 - z_t)^n, \ldots \tag{9.9}$$

(note that the remainder fraction is given by $(1 - z_t)^n$ after n steps).

When we move along the hyperbola the remainder fraction cannot be too small. It must necessarily be larger than s_0/s with s_0 of order $\langle \Gamma \rangle$. Therefore we obtain a formula for the typical multiplicity, n_t, in a Lund model fragmentation event:

$$(1 - z_t)^{n_t} \sim s_0/s \Rightarrow n_t \sim \frac{\log(s/s_0)}{\log[1/(1 - z_t)]} \tag{9.10}$$

This result can be interpreted geometrically; see Fig. 9.1. The length of the hyperbola is $\sim \tau_t \log(s/s_0)$ with τ_t the hyperbola parameter defined above. (Note that the notion of length, of course, corresponds to the invariant length in Minkovski space.)

If the hyperbola is cut up into pieces corresponding to particle masses then each piece will cover a typical rapidity gap Δy. In Fig. 9.1 one hadron at rest is shown. According to our findings in Chapter 7 the space size of such a yoyo-state is given by its mass.

We conclude that with a hadron density $1/\Delta y$ along the hyperbola we will obtain the same multiplicity formula as in Eq. (9.10) if we put

$$\Delta y = \log[1/(1 - z_t)] \tag{9.11}$$

9.4   The particle density in a general iterative cascade model

In order to understand the significance of the result in Eq. (9.11) we will
consider some properties of the iterative cascade fragmentation models,
which were mentioned in Chapter 7. For simplicity we consider the situation when there is only a single flavor and a single kind of meson.
The probability of obtaining the first-rank particle with a given energy-
momentum fraction z is $f(z)dz$. We note that f must be normalised to
unity:

$$\int_0^1 dz f(z) = 1 \tag{9.12}$$

We now define the *totally inclusive single-particle distribution* $F(z)dz$ as
the number of hadrons (irrespective of rank) with fractional energy-
momentum z. This function is not normalised to unity as is f in Eq. (9.12)
but, instead, to the total number of hadrons produced.

Inside the scaling cascade scheme this number is in general divergent. In
this subsection we will derive the behaviour of a general iterative cascade
model and in the next we will specialise to one particular shape of f and
perform some of the calculations in detail.

To investigate the properties of F we note that there is an integral
equation which relates F and f:

$$F(z) = f(z) + \int_0^{1-z} \frac{d\zeta}{(1-\zeta)} f(\zeta) F\left(\frac{z}{(1-\zeta)}\right) \tag{9.13}$$

The interpretation of the equation is that a hadron with z may be the
first-rank hadron in the jet (this is the first contribution $f(z)dz$ on the right-
hand side of Eq. (9.13)). After the first-rank particle has left a fraction
$1 - \zeta$ (with probability $f(\zeta)d\zeta$) the number of hadrons with z that occur
further down in the jet is $F(z/(1 - \zeta))dz/(1 - \zeta)$. This gives the integrand
in the second term of Eq. (9.13) (after division dz). We must sum over all
values of ζ compatible with the requirement that the argument of F is
between 0 and 1.

The equation can be solved by means of the moments method. We
obtain from Eq. (9.13)

$$M(r) = \int_0^1 z^r F(z) dz, \quad C(r) = \int_0^1 (1-z)^r f(z) dz,$$

$$m(r) = \int_0^1 z^r f(z) dz \Rightarrow M(r) = \frac{m(r)}{1 - C(r)} \tag{9.14}$$

which we leave to the reader to prove.

The normalisation condition in Eq. (9.12) implies that $C(0) = m(0) = 1$.
This evidently means that $M(r)$ diverges when $r \to 0$, which corresponds

to the normalisation equation for F. The reason for this divergence is that in Eq. (9.13) no provision is made for ending the cascade: there is no smallest value of z, or in other words the process is totally scaling. Instead one obtains a *rapidity plateau* (note that $y \simeq \log z$ implies $dz/z = dy$). After a rapidity region in the forward direction, the *fragmentation region* of the original quark, there will be a uniform distribution of hadrons in rapidity space. In practice this fragmentation region is about 1.5–2 rapidity units. This means that a large part of the energy is inside the fragmentation region. It is populated by the first few particles in rank but the density of particles is strongly fluctuating and dependent upon the flavor quantum number carried by the original color charge.

In the Lund model it is not sufficient to consider only the fractional energy-momentum along the jet, i.e. in one of the lightcone directions, as in iterative cascade models. There is also the energy-momentum along the opposite lightcone direction, for which we must account. This is the reason why in the last subsection we had to bring the plateau to an end by the request that we can use up the fractional energy-momentum only to the level s_0/s. In the integral equation in Eq. (9.13) the plateau will, however, go on forever.

The height of the plateau, i.e. the density of hadrons in the centre, can be calculated by simple means from Eq. (9.14). We may conclude, by expanding for small values of the moment parameter r, that $M(r) \to R/r$ where

$$\frac{1}{R} = -\left(\frac{dC(r)}{dr}\right)_{r=0} = \int_0^1 \log\left(\frac{1}{1-z}\right) f(z) dz \qquad (9.15)$$

which we again leave for the reader to prove.

We conclude that $F(z)dz$ behaves as $Rdz/z \equiv Rdy$ for small values of z, i.e. for rapidities far from the 'tip' of the jet. Thus the result in Eq. (9.11) is very general with Δy identified with $1/R$, i.e. the mean value of $\log[1/(1-z)]$ as calculated from the fragmentation function f.

We may in an intuitive sense identify Δy with the *mean loss of rapidity per produced hadron*. It is interesting to note that we again find a similar scaling energy-momentum distribution as for the virtual quanta in the MVQ and the partons in the PM in Chapters 2 and 5. In particular the result obtained in the Schwinger model for excitations by means of an external charged pair $\pm g$ leads to the result $R = 1$; cf. Chapter 6.

A detailed calculation of the inclusive distribution using a simple model

The method of moments is a very powerful mathematical technique but it may be difficult to understand the results on an intuitive level. We will therefore show by explicit calculation how the central plateau is built up by the contributions from the hadrons of different rank. The results of the calculation will also be useful further on, in Chapter 13.

We consider a very simple iterative cascade model with a constant fragmentation function, f, which then in order to be normalised as in Eq. (9.12) must equal unity. Although it is simple, it was used rather successfully at the beginning of the Lund model, [13], assuming then that all vector and pseudoscalar mesons were produced in accordance with their statistical weights. We now know that this is not the case. Further a constant distribution does not fulfil the requirements for a consistent fragmentation process listed at the beginning of Chapter 8.

A detailed study of the model is, however, instructive because it is straightforward to provide explicit results for the inclusive distributions of the nth-rank particles for all values of n. The first-rank particle is evidently distributed according to f. After it has taken z_1 (with the same probability for all z_1) the second-rank particle will take $z = z_2(1 - z_1)$, with a flat distribution for z_2 also.

This means that the inclusive distribution of the second-rank hadron is

$$D^{(2)}(z) = \int dz_1 dz_2 \delta(z_2(1 - z_1) - z)$$

$$= \int_0^{(1-z)} \frac{dz_1}{1 - z_1} = \int_z^1 \frac{dx_1}{x_1} = \log\left(\frac{1}{z}\right) \tag{9.16}$$

Using the same method we obtain for the nth rank hadron

$$D^{(n)}(z) = \int \left(\prod_{j=1}^{n} dz_j\right) \delta\left(z_n \prod_{j=1}^{n-1}(1 - z_j) - z\right)$$

$$= \int \left(\prod_{j=1}^{n-1} \frac{dx_j}{x_j}\right) \Theta\left(\prod_{j=1}^{n-1} x_j - z\right) \tag{9.17}$$

where Θ is the Heaviside function, equal to unity for a positive argument and vanishing elsewhere. We have also defined the obvious new variables $x_j = 1 - z_j$. In order to perform the integral we introduce $y_j = \log(1/x_j)$ and write, exchanging the product of the x_j for a sum of the rapidities y_j (the sum being introduced by means of a δ-distribution, $dy\delta(\sum y_j - y)$)

$$D^{(n)}(z) = \int dy\Theta[\exp(-y) - z] \int \left(\prod_{j=1}^{n-1} dy_j\right) \delta\left(\sum_{j=1}^{n-1} y_j - y\right) \tag{9.18}$$

We obtain a symmetrical integral (for $N = n - 1$), which is most easily solved by iteration:

$$I_N = \int \left(\prod_{j=1}^{N} dy_j\right) \delta\left(\sum_{j=1}^{N} y_j - y\right) = \frac{y^{N-1}}{(N-1)!} \tag{9.19}$$

We finally obtain by direct integration over y

$$D^{(n)}(z) = \frac{[\log(1/z)]^{n-1}}{(n-1)!} \qquad (9.20)$$

which is a nice and very satisfying result to derive! The following comments may be made.

- All but the first-rank hadron have a distribution in z which vanishes when $z \to 1$. Since $\log(1/z) = \log[1 + (1-z)/z] \simeq 1 - z$ when $z \to 1$ we find that the nth rank distribution will vanish like

$$(1-z)^{n-1}. \qquad (9.21)$$

The reason is evidently that there have already been $n-1$ earlier energy 'handouts'. The above result is a very general property of all physical systems, usually referred to as the *spectator relation*: if there are n basic constituents sharing a common energy and you require the inclusive distribution in energy for one of them its fraction usually behaves as in Eq. (9.21).

- The result (9.20) can be described as a distribution in $y = \log(1/z)$:

$$D^{(n)}(z)dz = dy \frac{y^{n-1}}{(n-1)!} \exp(-y) \qquad (9.22)$$

i.e. a Poisson distribution in rapidity. The distributions are evidently all normalised. This is exactly what was obtained in the external excitation model, derived from the Schwinger model in [39]; cf. Chapter 6.

- From the sum over all ranks we obtain the totally inclusive distribution, which, according to the predictions from the integral equation (9.13), corresponds to the particle density $R = 1$:

$$D(z)dz = \sum_{n=1}^{\infty} D^{(n)}(z)dz = dy = \frac{dz}{z} \qquad (9.23)$$

A useful exercise is to carry through the calculations above also for the case when f is exchanged for $f_a = (a+1)(1-z)^a$. Then one obtains $D_a(z) = (a+1)(1-z)^a/z$, i.e. a rapidity density equal to $R_a = a+1$.

In this way we can see in detail how the rapidity plateau occurring in the iterative cascade models is built up. From the properties of $D^{(n)}$ we conclude that the maximum of the distributions moves towards larger values in rapidity with increasing n; this is a useful exercise for the interested reader. Note, however, that an nth-rank hadron may very well have a larger z-value than the first-rank hadron (although with a small probability).

- For the model with a constant f the rapidity plateau evidently goes all the way from the tip and there is no evidence of a fragmentation region (but for the model with f_a there is such a region).

The reason why the model with a constant f works rather well is that if three times more vector mesons than pseudoscalars are produced as direct particles then the decay products from the vector mesons will move towards smaller z-values. This effect means e.g. that while the prediction for inclusive π^+ mesons for $z \to 1$ is $zD^{\pi^+} \simeq 0.1$ it becomes around 0.4 for $z \simeq 0.4$ because of the many π^+'s from the decays.

It is possible to derive many other analytic expressions, e.g. for the two-particle correlation functions, by the same means as for Eq. (9.13) and we refer to the original literature, [13]. We will not do it here because there are many kinematical complications. If this kinematics is included in the analytical equations the results become so complicated that it is in general much easier to take recourse to computer simulations.

It is very satisfying and highly recommended at this point for the reader to obtain a set of simple but useful and understandable distributions by the use of a Monte Carlo simulation program such as JETSET or HERWIG, just in order to appreciate the effects of really introducing kinematics!

9.5   The relationship between the vertex proper time and the momentum transfer across the vertex

We will now use the external-part formulas in a way proposed by Artru, [25]. We will start by showing that the quantity Γ, which in space-time has been related to the proper time of a vertex, in energy-momentum space can be interpreted as the invariant squared momentum transfer between the two jets produced by the appearance of the vertex.

Thus a vertex appearing at the space-time point $V = (x_+, x_-)$ (with $\Gamma = \kappa^2 x_+ x_-$) will divide the total system (see Fig. 9.2) into a right-moving jet with energy-momentum $P_r = (p_{+0} - \kappa x_+, \ \kappa x_-)$ and a left-moving jet with $P_l = (\kappa x_+, \ p_{-0} - \kappa x_-)$.

The situation depicted in Fig. 9.2 can be interpreted as if the original q-particle with $P_+ = (p_{+0}, 0)$ is transformed into the right-movers and the original $\bar{q}$-particle with $P_- = (0, p_{-0})$ is transformed into the left-movers. There is then evidently a momentum transfer in this process equal to $q \equiv P_+ - P_r = -(P_- - P_l) = \kappa(x_+, -x_-)$.

We note that in order to obtain positive masses for the two systems it is necessary that this momentum transfer is a spacelike vector, i.e. that the (Lorentz-)square of the vector is negative. The fact that it is the negative-lightcone component which becomes negative in our formula is due to

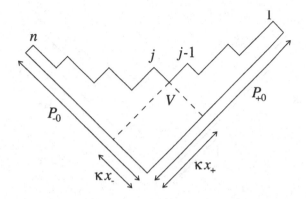

Fig. 9.2. The vertex $V = (x_+, x_-)$ divides the system into right-movers (hadrons $1, \ldots, j-1$) and left-movers (hadrons $j, \ldots, n$).

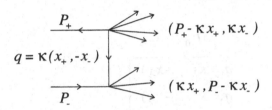

Fig. 9.3. The energy-momentum transfer between the right-movers produced by the original q and the left-movers produced by the original $\bar{q}$.

our choice to define the momentum transfer in the direction from right to left.

Pictorially we may describe the situation as in Fig. 9.3 in which there are two 'transformation points' where q becomes the rightmovers and $\bar{q}$ becomes the leftmovers and the momentum transfer q connects the two points. The result is evidently very similar to a simple Feynman diagram.

The system can be further partitioned. Each of the two new jets is naturally subdivided into smaller systems by means of internal vertices. This process can be continued until we reach the level of individual particles.

We may in this way relate the production process in the model, as shown in Fig. 9.4, to a particular *multiperipheral diagram* shown in Fig. 9.5 with the final-state hadrons coming out along a 'chain'.

There is a dual relationship between the picture we have had of producing the particles in space-time (cf. Fig. 9.4) and this kind of diagram.

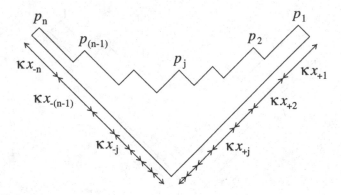

Fig. 9.4. The production of a multiparticle state in the Lund model. Note that the difference vectors between the space-time vertices correspond to the energy-momenta of the produced hadrons, while the positions of the vertices correspond to the energy-momentum transfers.

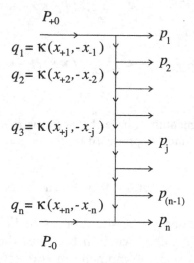

Fig. 9.5. The production process described in terms of momentum transfers in a chain.

The string production vertices (the space-time production points) are exchanged for a set of momentum transfers (the connecting *links* along the chain in the multiperipheral diagram). The invariant size of these momentum transfers (apart from the factor κ) corresponds in Figs. 9.2 and 9.5 to the distance from the origin to the space-time vertex points.

Models of a multiperipheral type have been under intense investigation

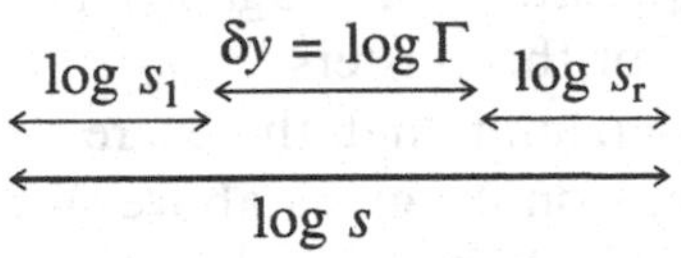

$$\underset{\overset{\longleftrightarrow}{\log s_1}}{\qquad} \underset{\overset{\longleftrightarrow}{\delta y = \log \Gamma}}{\qquad} \underset{\overset{\longleftrightarrow}{\log s_r}}{\qquad}$$

$$\overset{\longleftrightarrow}{\log s}$$

Fig. 9.6. The total range of rapidity $\sim \log s$ subdivided into the rapidity ranges of the right- and left-movers and the rapidity difference $\delta y \sim -\log \Gamma$.

since the 1950s and one notes the close resemblance to the Feynman diagrams for a two-body to many-body interaction. We will come back to such models in the next chapter after we have further developed our understanding of the Lund model fragmentation formulas.

We will meet the same kind of diagrams when we relate the lightcone singularities in deep inelastic scattering to the leading-log approximation formulas derived by the theorists working with Gribov, cf. Chapter 19.

We note that the momentum transfer divides the state so that the right-movers are, intuitively speaking, dragged apart from the left-movers. The proper measure for this effect is the rapidity difference between them.

We note that Γ can be expressed in terms of the total squared mass s of the original system and the squared masses s_l and s_r of the left- and right-moving systems through the equation (for the reader to prove):

$$s\Gamma = (s_l + \Gamma)(s_r + \Gamma) \approx s_l s_r \tag{9.24}$$

This means that for large-mass systems

$$-\ln \Gamma \simeq \ln s - \ln s_l - \ln s_r \tag{9.25}$$

The right-hand side of this expression is basically the rapidity difference δy between the right-movers and the left-movers. To see this we note that according to our results in the last subsection the available rapidity region for a system of mass $\sqrt{s}$ to deliver its particles is $\approx \log s$ (i.e. the length of the typical hyperbola). Then, as seen in Fig. 9.6, after having taken away the rapidity region inhabited by the left- and right-movers, we are left with a rather approximate measure of the rapidity difference δy. From the distribution in Γ, $H(\Gamma)$, the distribution in δy is

$$\sim d(\delta y) \exp[-(a+1)\delta y] \tag{9.26}$$

(we have neglected the slowly varying exponential). Thus we obtain a prediction for the (approximate) distribution of the rapidity gaps in our breakup process: when evaluated inclusively, it should be an exponentially decaying distribution with $a + 1$ as the exponential rate.

This is reminiscent of the formulas occurring in Regge-Mueller analysis, in which the parameter a plays the role of a Regge parameter. We will come back to this interpretation of a again later when we consider the internal decay properties of the clusters.

It is necessary to understand that there are several 'ifs and buts' in connection with the results on δy given above. We note that the rapidity difference between two neighbors in rank actually can be both positive and negative even if we know the rank-ordering. In order to see this consider the joint probability

$$P((\delta y)_n) = \int dz_1 f(z_1) \frac{dz_2}{1-z_1} f\left(\frac{z_2}{1-z_1}\right) \delta\left((\delta y)_n - \log\left(\frac{z_1}{z_2}\right)\right) \quad (9.27)$$

(the index n stands for rank neighbors). It is easy to manipulate the formula into

$$P((\delta y)_n) = N^2 \int_0^1 \frac{dz}{z} (1-z)^a \exp[bm^2 \exp(\delta y)_n]$$

$$\times \exp\left(\frac{-4bm^2}{z} \cosh\frac{(\delta y)_n}{2}\right) \quad (9.28)$$

This integral must be calculated by numerical methods. One obtains a generally smooth distribution with a maximum around $(\delta y)_n = 0$ but with an appreciable tail on both sides.

Thus *a lower-rank particle may be faster than a higher-rank particle*. It may even happen (with admittedly a small probability) that two particles close in rapidity may be far away in rank-ordering.

In a real experiment it is very difficult to observe rank-ordering because many directly produced particles are resonances which will decay quickly (mostly into pions but also into some kaons etc). Furthermore most of the particles in the final state contain u- and d-flavors and antiflavors. It is nevertheless known in many experimental situations that if one orders the observed particles in rapidity then the distribution in the rapidity gaps will be basically exponential (for large values of δy). There are, however, many different contributions to this distribution and it is not a useful way of determining the parameter a from such measurements.

10

The internal-part fragmentation formulas and their relations to the unitarity equations of a field theory; Regge theory

10.1   Introduction

In this chapter we will consider the decay properties of a cluster. We start to derive some results from the internal-part formulas, Eqs. (8.41) and (8.43).

I1   If we sum over all available states in the decay formulas of a cluster of squared mass s we obtain asymptotically, i.e. for large values of s, the behaviour $\sim s^a$. We will consider these state equations both for the case of a single species of flavor and meson and also for the case of many flavors and many hadrons in each flavor channel.

I2   At the same time we will derive the finite-energy version, f_s, of the fragmentation function f in Eqs. (8.16), (8.17). We will show that f_s tends rapidly towards f when s is larger than a few squared hadron masses (just as $H_s \rightarrow H$ according to the results of Chapter 9).

The method we will use is to derive a set of integral equations and then to solve them. In that way we will find that there are some necessary relationships between the parameters a, b and the normalisation parameters that constitute a set of eigenvalue equations for the integral equations.

The whole procedure is very similar to that for obtaining the *unitarity conditions for the S-matrix in a quantum field theory*. We will exploit these relationships by showing that the results obtained under I1 are just the same as are obtained for the *multiperipheral ladder equations* in a quantum field theory. Even the methods of constructing the integral equations are the same. One major result is that the parameter a in the fragmentation functions occurs in a similar way to the Regge intercepts in the Reggeon field theory.

It was Gribov who first understood that the unitarity equations of the S-matrix can be used to derive very general relations between the matrix

177

elements and the cross sections in any field theory and these considerations form the basis of his Reggeon field theory. Due to space limitations we will have to omit it from this book, but it is just as beautiful, simple and general as another part of the work he has initiated, which is presented in terms of the DGLAP equations in Chapter 19.

10.2   The decay properties of a cluster

We start by noting that the internal breakup distribution in Eq. (8.41) contains two factors:

$$dP_{int} = d\tau_n |\mathcal{M}|^2 \tag{10.1}$$

where the n-particle phase space volume, cf. Chapter 4, is given by

$$d\tau_n = \prod_{j=1}^{n} N d^2 p_{oj} \delta^+(p_{oj}^2 - m^2) \delta^2 \left(\sum_{j=1}^{n} p_{oj} - P_{rest} \right) \tag{10.2}$$

and the squared matrix element by the area-law suppression

$$|\mathcal{M}|^2 = \exp(-bA_{rest}) \tag{10.3}$$

The somewhat fancy notation is used in order to make a connection with Fermi's Golden Rule for quantum mechanical transitions (cf. Chapter 3): the probability for a particular transition is obtained by multiplying the square of the transition matrix element by the available number of states, i.e. the phase space volume of the final-state hadrons. (One should also multiply by a flux factor for the initial state but this is omitted here.)

The interpretation of the area suppression law as a squared matrix element will be provided in the next chapter.

It is evident, however, that the formulas contain two scales. One of these is the quantity b in the exponent. In the derivation of the formulas we noted that b must be the same for all the breakup vertices. Thus b must be flavor independent and so contain basic information about the color force field for which the string is used as a model.

We note, however, that the other parameter in the fragmentation function, a, has vanished from the expressions in Eq. (10.1) but that the normalisation constant N is still present. (If there are several values of a then the differences $a_{j-1} - a_j$ occur directly in the formulas, cf. Eq. (8.43).)

The normalisation constant N, which occurs together with each of the hadronic state factors, can be thought of as a scale factor for the hadronic states. In the simple picture used up to now, in which we have discussed a $(1 + 1)$-dimensional model, then N is, of course, dimensionless. In the actual $(3 + 1)$-dimensional world then N would need to have the same dimensions as b in order to obtain the correct dimensions of the cross sections, etc. It is by no means obvious that the scale determining the breakup properties of the string color field, i.e. b, is the same as the scale governing the density of the stable hadronic states.

The reason why the a-dependent factors have vanished from Eq. (10.1) is that this formula is an *exclusive expression* for the probability to produce just the particular set of n particles with energy-momenta $\{p_{oj}\}$, $j = 1,\dots,n$ and nothing else. Our earlier fragmentation formula for $f(z)$ (Eqs. (8.16), (8.17)) is an *inclusive expression*, i.e. it describes the probability of producing one particular meson independently of whatever will come before (or after); there is, however, an implicit assumption that in general there will be something more. We will now prove that it is this expectation that does in fact produce the a-dependent factors in the Lund model.

1 The case of a single a-parameter and a single hadron

We start by considering the probability for producing a first hadron with energy-momentum

$$(p_{o1+}, p_{o1-}) \equiv (u_1 W_+, m^2/u_1 W_+) \tag{10.4}$$

from the cluster with mass $\sqrt{s}$, independently of what comes after. We will use the same notation as in Chapter 8. Now we must pick out those properties of the expression that are u_1-dependent. Then we integrate and sum over everything else, keeping u_1 and s fixed.

Let us first note that if we sum over *everything*, including even the first-rank hadron, then the only thing that the expression can depend upon is the total squared mass s. (This is Lorentz invariance at work in a situation where the only Lorentz invariant is s.) Therefore we can define the function $g(s)$ as follows:

$$g(s) = \sum_n \int \prod_{j=1}^{n} N d^2 p_{oj} \delta^+(p_{oj}^2 - m^2) \delta \left(\sum_{j=1}^{n} p_{oj} - P_{rest} \right) \exp(-bA_{rest}) \tag{10.5}$$

$$s = P_{rest}^2 \equiv W_+ W_-$$

If we introduce the above parameter u_1 (by means of $\delta(p_{o1}^2 - m^2)d^2 p_{o1} = du_1/u_1$), then noting that the area A_{rest} in Fig. (10.1) can be subdivided in an obvious way to give $A_{rest} = \frac{m^2}{u_1} + A(2 - n)$ we obtain the result

$$g(s) = \int \frac{N du_1}{u_1} \exp\left(-\frac{bm^2}{u_1}\right) h \tag{10.6}$$

The quantity h is given by the same expression for the particles indexed 2 to n as that for the function $g(s)$ for all the particles. There is, however,

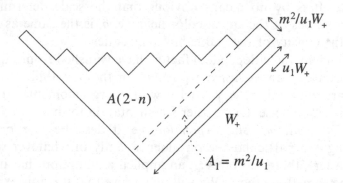

Fig. 10.1. The subdivision of the full cluster area into the area characteristic of the first-rank hadron, A_1, and the area of the remaining ones, $A(2-n)$.

a changed (u_1-dependent) value for the squared mass:

$$s_1 = (P_{rest} - p_{o1})^2 = (W_{+1} - u_1 W_{+1})\left(W_{-1} - \frac{m^2}{u_1 W_{+1}}\right)$$

$$= (1 - u_1)\left(s - \frac{m^2}{u_1}\right) \tag{10.7}$$

This is the squared mass for the hadrons 2 to n; we have taken away the first-rank particle, with a fixed value of u_1. Thus $h \equiv g(s_1)$. Combining these results we obtain an integral equation for g:

$$g(s) = \int \frac{N du_1}{u_1} \exp\left(-\frac{bm^2}{u_1}\right) g\left((1 - u_1)\left(s - \frac{m^2}{u_1}\right)\right) \tag{10.8}$$

To be precise, this integral equation is only valid if s is larger than the square of the single-particle mass. Further, the integration region does not extend all the way down to $u_1 = 0$ because the remainder mass must also be reasonably large.

Nevertheless we note that for large values of s there are solutions of a power character for g (here $\mathscr{C}$ is a constant):

$$g(s) \simeq \mathscr{C} s^a \text{ if } 1 = \int \frac{N du_1}{u_1}(1 - u_1)^a \exp\left(-\frac{bm^2}{u_1}\right) \tag{10.9}$$

Note that this is a requirement on a, i.e. there is a relation between the normalisation constant N, bm^2 and a. This requirement is, of course, nothing other than the original normalisation conditions for f (remember the discussion in Chapter 8).

We could in fact call this property (a form of) *unitarity*: there is a total probability equal to unity that something will happen in connection with the production process!

We find in this way that the fragmentation function for a finite-mass cluster is formally s-dependent, from the normalisation factor h/g:

$$f_s(u_1) = \frac{Ndu_1}{u_1} \exp\left(-\frac{bm^2}{u_1}\right) \frac{g((1-u_1)(s-\frac{m^2}{u_1}))}{g(s)}$$

$$\simeq f(u_1)\left(1 - \frac{m^2}{su_1}\right)^a \tag{10.10}$$

In practice the quantity m^2/u_1 is much smaller than s. Therefore we recover our starting expression in Eq. (8.16), i.e. the function f_s tends to f rapidly when $s \gg m^2$.

2 The case of several values of the a-parameter and several kinds of hadron

If there are several a-values, a_k, and several flavors and hadrons, we will for simplicity sum over all hadron and flavor indices in the formulas using the convention that the normalisation constants of the fragmentation functions are only nonzero when the hadron index and the flavor indices are compatible. We treat only the case of a single hadron for each flavor combination. The more general case can be inferred from that.

The major difference from the case of a single flavor and hadron is that the hadronic phase space volume is changed (the area-law suppression is the same but with the relevant mass values inserted):

$$d\tau_{0,\alpha} = \delta\left(1 - \sum_1^n u_j\right) \delta\left(s - \sum_1^n \frac{m_{j-1,j}^2}{u_j}\right)$$

$$\times \prod_1^n \frac{N_{f_{j-1},f_j} du_j}{u_j} (u_j)^{a_{f_{j-1}} - a_{f_j}} \tag{10.11}$$

with the convention that the first-rank hadron has the flavor $f_1 = 0$ and the last one the (anti)flavor corresponding to $f_n = \alpha$.

We will now also prescribe that the first-rank hadron should have the antiflavor corresponding to β and energy-momentum fraction u_1. We can use a division trick (this time with $1 - u_1$) in the δ-distributions to rearrange them as follows:

$$\delta\left(1 - \sum_1^n u_j\right) \delta\left(s - \sum_1^n \frac{m_j^2}{u_j}\right)$$

$$= \delta\left(1 - \sum_2^n \frac{u_j}{1-u_1}\right) \delta\left(s_1 - \sum_2^n \frac{m_j^2(1-u_1)}{u_j}\right) \tag{10.12}$$

We note firstly the occurrence of s_1 from Eq. (10.7) in the δ-distribution in the second line and secondly that in this way the u_j, $j = 2, \ldots, n$, all occur in the rescaled version $u_j/(1 - u_1) \equiv \zeta_j$. Introducing these rescaled ζ_j we obtain from the phase space factor

$$d\tau_{0,\alpha} = \frac{N_{0,\beta} du_1}{u_1} u_1^{a_0 - a_\alpha} \left(\frac{1 - u_1}{u_1} \right)^{a_\beta - a_\alpha} d\tau_{\beta,\alpha} \qquad (10.13)$$

The interpretation is that the phase space volume of a flavor-ranked string of hadrons, starting at flavor 0 and ending at antiflavor $\bar{\alpha}$, can be rearranged into a product with one factor for a single first-rank hadron with flavors $0, \bar{\beta}$ and the other for the phase space volume of the string $\beta, \bar{\alpha}$.

The area-law suppression factor can be rearranged just as in Eq. (10.6) and we conclude that there is a corresponding integral equation, as for the simpler case in Eq. (10.8):

$$g_{0,\alpha}(s) = \sum_{\beta \in \{f\}} \int du_1 f_{0,\beta}(u_1)(1 - u_1)^{-a_\alpha} g_{\beta,\alpha}(s_1) \qquad (10.14)$$

We have here introduced the fragmentation function for the first-rank hadron and also the quantities g_{f_j, f_k} for the total sum over all possible production contributions starting at flavor f_j and ending at (anti)flavor f_k. Note that the argument of $g_{\beta,\alpha}$ is the reduced squared mass s_1. We must sum over all possible antiflavors of the first-rank hadron. The interesting thing is that this equation has solutions g which, for large values of s, depend solely on the final flavor, a_α, i.e.

$$g_{0,\alpha} \simeq g_{\beta,\alpha} \simeq \mathscr{C}_\alpha s^{a_\alpha} \qquad (10.15)$$

as is easily seen using the result for s_1 in Eq. (10.7) (the factor $(1 - u_1)^{-a_\alpha}$ is compensated by the corresponding factor in $s_1^{a_\alpha}$). There is a requirement again corresponding to unitarity that

$$\sum_{\beta \in \{f\}} \int f_{0,\beta}(u_1) du_1 = 1 \qquad (10.16)$$

i.e. that the total probability is 1 that some flavor β is produced at the first vertex and thereby that there is always a first-rank hadron. The statement that the total sum over all possible productions should depend only upon the final flavor was deduced also in the discussion of the Γ-distribution of the final vertex, in the context of the external-part formulas in the last chapter. We note that Eq. (10.16) must be valid not only for the index 0 but for all the flavor indices in $\{f\}$.

If we analyse our results it is evident that we have repeatedly made use of the fact that *the Lund model fragmentation formulas have simple factorisation properties*. In the next section we will consider the unitarity

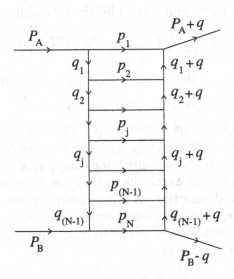

Fig. 10.2.   A Feynman diagram describing the scattering $(p_A, p_B) \to (p_A+q, p_B-q)$ with intermediate states and momentum transfers exhibited.

equations for the S-matrix in a quantum field theory. We will show that factorisation properties also in that case lead to very similar results for the correspondence to our state sum.

10.3   The relationship to the unitarity equations for the S-matrix in a quantum field theory

1   The AFS model

It has been known for a long time that it is possible to prove Regge asymptotic behaviour from the unitarity equations for the S-matrix in a quantum field theory. We will, in this and the following subsections, make use of a description similar to the one given in [28].

Amati, Fubini and Stanghellini (AFS), [1], formulated a set of integral equations based upon the so-called *ladder Feynman diagrams* or *multiperipheral ladder diagrams* (see Fig. 10.2). The starting point is that (the imaginary part of) the elastic scattering amplitude

$$T_{AB} \equiv T(p_A, p_B;\ p_A + q, p_B - q) \qquad (10.17)$$

describing the elastic scattering of the particles A, B with momentum transfer q must, owing to unitarity, fulfil the following equation:

$$\text{Im}(T_{AB}) = (1/2)\sum_N \int T_N(p_A, p_B; \{p_j\}) T_N^\star(p_A', p_B'; \{p_j\}) d\tau_N \quad (10.18)$$

Here the quantity $d\tau_N$ is the N-particle phase space volume and

$$p_A' = p_A + q \quad \text{such that} \quad p_A'^2 = p_A^2 \quad\quad (10.19)$$

(similarly for p_B'). The point is that all possible hadronic states (denoted $\{p_j\}$ in Eq. (10.18)), fulfilling energy-momentum conservation and capable of being produced from the incoming and outgoing states A, B, should be included in the sum and integral (the p_j being on the mass shell). We note the similarity to the description of a propagator in the Källén-Lehmann representation, where also all possible *intermediate states* occur.

The T (*transition*)-operator is related to the S-operator (which is defined in Chapter 3) in the usual way:

$$S = 1 + iT \quad\quad (10.20)$$

In the AFS model the amplitudes T_N are taken from the ladder diagrams, see Fig. 10.2:

$$T_N^{AFS} = \prod_{j=1}^N \lambda(q_{j-1}, q_j) D(q_j) \quad\quad (10.21)$$

Here $D(q_j)$ is the propagator for the momentum transfer $q_j = q_{j-1} - p_j$ ($q_0 = p_A$, $q_N = -p_B$, for $j = N$ the propagator is equal to 1) at the (Feynman) vertex $j - 1 \to j$ and $\lambda(q_{j-1}, q_j)$ is the corresponding vertex factor for producing the particle p_j.

We note for future reference that the intermediate N-particle state is in this way built up iteratively, with one particle being produced at a time along a ladder containing the propagators from vertex to vertex.

The expression on the right-hand side of Eq. (10.18), which from now on will be called *rhs*, will have a simple behaviour for a great many production models provided that:

U1 the amplitudes T_N fall off rapidly except when the energy-momentum transfers are small, i.e. $q_j^2 \lesssim m^2$ with m a typical mass size;

U2 there is no long-range order in the momentum transfers. Thus the amplitude T_N is independent of q_j and q_k if $|j - k| \gg 1$, i.e. the vertices j and k are far from each other in the production process;

U3 the amplitude is not large when the sub-energies $s_{j,j+1} = (p_j + p_{j+1})^2$ of neighboring pairs are large. This means in practice that there are no large rapidity gaps between the produced particles anywhere in the included chains.

These assumptions are at the basis of Gribov's Reggeon calculus, and all of them seem very natural. It should be noted, however, that in QCD the momentum transfers are in general larger than allowed in Gribov's basic assumptions so there is need for some caution in applying the rules to QCD.

In particular one can show from the assumptions U1–U3 that the equations will lead to Regge behaviour, i.e. that

$$\text{Im} \, T_{AB} \sim s^{\alpha} \tag{10.22}$$

The parameter(s) α will in general depend upon the quantum numbers of the particles A, B and also upon the squared momentum transfer $t = q^2$ in the process. The way to obtain the result in Eq. (10.22) is, in the AFS model, to make use of the following simple factorisation property of the amplitude in Eq. (10.21):

$$T_N^{AFS} = \lambda(p_A^2, q_1^2) D(q_1^2) T_{N-1}^{AFS} \tag{10.23}$$

Introducing this factorisation property one obtains immediately an integral equation for the right-hand side, $(rhs)^{AFS}$, of Eq. (10.18) ($q = 0$, $\mathcal{N}$ a numerical constant and $s_1 = (P_{tot} - p_1)^2$ as in Eq. (10.7)):

$$(rhs)^{AFS}(s) = \int \mathcal{N} dp_1 \delta(p_1^2 - m^2)$$
$$\times |\lambda(p_A^2, (p_A - p_1)^2)|^2 |D((p_A - p_1)^2|^2 (rhs)^{AFS}(s_1) \tag{10.24}$$

The similarity between this expression and the integral equation(s) for g and $g_{\alpha,\beta}$ in Eqs. (10.8) (10.15) are obvious. With a few manipulations one obtains the desired power behaviour in $s = P_{tot}^2$, the power-law parameter α_0 being determined from the eigenvalues of the equation, [1].

2   A detour into transverse dimensions

In this section we consider the extension of the results in Eq. (10.24) to nonzero values of the momentum transfer q. This is a preliminary to the extension of the Lund model formulas to a $(3 + 1)$-dimensional world. A second reason is that in this way there emerges a simple and intuitively appealing picture of the behaviour of multiparticle production models. We start with a brief discussion of the influence of the requirements U1–U3 on the results.

In order to make the integrals in the formulas for the function rhs convergent it is necessary to have a fast falloff in q_1^2, i.e. to make use of the requirement U1 above. It is possible to have more complex factorisation properties than in the simple AFS model, i.e. one may introduce short-range correlations between the vertices. The requirement U2 does, however, ensure the possibility of still writing integral equations (albeit, in this case, systems of integral equations).

The requirement U3 is needed for more subtle reasons and is only necessary when considering transverse dimensions also. (It is instructive to show that in a $(1 + 1)$-dimensional scenario the requirements U1 and U3 are equivalent and the reader is urged to do that.) The (general) requirement U3 is intended to solve a type of *transmission problem* by means of the *law of large numbers*.

Energy-momentum conservation at every (Feynman) vertex means, according to Fig. 10.2, that the individual momentum transfers at the jth cell of the ladder must fulfil

$$q'_j = q_j + q \qquad (10.25)$$

Therefore in order to transmit the momentum transfer q across the ladder it is necessary that neither q_j^2 nor $(q+q_j)^2$ should imply strong suppression of the vertex and propagator functions for the different steps. Therefore q^2 must be limited in size in accordance with the requirement U1 above. Actually, as we will see, the restrictions on q are essentially stronger because in a chain with n vertices there will be n requirements to accommodate.

Further, the momentum transfer q must be spacelike to keep the particles A and B on the mass shell. At high energies it is even necessary that q *should be almost transversely directed*, i.e. q should be almost orthogonal to the direction $\mathbf{p}_A + \mathbf{p}_B$ (from now on the *beam direction*). To prove this let us go to the cms of the two particles A and B; they have energy-momenta in a lightcone frame $p_A = (W, m^2/W, \mathbf{0}_t)$, $p_B = (m^2/W, W, \mathbf{0}_t)$. Then if q has large components along any of the lightcones, the mass-shell conditions for A and B, Eq. (10.19), cannot be fulfilled for large values of W and small values of q^2.

We will therefore use the approximation that $q \simeq \mathbf{q}_t$. It is useful to divide the hadronic phase space volume, $d\tau_N$, into transverse (t) and longitudinal (l) parts with respect to the beam direction:

$$d\tau_N = d\tau_{Nt} d\tau_{Nl} \quad \text{so that} \quad d\tau_{Nt} = \prod_{j=1}^{N} d^2 p_{tj} \delta \left(\sum_{j=1}^{N} \mathbf{p}_{tj} \right) \qquad (10.26)$$

We also observe from the ladder graph in Fig. 10.2 that the hadronic transverse momenta can be expressed as

$$\mathbf{p}_{tj} = \mathbf{q}_{t(j-1)} - \mathbf{q}_{tj}, \quad j = 1, \ldots, N \qquad (10.27)$$

This ensures that the transverse momentum conservation δ-distribution in Eq. (10.26) is fulfilled. We may evidently introduce as integration variables the transverse components of the momentum transfers q_j, $j = 1, \ldots, N-1$ instead of the corresponding components for the hadrons, p_j.

The next step is to introduce the (transverse) Fourier transforms of the transition amplitudes in Eq. (10.18):

$$T_N(p_A, p_B; p_j) = \int F_N(\mathbf{b}_j, p_{lj}) \prod_{j=1}^{N-1} \exp(i\mathbf{q}_{tj} \cdot \mathbf{b}_j) d^2 b_j \qquad (10.28)$$

(the index l is for longitudinal). There is a corresponding result for the amplitude $T_N^*(p_A', p_B'; p_j)$ in terms of (the complex conjugate of) the same function:

$$T_N^* = \int F_N^*(\mathbf{b}_j', p_{lj}) \prod_{j=1}^{N-1} \exp[-i(\mathbf{q}_{tj} + \mathbf{q}) \cdot \mathbf{b}_j'] d^2 b_j' \qquad (10.29)$$

An essential point is *the appearance of the q-dependence as a common factor in the above equation*:

$$\exp\left[-i\left(\sum_{j=1}^{N-1} \mathbf{b}_j'\right) \cdot \mathbf{q}\right] \qquad (10.30)$$

because if we now perform the integrals on the right-hand side of Eq. (10.18) we obtain a very pretty description:

$$(\text{rhs})(s, q^2) = \int \exp[-i\mathbf{b} \cdot \mathbf{q}] d^2 b \mathscr{F}(\mathbf{b}, s)$$

$$\mathscr{F} = \sum_N \int d\tau_{Nl} \prod_{j=1}^{N-1} (2\pi)^2 d^2 b_j \delta\left(\sum_{j=1}^{N-1} \mathbf{b}_j - \mathbf{b}\right) |F_N(\mathbf{b}_j, p_{lj})|^2 \qquad (10.31)$$

We have here repeatedly made use of the well-known Fourier distribution identity

$$\int d^2 q_t \exp\left[i(\mathbf{b} - \mathbf{b}') \cdot \mathbf{q}_t\right] = (2\pi)^2 \delta(\mathbf{b} - \mathbf{b}') \qquad (10.32)$$

to make the identification $\mathbf{b}_j = \mathbf{b}_j'$.

The whole mathematical game has been to introduce instead of the transverse momentum transfers $\mathbf{q}_{tj}$ their canonically conjugate correspondences, the *impact-space vectors* $\mathbf{b}_j$ of the different *links* in the ladder graphs. We find that *the full q-dependence* (in the transverse approximation) *is described by the Fourier transform with respect to the sum of all the individual impact-space vectors for the different links.*

The question then arises of the distribution of the sum of these individual impact-space vectors. Now we can again make use of the two requirements U1 and U2 above. Equation (10.28) was used to define the distributions F_N and we can of course invert it by means of the Fourier transform relations into

$$F_N(\mathbf{b}_j) = \int T_N(p_A, p_B; \mathbf{q}_{jt}, p_l) \prod_{j=1}^{N-1} \frac{d^2 q_{jt}}{(2\pi)^2} \exp[-i(\mathbf{q}_{jt} \cdot \mathbf{b}_j)] \quad (10.33)$$

Then using the properties of the Fourier transforms we can deduce that the two requirements U1 and U2 will have the following implications for the impact-space vectors $\mathbf{b}_j$.

$U^i 1$. The function(s) F_N are smooth and well-behaved distributions in $\mathbf{b}_j$ with, typically $\mathbf{b}_j^2 \simeq 1/m^2$, i.e. the inverse of the q_j^2.

$U^i 2$. The correlations $\mathbf{b}_j \, \mathbf{b}_k = 0$ if $|j - k| \gg 1$.

Now, finally, comes the basic use of the requirement U3: it means that the major contributions to the integrals come from situations very similar to the one in the Lund model, i.e. *there will be no major contributions to the integrals from small multiplicities*. If this were not the case then there would be many large rapidity gaps. In the Lund model picture there would then have to be many small values of the proper times of the vertices, i.e. we would be far from the usual hyperbola breakup.

(In fact, owing to U1, nor will there be major contributions from very large multiplicities, for which, in accordance with our findings for the Lund model, it is necessary to have large values of the proper time or equivalently of the momentum transfers Γ.) This *large-multiplicity requirement* is necessary because we are now going to make use of the law of large numbers, which (apart from some mathematical epsilontics) reads as follows.

• Consider the distribution of a quantity Σ, which is the sum of a (sufficiently) large number of *independent* stochastic variables:

$$\Sigma = \sum_{j=1}^{n} \rho_j \quad (10.34)$$

Each of the ρ_j is distributed in one way or another with a mean $\langle \rho_j \rangle$ and a variance σ_j^2. Then Σ is distributed according to a gaussian:

$$\frac{dP}{d\Sigma} = \frac{1}{\sqrt{2\pi n \sigma_0^2}} \exp\left[-\frac{(\Sigma - n\Sigma_0)^2}{2n\sigma_0^2}\right] \quad (10.35)$$

The centre and the width of the Σ-distribution are given by

$$n\Sigma_0 = \sum_{j=1}^{n} \langle \rho_j \rangle, \quad n\sigma_0^2 = \sum_{j=1}^{n} \sigma_j^2 \quad (10.36)$$

In our case, the impact-space vectors evidently will all, due to symmetry, have a vanishing mean value. Depending upon the length of the correlations, according to $U'2$ they can be subdivided into groups which are independent (we assume each to have a correlation length $\langle \mathbf{r}^2 \rangle$). If the number of such groups is $n(s) \gg 1$ for a given value of the squared cms energy s we may apply the law of large numbers.

In this way we obtain a very general result for the distribution $\mathscr{F}$ in Eq. (10.31), i.e. a gaussian distribution in the total impact-space vector $\mathbf{b}$.

The method we have used can in somewhat vague language be described as follows.

- We go over to impact-parameter space and consider the building up of the total impact parameter $\mathbf{b}$ as a Brownian motion along the chain with each impact parameter (or for short-range correlations each group of) impact-parameter vector(s) $\mathbf{b}_j$ contributing in a random way.

This is a very general feature of transmissions through many steps: they tend to randomise rather quickly and then only the general mean and variance are noticeable (together with the number of steps). Note that if there are major contributions from the fluctuations down to a small number of steps these statements may not be true.

3 Coming back to the Regge phenomenology

After having obtained a very general distribution for $\mathscr{F}$ we will go back to the unitarity equations, as well as to the representation in Eq. (10.31), to obtain

$$(rhs)(s, q^2) = (rhs)(s, q^2 = 0) \exp \left[\frac{-n(s) \langle \mathbf{r}^2 \rangle \, \mathbf{q}_t^2}{2} \right] \qquad (10.37)$$

i.e. we perform a straightforward gaussian integral.

For the case when the correlations, according to U2, are reasonably short-range one expects $n(s) = c \langle n \rangle$, i.e. this number should be proportional to the mean multiplicity at that energy.

In any of the models which fulfils not only U1 and U2 but also U3, this mean multiplicity will grow logarithmically in s. We therefore expect that the Regge parameters α in Eq. (10.22) are generally *linear functions of t* (note that $\exp(-C \ln s) = s^{-C}$). Conventionally these are written as

$$\alpha(t) = \alpha_0 + \alpha' t \qquad (10.38)$$

with $t = q^2 \simeq -\mathbf{q}_t^2$. Here α_0 is called the *intercept* and the parameter α' *the Regge slope*). This latter parameter should according to our formula be proportional to $\langle \mathbf{r}^2 \rangle /2$ and therefore depend both upon the value of the (average) transverse momentum and upon the size of the short-range correlations.

It is known phenomenologically that for most *Regge trajectories*, as the $\alpha(t)$-line in Eq. (10.38) is called, this parameter $\alpha' \sim 1$ $(\text{GeV}/c)^{-2}$. However, there is one trajectory, equipped with the vacuum quantum numbers, called the *Pomeron*. The Pomeron has an essentially smaller slope, $\alpha'_P \simeq 0.25$ $(\text{GeV}/c)^{-2}$.

In this way both the elastic and the total cross sections are dominated by power behaviour in s when they are expressed in terms of the matrix elements of T as

$$\frac{d\sigma_{el}}{dt} = \frac{|T_{AB}|^2}{16\pi s^2} \propto s^{2\alpha(t)-2}$$

$$\sigma_{tot} = \frac{\operatorname{Im} T_{AB}}{s} \propto s^{\alpha(t)-1} \tag{10.39}$$

The second line of Eq. (10.39) is of course nothing other than a statement that the sum over all states, like in Eq. (10.18), includes everything that can happen; the factor $1/s$ is the flux factor of the incoming state. For the first line of Eq. (10.39) it is necessary also to know something about the real part of T_{AB}, but fortunately the latter turns out to give only a small correction.

Similar power results for the high-energy behaviour are obtained from potential scattering models by rewriting the scattering amplitude using the Sommerfeld-Watson transform, which was originally developed for light-scattering, [46]. These features are, however, outside the scope of this book. One very important result in this connection is that the Regge trajectories $\alpha(t)$ also contain information on the bound-state spectrum of the potential. It is possible to show that $\alpha(t)$ is an analytical function of t in the potential scattering models. In particular, when the squared momentum transfer t (which in the case discussed up to now must be negative) is continued to positive values then

$$\alpha(t = m^2) = j \tag{10.40}$$

with j the angular momentum of the bound state with mass m.

Regge behaviour has also been proposed and investigated in more complex processes than elastic scattering, e.g. in charge exchange processes such as $\pi^- p \to \pi^0 n$ (p, n stand for proton and neutron, respectively). In this case the use of the *ρ-trajectory* has provided a very good description of all available experimental data. From the experimental analysis, this trajectory actually does exhibit the straight-line behaviour expected in Eq. (10.38), and with the constraint from Eq. (10.40) (that the ρ-meson spin is 1) the value of $\alpha_{0\rho}$ has been decided as being close to 0.5.

The ρ-trajectory is, in accordance with the flavor composition of ρ as a mixture of $u\bar{u}$ and $d\bar{d}$, related to the most common flavors. It should be more than a coincidence that the phenomenological values for the Lund model parameter a, which are obtained from studies that also include gluonic radiation, tend to demand a value just above 0.5.

11

The dynamical analogues of the Lund model fragmentation formulas

11.1 Introduction

The Lund model fragmentation formulas are based upon general principles such as causality, Lorentz covariance and confinement. Only classical probability concepts and semi-classical dynamical considerations have been used in the derivations in the earlier chapters. Nevertheless, the resulting formula for the decay of a (color- and flavor-connected) cluster has an appealing simplicity, similar to those obtained in other dynamical situations. It is the product of the n-particle phase space factor and an *area suppression factor* written as the square of a 'matrix element' $\mathcal{M}$:

$$dP_{int} = d\tau_n |\mathcal{M}|^2, \quad |\mathcal{M}|^2 \equiv \exp(-bA) \tag{11.1}$$

In this way we connect with formulas for multiparticle production cross sections in a quantum field theory and in Chapter 10 we have exhibited a wide class of models, the multiperipheral ladder models, with properties similar to the Lund model results. In this chapter we consider further dynamical analogies to the Lund model. We show that the *area suppression law* in Eq. (11.1) can be interpreted in at least two different ways stemming from field theory:

- in terms of a quantum mechanical tunnelling process (the decay of the vacuum in an external field); we will call this the *Schwinger way*,

- in terms of basic gauge-independent quantities, the Wilson phase operators, in a gauge field theory; in the same language this interpretation will be called the *Wilson way*.

Both these interpretations will lead to a discussion of the meaning of the parameter b in the Lund fragmentation model and of the behaviour of the color force field in the QCD vacuum. We will also show that the formulas can be interpreted in a statistical mechanics scenario as follows.

- The internal-part formula corresponds to the partition function for a gas with two-body interaction potentials. The coordinate space for the gas is then the longitudinal rapidity space of the cluster.

In this way another familiar phenomenological tool, the *Feynman-Wilson gas in rapidity space*, will occur in connection with the Lund model fragmentation formulas. Approximating the gas partition function according to the first nontrivial part of the virial expansion, we derive a relationship between the normalisation constant N, the parameters b and a and the particle density in rapidity space. The result can also be considered as an ideal gas law for the rapidity-space gas.

11.2 The decay of the vacuum in an external field

We now meet another example of the *law of the conservation of useful dynamics*. Again this is a case for which it is possible to obtain a closed expression in terms of elementary functions for a dynamically interesting situation. This time it is the reaction of the vacuum, defined as the state containing no quanta, to the onset of an external electromagnetic field. Quantum matter fields, such as an electron-positron ($e^- e^+$) field, which are coupled to the electromagnetic field, will then start to fluctuate.

In Chapter 4 we have shown that that the vacuum in a quantum field theory is, due to quantum fluctuations, very similar to a dielectric medium. If there is an external field stretching over macroscopic regions then these polarisation charges will be driven by the field. Therefore the original no-particle vacuum state, existing before the onset of the field, will break down into a new state.

The problem of the reaction of the vacuum was considered by Heisenberg and Euler in the 1930s, [79], by Schwinger in the 1950s, [100], and by many authors in the 1970s and the 1980s. In this section we will formulate the problem as a *tunnelling process*. We will be satisfied with using semi-classical considerations (similar to the ones presented in [64]) to derive a formula for the *vacuum persistence*, i.e. the probability that the vacuum does not decay.

1   *The tunnelling formulas*

In Chapter 6 we found that a classical particle with mass m which experiences a constant force field (force constant κ) will move in from far away towards the origin, 'bounce' at the classical turning point $|x_c| = m/\kappa$ and then go back outwards again.

The origin $x = 0$ is then defined by the requirement that the total energy of the particle, i.e. the sum of the kinetic and potential energies,

$$E = \sqrt{|\mathbf{p}|^2 + m^2} - \kappa x \tag{11.2}$$

vanishes, i.e. we choose the value $E = 0$ (note that you may place the origin wherever you like by suitable choice of E).

For a quantum mechanical particle, however, the wave function ψ will be oscillating for $x > x_c$ but it will not vanish at the classical turning point x_c. There will be an exponential tail for smaller values of x and this tail can be approximately calculated by means of the well-known WKB approximation (for details on the WKB approximation, cf. Merzbacher). A WKB solution to the wave function inside the classically forbidden region is, with p_ℓ (index ℓ for the momentum component along the x-direction) chosen to fulfil Eq. (11.2),

$$\psi(x) = \psi_c \exp\left[-i \int_{x_c}^{x} p_\ell(x')dx'\right], \quad x_c \geq x \geq 0 \tag{11.3}$$

$$p_\ell(x) = i\sqrt{E_\perp^2 - (\kappa x)^2}, \quad E_\perp = \sqrt{|\mathbf{p}_\perp|^2 + m^2} \tag{11.4}$$

The quantity ψ_c is the value of the wave function for $x = x_c$. We will come back to its significance later. The particle is assumed to move with a transverse momentum $\mathbf{p}_\perp$ in the force field along the x-axis (transverse meaning orthogonal to the x-axis) so that the classical turning point is given by $x_c = E_\perp/\kappa$. Note that both classically and quantum mechanically the transverse momentum $\mathbf{p}_\perp$ will be a conserved quantity.

We obtain for the integral in the exponent for the value $x = 0$:

$$R(p_\perp) = \frac{\psi(0)}{\psi_c} = \exp\left[-\left(\frac{\pi E_\perp^2}{4\kappa}\right)\right] \tag{11.5}$$

We may then consider the following dynamical problem.

- Consider an incoming particle and antiparticle, a $q\bar{q}$-pair, each connected to the constant force field (the forces are, however, oppositely directed as the q and $\bar{q}$ have opposite charges). Let us assume that they move in along the $\mp$ directions, have opposite transverse momenta $\pm\mathbf{p}_\perp$ and vanishing (total) energies. What is the probability $P(p_\perp)$ that their wave functions will overlap, as required for them to annihilate each other?

We note that such an annihilation process is allowed (all quantum numbers of the pair are conserved) and that a reasonable answer is given by the square of the overlap of their wave functions at the origin:

$$P(p_\perp) = |R^2|^2 = \exp\left[-\left(\frac{\pi E_\perp^2}{\kappa}\right)\right] \tag{11.6}$$

This turns out to be the right answer, and the question is investigated in more detail in e.g. [17]. If the wave function for q is $\psi(x)$, then that for $\bar{q}$ is $\psi(-x)$, and the factor ψ_c can at this point be thought of as a flux factor, i.e. the density of incoming particles.

This annihilation probability is equal to the production probability in a quantum field theory. Thus we have deduced the probability that a $q\bar{q}$-pair with opposite transverse momenta will tunnel out in the constant force field κ. We note in passing that the result in Eq. (11.6) is intrinsically of a nonperturbative origin, i.e. we cannot expand the result as a power series in the force field constant κ for small values of κ.

The result in Eq. (11.6) can be compared to the results of Heisenberg's indeterminacy relation. Then one would ask, what is the probability of obtaining a vacuum fluctuation such that a $q\bar{q}$-pair occurs at a separation $\Delta = 2x_c$ with transverse masses $E_\perp$?

The answer is given by the square of the free coordinate-space propagator $\Delta_F(x, m)$ evaluated for a spacelike value of $x = \Delta$ and $m = E_\perp$:

$$[\Delta_F(\Delta, E_\perp)]^2 \sim [K_1(E_\perp\Delta)]^2 \simeq \exp[-2(E_\perp\Delta)] = \exp\left(-\frac{4E_\perp^2}{\kappa}\right) \quad (11.7)$$

The function K_1 is the modified Bessel function of rank 1 (which is equal to Δ_F for a spacelike argument, according to section 6.3, Eq. (6.39)) and here we use a simple approximation for it.

The results in Eqs. (11.6) and (11.7) are essentially the same except that the factor 4 in the exponential for the free (i.e. the no-field) case replaces the factor π in the previous case, where there is a field which pushes and therefore makes it easier for the pair to tunnel out.

2 The vacuum persistence probability

We will now consider the persistence probability of the vacuum as defined in [40]. It is the probability that no tunnelling will occur for any spin (s), flavor (f) and transverse momentum ($\mathbf{p}_\perp$) at any place, i.e. for any value of $0 < x < L_x$ and any time $0 < t < T$, L_x and T being the extent of the field in longitudinal space and time. This quantity, which we will denote by $\mathscr{P}$, is evidently given by

$$\mathscr{P} = \prod_{s,f,\mathbf{p}_\perp,x,t} (1 - P) = \exp\left[\sum_{s,f,\mathbf{p}_\perp,x,t} \ln(1 - P)\right] \quad (11.8)$$

We will start by considering the sums over the longitudinal (L_x) and time (T) extents of the field. We have repeatedly observed that it takes a spatial region of the size $\Delta = 2E_\perp/\kappa$ to produce a pair. The lifetime δt of such a pair is evidently $\delta t = 2\pi/(2E_\perp)$ according to the indeterminacy principle. We conclude that each pair will need a space-time region of size $\Delta\delta t = 2\pi/\kappa$ for its production to be possible.

As the probability P is independent of x and t we conclude that the summation over x and t in the exponent of Eq. (11.8) will give the factor

$$\frac{L_x T}{\Delta\delta t} = \frac{\kappa L_x T}{2\pi} \tag{11.9}$$

In this way the possible pairs are as 'closely packed' as possible.

Next we consider the transverse extent $A_\perp$. In accordance with the discussion in Chapter 3, the number of plane wave solutions that can be fitted into such a transverse box is

$$\frac{A_\perp d^2 p_\perp}{(2\pi)^2} \tag{11.10}$$

We conclude that the sum in the exponent of Eq. (11.8) can be written as

$$\sum_{s,f,\mathbf{p}\perp,x,t} \ln(1-P) = \frac{\kappa L_x T A_\perp}{(2\pi)^3} \sum_{s,f} \int d^2 p_\perp \ln(1-P) \tag{11.11}$$

The integral over the transverse momenta is easily performed in terms of gaussian integrals if we expand the logarithm:

$$\ln(1-P) = -\sum_{n=1}^{\infty} \frac{1}{n} \exp\left[\frac{-n\pi(m^2 + \mathbf{p}_\perp^2)}{\kappa}\right], \quad \mathscr{P} = \exp(-\kappa^2 L_x T A_\perp \Pi),$$

$$\Pi = \frac{1}{4\pi^3} \sum_{s,f} \sum_{n=1}^{\infty} \frac{1}{n^2} \exp\left(\frac{-n\pi m^2}{\kappa}\right) \tag{11.12}$$

Comparing to Eq. (11.1) we find the Lund model area suppression law:

$$\mathscr{P} = \exp(-bA), \quad \kappa^2 L_x T = A/2. \tag{11.13}$$

This is the natural interpretation since the region A in the Lund formula is spanned in the longitudinal and time directions and is just the region over which the MRS persists, i.e. does not decay. Thus we identify $|\mathscr{M}|^2$ in Eq. (11.1) with $\mathscr{P}$ (but this obviously does not provide the phase of the matrix element $\mathscr{M}$; cf. section 11.3). The factor $1/2$ in Eq. (11.13) is due to our use of a lightcone metric $dA = \kappa^2 dx_+ dx_-$, which is double the usual metric $\kappa^2 dx dt$ (cf. Chapter 7).

At the same time *we have obtained a formula for the parameter b in terms of the transverse size of the force field, $A_\perp$:*

$$b = A_\perp \frac{\Pi}{2} \tag{11.14}$$

The quantity Π in Eq. (11.12) is, for a number n_f of massless spin $1/2$

particles (we will neglect the massive flavor contributions),

$$\frac{\Pi}{2} = \frac{n_f}{12\pi} \tag{11.15}$$

Although we have used semi-classical arguments for the evaluation of the persistence probability, the final result coincides with the one Schwinger wrote down for the production of e^+e^--pairs in a constant electric field. He obtained the same formula as Eq. (11.12) with the quantity κ replaced by $e\mathscr{E}$, i.e. the electronic charge times the electric field strength, which evidently is the force acting upon the electrons or positrons in constant external field. We are instead applying the formula to the color fields in QCD and to the production of $q\bar{q}$-pairs along the constant string force field when we compare to the Lund model area law.

If we use $n_f = 2$ (i.e. consider the u- and d-flavors to be massless and neglect the rest) we obtain the following value for the transverse radius $R_\perp$ ($A_\perp = \pi R_\perp^2$) of the force field:

$$R_\perp = \sqrt{6b} \simeq 0.55 \quad \text{fm} \tag{11.16}$$

using the phenomenological value $b \simeq 0.75$ GeV^{-2} which we have discussed before.

3 The relation between the parameter b and the fields and charges

We will next relate the transverse area $A_\perp$ to the charges of the $q\bar{q}$-pairs, which generate the fields. Although we will repeatedly make use of the analogous situation in the (abelian) QED field theory, we actually have in mind the more complicated color fields in the (nonabelian) QCD theory. We will therefore consider two different situations, one which we call the abelian setting and one that should be characteristic of a confining QCD vacuum.

In the abelian setting (where all fields and charges can be added in any order) the relation between the charge of a q-particle, which we will call g, and the electric field, $\mathscr{E}_1$, stemming from it is from Gauss's law (see Fig. 11.1)

$$2\mathscr{E}_1 A_\perp \equiv \mathscr{E} A_\perp = g \tag{11.17}$$

We note that the total electric field is $\mathscr{E} = 2\mathscr{E}_1$ as it obtains contributions which add up in between the charges but subtract to zero elsewhere. This corresponds to confinement in this case, i.e. there is a field only in between the charges. Thus for the abelian scenario all the fields arise from the charges connected to the string force field and there is no influence from the vacuum.

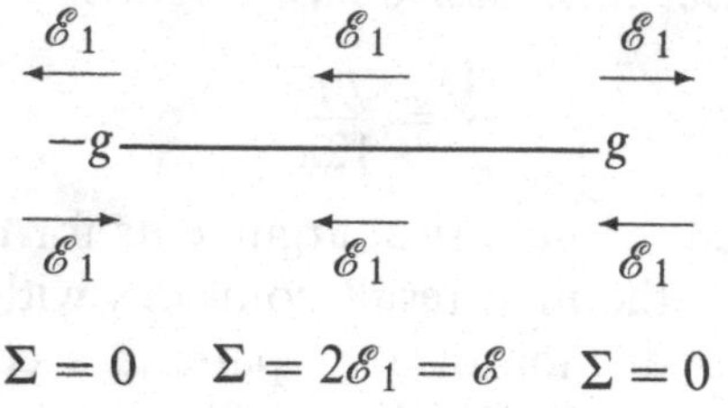

$$\mathscr{E}_1 \qquad \mathscr{E}_1 \qquad \mathscr{E}_1$$

$$-g \text{———————————} g$$

$$\mathscr{E}_1 \qquad \mathscr{E}_1 \qquad \mathscr{E}_1$$

$$\Sigma = 0 \quad \Sigma = 2\mathscr{E}_1 = \mathscr{E} \quad \Sigma = 0$$

Fig. 11.1.   The color electric field $\mathscr{E} = 2\mathscr{E}_1$ from a connected charged pair $\pm g$. There is no field outside the connected region between the charges (a simple-minded approach to confinement). The field strengths from the charges are shown above and below the string, and the contributions are summed in the last line.

The force κ_a (index a for abelian) between the $\bar{q}$ and the q is therefore

$$\kappa_a = g\mathscr{E}_1 = \frac{g\mathscr{E}}{2} = \frac{A_\perp \mathscr{E}^2}{2} \tag{11.18}$$

in accordance with the ordinary formulas for the energy density of an electric field. Note the difference from the Schwinger result where $\kappa \to e\mathscr{E}$ for a truly external field $\mathscr{E}$ (e the electric charge e being the electron). In our case we identify the external field with that spanned by the original $q\bar{q}$-pair at the endpoints of the MRS.

For consistency we note that if a $q\bar{q}$-pair is produced along the field then the same force and the same field relations are valid. In between the produced pair the two fields just compensate each other (to secure confinement) while the new endpoint fields take over in between the old and the new endpoints.

From these relations we obtain for the parameter $b = b_a$, introducing the ordinary coupling constant $\alpha = g^2/(4\pi)$,

$$b_a \kappa_a = \frac{\Pi A_\perp}{2} \kappa_a = \frac{\Pi g^2}{4} = \frac{n_f \alpha}{6} \tag{11.19}$$

There is (at least) one point in this discussion which is disturbing. *There is no reason why such an abelian field should be kept inside a thin transverse region.* It is well known that the electromagnetic fields in the abelian QED field theory do not behave like that. We will therefore consider a different scenario in which confinement is actually enforced by the properties of the QCD vacuum.

If we consider the color dynamical fields in QCD it is not obvious how to treat Gauss's law in Eq. (11.17). The electric field in this case is a color-8 operator while the charge is a color-3 (for the q) or a color-$\bar{3}$ (for the $\bar{q}$). The energy density along the string due to the color electric field should, however, have a color scalar, i.e. a color singlet meaning (although different 'ideologies' with respect to color dynamics may provide different numerical values).

We will now assume that for the QCD force fields the vacuum exerts a pressure on the fields and the charges (according to the bag model for hadrons mentioned in Chapter 6). There are, in the vacuum, gluonic field configurations, which compensate the fields from the charges outside the string region. In Chapter 6 we used the analogous picture of a color-superconducting QCD vacuum with a Meissner effect produced by these gluonic 'Cooper pairs'. When the field is built up this vacuum pressure must be overcome and thus the total work done in creating the field configuration in the vacuum is twice as large as for the abelian situation.

In this case the force on a charge is $\kappa \to \mathscr{E}^2 A_\perp/2 + BA_\perp = \mathscr{E}^2 A_\perp \equiv \kappa_b$, with $B = \mathscr{E}^2$ the bag pressure. The corresponding value for the parameter $b \equiv b_b$ is then, in terms of the squared field flux Ξ divided by the squared q- or $\bar{q}$-charge, which should have a meaning also in color dynamics,

$$b_b \kappa_b = \frac{n_f \alpha}{6} \Xi, \quad \Xi = \frac{(\mathscr{E} A_\perp)^2}{g^2} \tag{11.20}$$

One can argue in different ways at this point. One way, which is certainly not unreasonable, is to say that $\Xi\alpha \equiv \bar{\alpha}$, i.e. the effective coupling for gluonic emission along the field (cf. Chapters 16 and 17). Then the typical value, using $\bar{\alpha} = n_c \alpha$ (with the QCD value $n_c = 3$), for the strong coupling would be $\alpha \simeq 0.3$ (once again using $n_f = 2$, $b = 0.75$ GeV^{-2} and $\kappa \simeq 1$ GeV/fm). There are other ways to interpret Ξ but all of them will, within a factor of 2, provide a similar 'reasonable' result for the strong coupling. We will come back to these formulas later on, both when we consider another field theoretical analogy to the Lund model fragmentation distributions (in the next section) and when we have learned more about the way gluonic radiation 'resolves' the color force field, i.e. how we can treat the massless relativistic string as a model for the color force field in accordance with the Lund interpretation (cf. Chapter 17).

11.3 The Wilson loop exponential laws and gauge invariance

In chapter 2 we considered the invariance of the electromagnetic fields under gauge transformations. We will discuss the implications of gauge invariance for the matter fields in the first subsection below. We will then show how gauge invariance should constrain the production of $q\bar{q}$-pairs along the color force fields. We will find that the Lund model area suppression law is a natural consequence of these constraints.

These considerations will provide us with a possible phase for the matrix element $\mathcal{M}$ in Eq. (11.1), which we will later, in chapter 14, show to have significance in connection with the so-called Hanbury-Brown-Twiss (or Bose-Einstein) effect in multiparticle production.

1 The implications of gauge invariance for the matter fields

We have in chapter 2 considered the Maxwell equations and remarked that if we introduce the (four)vector potential A there is still a gauge degree of freedom. This means that we obtain the same electromagnetic fields if A_μ is changed as follows (with Λ an arbitrary function in space and time):

$$A_\mu(x) \rightarrow A_\mu(x) + \frac{\partial}{\partial x_\mu}\Lambda(x) = A_\mu(x) + \delta A_\mu(x) \qquad (11.21)$$

We will now show that *the gauge degrees of freedom also have significance for the matter fields which couple to the electromagnetic fields.*

We firstly note that the motion of a nonrelativistic particle with mass m and coordinate $\mathbf{x}$ will under the influence of a scalar (non-electromagnetic) potential V be described by the hamiltonian equations

$$\dot{p}_j = -\frac{\partial H}{\partial x_j}, \quad \dot{x}_j = \frac{\partial H}{\partial p_j} \qquad (11.22)$$

The dotted variables mean derivatives with respect to time. If hamiltonian h is independent of one of the coordinates x_j (i.e. the derivatives of h with respect to that coordinate vanish) then the first line of Eq. (11.22) will provide us with a constant of motion, p_j, which in general is equal to $m\dot{x}_j$. In other words, *the (mechanical) momentum is conserved* if there is no force along that direction.

Next we consider a charged particle (charge g) moving under the influence of a constant magnetic field $\mathcal{B}$ along the 3-axis : $\mathcal{B} = \mathcal{B}\mathbf{e}_3$. Possible vector potentials $\mathbf{A}$ to describe this field are given by e.g. the following two:

$$A_1^{(1)} = -\mathcal{B}x_2 \quad \text{or} \quad A_2^{(2)} = \mathcal{B}x_1 \qquad (11.23)$$

with the components not exhibited in the two cases vanishing. The equation of motion for the particle is

$$\frac{dm\dot{\mathbf{x}}}{dt} = g\dot{\mathbf{x}} \times \mathcal{B} \qquad (11.24)$$

from which we immediately obtain that

$$m\dot{x}_1 - g\mathcal{B}x_2 = c_1, \quad m\dot{x}_2 + g\mathcal{B}x_1 = c_2 \qquad (11.25)$$

with c_1, c_2 constants of motion. Consequently, in this case the mechanical linear momentum $m\dot{\mathbf{x}}$ is not conserved *but there is a combination of it and*

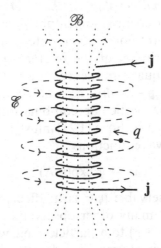

Fig. 11.2. A pointlike charge q is affected by the electric field $\mathscr{E}$ arising when a sudden current $\mathbf{j}$ causes a magnetic field $\mathscr{B}$ through the solenoid.

a field quantity with this property. A little reflection using Eqs. (11.23) will lead us to guess the right answer:

- for a particle moving under the influence of an electromagnetic field described by a four-vector potential A_μ, such that A_μ can be chosen to be independent of a coordinate x_j, the combination $m\dot{x}_j + gA_j \equiv p_j$ is a constant of motion.

As another example, which stems from the Feynman Lectures, consider a solenoid, Fig. 11.2. Suppose that we suddenly turn on a current through the wires and that there is a charged particle nearby. There will then be a sudden magnetic flux through the solenoid and correspondingly a *sudden buildup of the circumferential vector potential* **A**. Note that the relation between **A** and the magnetic flux means that the line integral around the solenoid of **A** is equal to the magnetic flux. There is also a sudden electric field

$$\mathscr{E} = -\frac{\partial \mathbf{A}}{\partial t} \tag{11.26}$$

which provides a force on the particle. This force is equal to the charge times the field so that there is an impulse during the buildup of the magnetic field in the solenoid corresponding to $-g\mathbf{A}$. *The difference between the mechanical momentum $m\dot{x}$ and that added by the impulse does not change,* i.e. once again we find that $\mathbf{p} = m\dot{x} + g\mathbf{A}$ is a conserved quantity.

This *principle of minimal coupling of the electromagnetic field*, that everywhere the mechanical energy-momentum p_μ is replaced by $p_\mu - gA_\mu$, i.e. a combination with the four-vector potential A_μ, is of fundamental importance for the description of the interaction between the field and the charged particles. In quantum theory, where the canonical momentum and energy are operators $\mathbf{p} \to -i\hbar\nabla$, $E \to i\hbar\partial/\partial t$ acting on wave functions ψ, the occurrence of the particular combination $p_\mu - gA_\mu$ means that the gauge transformations in Eq. (11.21) must be implemented as phase transformations on the wave functions:

$$\psi \to \psi \exp(ig\Lambda) \tag{11.27}$$

This means that the observable $|\psi|^2$ is unaffected by gauge transformations. The same goes for many other observables. As an antiparticle has the opposite charge ($g \to -g$) to a particle their wave function overlap at a single point (as in a current or charge density) will also be unaffected.

If we want to consider the overlap of the wave functions for a particle-antiparticle pair at different points, however, then the phase plays a role:

$$\rho(x_\mu, x'_\mu) = \psi(x_\mu)\overline{\psi}(x'_\mu) \to \rho \exp\{ig[\Lambda(x_\mu) - \Lambda(x'_\mu)]\} \tag{11.28}$$

This phase can evidently be expressed as a line integral of the vector δA_μ in Eq. (11.21) between the two points x'_μ, x_μ:

$$\int_{x'_\mu}^{x_\mu} ig\delta A_\mu dx^\mu \tag{11.29}$$

A general prescription by Schwinger for handling these situations is to endow the wave function overlap with a phase as follows:

$$\rho(x_\mu, x'_\mu) \to \rho \exp\left(i\int_{x'}^{x} gA_\mu dx^\mu\right) \tag{11.30}$$

We note that as long as there are no singularities in the four-vector potential A_μ then the line integral can be evaluated along any curve connecting the particle and antiparticle positions.

It is at this point worthwhile to note that the appearance of this phase difference, depending upon the vector potential A_μ, was predicted to be an observable result, e.g. for interference effects in charged particle motion, by Aharanov and Bohm, [3] in 1956. It also turns out to be an observable effect, [23] when there is a *singular potential*, i.e. when the wave function difference is connected over a region inside which there is a nonvanishing magnetic field flux.

We have up to now considered *abelian* gauge transformations, i.e. those transformations in which it is possible to add and subtract charges and fields in any order. For *nonabelian* gauge transformations it is necessary to generalise our notions to take into proper account the order in which the different quantities are added and multiplied just as for the exponent defining the *S*-operator in perturbation theory; cf. Chapter 3.

For QCD the quantity gA_μ is a matrix in color space and the line integral contains the multiplication of all the matrices at every point. It is said to be *path-ordered*. With some care it is possible to carry through essentially the whole discussion above for the nonabelian case also.

2 The application of gauge invariance to the string decay process in the Lund model

The production of hadrons in the Lund model occurs by means of a *q*-particle from one vertex (V_1) and a $\bar{q}$ from an adjacent vertex (V_2). Therefore the production matrix element should contain at least the factor

$$\gamma(V_1, V_2) = \exp\left(ig \int_{V_2}^{V_1} A_\mu dx^\mu\right) \tag{11.31}$$

in order to maintain gauge invariance. The next production will similarly involve $\gamma(V_2, V_3)$ and so on. Therefore a minimal requirement for gauge invariance is that the matrix element contains a factor (with $V_{n+1} = V_0$) that becomes an integral around the production area,

$$\prod_{j=0}^{n} \gamma(V_j, V_{j+1}) = \exp\left(ig \oint A_\mu dx^\mu\right) \tag{11.32}$$

This is a *Wilson loop operator* when it is evaluated in a quantum field theory state. Wilson's condition for confinement is that it should behave as

$$\left\langle s \left| \exp\left(ig \oint A_\mu dx^\mu\right) \right| s \right\rangle = \exp(i\xi\tilde{A}) \tag{11.33}$$

where $|s\rangle$ is the state and $\tilde{A}$ is the (space-time) area enclosed by the integration contour. The quantity ξ is a parameter whose real part is equal to the string constant κ, i.e. the force on the charges in the confining force field.

Wilson's suggestion has been studied by means of approximative calculations on a lattice and his area law has been confirmed for a number of confining situations. These calculations are, however, outside the scope of this book.

In order to understand the reason why one should obtain this result for the loop integral and also to see why the loop integrals are gauge-invariant we will consider the integral around a connected curve rearranged as in Fig. 11.3. In this way the integral is seen to be equivalent to a large number of integrals around smaller curves, which together make up the larger one.

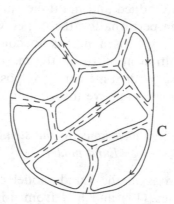

Fig. 11.3. The area inside the boundary curve C is subdivided into smaller areas defined by closed curves. If one sums over the contributions from them all only the original curve will be left since each of the internal parts of the curves is traversed twice, in opposite directions.

This is the construction used to prove Stoke's theorem: the line integral of a vector field around a connected curve equals the area integral of the rotation of the vector field. The area integration vector $d\mathbf{a}$ is defined by the right-hand screw rule from the direction of the curve C:

$$\oint_C \mathbf{A} \cdot d\mathbf{x} = \int_S (\nabla \times \mathbf{A}) \cdot d\mathbf{a} \tag{11.34}$$

If we identify $\mathbf{A}$ with the vector potential we obtain the gauge-independent magnetic field $\mathscr{B}$ in the integral and in this way the integral corresponds to the magnetic flux through the region surrounded by the curve C. There will consequently be a phase difference between wave functions describing motion on one side of the field and on the other side, according to Eq. (11.30), which is just the Aharanov-Bohm prediction mentioned above. Stoke's theorem can be extended to surfaces in a longitudinal space-time plane; one obtains

$$\oint_C g A_\mu dx^\mu = \int_{\tilde{A}} g \mathscr{E}_\ell dx_\ell dt \tag{11.35}$$

from the relationship given in Chapter 2 between the four-vector potential A_μ and the electric field $\mathscr{E}$ (the index ℓ stands for the direction along the field). Wilson's criterion for confinement implies that the surface integral of the electric force over $\tilde{A}$ should be proportional to the area of $\tilde{A}$. This is fulfilled when (the real part of) $g\mathscr{E}$ is a constant.

The description in Fig. 11.4 of Lund model fragmentation by a subdivision of the string persistence region in different ways is a direct realisation

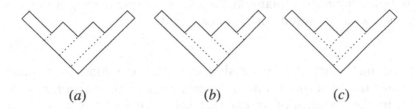

(a)   (b)   (c)

Fig. 11.4. Different ways to subdivide the string region in the fragmentation process, corresponding to different gauge choices: (a) the gauge $A_- = 0$; (b) the gauge $A_+ = 0$; (c) mixed gauge.

of choosing a gauge in different ways and should be compared with the Stoke's construction in Fig. 11.3.

For the case when the hadron yoyos are produced along the positive lightcone (Fig. 11.4(a)) the gluonic field can be choosen in the lightcone gauge $A_- = 0$. This means that there is no gluon-field transmission along the negative lightcone. The pairs can be thought of as produced by the gluons emitted along the positive axis from the original $\bar{q}$. In the case in Fig. 11.4(b) the corresponding gauge is $A_+ = 0$. The case exhibited in Fig. 11.4(c) corresponds to a mixed gauge condition, which is different in different parts of the string region.

3 The possible relationship to the Lund area law

At this point it is necessary to make a few clarifying remarks. We have already pointed out that the phase integral in Eq. (11.30) is independent of the curve choosen between the particle and antiparticle positions. This is, however, only true for non-singular fields and *the field we are working with is singular*. It is constant inside the string region and vanishes outside. The sudden change occurs along the curve along which we are integrating.

For the abelian case described above the force on a charge is $g\mathscr{E}/2$. It seems reasonable to identify the constant in the integral in Eq. (11.35) with this value. The field from the particle itself (i.e. from the particle on the contour) should not be counted in order to avoid self-interactions.

For the second case, in which the external vacuum fields contribute, we again expect to identify the constant with the true force on the charge. We will therefore use κ_b times the area $\tilde{A} = A/(2\kappa_b^2)$, where we have introduced the (lightcone metric) area A used in the Lund model. Then the real part of $\xi\tilde{A}$ equals $A/2\kappa$.

There should, however, also be imaginary contributions to ξ. If there is absorption the dielectricity ϵ changes from its vacuum value (which in our case is unity); cf. Chapters 2 and 4. This occurs because the vacuum itself is polarisable in a quantum theory and consequently we expect in accordance with the results in Chapter 4 that

$$\epsilon = 1 + i\eta \tag{11.36}$$

where the imaginary part η equals the absorption rate. In a quantum field theory this absorption rate is π times the pair production rate in the vacuum in the presence of an external field. For QCD with n_f massless flavors this becomes $\eta = n_f \alpha_s / 6$ (cf. the calculations of the imaginary part of the vacuum polarisation tensor in Chapter 4).

Therefore we expect that the quantity ξ in the Wilson area law contains both a real and an imaginary part,

$$\xi = \kappa\epsilon \tag{11.37}$$

We are, however, again in trouble with regard to the interpretation of this. For the abelian case there are no problems in relating the imaginary part of ϵ to the production of pairs because we are then only discussing a field stemming from the true charges. The energy density is $\kappa_a = \mathscr{E}^2/2$, the electric field being generated by the q- and $\bar{q}$-charges.

For the nonabelian case the force κ contains contributions not only from the true charges but also (the same amount) from the vacuum field pressure. On the one hand *the vacuum field should not be allowed to produce pairs, because that would mean that the vacuum state is not stable.* On the other hand when the pairs have been produced (by the true field) then the vacuum pressure may well push them apart during the tunnelling process.

We will then write for $\xi = \xi_b$,

$$\xi_b = \kappa_b \left(1 + i\frac{\eta}{2}\right) \tag{11.38}$$

The matrix element $\mathscr{M}$ in Eq. (11.1) can be identified, in the expression in Eq. (11.33), with the 'true' area A used in the Lund model fragmentation functions:

$$\mathscr{M} = \exp\left[\left(i\kappa - \frac{\kappa_I}{2}\right)\tilde{A}\right] = \exp\left[\left(\frac{i}{2\kappa} - \frac{b'}{2}\right)A\right] \tag{11.39}$$

The parameter b' is then equal to the b we obtained in the last section from Schwinger's persistence probability:

$$b' = \frac{\eta}{2\kappa} = \frac{n_f \alpha}{12\kappa} \tag{11.40}$$

Evidently the matrix element $\mathscr{M}$ will then fulfil the area law

$$|\mathscr{M}|^2 = \exp(-bA) \tag{11.41}$$

We now have, accepting the considerations above, no 'fudge-factor' in the b-value and we have also a well-defined phase for $\mathscr{M}$ from Eq. (11.39).

11.4   The fragmentation formulas and the partition functions for the Feynman-Wilson gas in rapidity space

1   Preliminaries and definitions

We will now rewrite the decay distributions for a finite-energy cluster in terms of the partition functions of statistical physics. In this way we may identify the area suppression law with the well-known Boltzman factor, i.e. the negative exponential of a state energy divided by the temperature.

To this end we introduce in the cms system the rapidity variables instead of the hadron momenta, p_{oj} (with the index o for 'observable'; for simplicity we consider only a single species of hadron with mass m):

$$(p_{oj+}, p_{oj-}) = m(\exp y_j, \exp(-y_j)), \quad W_{\pm} = \sqrt{s} = W \quad (11.42)$$

Then we obtain for the hadronic phase space volume element in Eq. (11.1)

$$\prod_{j=1}^{n} N dy_j \delta \left(W - m \sum_{j=1}^{n} \exp y_j \right) \delta \left(W - m \sum_{j=1}^{n} \exp(-y_j) \right) \quad (11.43)$$

The area A in the exponent in Eq. (11.1) can be expressed in terms of the rapidities in the following way (see Fig. 11.5):

$$A = \left(\sum_{j=1}^{n} p_{oj-} \right) \left(\sum_{k=j}^{n} p_{ok+} \right) = m^2 \sum_{j=1}^{n} \sum_{k=j}^{n} \exp(y_k - y_j) \quad (11.44)$$

This corresponds to summing systematically the partial areas corresponding to the different particles starting from the positive and going towards the negative lightcone. (The same result is of course obtained by going the opposite way.)

In this way we have exhibited the two-particle correlations explicitly. We note that the area looks very much like a sum of 'two-body potentials', $V(y_j - y_l)$, in the rapidity differences. This is the way we are going to treat the expression and we define a 'partition function' Z by

$$Z = \sum_{n} Z_n = \sum_{n} \prod_{j=1}^{n} N dy_j \delta(\cdots) \delta(\cdots) \exp \left[-\frac{\sum V(y_j - y_k)}{kT} \right] \quad (11.45)$$

the exponential factor being given by bA with A written as in Eq. (11.44). Z essentially has the properties of a partition function if the particles are imagined as making up a gas in rapidity space and interacting via the exponential two-body potentials. The hamiltonian in that case is

$$H = \sum T_j(\pi_j) + \sum V(y_j - y_k) \quad (11.46)$$

and the phase space volume element is $\prod dy_j d\pi_j$, the quantities π_j being the 'momenta' canonically conjugate to the 'coordinates' y_j.

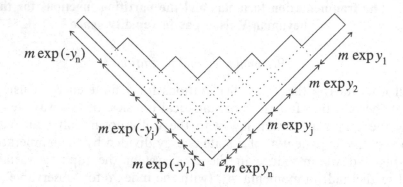

Fig. 11.5. The fragmentation area partitioned into two-particle regions in order to understand how the correlations are produced.

The kinetic energy factors T_j (which correspond to gaussian integrals for a nonrelativistic gas and can be expressed in terms of Bessel functions for the relativistic case) are integrated out in Eq. (11.45) and incorporated into the constants N. These factors then play the role of (the exponentials of) 'chemical potentials' or 'fugacities'.

We note that the potentials V correspond to interactions also between particles distant in rank. Therefore the interaction term is principally of a long-range character. But, owing to the exponential falloff ($y_j \gg y_k$ if $j \ll k$), it is in practice a good approximation to keep only a few of the near-neighbor terms if the gas is dilute.

The two δ-distributions contain the requirement that the the 'gas volume' should be of the order of $\log s$. To see this we may integrate out the rapidities of the first and the last particles in rank to obtain

$$dy_1 dy_n \delta(\cdots)\delta(\cdots) \simeq \frac{1}{s}, \quad y_1 \simeq -y_n \simeq \log(\sqrt{s}) \qquad (11.47)$$

We may in this approximation choose a number s_0 in such a way that

$$\Delta Y \equiv y_1 - y_n = \log(s/s_0) \qquad (11.48)$$

and assume that all the particles are kept inside this rapidity 'volume'.

If we order the particles in rank, $y_1 > y_2 > \cdots > y_n$, the phase space volume is (the result is most easily obtained by iteration, using the translational invariance and the similarity to the symmetrical integral in Eq. (9.19))

$$\int_{y_n}^{y_1} N dy_2 \int_{y_n}^{y_2} N dy_3 \cdots \int_{y_n}^{y_{n-2}} N dy_{n-1} = \frac{(N\Delta Y)^{n-2}}{(n-2)!} \qquad (11.49)$$

This approximation of 'well-orderedness' in rank and rapidity may seem drastic, in particular when we remember the results in Chapter 9 that two rank neighbors may well have different rapidity order.

The latter result is true as a local statement, i.e. it may well happen that a few, $m \ll n$, pairs are not well-ordered. But if many pairs, $m \sim n$, are not well-ordered then many of the exponential potentials will be strongly increasing, i.e. the area suppression in the Lund model will make these contributions small. We are evidently invoking the same arguments as those which were used for the multiperipheral models in Chapter 10.

We next turn to the exponential and the sum of the potentials. We will momentarily go back to the usual Lund model notions and observe that in order to be on the mass shell a particle p_j produced between two vertices (or rather between two momentum transfers q_j, q_{j-1} as in the multiperipheral chains in Fig. 10.2) must fulfil

$$m^2 = p_j^2 = (q_j - q_{j-1})^2 = 2Q_j Q_{j-1} \cosh(y_j - y_{j-1}) - Q_j^2 - Q_{j-1}^2 \quad (11.50)$$

in terms of the momentum transfer sizes $\Gamma_j \equiv Q_j^2 = -q_j^2$ and the rapidities y_j of the vertices. Expanding these formulas up to second order in the differences $\delta Q_j = Q_j - Q_{j-1}$ and $\delta y_j = y_j - y_{j-1}$ we obtain

$$m^2 = (\delta Q_j)^2 + Q^2 (\delta y_j)^2, \quad Q = \frac{Q_j + Q_{j-1}}{2} \quad (11.51)$$

Such a formula occurs in statistical analysis when there are independent variations in the quantities δQ_j and $Q \delta y_j$. We interpret the result to imply that the particles gather along a hyperbola with size parameter Q. They may 'twist and turn' independently along the hyperbola ($Q\delta y$) and transversely to it (δQ); cf. also the results in Chapter 18.

If the particles are distributed along this hyperbola there will be a relation between Q and the number of particles (c describes the relative weight between the transverse and longitudinal variations along the hyperbola, with $c = \sqrt{2}$ if they are equal):

$$m \simeq cQ(y_j - y_{j-1}), \quad nm \simeq cQ\Delta Y \quad (11.52)$$

The area below such a hyperbola is

$$A = Q^2 \Delta Y \simeq \frac{n^2 m^2}{c^2 \Delta Y} \quad (11.53)$$

2   An equation of state for an (almost) ideal rapidity gas

We will now use the results for the phase space factor and the area suppression we derived in the last subsection. Thus we obtain for the term Z_n in Eq. (11.45), using the Stirling approximation for the factorial in the denominator of Eq. (11.49) and assuming $n \gg 1$:

$$Z_n = \frac{\exp \Phi_n}{s}$$

$$\Phi_n = n \log(N\Delta Y) - n \log n + n - \frac{n^2 b m^2}{c^2 \Delta Y} \tag{11.54}$$

As a function of n the the exponent Φ_n has a maximum for

$$n \equiv \bar{n} = R\Delta Y \quad \text{with} \quad \Phi_{\bar{n}} = \left(R + \frac{R^2 b m^2}{c^2} \right) \Delta Y \tag{11.55}$$

where the parameter R is determined from

$$\frac{2Rbm^2}{c^2} = \log \left(\frac{N}{R} \right) \tag{11.56}$$

In this way we obtain as a result for $Z \simeq Z_{\bar{n}}$ that as a function of the squared cms energy s it is

$$Z \sim s^a, \quad a = R - 1 + \frac{R^2 b m^2}{c^2} \tag{11.57}$$

The parameter R evidently corresponds to the density of particles in rapidity and it is worthwhile to relate it to our earlier results in Chapter 9, cf. Eq. (9.15). In that case we have derived that the inverse density is given by the average value of $\log 1/(1 - z)$, the average taken over the fragmentation function $f(z)$. We remember from Eq. (9.11) that $\langle \log 1/(1 - z) \rangle$ corresponds to a typical rapidity difference in the cascade. For the Lund model fragmentation function it is straightforward to derive a formula for this rapidity difference:

$$-N \frac{\partial}{\partial a} \left(\frac{1}{N} \right) = N \int \frac{dz}{z} \log \left(\frac{1}{1 - z} \right) (1 - z)^a \exp(-bm^2/z)$$

$$\equiv \langle \log 1/(1 - z) \rangle \tag{11.58}$$

where we have used the normalisation condition

$$\frac{1}{N} = \int \frac{dz}{z} (1 - z)^a \exp(-bm^2/z) \tag{11.59}$$

Therefore we need the logarithmic derivative of the normalisation constant with respect to the parameter a with the parameter b fixed. From Eqs. (11.56) and (11.57) it is straightforward to show that

$$\frac{\partial R}{\partial a} + 2R \frac{\partial R}{\partial a} \frac{b m^2}{c^2} = 1$$

$$\frac{\partial \log N}{\partial a} = \frac{\partial \log R}{\partial a} + 2 \frac{\partial R}{\partial a} \frac{b m^2}{c^2} = \frac{1}{R} \tag{11.60}$$

where we have used the result of the first equation to obtain the expected result in the second one. Note that the introduction of R in Eqs. (11.55) and (11.56) stems from a 'global' result, i.e. it is obtained from (an approximation of) the total area law for the average state, while the particle density as defined by Eq. (11.58) is a local result, defined from the fragmentation function for a single particle in the Lund model.

The identification of Z with the maximal term in the sum is a good approximation when n is very large (as in an ordinary gas). In our case we can estimate not only the largest contribution to Z but also the width of the 'multiplicity distribution', i.e. $P_n = Z_n/Z$.

This is obtained if we expand the exponential Φ_n in a Taylor series to second order in n (treating n as a continuous variable):

$$\Phi_n \simeq \Phi_{\bar{n}} + (n - \bar{n})\frac{d\Phi}{dn}\bigg|_{n=\bar{n}} + \frac{(n - \bar{n})^2}{2}\frac{d^2\Phi}{dn^2}\bigg|_{n=\bar{n}} \qquad (11.61)$$

The first-order term vanishes due to the choice of $\bar{n}$ as the maximum value and we obtain as an approximation for Φ_n to second order:

$$\Phi_n = \Phi_{\bar{n}} - \frac{(n - \bar{n})^2}{2\mathcal{V}} \qquad (11.62)$$

In this gaussian approximation we can identify the width $\mathcal{V}$ of the distribution, i.e. the inverse of the coefficient of $(n - \bar{n})^2$, to be $\overline{(n - \bar{n})^2}$, the variance in n:

$$\mathcal{V} = \overline{n^2} - (\bar{n})^2 = \frac{\bar{n}c^2}{c^2 + 2bm^2R} \qquad (11.63)$$

We conclude that the multiplicity width in the Lund model should be somewhat more narrow than the predictions from a Poissonian distribution (where the width is $\mathcal{V}_p = \bar{n}$). This is also true if we produce a single species of hadron and neglect transverse momentum fluctuations.

If these are taken into account together with the various hadrons, resonance decays and gluon radiation etc. occurring in the experiments then the multiplicity width (for the decay of a single string) behaves rather like $\bar{n}^2$ as we will see later. The most essential contribution for large energies is that of the gluon radiation (but we are then no longer in a single space-dimensional setting).

There is, according to statistical mechanics, a simple relationship between the grand canonical partition function and the properties of the gas:

$$\log Z \simeq \Phi_{\bar{n}} = \frac{PV}{kT} \qquad (11.64)$$

Our treatment in Eq. (11.55) basically corresponds to the first two terms in the virial expansion in the particle density $R = \bar{n}/V$, where $V = \log(s/s_0)$

(although our treatment of the fluctuations around the mean hyperbola by means of the single parameter c is probably too cavalier).

It is of course possible to consider the virial expansion to higher orders and to calculate different quantities for the rapidity gas, such as its entropy etc., but that will be left for the interested reader.

12

Flavor and transverse momentum generation and the vector meson to pseudoscalar meson ratio

12.1 Introduction

In this and the next chapter we are going to extend the Lund fragmentation model in several different directions in order to make it into a realistic model for the production of different kinds of hadron.

In the first section we start by investigating the classical motion of a string when it contains a $q\bar{q}$-pair having both mass and momentum components transverse to the string direction. One reason for doing this is to get an insight into the classical motion of a confined object. From this exercise we will learn that there are modes of motion of the massless relativistic string (the MRS) which contain much richer dynamics than that of a linearly rising potential. In this case the different parts (called 'segments') of the string will be found to move with respect to each other. We will meet similar examples later on in which our experience from this investigation will be useful.

It is possible to continue the investigation into the properties of the wave functions for this situation, [17]. Although we will present a few steps in that direction below a more complete treatment necessitates considerable mathematical machinery. The main result is, however, essentially the same as that obtained from the simple WKB-approximation, which was presented in Chapter 11.

The production of heavy quark masses and transverse momentum in a string field is intrinsically non-perturbative and leads to a gaussian suppression in both the quantities. The phenomena are governed by the string constant κ, i.e. the available energy per unit length along the force field. We will discuss a general process for transverse momentum generation and afterwards show its close resemblance to Brownian motion.

The results will necessitate a few phenomenological remarks. In particular it will become evident that heavy quark flavors like charm and

213

bottom will never be produced during the fragmentation process of the Lund model. We briefly consider the pion-to-kaon ratio and its significance for measuring the strange-to-up and strange-to-down flavor ratios in multiparticle reactions.

We will after that consider the vector-to-pseudoscalar rate in a fragmentation process. The tunnelling process, which we deal with in the first section, also has implications for the relative rate of final-state hadrons. The produced $q\bar{q}$-particles have up to now been treated as if they were freely moving outwards after their production. This is evidently not the case in connection with the Lund model fragmentation process. In this process *they are tunnelling into bound-state hadronic wave functions*.

The classical turning point for the potential, i.e. the point where the kinetic energy vanishes for classical motion, is also the place where the virtual $q\bar{q}$-pair will come onto the mass shell after the production. The pair production rate is directly proportional to the squared wave function at this point. This is the place where the bound-state wave function starts to play a role for the produced q and $\bar{q}$. We will show that, depending upon the properties of the bound-state interaction, different kinds of hadron may have a different size of wave function at the turning point.

The spin-spin interaction between the constituents, which is different for vector mesons and pseudoscalar mesons, implies that the vector meson production rate should be suppressed compared with the pseudoscalar rate. In other words *it is more difficult to tunnel out into a vector meson state than into a pseudoscalar meson state*.

We consider a simple model for the phenomenon (a 'one-dimensional bag'). Then the relative rate behaves as a power law in the masses. The simplicity of the model prevents it from being a reliable tool for quantitative prediction of the suppression. But it provides a useful method of finding the wave functions for the tunnelling process, which we omitted before. We will also find qualitative agreement between the phenomenology and and the experimental observations.

We end by pointing out another problem, related to the production rate of η' particles. This also provides an opportunity to discuss the assignment of projection probabilities for the different flavor states in the Lund model.

12.2   The classical transverse motion of a string

The classical motion of a string when one of the endpoints has a momentum component transverse to the string direction is more complex than the modes of the MRS we have encountered up to now. The string field no longer behaves as a classical potential because different segments of the string are moving relative to each other.

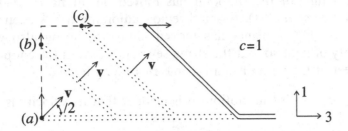

Fig. 12.1. The motion of a (massless) endpoint particle initially possessing transverse momentum together with the adjoining string piece. The times (*a*), (*b*) and (*c*) correspond to snapshots of the situation, as described in the text.

The transverse momentum of the endpoint is a very transitory property because of the interaction between the string and the particle at the endpoint, which will lead to a *transfer of the momentum from the endpoint particle to a region of the string in its neighborhood*. The size of this string segment is proportional to the size of the transverse momentum if the endpoint particle is massless and somewhat more complicated if the endpoint is a massive particle.

This means that transverse momentum is not a conserved property for such a particle, when it is part (and also generator) of a confining force field. But it is not possible to distiguish between the particle and the neighboring force field unless one introduces some measurement prescription.

We therefore feel confident in treating the particle quantum mechanically as an entity (although we are then also incorporating some part of the field in its neighbourhood). The major difference from the $(1 + 1)$-dimensional scenario we have discussed before is that a segment of the string containing transverse momentum will seem to have a larger longitudinal size, proportional to its transverse mass, rather than its ordinary mass.

1   Transverse string motion with a massless endpoint particle

Consider the situation depicted in Fig. 12.1(*a*) as a given starting point for the motion. The endpoint is assumed to have longitudinal momentum $k_3 = 0$ and a transverse momentum component $k_1 \neq 0$.

We assume that it is connected to a string stretched along the 3-direction, which is sufficiently long that we do not need to consider the other endpoint and its motion. We start with the case when the endpoint particle q is massless; we will discuss a set of snapshots of the motion in time.

- After a time δt the endpoint has moved out along the 1-direction a distance δt ($c = 1$). A small region adjoining it has been affected (Fig. 12.1(b)). A signal has moved along the string also with the velocity of light so that the string segment between the endpoint and the rest of the string has a (geometrical) length equal to $\sqrt{2}\delta t$.

This segment is also in motion; it is evident that its velocity is

$$v = \frac{1}{\sqrt{2}} \tag{12.1}$$

This velocity, as in the transverse motion of the yoyo mode in Chapter 6, is related to the half angle $\theta/2$; in this case $\theta = \pi/2$:

$$v = \cos(\theta/2) = 1/\sqrt{2}$$
$$\gamma(v) = \frac{1}{\sqrt{1 - v^2}} = \frac{1}{\sin(\theta/2)} = \sqrt{2} \tag{12.2}$$

This means that the string segment will have the energy-momentum

$$e = \kappa\sqrt{2}\delta t\,\gamma(v) = 2\kappa\delta t$$
$$p_1 = \kappa\sqrt{2}\delta t\, v\gamma(v) \cos(\theta/2) = \kappa\delta t \tag{12.3}$$
$$p_3 = \kappa\sqrt{2}\delta t\, v\gamma(v) \sin(\theta/2) = \kappa\delta t$$

Consequently

- the endpoint has lost $\kappa\delta t$ both in energy and in the 1-component of its momentum, in accordance with the ordinary equations of motion. The string segment has picked up these quantities together with the energy-momentum which was in the string behind the moving signal-corner.

In this way we make use of the same considerations of local energy-momentum conservation as in Chapter 6 when we traced the transverse motion of the yoyo-state.

- This part of the motion continues until the endpoint q has lost all its original energy-momentum along the 1-direction. At that time it will start to move in the 3-direction (Fig. 12.1(c)). From now on, until the q reaches the other endpoint *the string segment serves only as a convenient transporter of energy-momentum to the endpoint q.* Thus the q-particle will gain energy-momentum along the 3-axis just as if it were a particle joined to an elastic cord, i.e. to the MRS.

In fact the string segment now picks up energy-momentum at the other end (from the remaining string), at the rate of κ per unit time and length, and delivers the same amount to the q. The string segment is in this

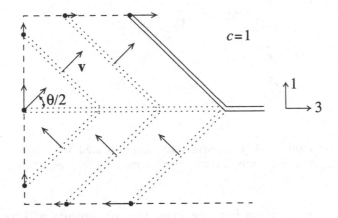

Fig. 12.2. The full motion when a q-particle comes in and leaves again. The q and the adjoining string segments share the transverse momentum in different proportions at different times.

way also moving on just like a rigid pole (but without changing its total energy-momentum).

In Fig. 12.2 we exhibit the full motion of a q-particle which comes in, dragging out the string behind it, and then turns around and leaves outwards again in accordance with the discussion above.

This corresponds to the motion discussed in Chapter 6, when a classical particle comes in and rebounds from the classical turning point. It is useful to trace the part of the motion before that discussed above, i.e. the motion inwards; we will leave it to the reader to understand the details by making use of local energy-momentum conservation and the fact that the endpoint particle moves with the velocity of light.

2   Transverse string motion with a massive endpoint particle

Before we display the above motion in a space-time diagram we will briefly consider the changes which occur when a massive q-particle (mass μ) is at the endpoint. The starting situation is the same as before but this time the endpoint is no longer moving with the velocity of light and consequently it can no longer go straight out along the 1-axis.

It will instead follow a hyperbola and, using the parameter $a(t)$ indicated in Fig. 12.3, we can write out the energy-momentum conservation equations. We use the indices p for the particle and s for the segment connecting the particle to the rest of the string. Let us suppose that the starting value of the particle's energy is $E_\perp = \sqrt{k_1^2 + \mu^2}$. Then the

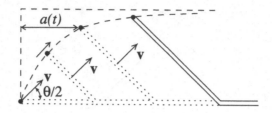

Fig. 12.3. The motion of a massive particle connected to a string when the particle initially possesses only a transverse momentum component.

equations for the conservation of energy and momentum will be

$$e_s(t) + e_p(t) = E_\perp + \kappa t, \quad e_s(t) = 2\kappa(t - a)$$
$$p_{3s} + p_{3p} = \kappa t, \quad p_{3s} = \kappa(t - a) \tag{12.4}$$
$$p_{1s} + p_{1p} = k_1, \quad p_{1s} = \kappa(t - a)$$

in accordance with the results in Eq. (12.3). We have used the fact that a string segment of a given size and with the transverse velocity given in Eq. (12.1) will have the same relation between size and energy-momentum as in the massless case. The string segment does not know that it is connected to a massive particle this time!

The requirement that the endpoint particle should be on the mass shell, i.e. $e_p^2 - p_{3p}^2 - p_{1p}^2 = \mu^2$, provides equations for the quantity $a = a(t)$ and also for the hyperbola along which the particle will move,

$$(\kappa x_{1p} - E_\perp)(\kappa x_{3p} + E_\perp - k_1) = -E_\perp(E_\perp - k_1) \tag{12.5}$$

(the coordinates x_{1p}, x_{3p} being defined with respect to an origin at the turning point). Note that when $\mu = 0$ we obtain the orbits shown above in Fig. 12.2. We leave it to the reader to prove Eq. (12.5) and to obtain the equation for $a(t)$.

In this way it is obvious that as long as we are not resolving the motion on a scale corresponding to the value of $\kappa(t - a(t)) - k_1$, i.e. $E_\perp - k_1$ (valid at large values of t) we do not know whether there is a massless or massive particle at the endpoint.

This is shown in a projection onto the tx_3-plane (i.e. the plane where the remainder of the string is dwelling) of the motion of a massless and of a massive endpoint particle (Fig. 12.4). We note in particular that the distances of closest approach (to the vertex of the hyperbola) in the two cases are given by $k_\perp = k_1$ and $E_\perp > k_1$ respectively.

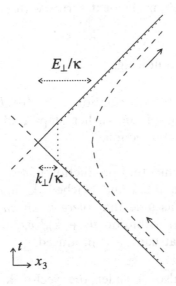

Fig. 12.4. The space-time diagram of the motion of an endpoint particle with a transverse momentum at the endpoint of a string, projected onto the plane spanned by the time and logitudinal direction of the string. The dotted line shows a massless particle; at closest approach it is at a distance $k_\perp/\kappa$ from the hyperbola focus. The broken line shows a massive particle; at closest approach it is at a distance $E_\perp/\kappa$ from the focus.

12.3 A general process for transverse momentum generation

1 Preliminaries

With the same methods used to derive the longitudinal fragmentation function we will now exhibit a general method for endowing the produced particles with transverse momentum. This leads to a correlation length in the model (further discussed and determined in Chapter 18). We will relate the result via the Langevin equation to Brownian motion; this is the Ornstein-Uhlenbeck process, [109].

We will assume that there is a set of hadrons produced along the positive lightcone direction; hadron j possesses transverse momentum $\mathbf{p}_j$. After n steps the total transverse momentum is $\mathbf{k}$:

$$\sum_1^n \mathbf{p}_j = \mathbf{k} \qquad (12.6)$$

We now take one more step and produce a hadron with transverse momentum $\mathbf{p}$, thereby reaching the next vertex with transverse momentum

$$\mathbf{k}' = \mathbf{k} + \mathbf{p} \tag{12.7}$$

We further assume the following.

- After many steps the distribution in $\mathbf{k}$, $f(\mathbf{k})d^2k$, 'saturates' and becomes independent of the earlier steps and there is no longer a preferred transverse direction.

- The transverse momentum production of the next hadron then only depends upon the the value reached, $\mathbf{k}$, and on the step, $\mathbf{p}$. We will in particular assume that *there is an anticorrelation* so that it depends upon the distribution $g(\mathbf{p}+\gamma\mathbf{k})d^2p$, γ being a positive number determined by what hadron is produced.

We may evidently also consider the vector $\mathbf{k}'$ as the result of the production of a set of hadrons along the negative lightcone with transverse momenta $\mathbf{l}_j$, thereby reaching the point

$$\mathbf{k}' = - \sum_{j=n+1}^{N} \mathbf{l}_j \tag{12.8}$$

(the minus sign is necessary to conserve the total transverse momentum, cf. Eq. (12.7) and we assume that there are N particles produced). The probability of doing this is again given by the saturating distribution $f(\mathbf{k}')d^2k'$. The next step from $\mathbf{k}'$ to $\mathbf{k}$, with $\mathbf{k} = \mathbf{k}' - \mathbf{p}$, thereby producing the hadron with transverse momentum $\mathbf{p}$, is given by $g(-\mathbf{p} + \gamma\mathbf{k}')d^2p$.

2 The resulting distributions

If we equate the two probabilities for producing a hadron with transverse momentum $\mathbf{p}$ we obtain (exchanging $\mathbf{p}$ for $\mathbf{k}' - \mathbf{k}$):

$$f(\mathbf{k})g(\mathbf{k}' + (\gamma - 1)\mathbf{k})d^2kd^2k' = f(\mathbf{k}')g(\mathbf{k} + (\gamma - 1)\mathbf{k}')d^2kd^2k' \tag{12.9}$$

After removing the differentials we may take logarithms of the functions (putting $\log f = F$, $\log g = G$) and then partial differentials of the result with respect to the same components of $\mathbf{k}$ and $\mathbf{k}'$. We note the similarity to the methods used in Chapter 8 to derive the longitudinal fragmentation function. Once again one set of functions vanishes, this time the f's.

We are then as in that case left with a second-order differential equation:

$$(\gamma - 1)\ddot{G}(\mathbf{k}' + (\gamma - 1)\mathbf{k}) = (\gamma - 1)\ddot{G}(\mathbf{k} + (\gamma - 1)\mathbf{k}') \tag{12.10}$$

Therefore if $\gamma \neq 0, 1$ we conclude that the two sides must be equal to the same constant, to be called $-4\beta/[\gamma(2-\gamma)]$. We obtain directly the result

$$G(\mathbf{P}) = \log N - \frac{4\beta\mathbf{P}^2}{\gamma(2-\gamma)} \quad \rightarrow \quad g(\mathbf{P}) = N \exp\left[-\frac{4\beta\mathbf{P}^2}{\gamma(2-\gamma)}\right] \quad (12.11)$$

We have here invoked (euclidean) invariance, i.e. the assumption that there is no preferred direction. Therefore there is no linear term in the vector arguments.

The result for $F(f)$ is obtained by noting that if we introduce the result for $G(g)$ in Eq. (12.9) we may gather up the contributions depending upon $\mathbf{k}$ and $\mathbf{k}'$ on each side. The two sides must therefore again be equal to the same constant, $\log N'$:

$$F(\mathbf{P}) = \log N' - 4\beta\mathbf{P}^2 \quad \rightarrow \quad f(\mathbf{P}) = N' \exp(-4\beta\mathbf{P}^2) \quad (12.12)$$

The constants N, N' are evidently normalisation constants, while β, γ are dynamical quantities.

The result can be written in different ways. One interesting way is to write it as a squared matrix element (defining $\gamma \equiv 1 - \exp(-\tau)$):

$$f(\mathbf{k})g(\mathbf{k}' - \exp(-\tau)\mathbf{k}) = |\mathcal{M}|^2, \quad \mathcal{M} = \langle 0|\mathbf{k}'\rangle \, p_\tau(\mathbf{k}', \mathbf{k}) \, \langle \mathbf{k}|0\rangle \quad (12.13)$$

We have introduced the harmonic oscillator ground states (Chapter 3) in the momentum representation:

$$\langle \mathbf{k}|0\rangle = C \exp\left(-\beta\mathbf{k}^2\right) \quad (12.14)$$

The symmetrical function $p_\tau(\mathbf{k}', \mathbf{k})$ is equal to the *transition matrix* from the state $\mathbf{k}$ to the state $\mathbf{k}'$, cf. [83] and Eq. (12.24) below:

$$p_\tau(\mathbf{k}', \mathbf{k}) = \langle \mathbf{k}'| \exp(-H\tau) |\mathbf{k}\rangle \quad (12.15)$$

with H the harmonic oscillator hamiltonian in terms of the canonically conjugate variables $\mathbf{k}, \mathbf{b}$:

$$H = \frac{\sqrt{\beta}\mathbf{k}^2}{2} + \frac{\mathbf{b}^2}{2\sqrt{\beta}} \quad (12.16)$$

This brings out the symmetry of the results with respect to the transverse momentum and the impact parameter, $\mathbf{b}$. At the same time we note the similarity to Feynman's path-integral formulation of quantum mechanics, although the result refers to an imaginary time τ. The imaginary-time formalism fits into a statistical physics scenario and we will now show the close relationship between the results above and the velocity distribution of a particle undergoing Brownian motion.

It is of interest to note that motion in spacelike directions (i.e. in impact-parameter space) is formally equivalent to an imaginary-time formalism.

This is evident from the considerations of Chapter 2, i.e. in spacelike directions the proper time τ becomes $i\sqrt{\mathbf{b}^2 - t^2}$.

3   The relation to Brownian motion: the Ornstein-Uhlenbeck process

Another way to understand the results in Eqs. (12.11) and (12.12) is to consider a seemingly different problem, the motion of a Brownian particle, mass m, under the influence of a friction force proportional to the velocity, $-m\rho v$, and a *gaussian random force*, mR. This is expressed by the Langevin equation:

$$\frac{dv}{dt} = -\rho v + R \tag{12.17}$$

In this way v is obtained as a *stochastical variable* defined by R. We assume that there is an ensemble of states on which measurements may be made. The ensemble average of a measurement of a dynamical variable a will be denoted $\langle a \rangle$.

We may evidently write as a general solution for Eq. (12.17):

$$v(t) = v(t_0)\exp(-\rho(t - t_0)) + \int_{t_0}^{t} dt' R(t')\exp(-\rho(t - t')) \tag{12.18}$$

The gaussian randomness assumption on R means that only 'white' noise is included. This means that R has a vanishing average value and only contains equal-time correlations; with I_R a constant we have for the ensemble averages

$$\langle R(t) \rangle = 0 \tag{12.19}$$
$$\langle R(t)R(t') \rangle = 2\pi I_R \delta(t - t')$$

From this we conclude that the mean value and the *time correlation function* of v are given by

$$\langle v(t) \rangle = \langle v(t_0) \rangle \exp[-\rho(t - t_0]$$
$$\langle v(t)v(t') \rangle = \frac{\pi I_R}{\rho}\exp(-\rho|t - t'|) \tag{12.20}$$

We have then assumed that the distribution is 'thermalised' at the starting time t_0 so that, according to the Maxwell velocity distribution,

$$\left\langle v^2(t_0) \right\rangle = \frac{\pi I_R}{\rho} = \frac{kT}{m} \equiv \frac{1}{2\beta} \tag{12.21}$$

It is then also the same for all times, $\left\langle v^2(t_0) \right\rangle \equiv \left\langle v^2(t) \right\rangle$, but there is an exponentially falling correlation between the value of v obtained at one time and at another. The correlation function depends only upon the time difference and therefore the stochastic process is called a *stationary stochastic process*. The variance of $v(t)$ is

$$\left\langle (v(t) - \langle v(t) \rangle)^2 \right\rangle = \frac{[1 - \exp(-2\rho(t - t_0))]}{2\beta} \tag{12.22}$$

If we define $u(t) = v(t) - \langle v(t) \rangle$ then $u(t)$ considered as a stochastical variable will have the particular *gaussian property that all higher-order correlation functions are determined by the two-point correlations*:

$$\left\langle \prod_{j=1}^{2n} u(t_n) \right\rangle = \sum_{perm} \prod_{j \neq k} \langle u(t_j) u(t_k) \rangle \tag{12.23}$$

i.e. they only contain all possible two-point correlations in time (the notation '*perm*' means all possible permutations of $1, \ldots, 2n$). The expectation value of an odd number of u's vanishes.

It is no coincidence that the Wick rearrangements of a time-ordered product of free fields, as discussed in Chapter 3, lead to the same results. The vacuum expectation values of free fields are just like the gaussian processes we consider here and the Feynman propagator corresponds to the correlations between two space time points.

The *transition probability* $P(v_0, t_0 | v, t)$ from the value v_0 at t_0 to v at t is then given by the gaussian distribution

$$P(v_0, t_0 | v, t) = \left\{ \frac{\beta}{\pi[1 - \exp(-2\rho\delta t)]} \right\}^{d/2} \exp \left\{ -\frac{\beta[v - v_0 \exp(-\rho\delta t)]^2}{1 - \exp(-2\rho\delta t)} \right\}$$

$$\int d^d v_1 P(v_0, t_0 | v_1, t_1) P(v_1, t_1 | v_1, t_1) = P(v_0, t_0 | v, t), \quad t_0 \leq t_1 \leq t \tag{12.24}$$

Here we have introduced the number of dimensions, d, in which the process goes on and written $\delta t = t - t_0$ for the time difference.

It is not difficult to recognise the distribution g (with $\gamma = 1 - \exp(-\rho\delta t) \equiv 1 - \exp(-\tau)$) in the transition probability for the change in velocity of a Brownian particle in thermal equilibrium under the influence of gaussian white noise; see Eq. (12.11).

4   Concluding remarks

The formulation of transverse momentum generation above may seem abstract. It is obvious, however, that it corresponds to the possibility of producing a pair at every vertex with transverse momenta $\pm \mathbf{k}$. Together the q (with transverse momentum $-\mathbf{k}$) and the $\bar{q}$ (with $\mathbf{k}'$) from adjacent vertices will then combine to give a hadron with transverse momentum $\mathbf{p} = \mathbf{k}' - \mathbf{k}$.

From this interpretation we may, using the tunnelling arguments in Chapter 11, identify the scale parameter, β, in terms of the string tension κ (note that the factor 4 in the definition of β in Eq. (12.11) is justified if we compare with the tunnelling process in Eq. (11.5) and also with the transition matrix elements in Eqs. (12.13) and (12.16)):

$$\beta = \frac{\pi}{4\kappa} \tag{12.25}$$

In the present Lund model the possibility of a correlation between the vertices for transverse momentum generation is not used, i.e. we have always put $\gamma = 1$. This corresponds to a large 'friction' coefficient, i.e. the 'memory' is very short.

There is, however, one kind of hadron, the pions, which have very small mass and for which, therefore, the two production vertices are very close. There are several indications that a proper treatment of the pions directly produced along the string actually does require a correlation, which then will diminish the transverse momentum of the final hadron. Note that the inclusive distribution of the hadron $\mathbf{p}$ is

$$\int d^2k f(\mathbf{k}) g(\mathbf{p} + \gamma\mathbf{k}) \quad \propto \quad \exp\left(-\frac{2\beta\mathbf{p}^2}{\gamma}\right) \tag{12.26}$$

This means that the mean transverse momentum of the hadron is

$$\sqrt{\langle \mathbf{p}^2 \rangle} = \sqrt{\frac{2\gamma\kappa}{\pi}} \tag{12.27}$$

which diminishes with γ. We will return to these features in Chapter 18.

12.4　The phenomenological implications of the tunnelling process

1　The production of heavy flavors

The results derived above are compatible with the WKB results, i.e. they are equivalent to Schwinger's result for the decay of the no-particle vacuum in the presence of an external electric field. We obtain for the $q\bar{q}$ production rate, with μ the mass and $\pm\mathbf{k}_\perp$ the transverse momenta of the pair,

$$dP \simeq d^2k_\perp \exp(-\pi E_\perp^2/\kappa), \quad E_\perp^2 = \mathbf{k}_\perp^2 + \mu^2 \tag{12.28}$$

The result in Eq. (12.28) has several different consequences.

The first is related to the relative abundance of different flavors in the fragmentation process. It is difficult to obtain precise mass-values for the unobservable $q\bar{q}$-particles but it is possible to obtain estimates.

If we start with the heavy flavors, i.e. charm, c, and bottom, b, then there are potential models for the bound states of the $c\bar{c}$-states, J/Ψ and its relatives, and likewise for the corresponding $b\bar{b}$-states, Υ etc., [55].

These authors make use of nonrelativistic kinematics and potential terms containing Coulomb contributions, a linearly rising confining potential and also spin-spin and angular-momentum-spin interactions. In this way they obtain $\mu_c \sim 1.5$ GeV and $\mu_b \sim 5$ GeV (in both cases with a spread of a few hundred MeV).

Using the value $\kappa \simeq 0.2$ GeV^2 we find numbers smaller than 0.3×10^{-11} for the c-flavor production and vanishingly small numbers for the corresponding b-production. As the lighter-flavor masses are at most in the range of a few hundred MeV with small suppression we conclude that *c- and b-flavors are never produced in a fragmentation process*. The available energy density (κ) means that we need a string with a size at least around 3 fm to obtain a $c\bar{c}$-pair. At that point the lighter flavors have already supplied tremendous amounts of possible string-breakers.

2 The production of light flavors

The relative rates of u-, d- and s-flavor production are much more difficult to estimate. The reason is that there are different ways of attributing mass-values to these light flavors. In the classical treatment of the motion of a string with transverse momentum at the endpoint, we found that there is a certain region of the string close to the endpoint particle that exchanges energy-momentum with the particle. Thus it is difficult to distinguish between the particle and the field around it, at least in stable, long-time situations.

There is a corresponding notion in phenomenological models for the hadrons, i.e. the 'constituent quark mass' whereas in fast-production situations one talks of the 'current quark mass'. A quark-parton, which is exposed to an external probe acting very quickly in an almost pointlike way, is a current quark and one then expects to be able to neglect the field surrounding the quark charge. In the corresponding bound-state (stable) situation a constituent quark is always part of the bound-state force field and therefore 'heavier' than the 'bare' current quark.

If we compare the masses of the ρ-, K^*- and ϕ-mesons, which contain zero, one or two s- and/or $\bar{s}$-flavors, we may tentatively assign a mass difference $\mu_s - \mu_u \simeq 120$ MeV to the constituents. We are then referring the whole mass difference to the quark masses and are using a linear interpolation between the meson masses (there have been suggestions that the mass formulas should be quadratic for the mesons but we will not consider the reasons for such complications).

If we further assume that both the K^*- and the K-mesons are 'normal' with respect to chiral symmetry breaking (which strongly affects the mass of the π- and the ρ-mesons) then we could expect that the constituent u-quark mass is around 330 MeV. Then the ratio between the s- and the u-flavor rates should be close to 0.25 according to Eq. (12.28).

A change of κ to 0.24 GeV^{-2} or a reduction in the u-quark mass to 260 MeV will give a number close to 0.3. This seems to be the preferred relative fraction among the Monte Carlo users of the Lund model (i.e. the JETSET simulation program [105]), although 0.25 is not ruled out. In order to observe the ratio it is also necessary to take into account a few simple kinematical properties (cf. below in subsection 4).

3 Transverse momentum generation

We will later learn that the major contributor to the transverse momentum spectrum of the hadrons in high-energy interactions is the gluonic radiation. The transverse momentum we obtained in Eq. (12.27) should rather be looked upon as a quantum mechanical fluctuation in the ground state of the string (a zero-point fluctuation). These fluctuations imply that any primary meson should come out with a gaussian transverse momentum spectrum of width (the result stems from Eqs. (12.26) and (12.25))

$$\left\langle \mathbf{p}_\perp^2 \right\rangle = \frac{2\gamma\kappa}{\pi} \simeq (0.35)^2 \gamma (\text{GeV}/c)^2 \tag{12.29}$$

We have inserted the word 'primary' above in order to distinguish between the hadrons that are produced directly in the fragmentation process and those which are actually observed. About half the primary particles are resonances and from the tables of the Particle Data Group one finds that they decay afterwards into many π's and K's. These decay products therefore constitute a large part of the observed charged particles.

The general preference of Monte Carlo users seems to be an average $\langle p_\perp \rangle \sim 0.4$ GeV/c, which is a little above Eq. (12.29) (with $\gamma = 1$). The difference can be easily attributed to soft gluonic radiation.

If we go back to the result in Eq. (12.28) then we may use the impact vector description from Chapter 10 to obtain an idea of the transverse width inside which the $q\bar{q}$-particles are produced. The matrix element becomes

$$\sim \int d^2 p_\perp \exp\left(i\mathbf{p}_\perp \mathbf{b} - \frac{\pi \mathbf{p}_\perp^2}{2\kappa} \right) \simeq \exp\left(-\frac{\kappa \mathbf{b}^2}{2\pi} \right) \tag{12.30}$$

Thus the average impact-vector size is $\left\langle \mathbf{b}^2 \right\rangle = \pi/\kappa \simeq (0.8)^2 \text{fm}^2$, a number almost twice as large as the transverse radius value we obtained from the phenomenology of the b-parameter in the Lund model in Chapter 11. However, we are not discussing the same quantity when we refer to the transverse size in connection with the *b*-parameter as when we refer to the size provided by the transverse momentum fluctuations in the string fragmentation process.

For the transverse momentum fluctuations it is not only the size of the string field but also the localisation of its centre which is of interest. Thus if we localise a quantum mechanical object very well in coordinate space then the wave function will evidently contain a large spread of momentum components in the dual space. The size 0.8 fm corresponds more to how well localised the string is in transverse directions than to the size of the emission region.

Note that the transverse momentum generation also influences the longitudinal distributions because in this case we must use the transverse mass $m_\perp = \sqrt{\mathbf{p}_\perp^2 + m^2}$ instead of the ordinary mass m. The reason is the relation $p_+ p_- = m_\perp^2$, which implies that we can conserve energy-momentum in the string plane only by this exchange.

4 The mother–daughter relation

There is another aspect of the resonance decays. The π's and K's from the decays populate phase space in different ways. There is a well-known kinematical property, usually referred to as the 'mother–daughter relation'.

Consider a resonance decay in the rest frame of the resonance. Obviously the decay products will go out in different directions in order to conserve the momentum. Their velocities are given by the mass of the particles, $\sqrt{1 - 4m^2/M^2}$ (cf. Chapter 4) when a resonance with mass M decays into two particles with the same mass m. This translates into a rapidity difference of the order of 1 to 2 units. (For $\rho \to \pi\pi$ decay, which is an extreme case because of the smallness of the π mass, one obtains, after angular averaging, around 1.5 units.)

Therefore the 'daughters', i.e. the decay products, tend to have roughly the same rapidity as the 'mother' so that they will have momenta which are proportional to their masses. In the decay of a K^* to a $K\pi$ this means that the (much) lighter π-particle in general takes a much smaller share of the mother's momentum than the K-particle. The π's also occur much more often as decay products than do other particles. The large amount of π's and the smaller correlation to their 'heritage' means that the π's occur in a rather uncorrelated fashion in the final states. They often have small momenta, in particular transverse to the jet.

This means that the central parts of rapidity space and the small transverse momentum region contains many more π's than K's and the ratio K/π is much smaller than the s/u ratio discussed above (typically below 0.1). If we consider the ratio K/π as a function of transverse momentum then it grows quickly with transverse momentum size and reaches for $p_\perp \simeq 0.4$ GeV (which corresponds to the average transverse momentum) the value 0.3.

We note, however, that if a primary π is produced with a large transverse momentum then there is a smaller difference from the production of a directly produced K because they will then have similar transverse masses.

12.5   Vector meson suppression

1   Preliminary remarks

At first sight one may guess that the ratio of the vector mesons (in the $J^{PC} = 1^{--}$ nonet; we use the usual notation with J the spin, P the parity and C the charge conjugation quantum number, cf. the Particle Data Group tables) to the pseudoscalar mesons (in the $J^{PC} = 0^{-+}$ nonet) should be 3 : 1. This is a purely statistical result corresponding to the number of states (three spin states for a vector as compared to the single spin 0 state for a pseudoscalar). The numbers of isospin and strangeness states are of course the same because the vector mesons and the pseudoscalars are both nonets in SU(3)-flavor.

This ratio is not in accordance with the results of e^+e^- annihilation experiments. Although it is difficult to disentangle the vector mesons (there are many possible combinations of K's and, particularly, of π's in a multiparticle final state) the general consensus for the PEP–PETRA energy region (20–40 GeV) is that the ratio of vector mesons to pseudoscalar mesons is in fact about 1 : 1 or maybe even smaller.

There is a good dynamical reason for this disagreement and we will now consider it in some detail. The vector meson wave functions are actually more difficult to tunnel into because they are smaller at the point, the classical turning point, where the produced q and $\bar{q}$ start to notice their final fate. The ensuing model should be applicable to all relative yields in which two produced hadrons, with similar quantum numbers, for some reason, e.g. due to constituent interactions, have different masses.

The q- and $\bar{q}$-particle stem from two adjacent vertices and there is an attractive force from the string which will bind them together. There are also, however, spin-dependent forces which will act in opposite ways in a pseudoscalar state and a vector state. The spin-spin correlation between particles 1 and 2 is positive for vectors and negative for pseudoscalars (note that the eigenvalue of $\mathbf{S}^2$ is $s(s+1)$):

$$\mathbf{S}_1 \cdot \mathbf{S}_2 = \tfrac{1}{2}\left[(\mathbf{S}_1 + \mathbf{S}_2)^2 - \mathbf{S}_1^2 - \mathbf{S}_2^2\right] = \begin{cases} \tfrac{1}{4} & \text{for vectors} \\ -\tfrac{3}{4} & \text{for pseudoscalars} \end{cases} \qquad (12.31)$$

It is a very general property that all physical systems try to 'economise' with the energy. This means that in a system with a hamiltonian H (H_0 provides the space structure of the $q\bar{q}$ interaction),

$$H = H_0 + g\mathbf{S}_1 \cdot \mathbf{S}_2 \qquad (12.32)$$

the constituents of a vector meson will try to avoid regions with positive g-values because the interaction energy of the state is increased in these regions and avoidance therefore implies a total decrease in the state energy. In contrast, for the pseudoscalar states it is advantageous for the constituents to be in regions with positive g-values. Therefore *we expect that the wave functions of the pseudoscalars will be concentrated in regions with g positive while the vector states will behave in the opposite way.*

This type of spin-dependent force occurs in a state with color-force binding, because of gluon exchange. It turns out that the effect has a very short-range character in coordinate space. A simple model for it is a positive contact form, which is very large when the constituents are close and vanishing when they are apart (x_j, $j = 1, 2$ are the coordinates of the particles and $\alpha > 0$ the effective coupling):

$$\alpha\delta(x_1 - x_2)\mathbf{S}_1 \cdot \mathbf{S}_2 \qquad (12.33)$$

For such an interaction it is evident that in one space dimension the vector state, where the q- and the $\bar{q}$-particle try to stay apart, will be larger in size than the corresponding pseudoscalar state, where they would like to stay close together. A more spread-out vector meson wave function, which is still normalised to unity over the region, will be smaller at the classical turning point.

For real, three-space dimensional, bag models of the hadrons there is a corresponding effect, although the bag radii in this case are similar. Nevertheless the constituents in a vector meson bag move close to the outskirts of the bag so that they stay apart as much as possible while the pseudoscalar bag constituents tend to stay close together at the centre.

2 A one-dimensional bag model

We will now estimate the ratio, mentioned at the beginning of the last subsection by solving for the eigenstates of the hamiltonian in Eq. (12.32). For simplicity we will use nonrelativistic kinematics for H_0 (with equal masses μ for the q_1 and the $\bar{q}_2$):

$$H_0 = 2\mu + \frac{p_1^2}{2\mu} + \frac{p_2^2}{2\mu} + \kappa|x_1 - x_2| \equiv 2\mu + \frac{P_{cms}^2}{4\mu} + \frac{p^2}{\mu_r} + \kappa|x - x_0| \quad (12.34)$$

Here the cms motion of the pair has been omitted and we choose for the cms momentum eigenvalue $P_{cms} = 0$. The relative coordinates $x - x_0$ and

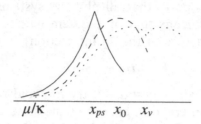

Fig. 12.5. The shapes of the wave functions corresponding to tunnelling into a vector meson (dotted line) and into a pseudoscalar meson (solid line) as well as into the Airy function ψ (broken line) mentioned in the text.

momentum p are introduced together with the reduced mass $\mu_r = \mu/2$. We will now construct the wave functions for the ground states in the pseudoscalar case, the vector case and the case when $\alpha = 0$ in Eq. (12.33), denoted $j = ps, v, 0$, respectively.

The linear potential evidently vanishes at $x = x_0$ and we note that for $x < x_0$ we have the differential equation

$$\left[-\frac{1}{2\mu_r} \frac{d^2}{dx^2} + \kappa(x_0 - x) \right] \Phi_j(x) = (E_j - 2\mu)\Phi_j(x) \qquad (12.35)$$

For all wave functions Φ_j we choose a value of $x_0 \equiv x_j$ such that the left-most classical turning point (classically defined by $p = 0$ and quantum mechanically by an inflection point in the wave function) is at the same point $x = \mu/\kappa$. This means that for all cases when $x < x_j$ we have

$$\left(-\frac{1}{2\mu_r} \frac{d^2}{dx^2} - \kappa x \right) \Phi_j(x) = -\mu\Phi_j(x), \quad x_j = \frac{E_j - \mu}{\kappa} \qquad (12.36)$$

This means that the wave functions for all j have the same behaviour in this region. The differential equation can be solved and one obtains a function, known as an Airy function (which we denote $\psi(x)$). It decreases exponentially along the negative x-axis. We do not need its properties but it is shown in Fig. 12.5.

The different wave functions do not need to be normalised in the same way and we therefore introduce the normalisation constants N_j:

$$\Phi_j(x) = N_j\psi(x), \quad x < x_j \qquad (12.37)$$

From the symmetry of the problem we deduce that

$$\Phi_j(x) = N_j\psi(2x_j - x), \quad x > x_j \qquad (12.38)$$

thus ensuing continuity at $x = x_j$.

For the case when the spin-spin interaction vanishes ($j = 0$) the eigen-value $E_0 = \kappa x_0 + \mu$ is determined from the requirement that the first derivative at $x = x_0$ should be continuous, which implies

$$\frac{\Phi_0}{dx}(x \equiv x_0) = \frac{d\psi}{dx}(x_0) = 0 \qquad (12.39)$$

For the two other cases we obtain a discontinuous first derivative by integrating the differential equation over the region including the point x_j:

$$\lim_{\epsilon \to 0} \frac{1}{2\mu} \left[\frac{d\Phi_j}{dx}(x_j + \epsilon) - \frac{d\Phi_j}{dx}(x_j - \epsilon) \right] = \alpha \mathbf{S}_1 \cdot \mathbf{S}_2 \Phi_j(x_j) \qquad (12.40)$$

This means

$$-\psi'(x_j)/[\mu\psi(x_j)] = \begin{cases} -3\alpha/4 & \text{for } j = ps \\ \alpha/4 & \text{for } j = v \end{cases} \qquad (12.41)$$

We conclude that for the pseudoscalar (vector) case the ratio of the wave function and its derivative (known as the logarithmic derivative) is positive (negative) while for the case $j = 0$ it vanishes.

The function ψ starts out very small, for large negative x-values, and then increases towards a maximum at the value $x = x_0$; afterwards it decreases. Therefore we conclude that the value $x = x_{ps}$ must be on the upward slope, and the value $x = x_v$ on the downward slope, in order that the logarithmic derivatives should have the right signs.

Thus we obtain without much effort that

$$x_{ps} < x_0 < x_v \qquad (12.42)$$

which is basically what we set out to prove, namely that the vector meson state in this one-space dimensional model is essentially larger than the pseudoscalar state in extension.

It is also rather easy to estimate the relative size of the wave functions at the classical turning point, because this is given by

$$\frac{|\Psi_v|^2}{|\Psi_{ps}|^2} = \frac{|N_v|^2}{|N_{ps}|^2} \qquad (12.43)$$

i.e. by the normalisation condition for the integrals

$$1 = |N_j|^2 \left(\int_{-\infty}^{x_j} dx |\psi(x)|^2 + \int_{x_j}^{\infty} dx |\psi(2x_j - x)|^2 \right) \qquad (12.44)$$

As the functions are subject to exponential rapid decrease outside the classical turning points and vary reasonably slowly in between these points we can estimate that the ratio in Eq. (12.43) is

$$\frac{|N_v|^2}{|N_{ps}|^2} \simeq \frac{x_{ps}}{x_v} = \frac{E_{ps} - \mu}{E_v - \mu} \qquad (12.45)$$

A more detailed numerical investigation of the ratio tells us that Eq. (12.45) is a very good approximation if the ratio is larger than $\simeq 0.2$. For smaller values of $E_{ps} - \mu$ the ratio in the first line of Eq. (12.45) levels out and for a vanishing $E_{ps} - \mu \cdot$the ratio of the normalisation constants becomes $\simeq 0.12$. The question whether there should be one factor for the q and another for the $\bar{q}$ is probably of little interest because of the general simplicity of the model.

Phenomenologically the Lund model has been successful with the following suppression rates:

$$\frac{\rho}{\rho + \pi} \simeq 0.5, \quad \frac{K^*}{K^* + K} \simeq 0.6, \quad \frac{D^*}{D^* + D} \simeq 0.75 \qquad (12.46)$$

We note that there is a clear tendency that the closer the masses of the vectors and the pseudoscalars, the closer we come to the statistical value $3 : 1$.

3   The η' puzzle

In order to exhibit the way in which the Lund model distributes the probabilities for different flavor configurations and also to mention a current phenomenological problem we will discuss the assignment of probabilities to the isoscalar states η and η'. Recently some doubt has appeared, [51], about the $\rho/(\rho + \pi)$ ratio – it should probably be even smaller than we predicted in the earlier section, maybe in the range 0.3–0.4. It turns out, however, that the η' has a large branching ratio for the decay

$$\eta' \to \rho + \pi \qquad (12.47)$$

and in this way the number of η'-particles will influence the observed ρ-signal. We will now show that according to the Lund model rules the number of η'-particles must be enhanced if we suppress the directly produced ρ-particles. But then we will get back a large ρ-rate through the decay channel in Eq. (12.47). To get around the problem diminishing the ρ-rate we must therefore also suppress η'. The practical effect is then that we give an overall enhancement to the π-mesons!

According to the philosophy of the model all states should be populated according to the probability of projecting out a given flavor composition upon them. The η and η' in the pseudoscalar nonet (from now on called PS) play the same role as the ϕ and the ω for the vector (V) mesons.

In more detail, a state with the third component of isospin vanishing is, in SU(3)$_f$ (with f for flavor) either the $I_3 = 0$ component, ($|0_j\rangle$), of the isovector (ρ_0 for the V and π_0 for the PS) or the total singlet state ($|1_j\rangle$) or octet state ($|8_j\rangle$) in either nonet; $j = PS, V$. Knowing that

$|0_j\rangle = (|u\bar{u}\rangle - |d\bar{d}\rangle)/\sqrt{2}$ (compare with the spin 1 states built from spin 1/2!) we may write for the other states

$$|1_j\rangle = \frac{|u\bar{u}\rangle + |d\bar{d}\rangle + |s\bar{s}\rangle}{\sqrt{3}}, \quad |8_j\rangle = \frac{|u\bar{u}\rangle + |d\bar{d}\rangle - 2|s\bar{s}\rangle}{\sqrt{6}} \quad (12.48)$$

to obtain orthogonal combinations in the three flavors u, d, s. There is, however, one further degree of freedom, called 'mixing', meaning that the true observed states may be mixtures of the singlet and octet states:

$$|h_{j1}\rangle = |8_j\rangle \cos\theta_j - |1_j\rangle \sin\theta_j$$
$$|h_{j2}\rangle = |8_j\rangle \sin\theta_j + |1_j\rangle \cos\theta_j \quad (12.49)$$

The vector nonet states are 'simple' from the point of view of mixing. The ϕ-particle is almost a pure $s\bar{s}$-state, as it decays almost exclusively into a $K\bar{K}$-pair. Therefore the V mixing angle $\theta_V \sim 0.62$ so that $\tan\theta_V \simeq 1/\sqrt{2}$. Then the ω-particle is the combination $(|d\bar{d}\rangle + |u\bar{u}\rangle)/\sqrt{2}$. But the pseudoscalar nonet states η and η' are more complex flavor states: different authors (cf. the Particle Data Group tables) assign different mixing angles θ_{PS} from -0.17 to -0.40.

Now suppose that we start with a u (or d) and pick up from the next vertex a $\bar{u}$ (or $\bar{d}$). Then we may produce either a vector state (ρ_0 or ω) or a pseudoscalar state (π_0, η or η') according to the assignments given above. If we suppress the vectors we enhance the pseudoscalars and therefore the decay channel in Eq. (12.47) will also be enhanced!

We will in Chapter 14 again find reasons for suppressing the η'-rate (which actually is not very well known from direct measurements). One possible clue to such a dynamical suppression is the very large masses of the η and η'. There have been different models suggested (mostly built upon the possibility that the vacuum exhibits a dynamical 'chiral symmetry breaking', which we will have no space to consider in this book).

In such models there will necessarily be some mechanism which makes the isoscalar states more massive than the corresponding isovector ones in the PS nonet. This will then, in all cases, have the further effect that we obtain a suppression factor similar to the one we found in our one-dimensional bag model, the spin interaction being exchanged for the mechanism that makes the isoscalars more massive.

13

Heavy quark fragmentation
and baryon production

13.1   Introduction

In this chapter we are going to consider a few further phenomena that should be included in a realistic model for hadron production.

We start by considering heavy flavor fragmentation. There should be no production of heavy flavors in the fragmentation process itself because of the very strong suppression from the tunnelling process. Heavy quark jets will nevertheless occur when the heavy flavor is produced in a process where there is a large energy concentration, e.g. in an e^+e^- annihilation process. Then the first-rank hadron in the jet contains the heavy flavor and such a hadron will, in general, have a larger mass than the ordinary hadrons, which are made up from the lighter quark flavors, u, d and s. We have seen (cf. Chapter 9) that for the usual Lund fragmentation function a larger-mass hadron will have a 'harder' z-spectrum, i.e. the typical value of the fragmentation variable z will be closer to unity.

We will consider a number of different models, both those that tend to give $1 - z \propto 1/M$ and those that give $1 - z \propto 1/M^2$ for the first-rank hadron with large mass M. We will also consider a rather different treatment which leads to the so-called Peterson formula, [99], for heavy quark fragmentation. The basic idea is to make use of the wave functions obtained in a lightcone-dynamical scenario.

After that we will continue with a discussion of baryon-antibaryon ($B\bar{B}$) production. A baryon, or at least a baryon resonance state, may well have a more complex structure than that which a $(1 + 1)$-dimensional dynamical scenario can provide. The number of parameters which occur for the description of $B\bar{B}$-production in the Lund model is rather large. The number of baryon states is eight if we count the ones in the $J^P = (1/2)^+$ octet (we use the usual notation with J the spin and P the parity of the states) (N, Λ, Σ, Ξ) and in the $J^P = (3/2)^+$ decuplet (Δ, Σ^*, Ξ^* and Ω^-).

The number of parameters to describe their yields is, however, basically seven! There are good reasons for complaints about the predictive powers of the model.

The two models we will discuss, the *diquark model* and the *popcorn model*, have, however, some endearing qualities. If we are only interested in describing the ordinarily observed baryon states, i.e. the nucleons N (proton and neutron) and the Λ-particle, then their rates are determined by a single number, the baryon-to-meson rate. This is so if we use an SU(6) (flavor-spin) symmetrisation of the wave functions.

Both models account for a strong increase with energy in the baryon-to-meson ratio (which is not only a kinematical effect) as well as an increased baryon fraction in gluon jets. This will be discussed after the introduction of the Lund gluon model (see Chapter 15). Also both models exhibit a string 'drag' effect in the sense that the B and $\bar{B}$ in a pair tend to go in different directions along the string. This is due to the correlation between flavor and color, i.e. a q has color and flavor and is therefore dragged by the string in the opposite direction to a $\bar{q}$, which has anticolor and antiflavor. This was experimentally observed early on by the TPC group at PEP.

The difference between the models is mostly related to the transverse momentum correlations. In the diquark model the B and $\bar{B}$ are neighbors in rank and therefore contain stronger transverse momentum correlations than $B\bar{B}$-production in the popcorn model, where there may be mesons produced in between. The experimental data tend to confirm the popcorn scenario.

We will end the chapter with a brief discussion of a different use of the Lund model fragmentation formulas in the way suggested by a group in UCLA, [38]. The basic idea is to make use of the Lund model area law to determine the *relative rates for different kinds of hadrons*. In this approach there is no use of the probabilities for producing different $q\bar{q}$-pairs. Instead it is the fact that the hadrons have different masses, and consequently will effectively use up different areas in the Lund model fragmentation formulas, that provides the relative probabilities. The model contains an intriguing picture, which, for some reason that is not understood, seems to mirror rather well the observed rates and spectra for different hadrons.

13.2 Heavy quark fragmentation

In Chapter 12 we learned that, at least within a tunnelling scenario with an available energy per unit length equal to the string tension $\kappa \simeq 1$ GeV/fm, there is no heavy quark production along the string.

There is nevertheless the possibility that a heavy quark pair $q_h, \bar{q}_h$ is produced initially in e.g. an e^+e^- annihilation event, if there is a large energy concentration available from the annihilating pair. Later, we will also discuss a particular 'hard' process in which a gluon may split up into a $q\bar{q}$-pair. For this we will once again need the possibility of fragmenting strings containing heavy quarks.

We will start with a few introductory remarks due to Bjorken. After that we will consider three different scenarios.

In the first we will use the Lund fragmentation function based upon the usual area suppression.

In the second we will use a prescription first proposed by Bowler, [32]. He noted that the area spanned by a heavy quark, moving until it meets its 'light' partner (from the $q\bar{q}$-pairs usually produced) to form a hadron, is smaller than that used in the Lund model formula. A heavy particle connected to a string moves along a hyperbola, while a massless quark moves along the lightcone which is the asymptote of the hyperbola. Therefore the area spanned is smaller for the heavy particle. We note that in the interpretation we used for the area law in Chapter 11, it was actually the area in space-time which should occur in the area law. We will show that a correction of the type Bowler advocates leads to a different version of the Lund model fragmentation formulas, in which there are different values for the fragmentation parameter a for the light and for the heavy flavor.

We will then pursue a very different approach to fragmentation and derive the Peterson formula, [99]. In this approach the basic idea is to build up a wave function for the final state from the lightcone dynamics that we sketched in Chapter 3. Such a wave function is based upon the off-shell nature of the state. This leads to a simple one-parameter formula for the distribution of the first-rank hadron in a heavy quark jet, which has been used rather successfully.

1 Bjorken's remarks

The following arguments for an average cascade correspond to the essence of Bjorken's ideas. Suppose that we consider an ordinary quark jet with, for simplicity, a single kind of hadron (mass m). The first-rank hadron will have rapidity y_l and the rest will have rapidities $y_l - \delta y, y_l - 2\delta y, \ldots$. This means that the total energy W is

$$m \exp y_l \sum_{j=0} \exp(-j\delta y) = \frac{m \exp y_l}{1 - \exp(-\delta y)} = W \qquad (13.1)$$

We conclude that the first-rank particle will take a fraction of the total energy

$$z_l = \frac{m \exp y_l}{W}, \quad z_l + \exp(-\delta y) = 1 \tag{13.2}$$

This is equivalent to the results in Eq. (9.11) in connection with the iterative cascade models.

Next we consider a heavy quark jet and assume that the first-rank hadron has (large) mass M at rapidity y_h. *All the remaining ones should, however, behave as before*, i.e. have the 'ordinary' average rapidities $y_h - \delta y, y_h - 2\delta y, \ldots$ This means that Eq. (13.1) is exchanged for

$$M \exp y_h \left[1 + \sum_{j=1} \frac{m \exp(-j\delta y)}{M}\right] = M \exp y_h \left[1 + \frac{m(1 - z_l)}{M z_l}\right] = W. \tag{13.3}$$

Therefore the first-rank particle in a heavy quark jet should take a fraction of the total energy

$$z_h = \frac{M \exp y_h}{W} = \frac{1}{1 + m(1 - z_l)/M z_l} \simeq 1 - \frac{m_0}{M} \tag{13.4}$$

The whole discussion is evidently an instructive demonstration of the difference between rapidity and energy-momentum. From a knowledge of ordinary quark jets we may guess that the characteristic mass m_0 should be of the order of 1 GeV.

2 Ordinary Lund area suppression versus a more literal interpretation

The Lund model fragmentation function for the production of a hadron with flavors α, β and with a mass m is given by

$$f_{\alpha \to \beta}(z)dz = \frac{N dz}{z} z^{a_\alpha - a_\beta} (1 - z)^{a_\beta} \exp\left(-\frac{bm^2}{z}\right) \tag{13.5}$$

It is mostly used with the same value for the parameters, $a_\alpha = a_\beta \equiv a$. For this case we found in Chapter 9 that for a commonly used value of $a = 0.5$ there is a maximum value of f when the fragmentation variable $z = 1 + bm^2 - \sqrt{1 + (bm^2)^2}$.

Therefore in this case the correspondence to z_h in Eq. (13.4) will depend upon the mass of the first-rank hadron according to

$$z_{1,oL} \simeq 1 - \frac{1}{bM^2 + \sqrt{1 + (bM^2)^2}} \tag{13.6}$$

(the notation oL stands for 'ordinary Lund'). For large values of M it behaves as $1 - (\mu_0/M)^2$ instead of linearly as in Bjorken's guess.

Phenomenologically the behaviour in Eq. (13.6) seems to be too stiff, i.e. to predict values of z that seem too large although they are not

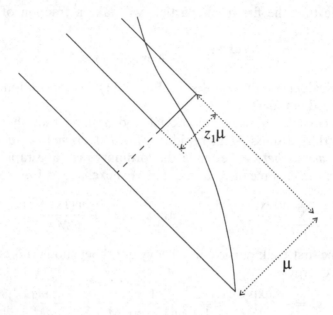

Fig. 13.1. The motion under the influence of the string tension of a heavy flavor parton in the coordinate system in which the parton was originally at rest.

actually excluded by the experimental data. We will now consider a modification introduced by Bowler (in the context of the Artru-Menessier-Bowler model; see Chapter 8).

We will go to a frame where the (original) q_h is initially at rest (see Fig. 13.1). According to the equations of motion for a heavy quark it will start to be dragged away by the string field along the hyperbola

$$(\kappa x_+ - \mu)(\kappa x_- + \mu) = -\mu^2 \tag{13.7}$$

(cf. Chapter 6 and note that in the rest frame of the q_h the total $p_+ = \mu$).

We will assume that a first-rank hadron with mass M is produced, from q_h together with a massless $\bar{q}_1$ stemming from the first vertex, with energy-momentum fraction z_1 as indicated in Fig. 13.1. We also find that the negative lightcone component of the first vertex x_{-1} is, from the mass requirement, given by

$$\kappa x_{-1} = \frac{M^2}{z_1 \mu} \tag{13.8}$$

Bowler then assumed that *only the area between the broken line and the orbit of the heavy quark* (Fig. 13.1) *should be counted in connection with an area suppression law*, i.e. he suggested the region

$$A_B = \kappa^2 \int_0^{x_{-1}} dx_- \int_0^{x_+(x_-)} dx_+ \qquad (13.9)$$

where the quantity x_+ (x_-) is calculated from Eq. (13.7). The integrals are easily performed using Eqs. (13.7), (13.8) and we obtain for Bowler's area

$$A_B = \frac{M^2}{z} - \mu^2 \left[1 + \log \left(\frac{M^2}{\mu^2 z} \right) \right] \qquad (13.10)$$

If we introduce Bowler's area into the fragmentation function in Eq. (13.5) (with a single value of a) then we obtain:

$$f_B = \frac{N'dz}{z} z^{a_h - a}(1 - z)^a \exp \left(-\frac{bM^2}{z} \right) \qquad (13.11)$$

where we have incorporated the difference between Bowler's area and the usual area as a change in the normalisation constant $N \to N'$ and as a new a-parameter

$$a_h = a - b\mu^2 \qquad (13.12)$$

characteristic of heavy quark fragmentation.

This result is interesting for several different reasons. Firstly it is evident that the spectrum will become softer than the one obtained in Eq. (13.6) because in Eq. (13.11) there is a negative z-power. Due to the very strong suppression in the exponent there is, of course, no trouble from this term in the normalisation of f_B.

An observational problem is that some of the b-flavored particles will decay into c-flavored ones. Therefore the observed c-flavored mesons have a different spectrum from that obtained using any of the formulas discussed above. Also, ordinary Lund model fragmentation looks very similar to the observed data, e.g. for D^*-mesons, because of this decay contribution. Only when the experimentalists have been able to disentangle the corresponding B-meson signal will it be possible to know whether any of these alternatives is correct.

A second reason is related to the findings in Regge theory. We have in an earlier discussion (Chapter 10) related the parameter a to the Regge intercepts and we see from the result in Eq. (13.12) that there should be a smaller intercept if there are larger-mass constituents involved. This is in accordance with the early phenomenological findings (cf. [46]).

3 A different approach based upon a wave function from lightcone dynamics

The Feynman propagator, in the contexts in which have met it, contains the notion of the *virtuality of a state*. In particular the free propagator describes the way in which a quantum fluctuation may develop in space-time according to Heisenberg's indeterminacy relation. We will find a similar description of a decaying state, i.e. a resonance, in connection with the Breit-Wigner form factor, in Chapter 14. In that case the correspondence to the Feynman limiting $\epsilon \to 0$ in Eq. (3.81) has a size corresponding to the inverse lifetime of the state.

A bound state in quantum theory is often described in terms of an exponentially falling wave function in space. This corresponds in energy-momentum space to an inverse power in the momentum. Our experience of form factors, which corresponds to the description of extended distributions in space, cf. the example of the elastic proton form factors in Chapter 5, confirms this general inverse-power behaviour in energy-momentum space.

In [37] there is a treatment of the possible bound-state wave functions using the lightcone formalism for the Feynman rules that we considered in Chapter 3. A possible wave function for a $c\bar{q}_l$-state (index l for light so that the state corresponds to a D-meson) is deduced from the distribution

$$
\begin{aligned}
|\psi|^2 &\sim N \frac{\delta(1 - x_1 - x_2)}{(M_D^2 - m_c^2/x_1 - m_l^2/x_2)^2} \\
&= N' \frac{\delta(1 - x_1 - x_2)}{(1 - 1/x_1 - \epsilon/x_2)^2}
\end{aligned}
\tag{13.13}
$$

for the energy-momentum fractions x_j of the two constituents. Here N is a normalisation constant and the masses M_D, m_c, m_l correspond to the D, the c-quark and the light quark, respectively. In the second line it is assumed that the mass $M_D \simeq m_c$ is divided out and that $\epsilon = m_l^2/m_c^2$.

In [99] the authors take a step further and assume that the gross features of the amplitude for a heavy quark Q to fragment into a hadron $H = Q\bar{q}$ and a light quark q should be determined in a similar way. This means that it is the value of the energy transfer, ΔE, in the breakup process which will determine the distribution. Therefore the amplitude should behave basically as $(\Delta E)^{-1}$, ΔE being proportional to the denominator in Eq. (13.13).

They include a factor $1/z$ from the longitudinal phase space, perform the integral over the δ-distribution to get $x_1 = z$, $x_2 = 1 - z$, and obtain the shape

$$
D_Q^H(z) = \frac{N}{z[1 - 1/z - \epsilon/(1 - z)]^2}
\tag{13.14}
$$

as a possible fragmentation function for a heavy quark into a first-rank hadron. The formula evidently contains only one parameter and has been used extensively and successfully with reasonable values of ϵ.

This means that it is possible to use a very different approach to fragmentation than that usually advocated in this book. The lesson is that although only experimental data can distinguish between possible theoretical developments, it is actually quite difficult to observe such differences since very different functional shapes give closely similar predictions for the experimental data.

13.3 Baryon-antibaryon production

1 Preliminaries

We will make use of two general assumptions in $B\bar{B}$-production.

B1 We will only be concerned with the baryons in the octet, $J^P = (1/2)^+$, and decouplet, $J^P = (3/2)^+$, representations of SU(3) flavor and we will neglect all other production channels.

B2 The octet and decouplet can be made into a 56-representation (with completely symmetrical wave functions) of the group SU(6) in combined flavor-spin assignment. We will assume that all wave functions of the constituents are determined in accordance with the SU(6) requirements and that only the states in the 56-representation are actually produced.

The first assumption is one of economy. We know that the higher baryon resonances occur in only very tiny fractions in those exclusive channels that have been studied.

The second assumption is very basic in the Lund model. The production mechanism in the model is determined not only by the fragmentation probability (e.g. the area suppression law) but also by the available number of states (e.g. the phase-space factors).

Although it is known that SU(6) symmetry is rather badly broken (the different states do not have the same masses and here also we will break the symmetry in a similar way) we will insist on a projection to completely symmetric flavor-spin states for the baryons. If we did not use this requirement then the probability of picking up three u-quarks at random (making a Δ^{++}-baryon) would be one chance in 27 (assuming only u, d and s are produced, each with, for simplicity, the same probability). But the probability of picking up a u-quark, a d-quark and an s-quark at random, producing any of the three states Λ, Σ^0 or Σ^{*0}, would be six

times larger, by pure combinatorics. Such predictions would be very much
in conflict with the experimental findings.

The requirement, of a totally symmetric baryon flavor-spin wave function, means that if we start by producing e.g. an (effective) diquark $(ud)_1$
(corresponding to a state with spin = isospin = 1) and add an s-quark
we have a probability of only $1/2$ of obtaining a symmetric state; of
this $1/3$ will correspond to a decouplet (Σ^{*0}) and $1/6$ to an octet (Σ^0).
The remaining $1/2$ corresponds to the production of a state outside the
56-representation.

2   The diquark production model

We will next outline a possible production mechanism for effective diquarks. We do not believe that diquarks, although we may equip them
with different internal quantum numbers like color, spin and isospin, are
basic quanta of the QCD force field. A diquark is a qq-state, which from
a color point of view behaves as a $\bar{3}$, i.e. effectively as a $\bar{q}$-charge.

It is, however, reasonable that even an extended charge may have an
effective coupling to the color force field in the same way as a heavy $\bar{q}$.
The density of virtual q-particles may be sufficiently large to make the
probability of finding a partner in a color-$\bar{3}$ diquark state approximately
unity. Then the diquark may, together with an antidiquark (a $\bar{q}\bar{q}$-pair)
tunnel out like a heavy quark pair. This is the basic production mechanism
in the diquark model.

In this way we obtain the following tunnelling properties.

- All hadrons have the same transverse momentum production mechanism.

- A $B\bar{B}$-pair, sharing a $(qq, \bar{q}\bar{q})$-pair, are neighbors in rank and are
  therefore sufficiently close in rapidity for correlation studies.

- As the tunnelling probability predicts a fast falloff for (effective)
  color charge masses above $\sqrt{\kappa/\pi} \simeq 0.25$ GeV a small difference in
  mass means a large change in probability.

The last property implies that the strange diquarks (and antidiquarks)
are strongly suppressed compared to the non-strange ones. Λ-particles
contain an s-quark and a $(ud)_0$-diquark (spin = isospin = 0), and also
(small) $(d + us)$-components and so on. Due to the suppression of strange
diquarks a Λ will not be produced very often with a $\bar{\Lambda}$ (which may occur
for the latter diquark combination) but instead with an $\bar{N}$ (antiproton
or antineutron, which contains only non-strange diquarks) and a K or
a K^* to compensate the strangeness. However, if we do find a $\Lambda\bar{\Lambda}$-pair

in an event then they are almost always rank neighbors in the diquark production model.

In order to make calculations in the model it is necessary to determine the 'masses' of the diquark pairs. The lightest diquark masses, obtained by consideration of the masses of, and mass differences between, the baryons may well be around 0.4 GeV. This would set the overall ratio of the production of diquark pairs to quark pairs at around 0.1. Such a production ratio of $B\bar{B}$ to mesons seems from the experiments to be in the right range (see the discussion after the popcorn model has been presented).

The remaining parameters in this model, $(ud)_1/3(ud)_0 \simeq 0.05$, the suppression ratio of strange diquarks $(us)(d)/(ud)(s) \simeq 0.4$ and the suppression ratio of double-strange diquarks $(ss)_1/(ud)_0 \simeq 10^{-3}$ are at best fitted parameters. The decouplet-to-octet ratio has been kept at unity but could of course be changed if future experiments should require it.

It is the diquark-to-quark production ratio which is the main parameter for the relative ratio of Λ's, protons and neutrons. If one were to decrease e.g. the ratio $(ud)_1/3(ud)_0$ then the increase in directly produced Λ's would compensate the decrease in Λ-production through the channels $\Sigma \to \Lambda + X$, $\Sigma^* \to \Lambda + X$, etc. (X is any other decay product).

Such a change would, however, produce a different amount of e.g. Δ's in a jet. At present a thorough search is being conducted for the resonance content in e^+e^- annihilation events. It will be interesting to see whether one can understand the spectra from these simple considerations (although we necessarily have five parameters for the detailed content already!).

Let us consider some of the consequences of the present scheme. For 30 GeV e^+e^- annihilation events, about 30% of these will contain one $B\bar{B}$-pair and about a further 6% two pairs. About half of the $B\bar{B}$-pairs will decay into $p\bar{p}$-states. Therefore in about 25% of the observed $p\bar{p}$-combinations the p stems from one original pair and the $\bar{p}$ stems from another. This evidently waters down the correlations stemming from the fact that the $B\bar{B}$-pairs are produced as rank neighbors.

For a directly produced $p\bar{p}$-pair in a quark jet, the $\bar{p}$ is about half a unit in rapidity behind the p (in a q-jet it is evident that we will first produce the baryon and only afterwards, i.e. at lower rank, the antibaryon). Regarding the analysis of an event, however, the quark and the antiquark directions are not defined a priori.

If we use one of the customary axes for an event, the thrust axis, cf. Chapter 15, to study the rapidity difference $|y_p - y_{\bar{p}}|$ we end up with an average distance around 1.3, which is evidently far away from the primary result. The transverse momentum correlations are also watered down and we find that

$$\frac{\langle \mathbf{p}_{\perp p} \cdot \mathbf{p}_{\perp \bar{p}} \rangle}{\langle \mathbf{p}_{\perp}^2 \rangle} \simeq 0.3 \tag{13.15}$$

There are some particular features of $B\bar{B}$-production found in connection with events where there is gluon radiation. These features are discussed in Chapter 15.

3 *The popcorn mechanism*

The gluon radiation influences the transverse momentum correlations. When gluons are included the correlation parameter in Eq. (13.15) becomes only half as large as in the diquark production model. But, according to the experiments, this is still a bit too big.

The tunnelling out of non-fundamental quanta such as the diquark-antidiquark pairs is also a less pleasing aspect of the model. Therefore we have produced another model within the Lund scheme, the *popcorn model*. This is based upon an idea first advocated in [40], i.e. that the baryons are produced in a stepwise manner, 'popping out'.

Unfortunately the popcorn model necessarily increases the number of parameters for the detailed content of the baryon species to seven, although it is still the same for the proton, neutron and Λ-particle.

A general fact about the number of parameters in a model is that even if they are not in practice used (an example is the probability of producing the decouplet baryons more or less often than the octet baryons) they are still parameters! This general definition of the notion 'parameter' means 'a possible degree of freedom not fixed by the present dynamics'.

The popcorn mechanism diminishes the correlations between the B and $\bar{B}$ in a pair (they will no longer necessarily be neighbors in rank) but it also provides a scenario in which massive but loosely bound 'diquarks' are produced, without invoking the tunnelling scenario for diquarks *per se*.

In order to understand this consider Fig. 13.2. We will assume that there are space-time regions, such as A, B and C in the figure, where there are fluctuations with the 'wrong' colors in the field. In simple language the original $q_0\bar{q}_0$-pair may start out as a color $r\bar{r}$-pair (although we remember that the state is actually a coherent color state). The field is then a $r\bar{r}$-colored one; 'wrong' means that there may be fluctuations inside regions of the field which are $g\bar{g}$ or $b\bar{b}$).

Then in such a region we would see the color combination $rg = \bar{b}$ in one direction and the combination $\bar{g}\bar{r} = b$ in the other (if the colors happen to combine in this particular way). Under these circumstances we find that the existence of the color fluctuation does not change the energy

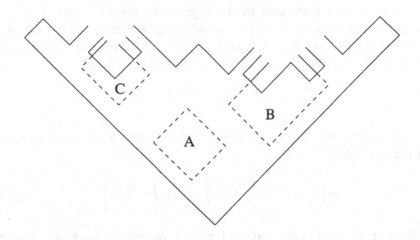

Fig. 13.2.   A space-time description of a string field with color fluctuations having disallowed colors. For the regions B and C this leads to $B\bar{B}$ pair production situations, in B with a meson in between.

density in the string field; it is still κ-characteristic for a $3\bar{3}$ color field in the region with the wrong colors. The force direction changes, however, to the opposite one. This implies that:

- the wrongly colored charges are pulled with the same magnitude of force in the two directions, i.e. there is no net force on them;

- one or more $b\bar{b}$-colored $q\bar{q}$-pairs may tunnel out within the region. If that happens then we evidently obtain a qqq-state at one end and a $\overline{qqq}$-state at the other, i.e. a $B\bar{B}$-pair is produced. There may also be one or more mesons produced in the $b\bar{b}$-part of the field, i.e. in between the B and $\bar{B}$.

The fact that the two 'wrongly' colored particles, q_1 and $\bar{q}_1$ (each assumed to have transverse mass μ_1) just float around without any force upon them means that we may estimate the probability for them to be at a (spacelike) distance x_1 apart by the same method as in Chapter 11, i.e. using the Heisenberg indeterminacy relation we obtain

$$|\Delta_F(x_1, \mu_1)|^2 \simeq \exp(-2\mu_1|x_1|) \tag{13.16}$$

If there is a pair $q_2\bar{q}_2$ (with transverse masses μ_2) produced in between q_1 and $\bar{q}_1$ then the probability for the particles in the pair to tunnel the distance x_2 is suppressed by the tunnelling factor

$$\exp\left(-\frac{\pi\mu_2|x_2|}{2}\right) \tag{13.17}$$

If no other pair is produced in the region the resulting pair B, $\bar{B}$ will be neighbors in rank. The minimum distance to which the quarks must separate to come onto the mass shell is related to the effective transverse mass, μ, of the 'diquark' $q_1 q_2$ and the antidiquark $\bar{q}_1 \bar{q}_2$ by

$$|x_1| = |x_2| = \frac{2(\mu_1 + \mu_2)}{\kappa} = \frac{2\mu}{\kappa}. \tag{13.18}$$

Then we obtain for the total probability for such a situation to occur, from Eqs. (13.16), (13.17),

$$\exp\left(-\frac{4\mu_1\mu + \pi\mu_2\mu}{\kappa}\right) \sim \exp\left(-\frac{\pi\mu^2}{\kappa}\right) \tag{13.19}$$

Therefore if we stretch the value of π to 4 we obtain back the ordinary tunnelling formula for the diquark-antidiquark! Thus in the popcorn model we obtain a similar suppression to that in the diquark production model for the heavier 'diquarks'.

For a meson to be produced, we need an even larger color fluctuation between q_1 and $\bar{q}_1$. We need an extra piece of the order M/κ to produce a meson with transverse mass M, which leads to an extra suppression factor

$$\exp\left(-\frac{2\mu_1 M}{\kappa}\right) \tag{13.20}$$

The mass of a mesonic system increases quickly with the multiplicity. It is not difficult to convince oneself that, with an average rapidity distance $\delta y \simeq 1$ between subsequent mesons, we obtain for the mass M_n of an n-particle system

$$M_n \simeq M \exp\left[\frac{(n-1)\delta y}{2}\right] \tag{13.21}$$

Therefore the probability of producing more than a single meson in between the $B\bar{B}$-pair will be small. The heavier vector mesons, in particular, may be even more suppressed than the pseudoscalars.

From Eqs. (13.20), (13.21) we may evidently estimate the average number of mesons in between as well as the multiplicity distribution. We obtain that in the mean about 0.5 mesons should be included.

We will therefore introduce the probability factor $(BMB) = 0.5$ of producing a single meson in between the $B\bar{B}$-pair and will neglect the larger multiplicities. Such larger multiplicities are nevertheless indirectly included by allowing the production of both pseudoscalar mesons and vector mesons (which afterwards decay into two or more pseudoscalars) in the same way as in any other part of the string field.

13.4 A different use of the Lund model formulas, the UCLA model

A group at UCLA has proposed, [38], the use of the Lund model formulas in a way different from the one we have presented up to now. They have been quite successful in interpreting the area law as a density for the final-state particle ratios also.

The prescription is the following. They make use of the ordinary projection coefficients, the Clebsch-Gordon coefficients, between a given $q\bar{q}$-state and a hadron state. These are used also in the normal version of the Lund model but then there is (as we have described in Chapter 12) extra suppression of the s-flavor and no production of c- and b-flavors in the fragmentation.

All are allowed in the UCLA version. It is, however, noted that if one produces e.g. an $s\bar{s}$-flavored pair then there must be two strange particles produced, with correspondingly larger masses than mesons that are composed of the u- and d-flavors. Suppressions of this kind can be determined by a few iterations of the basic method.

They then use the results in Chapter 10, [19], according to which every hadron with mass m is given stochastically a value of z according to the fragmentation function

$$f(z, m^2) = N \frac{(1-z)^a}{z} \left(1 - \frac{m^2}{zs}\right)^a \exp\left(-\frac{bm^2}{z}\right) \tag{13.22}$$

The authors use as a finite-energy correction the term we derived in Eq. (10.10). The basic assumption is that N, a and b are the same for all particles. The relative normalisation for different species is given by the integral of f. There are a set of extras, however, and we will mention a few.

They include a transverse momentum generation mechanism, which is an approximation of the mechanism we presented in Chapter 12. There is a claim that the approximation is good, [38].

Further, the method contains the Bowler implementation of heavy quark fragmentation in accordance with the description we gave in subsection 2. They formulate it in such a way that they can keep their general normalisation constant N by writing

$$z_{eff} = \frac{z}{1 - (\mu^2 z/m^2)[1 + \log(m^2/\mu^2 z)]} \tag{13.23}$$

Finally they use the popcorn mechanism for $B\bar{B}$ production, allowing any number of mesons in between. They provide each meson produced in this way with a factor $\exp(-\eta m)$, i.e. an exponential suppression proportional to the mass.

This means that in total they use the distribution

$$f(z_{eff}, m^2_{\perp,D})dzd^2k \times \text{Clebsch-Gordon coefficients}$$
$$\times \text{popcorn distribution} \qquad (13.24)$$

in order to generate a cascade.

In such a fitting-scheme it is of course essential to describe the data well and this method seems to do so although, as of yet, no basic reason why it should work has been put forward.

14

The Hanbury-Brown-Twiss
effect and the polarisation
effects in the Lund model

14.1 Introduction

In this chapter we consider two different observables, which, within the
Lund model, have some bearing upon the confinement properties of
QCD.

We start with the Hanbury-Brown-Twiss effect (the HBT effect) or, as
it is also called, the Bose-Einstein effect. It originated in astronomical
investigations, [78], where one uses the interference pattern of the photons
to learn about the size of the photon emission region, i.e. the size of the
particular star which is emitting the light.

The Goldhabers, [65], found and used in the same way as HBT a
correlation pattern among the produced pions when they investigated
proton-antiproton annihilation reactions close to the threshold (i.e. when
the annihilation occurs at very low relative velocities, so that the total
energy is essentially twice the proton rest mass). Photons and pions have
in common that they are bosons, which means that they thrive on being in
the same state. The HBT effect can be described as an enhancement of the
two-particle correlation function that occurs when the two particles are
identical bosons and have very similar values of their energy-momentum.

The size of the emission region obtained from these experiments in
hadronic physics seems to be essentially the same in almost any kind of
interaction. One obtains a radius of the order of 1 fm, which is a very
reasonable size. The extraction of this size as well as the finer details are,
however, still under intense discussion, [93], because it is very difficult
to determine the relative energy-momentum of the high-energy particles
to the necessary precision. We first discuss the idea behind the chaotic
interference pattern which is at the basis of the HBT effect. After that
we consider the reason why the source size should be similar for the
above-mentioned annihilation reaction at rest (which then is a 'low-energy

reaction') and in a truly high-energy hadronic reaction when the decay products move apart with large velocities.

According to the way we have described the final-state particle production in the Lund model the particles emerge over a large (longitudinal) space-time region, which increases with the energy. Nevertheless it turns out that *it is only the local proper times for the production vertices which play a role in the correlation measurement.* All the proper times are essentially the same for the production vertices in the Lund model and it is this value that will determine the HBT effect. It may very well be the same size in the Goldhaber annihilation measurements also.

We will discuss the implications of the Lund string model for the phenomenon. We will in particular show that the matrix element $\mathcal{M}$, which we derived in Chapter 11, provides a precise prediction, [22], for the HBT effect in an e^+e^- annihilation event. This prediction is in good agreement with the data but there is nevertheless one problem left. In the string model for the HBT effect presented in [22] it is necessary to neglect the fact that a large amount of the pions are decay products of resonances. From ordinary quantum mechanical considerations one expects that the wave functions of the produced pions are affected if the pion stems from such a decay. It is not possible to explain the observed HBT effect if one removes the pions which stem from long-lived resonances or modify their wave functions in accordance with the expectations from the resonance wave functions. It was, however, pointed out by Bowler, [34], that the main problem relates to the Lund model rate for η'-mesons. If that rate is decreased then the predicted HBT effect from [22] is restored to almost the same size even if decay pions are included, i.e. it is essentially in agreement with experiment.

The HBT effect is, as mentioned, also seen in other reactions beside e^+e^- annihilation events. For the case of DIS, deep inelastic scattering, there is also a single string in the Lund model scenario; cf. Chapter 20. Therefore all the considerations in our discussion of e^+e^- annihilation reactions are also relevant to this case. However, for hadro-production and for large gluon activities in general it is not evident how to treat the HBT effect within a model of the Lund type. We will not, therefore, in this book comment upon this topic for hadronic reactions, due to the lack of a sufficiently structured model with which to investigate the effect in these reactions.

The second subject in this chapter is the polarisation properties which are observed in high-energy processes. The momentum distributions of the final-state hadrons are to a large extent governed by longitudinal phase-space size and therefore polarisation properties offer a tool for gaining insight into the 'other dimensions' of the hadronisation process.

Actually, polarisation effects have always been expected to die away at large energies because it has been hoped that for 'asymptopia' there would be only a single production amplitude, which would dominate the processes. Under such circumstances there would be no polarisation because, according to conventional models, such a phenomenon necessitates interfering production amplitudes.

Nevertheless, even at the largest energies available there are strong polarisation effects noticeable in inclusive Λ-particle production. This cannot be explained from hard scattering processes as calculated in QCD and therefore it should stem from the soft confining interactions. We also note that the observed polarisation is not a large-angle phenomenon. It seems to saturate already at transverse momenta of the order of 1 GeV/c.

It is interesting to note that polarisation will come out naturally in a confined production scenario like the Lund model. An intuitive argument is that in a confined field there is always a local direction, the force direction $\mathbf{n}$, i.e. the direction of the flux of the color electric field. This means that, if we have a final-state particle which moves outwards with a momentum $\mathbf{p}$ not parallel to the direction $\mathbf{n}$, there is a nonvanishing axial vector, $\mathbf{a}$, obtained from the vector product $\mathbf{a} = \mathbf{n} \times \mathbf{p}$. A general experience of physics in any context is that wherever a possibility exists, Nature makes use of it. In this case it means that there is the possibility of a scalar coupling, in the overall hamiltonian, between the spin vector $\mathbf{S}$ of the quantum and the axial vector $\mathbf{a}$. (Note that all angular momentum vectors have an axial character.)

We will provide a simple semi-classical model, [9], in which this is a very noticeable effect. We will show the difference between a confined scenario and the production of e^+e^--pairs in an external electromagnetic field. We have used this picture before to illustrate various features. Polarisation properties are one of the few cases where there are major differences between the the dynamics of QED and of QCD. The other cases considered in this book are the behaviour of the running coupling constant, Chapter 4, and the growth of the phase space in multigluon emissions, Chapter 17.

We will be content to apply the model to the polarisation properties of the Λ-particle in baryon fragmentation. It is possible to provide many more predictions using the model but that would mean that we would need a more elaborate formalism.

14.2 The Hanbury-Brown-Twiss effect

1 The classical argument, coherence and chaos

The arguments in this subsection are based upon the discussion in [33]. We will consider the production of pions from a set of sources localised at different positions $\mathbf{x}_j$. Each of them will have some time-dependent wave function $f_j(t_j)$. Then the total amplitude for emission of a pion with energy-momentum $(\omega_1, \mathbf{k}_1)$ is given by

$$A_1 = \sum_j f_j(t_j) \exp[i(\mathbf{k}_1 \cdot \mathbf{x}_j - \omega_1 t_j)] \tag{14.1}$$

The joint amplitude for emission of pions with energy-momentum vectors $(\omega_\ell, \mathbf{k}_\ell)$, $\ell = 1, 2$, is then evidently given by the double sum

$$A_{12} = \sum_j f_j(t_j) \exp(-ik_1 x_j) \sum_\ell f_\ell(t_\ell) \exp(-ik_2 x_\ell) \tag{14.2}$$

where we have used four-vector notation in the complex exponentials. We note that this corresponds to a totally symmetric amplitude. This is necessary because the pions are bosons.

According to quantum mechanics the emission probability is proportional to the square of the amplitude, $P_{12} \propto |A|_{12}^2$, i.e.

$$P_{12} \propto \sum_{i,j,k,\ell} \exp(-ik_1 x_j) \exp(-ik_2 x_\ell) \exp(ik_1 x_i) \exp(ik_2 x_k)$$

$$\times f_j(t_j) f_l(t_\ell) f_i^*(t_i) f_k^*(t_k) \tag{14.3}$$

The basic idea in the HBT effect is to assume that the wave functions of the sources are wildly fluctuating so that there are only contributions to the sums above if

$$j = i, \ell = k \quad \text{or} \quad j = k, \ \ell = i \tag{14.4}$$

This is called the *chaotic limit* and we then obtain, writing $\rho_j = |f_j|^2$ for the source densities and exchanging the sums for integrals,

$$P_{12} \to \int dx_j dx_\ell \rho_j \rho_\ell [1 + \exp(iq\Delta x)] \tag{14.5}$$

We have here introduced the notation

$$q = k_1 - k_2, \quad \Delta x = -x_j + x_\ell \tag{14.6}$$

The result is evidently that

$$P_{12} = R^2(0) \left(1 + \frac{|R(q)|^2}{R^2(0)}\right) \tag{14.7}$$

in terms of the Fourier transform of the sources

$$R(q) = \int dx \exp(iqx) \rho(x) \tag{14.8}$$

If we calculate the one-particle yield in the same chaotic limit we obtain

$$P_1 = \sum_{ij} \exp(-ik_1 x_j) \exp(ik_1 x_i) f_j(t_j) f_i^*(t_i) \to \int dx \rho(x) = R(0) \tag{14.9}$$

and we note that $R(0)$ is a real, positive number. We conclude that the normalised two-particle correlation function in this case will be

$$C_{12}^{HBT} = \frac{P_{12}}{P_1 P_2} = 1 + |\mathscr{R}(q)|^2 \qquad (14.10)$$

where $\mathscr{R}$ is the normalised Fourier transform of the source densities,

$$\mathscr{R}(q) = \frac{\int \rho(x) dx \exp(iqx)}{\int \rho(x) dx} \qquad (14.11)$$

In this way we measure by means of the two-particle correlations something very similar to a *form factor of the source*. We conclude, just as we did for form factors, that the Fourier transform is sensitive to $(-q^2)$-values larger than the inverse squared length scale of the source. In principle it should even be possible to deduce the detailed shape of the source by performing the inverse Fourier transform. However, there is not only the problem that we lack a knowledge of the phases; it is also a sad fact that it is difficult to obtain sufficiently precise experimental data to distinguish between even very different assumptions on the general shape of the source.

The only thing upon which all experiments seem to agree is that there is one size-scale, of order 1 fm. There is no noticeable change in the HBT effect for larger values of $-q^2$ than those corresponding to this scale. But it is not known whether there are in addition *larger* size-scales in space-time (i.e. smaller in energy-momentum space) because to see this we would need precise measurements down to very small relative energy-momentum vectors q.

The HBT effect discussed above stems from the squaring of the (symmetrised) amplitude in Eq. (14.3) and the neglect of all contributions which do not fulfil the conditions in Eq. (14.4). Let us assume that the sums in Eq. (14.2) converge to a regular function F:

$$A_{12} \rightarrow \int dx_j f(x_j) \exp(-ik_1 x_j) \int dx_l \exp(-ik_2 x_l) f(x_l) \rightarrow F(k_1) F(k_2). \qquad (14.12)$$

Using the same arguments as before we find that the single rate is then $|F(k)|^2$ and the 'double' rate is $|F(k_1)|^2 |F(k_2)|^2$. This means that in this case we simply obtain the result

$$C_{12}^{coh} = 1 \qquad (14.13)$$

without the second, chaotic, term which occurs in Eq. (14.10)! This second limit is called the *totally coherent limit*. The term 'coherent' has been introduced because this is the result if we use the coherent states in a field theory coupled to external sources (this case is considered in Chapter 3). There the probability for emission of one or two quanta with given

energy-momenta depends only upon the square of the Fourier transform
of the external current density. This corresponds to the source density
discussed above. There are no chaotic phases in this case.

In Chapter 6 we discussed a simple model for particle production, [39].
In that model the final-state particles stem from the application of an
external current to the Schwinger model; then this particular coherent
state is obtained as a description of the reaction of the quantised dipole
density field. This led to early predictions that there should be no HBT
effect in a simple particle-production process such as e^+e^- annihilation.
The fact that there is and that it can be explained within the Lund model
is an explicit proof that there are basic dynamical differences between the
Schwinger model with an external source and the Lund model.

There has been intense theoretical discussion of whether the sources
in high-energy particle physics are *partially chaotic*, meaning that we
might have a scenario which is in between the chaotic HBT and the
coherent-state results. We will not develop this discussion here; we refer
the interested reader to the reviews [93].

Before we turn to the Lund model interpretation we will comment
upon the effects arising when the production regions are in large relative
motion, which they evidently are in the Lund model as well as in any
other relativistically covariant multiparticle production scenario.

2   The effect of moving sources on the HBT effect

The discussion of the HBT effect given above is sufficient for its application
in astronomy where there are, very probably, many photon emission
sources with (relative) chaotic phases. But they are all at rest or at least
moving slowly with respect to each other. This is not the case for high-
energy multiparticle production.

In the Lund model we have learned that the particle production structure is that, in the mean, all particles are produced after a certain proper
time in the local rest frame. Thus the particle production points are scattered around a hyperbola in space-time $t^2 - x^2 = \tau^2$. The full particle
production region has a large longitudinal extension, $L \sim \sqrt{s}/\kappa$, for the
production sources, although each vertex is governed by τ.

There is, however, also a strong correlation between the particle production points $x \sim \gamma(v)\tau$ and the momentum $p \sim mv\gamma(v)$, where we
use the usual notation for velocity v and the Lorentz contraction factor $\gamma^{-1} = \sqrt{1 - v^2}$. This means that particles from distant parts of the
production region will typically exhibit large momentum differences. Consequently the probability of finding particles from opposite ends of a
two-jet event with momenta less than $1/L$ (which is necessary to obtain
significant interference effects) should be negligible. Therefore the length

scale measured by the HBT effect in this case is not L; *it is instead the distance apart of production points for which the momentum distributions of the produced particles will still overlap.*

In order to get an estimate of the length scale inside which a particular source will deliver its final-state particles let us assume that we have a decay distribution which is completely isotropic in the rest frame of the source. We define the rapidity with respect to some axis. We will neglect the rest masses of the decay products, so that a particle produced along a direction at an angle θ to the rapidity axis will have the rapidity

$$y = \frac{1}{2} \log \left(\frac{E + p_\ell}{E - p_\ell} \right) = \log \cot \frac{\theta}{2} \tag{14.14}$$

where we have used $p_\ell = E \cos \theta$. Expressed in terms of this (pseudo)-rapidity variable we find the isotropic angular distribution

$$\sin \theta \, d\theta d\phi \quad \rightarrow \quad d\phi dy \frac{1}{\cosh^2 y} \tag{14.15}$$

The angle ϕ is the azimuthal angle around the rapidity axis. Thus in this case a typical particle source will produce particles inside a rapidity region with a width around 0.7 rapidity units. We conclude that the particle distributions from sources moving with a rapidity difference Δy will overlap reasonably well as long as $\Delta y \leq 1$.

Therefore this exercise shows that the HBT effect actually must correspond to the measurement of a source size of the order of the local proper time scale, i.e. $\tau \ll L$. In particular the distributions should hardly look more elongated in the longitudinal than in the transverse direction with respect to the main axis and the measured distributions should be independent of the cms energies involved.

3   The interference effect in the Lund model

In Fig. 14.1 we exhibit again the breakup of the Lund string into many final-state yoyo-hadrons, which stem from the combination of q- and $\bar{q}$-particles from adjacent vertices. In the same figure we show (by a broken line) the production of the very same final state but in this case the two particles, denoted 1 and 2, have been interchanged.

If these two particles are identical bosons then the amplitudes corresponding to the two possibilies shown in Fig. 14.1 will interfere according to quantum mechanics. We have up to now considered only the probabilities, and not the amplitudes, for the production processes in the Lund model. But, in connection with the discussion in Chapter 11 of the Wilson loop-integral analogy to the production process, we did provide a tentative

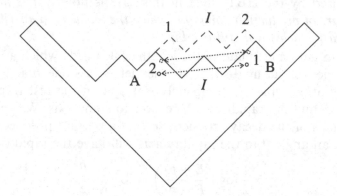

Fig. 14.1.   The breakup situation when two identical bosons, 1 and 2, are produced in the Lund model together with an *intermediate* state I and a set of hadrons outside the regions between the points A and B. The distance between the centres of the yoyo-hadrons is shown by the dotted arrows.

matrix element of the following form:

$$\mathcal{M} = \exp(i\xi A), \quad \xi = 1/2\kappa + ib/2 \tag{14.16}$$

The area A is as usual the space-time region (in energy-momentum space units) swept out before the string breaks. Note that throughout we use the lightcone-metric area just as we did in the derivation of the Lund model fragmentation functions. This means that it is a factor of two larger than the 'true' area and we have corrected for this with the factor $1/2$ in the real part of the parameter ξ.

We note in particular that *the areas are not the same in the two cases*.

Thus for the configuration shown by solid lines in Fig. 14.1 we have an area A_{12} in the matrix element $\mathcal{M}_{12}$ and for the one that includes the broken lines we have an area A_{21} in the matrix element $\mathcal{M}_{21}$. The production of the state with two identical bosons 1 and 2 must then be described by the symmetrised matrix element

$$\mathcal{M} = \mathcal{M}_{12} + \mathcal{M}_{21} \tag{14.17}$$

From this we conclude that the probability will contain the factor

$$\begin{aligned}
|\mathcal{M}|^2 &= |\mathcal{M}_{12}|^2 + |\mathcal{M}_{21}|^2 + 2\operatorname{Re}\mathcal{M}_{12}^*\mathcal{M}_{21} \\
&= [\exp(-bA_{12}) + \exp(-bA_{21})](1 + \mathcal{H})
\end{aligned} \tag{14.18}$$

with

$$\mathcal{H} = \frac{\cos(\Delta A/\kappa)}{\cosh(b\Delta A/2)} \tag{14.19}$$

These formulas are proved by straightforward algebra from the area law in Eq. (14.17).

The result is evidently that we may obtain the final state either through the channel '12' or through the channel '21'. *But the interference between the two situations will result in a multiplicative enhancement factor* corresponding to the term $1 + \mathcal{H}$, which depends upon the area difference ΔA between the two space-time breakups. This area difference will depend not only upon the two particles 1 and 2 but also upon the state produced in between, which we have denoted by I in the figure. We note that the area difference exactly vanishes if the energy momentum vectors p_1 and p_2 are equal but grows rapidly from zero with the mass of the state I.

Before we provide formulas for ΔA we note that if two identical *charged* pions are produced then it is necessary to have the state I in between in order to compensate the quantum numbers. Thus if $1, 2$ are positively charged pions then it is necessary to compensate by a state of negative charge, and vice versa if they are negatively charged pions. *If two neutral pions are produced, however, then there is no such requirement.* Consequently in an ideal world where it would be just as possible to make measurements on neutral pions as on charged ones we would obtain by straightforward means a smaller area for the neutral pions in general, and the model could thus be easily checked. Up to now this has not been possible because it is very difficult to disentangle a signal from two neutral pions with sufficient precision. They each decay predominantly to a two-photon state, and it is very difficult to pick out four photons with sufficient precision in multiparticle surroundings.

We will provide two different formulas for the area difference ΔA. The first one corresponds to an energy-momentum description:

$$\Delta A = |p_1 E_2 - p_2 E_1 + (p_1 - p_2)E_I - (E_1 - E_2)p_I| \qquad (14.20)$$

in easily understood notation. This is the true area, i.e. without the use of the lightcone metric and relevant to the result in Eq. (14.19). We note that it will vanish when the energy-momentum vectors of the bosons are equal and that it will grow quickly with the intermediate-system mass.

Another form that is interesting is obtained by rewriting ΔA as

$$\frac{\Delta A}{\kappa} = |(p_1 - p_2)\delta x - (E_1 - E_2)\delta t| \qquad (14.21)$$

and it is easy to construct the space and time differences $\delta x, \delta t$ which will fulfil Eq. (14.21):

$$\kappa \delta t = p_I + \frac{p_1 + p_2}{2}$$
$$\kappa \delta x = E_I + \frac{E_1 + E_2}{2} \qquad (14.22)$$

These are the space-time difference vectors between the centres of the two-particle yoyos during the cycle when, according to the Lund model interpretation, they are produced. These points are indicated in the figure and we note that although they do not coincide in the two cases their difference vector is, of course, the same.

With this interpretation of the area difference the result for the correlation term $\mathscr{H}$ in Eq. (14.19) is in a very natural way related to the chaotic correlation $\mathscr{R}$ which we obtained in Eq. (14.11); the phase difference between the two 'production points' occurs weighted by a denominator.

We can also easily understand that this phase difference is, in the Lund string model, related to the fact that there must in general be something else produced in between the pair. This intermediate state, called I above, is needed in order to conserve the quantum numbers in the production process. In this interpretation *the HBT effect measures the region inside which the quantum numbers of the production process are compensated.*

4   The introduction of transverse momentum

Before we compare the model to the experimental data it is necessary to account for transverse momentum generation and for the fact that many particles are not directly produced but come from the decay of resonances.

By means of the tunnelling mechanism described by WKB methods in Chapter 11 (cf. also Chapter 12), we should in the Lund model introduce the real factors

$$\exp[-\pi\mu_\perp^2)/(2\kappa)] \qquad (14.23)$$

in the matrix element at each production vertex. The quark-mass factors will be the same. But it is necessary to generate different transverse momenta for the two cases at the two vertices adjoining the state I, in order to obtain the same states. In order to see this we note that the transverse momentum $\mathbf{k}_j$ generated at the vertex j can be expressed in terms of the transverse particle momenta $\mathbf{p}_\perp$ as

$$\mathbf{k}_j = \sum \mathbf{p}_\perp \qquad (14.24)$$

The sum runs over all particles from one end of the string to the production point of the given q or $\bar{q}$. When the two particles are exchanged and there is a nonvanishing transverse momentum vector difference, $\mathbf{p}_{\perp 1} - \mathbf{p}_{\perp 2} \neq \mathbf{0}$, then this will result in changes in the transverse momentum generated at the q and the $\bar{q}$.

This means that the denominator term in $\mathscr{H}$ from Eq. (14.19) will change as follows:

$$\cosh(b\Delta A/2) \quad \rightarrow \quad \cosh(b\Delta A/2 + \delta_\perp) \qquad (14.25)$$

where

$$\delta_\perp = \frac{\pi\Delta(\sum \mathbf{k}^2)}{2\kappa}$$

and $\Delta(\sum \mathbf{k}^2))$ means the necessary changes in the sum. Note that the numerator stems entirely from the imaginary phases.

5 The resonance decay problems

A much more involved problem is the treatment of the particles which stem from the decay of directly produced resonances. We will briefly discuss what one should expect from a phenomenological treatment of the resonances in terms of Breit-Wigner form factors.

A decaying state with mass m must have in its own rest frame a wave function ψ satisfying

$$\psi \sim \exp(-imt - t\Gamma/2) \text{ so that } |\psi|^2 \sim \exp(-\Gamma t) \qquad (14.26)$$

if we are to obtain the well-known exponential decay law with lifetime $1/\Gamma$. This means that such a state behaves as if it has a complex mass $m - i\Gamma/2$, which in the limit $\Gamma \to 0$ corresponds to Feynman's prescription for the propagator, as described in Chapter 3.

Accordingly, one describes the propagation of such a state as a solution of the Klein-Gordon (or any other relativistically covariant) equation with this mass value inserted. We start by assuming that the resonance will be produced at the space-time point $x_R \equiv (t_R, \mathbf{R})$ with a certain production amplitude $f(x_R)$. We further assume, for simplicity, that it will decay to a two-particle state with energy-momentum vectors $(\omega_j, \mathbf{k}_j)$, $j = 1, 2$, at the space-time point $x_1 \equiv (t_1, \mathbf{r}_1)$.

There will be a decay amplitude for this, which we obtain by a coherent sum over all space-time points for the wave functions. For simplicity we assume that the decay products are described by plane wave solutions. The amplitude for the propagation and decay is then

$$M = \int dx \mathscr{M}(x),$$
$$\mathscr{M}(x) = f(x_R)g_R(t, \mathbf{r}) \exp[-i(k_1 x_1 + k_2 x_1)] \qquad (14.27)$$

Here $x = x_1 - x_R \equiv (t, \mathbf{r})$ and the (radially symmetric) propagation solution to the Klein-Gordon equation for the resonance is

$$g_R(x) = \frac{1}{r} \exp[i(k_R r - \omega_R t)] \qquad (14.28)$$

According to the mass assumption above we have the following relation between the (complex) momentum k_R and the energy ω_R of the decaying resonance:

$$k_R^2 = \omega_R^2 - (m - i\Gamma/2)^2 \tag{14.29}$$

The t-integral yields energy conservation,

$$\omega_R = \omega_1 + \omega_2 \tag{14.30}$$

and the angular integral over $d\Omega$, in the space differential $d^3r = r^2 dr d\Omega$, will yield for the decay-product plane waves

$$\int d\Omega \exp[-i(\mathbf{k}_1 + \mathbf{k}_2) \cdot \mathbf{r}] = \int d\phi d\theta \sin\theta \exp(-i|\mathbf{k}_1 + \mathbf{k}_2|r \cos\theta)$$

$$= \frac{4\pi}{|\mathbf{k}_1 + \mathbf{k}_2|r} \sin(|\mathbf{k}_1 + \mathbf{k}_2|r) \tag{14.31}$$

Finally the integral over r, now combined with Eq. (14.31), will be proportional to

$$\frac{1}{2|\mathbf{k}_1 + \mathbf{k}_2|} \left(\frac{1}{k_R - |\mathbf{k}_1 + \mathbf{k}_2|} + \frac{1}{k_R + |\mathbf{k}_1 + \mathbf{k}_2|} \right)$$

$$= \frac{1}{\omega_R^2 - (m - i\Gamma/2)^2 - (\mathbf{k}_1 + \mathbf{k}_2)^2}$$

$$= \frac{1}{M_{12}^2 - (m - i\Gamma/2)^2} \tag{14.32}$$

This is the well-known Breit-Wigner form factor, which relates the squared mass of the final two-particle state, $M_{12}^2 = (k_1 + k_2)^2$, to the complex mass of the decaying resonance. We have used the relations in Eqs. (14.29) and (14.30) in the last two lines and have left out a set of constant factors along the way, together with the remaining production amplitude factor $f(x_R) \exp[-i(k_1 + k_2)x_R]$.

There is evidently a close relationship between the Breit-Wigner distribution and the Feynman propagator in energy-momentum space. This means that the distribution in mass for the final-state particles will be proportional to $|M|^2$ and thus to

$$\frac{1}{(M_{12}^2 - m^2 + \Gamma^2/4)^2 + m^2\Gamma^2} \tag{14.33}$$

When we consider the correlation between a pion stemming from this kind of decay and one stemming from direct production it is necessary to symmetrise the wave functions etc. Bowler, [35], has done this for us and for the details we refer to his treatment of both this and a number of other final-state corrections to the HBT effect.

The result of such considerations is that if we have a 'spectator state' from the resonance decay, which with Bowler's notation we will call 3, together with two interfering bosons 1 and 2 then if the width of the resonance Γ in the formula above fulfils

$$\Gamma < |M_{13} - M_{23}| \qquad (14.34)$$

the interference effects will vanish. There is consequently no HBT enhancement effect for the decay products from sufficiently long-lived resonances. The question is then what we mean by 'long-lived'. To ascertain this, we investigated three different situations in [22], where the original Lund interpretation occurs.

1 Only charged or neutral pions are produced within the Lund model scenario. This is evidently not in accordance with the experimental observations for e^+e^- annihilation reactions but it does actually provide a reasonably good description of many features of the final states.

2 There is the usual mixture in the Lund model of stable and unstable particles, including strange particles and baryons. The matrix elements are in each case evaluated for the stable final particles, ignoring the fact that some of them come from resonances.

3 The decay products of a resonance with four-vector energy-momentum k, mass m and width Γ are allowed to contribute to the HBT effect only if $kq \leq m\Gamma$. Here q is the the four-momentum difference vector.

In [22] we compared the data from the TPC collaboration at SLAC-PEP to three different cases obtained by a Monte Carlo simulation of the Lund model predictions, with a weighting of each event by the factor $1 + \mathcal{H}$ in Eq. (14.19). It was found, firstly, that cases 1 and 2 above coincided with each other from all practical points of view and also with the data. It turns out that the results are essentially only sensitive to the numerator in $\mathcal{H}$, i.e. there is a quick falloff in the cosine function for $\kappa \simeq 0.2 - 0.3$ GeV^2, which we have been using in the Lund model. The hyperbolic cosine in the denominator only takes over after the cosine function has gone down to zero. The reason for this is that the b-parameter in the Lund model is essentially smaller than the scale provided by κ.

Case 3 is, however, very far from the data and the predicted HBT effect is very small. There are, according to the Particle Data Group tables, a set of long-lived resonances which may affect the results. Bowler, [34], has shown that the major problem is actually the rate of produced η'-mesons in the Lund model. Remember that in Chapter 12 we have already presented some problems related to the rate of the η'-particles in the Lund model.

Bowler found that around 40% of the like-sign pion pairs which come out with Q-values ($Q^2 = -q^2$) below 0.2 GeV stem from η'-decays. Further, the pions from other decays of long-lived resonances do not in general populate this region at all. This means that at the q-value range, where it matters, the long-lived η'-decay products really play a very large role. Bowler then questions the Lund rate of η'-production and as we mentioned in Chapter 12 the model predictions may well be wrong. Bowler finds that if there were a strong suppression of η'-particles in the production process and if instead the observed pions were directly produced then almost the same HBT effect as for the cases 1, 2 above is also predicted in case 3.

14.3   The polarisation effects in the Lund model

1   The dynamical idea

We will start with a semi-classical explanation for the existence of a large polarisation effect in the Lund model. We consider the production of a $q\bar{q}$-pair at a vertex and assume that the particles each have mass m and are tunnelling out with compensating transverse momenta $\pm\mathbf{k}_t$.

In order to conserve the energy they will appear on the mass shell at the relative distance $2l = 2m_t/\kappa$, where $m_t = \sqrt{m^2 + \mathbf{k}_t^2}$, as we have discussed several times before. In this way both the energy and the momentum are conserved. *But the angular momentum is not conserved.*

From Fig. 14.2 we immediately conclude that the orbital angular momentum of the pair state is equal to

$$L = 2k_t l = \frac{2k_t m_t}{\kappa} \tag{14.35}$$

in the direction along a unit vector determined by the vector product $l \times \mathbf{k}_t$; the direction of the force field, l, is then defined to go from the produced $\bar{q}$ to the q, q having the transverse momentum $\mathbf{k}_t$.

The size of L can by the usual tunnelling formulas be estimated from Eq. (14.35) to be very close to unity for an average transverse momentum size. Therefore the effect cannot be small for this average situation. It is also reasonable to assume that the force field, unless there are local excitations, should contain no angular momentum. So, the only way in which this increase in the orbital angular momentum can be compensated is if the combined spin of the produced pair S equals 1 (they are spin 1/2 particles) and if the transverse component of $\mathbf{S}$ is oppositely directed to the vector $\mathbf{L}$.

This means in spectroscopical notation that such pairs are produced in a state with the assignment $^3P_0$, meaning that a triplet spin state $S = 1$ combines with an orbital angular momentum state $L = 1$ to give

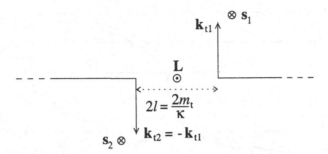

Fig. 14.2. A $q\bar{q}$-pair is produced with oppositely directed transverse momenta at a typical distance for a force field with finite energy density κ. The vector **L** points transversely outwards while the compensating spin vectors s_1, s_2 point inwards.

a state with total angular momentum $J = 0$. It turns out that owing to the intrinsic parity of the $q\bar{q}$-pair (which results in pseudoscalar spin 0 mesons) this state exactly corresponds to the quantum numbers of the vacuum ($L = 1$ states have negative parity).

The model also contains, however, predictions for the relative spin direction from a knowledge of the force field direction and the transverse momentum of the q or $\bar{q}$ with respect to the field. We will later show the consequences in connection with Λ-polarisation in a baryon fragmentation region.

2 The corresponding situation in QED

It is of some interest to note that there will be a very different result for the production of an e^+e^--pair in an external electric field.

To see this we assume that we have exactly the same production situation in QED as the one described above; let us also for the sake of argument assume that the pair will be polarised in the same way as above. Then *in QED this polarisation will not be conserved.*

The reason is that when the charges separate in the external field with momenta transverse to the field direction, see Fig. 14.3, then each of the charged particles will be accelerated along the electric field. But they will also cross the electric field lines, which means basically that there will be a torque working on the spin of the particles. Therefore the field will quickly take back the possible spin and kill the polarisation effects.

In order to discuss this effect in detail we consider the equation of motion for a spin vector in the particle's rest frame. In this frame the field, which was a constant electric field in the frame where the particle

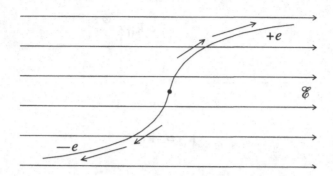

Fig. 14.3. The motions of an oppositely charged $\pm e$ pair in an external field $\mathscr{E}$.

was produced, is not purely electric. (The following discussion partly uses arguments from Jackson's book.)

The effect was noticeable for the fields we exhibited in a moving frame in connection with the method of virtual quanta in Chapter 2. There we found that a magnetic field was induced:

$$\mathscr{B} = -\mathbf{v} \times \mathscr{E} \tag{14.36}$$

and this is true in general ignoring correction terms of order v^2. Here also in the rest frame of the electron there is an induced magnetic field of this size.

A particle with spin $\mathbf{s}$ also has a magnetic moment $\boldsymbol{\mu}$ proportional to the spin vector:

$$\boldsymbol{\mu} = \frac{ge}{2m}\mathbf{s} \tag{14.37}$$

with the g-factor (as normal for a Dirac particle) equal to 2. (We will in the next section find the Thomas-precession correction to this result.)

Therefore there is an equation of motion for the spin

$$\frac{d\mathbf{s}}{dt} = -\boldsymbol{\mu} \times \mathscr{B} \tag{14.38}$$

which corresponds to an extra term in the hamiltonian

$$H' = -\boldsymbol{\mu} \cdot \mathscr{B} = \boldsymbol{\mu} \cdot (\mathbf{v} \times \mathscr{E}) \tag{14.39}$$

This means that in order to minimise the energy, the magnetic moment and therefore also the spin should be directed oppositely to $\mathbf{v} \times \mathscr{E}$.

This is exactly the opposite result to that obtained from the simple model described above. There we required that the spin should be oppositely directed to the 'produced' orbital angular momentum, which is directed along the direction $\mathbf{l} \times \mathbf{k} \propto \mathscr{E} \times \mathbf{v}$.

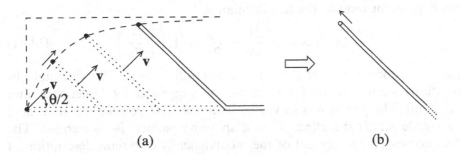

(a)   (b)

Fig. 14.4. The motion of a particle attached to a string (*a*) before and (*b*) after a boost to the rest system of the string piece adjoining the endpoint.

In the Lund model, however, there will be no such effect in a confined string situation. If we go back to the (admittedly classical) picture of the transverse motion of a particle attached to a string in Chapter 12 then we may utilise a frame in which the adjoining string piece is at rest, see Fig. 14.4.

This means that we are boosting in the transverse direction of the string piece, along the angle $\pi/4$ in Fig. 14.4(*a*), with velocity $v = \cos(\pi/4) = 1/\sqrt{2}$. In this frame we will find that the endpoint particle is simply moving outwards along a straight string, i.e. *there is always only a color electric field* acting on the particle. Therefore in a confined scenario of the Lund model type we do not have the torque on the spin discussed above.

This is actually the reason why we did not present calculations of how spin 1/2 particles would tunnel out of a confining force field. For particles with spin it is not sufficient to choose a potential that describes a purely electrostatic external field if we want to account for the relation between the force field and the particle in a confining situation. It is necessary to define a more general potential in such a way that *in the rest frame of the particle* the field is electrostatic.

3 *The Thomas-precession effect and a different model*

The result in Eq. (14.39) is actually identical to the so-called spin-orbit coupling in spectroscopy. For that case we write the following formula for the electric field in an atom as

$$e\mathscr{E} = -\frac{\mathbf{r}}{r}\frac{dV}{dr} \tag{14.40}$$

in terms of a spherically symmetric potential $V(r)$. This means that the spin-dependent term in the hamiltonian is

$$\frac{g}{2m^2}\mathbf{s}\cdot(\mathbf{r}\times m\mathbf{v})\frac{1}{r}\frac{dV}{dr}=\frac{g}{2m^2}\mathbf{s}\cdot\mathbf{L}\left(\frac{1}{r}\frac{dV}{dr}\right) \tag{14.41}$$

The only problem is that the g-factor according to all experimental observations should equal 1 and not 2 as suggested by Uhlenbeck and Goudsmit. The puzzle was solved by Thomas, who pointed out that there is a subtle relativistic effect when a spinning particle is accelerated. The effect comes very nicely out of the relativistically covariant description of a spinning particle in the Dirac equation.

We will not derive the Thomas effect in detail because we would then need an extensive formalism for $(3+1)$-dimensional Lorentz transformations. It is done in Jackson's book and we refer readers interested in the details to this. It is, however, a purely kinematical effect. The spinning particle may, in its own rest frame, have any spin vector direction. For the observer who is accelerated with respect to this rest frame there will be a bias in the direction of the coordinate system in the rest frame relative to that in the observer's frame.

Suppose that the observer adjusts his coordinate axes to coincide with those in the particle's rest frame at a time, t, when the particle has a certain velocity $\mathbf{v}(t)$. Then after a moment dt the particle will have the velocity $\mathbf{v}(t+dt)=\mathbf{v}(t)+d\mathbf{v}$.

Therefore when the observer compares the axes after the time increment dt he will have to make a Lorentz boost along the new direction. As we have said in Chapter 2, Lorentz boosts in different directions do not commute. In other words $L(\mathbf{v})L(d\mathbf{v})\neq L(\mathbf{v}+d\mathbf{v})$, where L is the boost operator, unless $\mathbf{v}$ and $d\mathbf{v}$ are parallel.

Consequently, the coordinate axis, and also the spin direction in the rest frame, will seem for the external observer to be rotating at a rate given by the Thomas angular velocity:

$$\boldsymbol{\Omega}_T=\frac{\gamma^2}{1+\gamma}\frac{d\mathbf{v}}{dt}\times\mathbf{v} \tag{14.42}$$

In the nonrelativistic approximation $\gamma=1$, which is relevant for an electron in an atom (an atomic electron moves with an average velocity equal to $\alpha\simeq 1/137\ll 1$) the acceleration will be given by the force in Eq. (14.40) and therefore the angular velocity will be

$$\boldsymbol{\Omega}_T=\frac{-1}{2m^2}\mathbf{L}\left(\frac{1}{r}\frac{dV}{dr}\right) \tag{14.43}$$

which will give an effective hamiltonian term equal to

$$H_T=\boldsymbol{\Omega}_T\cdot\mathbf{s}=\frac{-1}{2m^2}\mathbf{s}\cdot\mathbf{L}\left(\frac{1}{r}\frac{dV}{dr}\right) \tag{14.44}$$

This is of the same type as the spin-orbit coupling and together they will change the effective g-factor to $g - 1$, i.e. to the observed value.

In this way it is possible to obtain a spin effect that stems from the acceleration of the particle in a direction not parallel to its momentum vector $\mathbf{k}$. Thus if we imagine that after it has been produced the particle is accelerated along the force field direction l then there should be a Thomas precession effect with an effective hamiltonian

$$H'_T = h\mathbf{s} \cdot (l \times \mathbf{k}) \tag{14.45}$$

Here h is a positive-definite coupling constant equal to the force.

Now we do have a favorable situation for the particle to choose its spin in the direction opposite to the vector $l \times \mathbf{k} = l \times \mathbf{k}_t$, which is exactly in accordance with the prediction of the simple Lund polarisation model. In this case it would be the final-state interaction, i.e. the acceleration of the particle into the the final hadronic state, which would produce the polarisation effect, rather than the pair production mechanism, as in the Lund model explanation.

It may seem like magic, because there is really no force on the spin itself. It is instead an observational bias that produces the effect. It has nevertheless been suggested as a possible model to explain polarisation effects in hadronic production processes, [50].

4   The observable consequences

It is possible to make a large number of predictions from the simple model we discussed in subsection 1. We will be content, however, to discuss the results for Λ-particle polarisation in a baryon fragmentation region. The Λ-particle is, in some sense, one of the very few unqualified gifts which Nature has bestowed upon high-energy physicists, at least those interested in polarisation physics. Almost every other tool for observation contains very many complications. The reasons why the Λ is so nice are two-fold.

Firstly the Λ-particle decays via weak interactions to a nucleon and a pion. Weak interactions do not conserve parity. Consequently the Λ, through its decay, exhibits an asymmetry in the distribution of the angle between the nucleon and the pion which is directly related to its spin direction. And this asymmetry is large!

Secondly, the structure of a Λ-particle is rather simple. It can be described essentially as a state composed of a diquark $(ud)_0$, the index 0 denoting that the pair has spin and isospin equal to 0, and an s-quark. From this structure we conclude that it is the spin of the s-quark which determines the spin of the Λ-particle.

Thus the observation of Λ-polarisation reveals the direction of the s-quark's polarisation. If a Λ-particle is observed with a large fractional energy-momentum in the fragmentation region of a proton then it is most probably composed of a $(ud)_0$ diquark stemming from the original proton and a produced s-quark. If we imagine that the s-quark has been produced according to the Lund model prescription backwards along a string adjoining the $(ud)_0$ diquark (cf. the discussion of baryon fragmentation in Chapter 20) then the model of subsection 1 can be applied. It is only necessary to relate the transverse momentum of the produced s-quark to the transverse momentum of the observed Λ.

In the original paper, [9], we made two assumptions. The first was that the polarisation $\mathscr{P}$ of the produced s-quark will increase with the orbital angular momentum L of the $s\bar{s}$-pair and we chose the simple relation

$$\mathscr{P} = \frac{L}{L + \beta} \tag{14.46}$$

with the parameter $\beta \sim 1\text{--}2$. We further assumed that both the original diquark $(ud)_0$ and the produced s-quark had gaussian distributions of their transverse momenta, with widths σ_{qq} and σ_q, respectively. For the s-quark this can be justified from the tunnelling mechanism and for the diquark from the Fermi motion in the original baryon state.

This assumption means that the correlation between the momentum of the final-state Λ-particle, $\mathbf{p}_t$, and that of the s-quark, $\mathbf{k}_t$, will be

$$\left\langle \mathbf{k}_t \cdot \frac{\mathbf{p}_t}{|\mathbf{p}_t|} \right\rangle = \frac{\sigma_q^2}{\sigma_q^2 + \sigma_{qq}^2} |\mathbf{p}_t| \tag{14.47}$$

The resulting polarisation for the Λ-particle then agrees very well with the results of the ISR-data, see [9] and [58].

There must be corrections to the results for smaller values of the fragmentation variable z, i.e. the fraction of the original baryon energy-momentum carried by the Λ-particle. There are a set of possible channels that produce a Λ-particle in a baryon fragmentation region, according to the Lund model, cf. Chapter 20. It is then possible to predict the behaviour of the polarisation also for smaller values of the ratio z (or the Feynman variable x_F) in the fragmentation region of the proton and also to use the same mechanism for other hyperons, i.e. strange baryons.

The resulting predictions have been repeatedly confirmed. It is interesting that the polarisation effects are also found in states of a diffractive nature, [106]. Whether the dynamical mechanism for producing polarisation is the one proposed in the Lund model, i.e. the produced states come out with polarisation, or whether it is an effect of final-state Thomas precession is a question that we must leave open until more data on resonance hyperon polarisation become available.

15

The Lund gluon model, its kinematics and decay properties

15.1   Introduction

In this chapter we consider the way in which gluons are introduced into the Lund string fragmentation model, [7], [18], [104]. They are treated as *internal excitations on the massless relativistic string* (the MRS) similar to a sudden 'hammer hit' on an ordinary classical string. Thus they will be initially well localised in space-time. But we will find that they quickly disperse their energy-momentum to the surrounding string. This property means that *the gluon excitation disappears and reappears periodically as a localised energy-momentum-carrying entity during the string cycle*.

We will start as usual with a classical mechanics scenario and study some simple modes of motion of the MRS in order to get acquainted with the notion of an internal excitation. We start with the mode which has acquired the poetical name of 'the dance of the butterfly'. It certainly does exhibit the grace and the beauty that goes with this name. After a brief snapshot description of the appearance of this mode in space coordinates we proceed to a description in space-time. This will lead us to the general equations of motion for the MRS and to an understanding of the way the string is built up in terms of moving wave fronts.

After that we consider more complex modes, although there is no reason to go into too many details. The intention is simply to provide a sufficient understanding of the basics in this kind of string motion in order to make it possible to understand the way a string which has aquired a bend will fragment.

One property which is both useful and rather easily understood is the fact that *the space-time surface spanned by the string is a minimal surface*. We will spend some time considering this notion. We will also stress the notion of *infrared stability*, which is closely related to the minimal surface properties. This means that a small disturbance, such as a small or

269

collinear gluon excitation, does not change the string surface more than in a correspondingly small and local way.

After a brief description of the way fragmentation is handled (the whole process is a direct generalisation of the way a 'straight' string decays) we will turn to the consequences. We consider the correspondence in the bent string to the mean hyperbola decay, which was typical for the simple straight $q\bar{q}$-string in Chapter 9. We will find a noticeable similarity between the $(1 + 1)$-dimensional scenario and the multidimensional twistings and bends of the general string state.

In particular it is possible to generalise the rapidity variable for the simple straight string to a new variable which we have called λ in [48]. After we have introduced the cross section for gluon emission in Chapters 16 and 17 we will show how to calculate in an analytic form the properties of the λ-distribution and related variables. These distributions are governed by irregularities related to the so-called anomalous dimensions of QCD.

Within this pictorial scenario of gluon emission in the Lund model it is easy to understand both the increase in the multiplicity and the local properties of transverse flows. We will continue with a few remarks on *heavy quarkonia decays*. States such as the J/Ψ and Υ can, owing to their quantum numbers, decay only into three (or more) gluon final states. Such states are described in the Lund model by means of closed strings, the gluon excitations pulling out the string. Therefore there are differences between the decay of such a quarkonia state and that of a corresponding state with almost the same mass but outside the resonance (referred to as 'in the continuum'). We will explain the reason why there are more particles produced at the resonance, with its closed string state, than in the continuum with an open-ended string ending on a $q\bar{q}$-state.

15.2   The dance of the butterfly

1   Snapshots of the motion

In this section we will use some of the results from earlier investigations of string motion (cf. Chapter 6), in particular local energy-momentum conservation due to causality (cf. also Chapter 12). We consider the situation depicted in Fig. 15.1, where at the start the two endpoint particles q and $\bar{q}$ both have momentum $k_\perp$ along the same direction transverse to the connecting (straight) string (length $2l \geq 2k_\perp/\kappa$). As before we describe the ensuing motion in terms of a few snapshots in time.

A   After a time δt both endpoints have moved outwards in a straight line, in the same way as for the motion described in Chapter 12 (in that case only one endpoint moved transversely). There are two string segments, starting out with velocities $v = \cos(\pi/4) = 1/\sqrt{2}$, as indicated in Fig. 15.1. These have energies $2\kappa\delta t$, transverse momenta $\kappa\delta t$ upwards and compensating longitudinal momenta $\pm\kappa\delta t$.

The endpoints have each lost $\kappa\delta t$ in energy and transverse momenta and the remaining straight string is also $2\delta t$ shorter. In this way we account for all the available energy-momentum just as in Chapter 12. This part of the motion will continue until all the original energy and transverse momenta of the q and $\bar{q}$ have been used up.

B After that the q and $\bar{q}$ will start to move towards each other, each gaining energy and (oppositely directed) momentum at a rate of κ. The two connecting string segments continue inwards as two fronts, each with energy $2k_\perp$ and transverse momentum $k_\perp$, and with longitudinal momenta $\pm k_\perp$ respectively.

During this phase they serve as 'transporters' of energy-momentum to the two endpoints. More precisely they pick up energy from the remaining straight string piece, thereby gaining energy at the lower end and losing it to the endpoint particles at the upper end. This part of the motion continues until the fronts meet at a time $t = l$ and it again coincides with the results of Chapter 12.

The next part of the motion is, however, both surprising and beautiful.

C After a time δt the meeting point of the fronts has, for purely geometrical reasons, moved upwards by a distance δt and each front is now $\delta t \sqrt{2}$ shorter. Therefore each front has lost energy $2\kappa\delta t$, transverse momentum $\kappa\delta t$ and longitudinal momentum $\pm\kappa\delta t$. The two endpoint particles have gained energies $\kappa\delta t$ and longitudinal momenta $\pm\kappa\delta t$ (i.e. the joint longitudinal momentum loss vanishes).

The remaining energy and transverse momentum $2\kappa\delta t$ *are gathered at the meeting point of the fronts.* This point has moved upwards with the velocity of light (i.e. the same velocity as the endpoints) and is evidently gaining energy-momentum at a rate 2κ! This part of the motion will continue until the two endpoints $q, \bar{q}$ and the internal excitation, called g, meet at a transverse distance $k_\perp/\kappa$ from the starting position of the system.

D After another time period δt, the q and $\bar{q}$ have passed each other and are now at a distance $2\delta t$ and moving outwards. The g continues upwards and there are two new string segments connecting the three particles. Each segment moves with velocity $v = 1/\sqrt{2}$, as indicated in Fig. 15.1, and therefore has energy $2\kappa\delta t$, transverse momentum $\kappa\delta t$ and longitudinal momentum $\pm\kappa\delta t$.

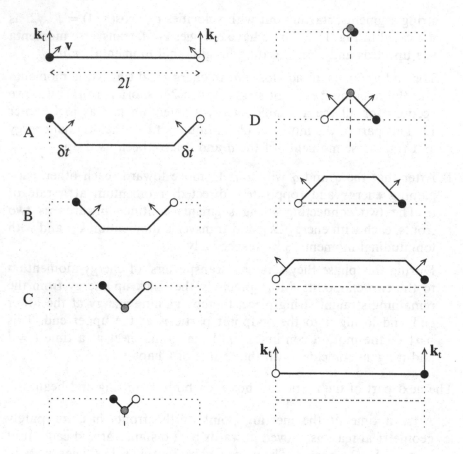

Fig. 15.1. The butterfly-dance mode of the MRS with the velocity v of the string segments and the different situations described in the text exhibited. The quark (antiquark) is denoted as an open (solid) circle and the gluon when it appears as grey. The arrows denote the directions of motion of the particles and the string segment fronts and the orbits of the quark and antiquark are shown as dotted.

> The q and $\bar{q}$ have each lost the energy $\kappa\delta t$ and the longitudinal momentum component $\kappa\delta t$. *The remaining energy and momenta in the moving string segments stem from the internal g-excitation, which has lost $2\kappa\delta t$ both in energy and transverse momentum* (it is useful for the reader to calculate the amount of energy-momentum in the fronts and thereby the amount which must stem from the g-excitation).

The internal excitation, the g, is evidently connected to the string in the same way as the q and the $\bar{q}$, except that the string tension acts with a force 2κ on the g. This can be understood intuitively from the fact that there are two connecting segments to the g and only one to each of the

q and $\bar{q}$; we will see this property in more detail later. The g-excitation encountered in this way evidently has particle properties, i.e. it carries energy-momentum in a local way and *in the Lund model it is used as a model for a gluon*, just as the energy-momentum-carrying endpoints are used as models for a quark and an antiquark.

The motion described under A–C above takes a time $l + k_\perp/\kappa$ and corresponds to a quarter of the full cycle of the butterfly dance. The motion following this is depicted in Fig. 15.1 for the next quarter-cycle also. It can easily be extrapolated from what we already know. After half a cycle we are back in the starting situation except that the q and $\bar{q}$ have changed places. It takes another half period before they are back in their original positions.

We note that the total energy E is $2\kappa l + 2k_\perp$ and the total momentum P is $2k_\perp$ and is transversely directed. The total period of motion before we are back in the starting position is therefore, as usual, $2E/\kappa$ and we also note that the system has moved a distance $2\mathbf{P}/\kappa$ during this period.

The g-particle is evidently only present as a point particle during $4k_\perp/\kappa$ of the full period. During the remaining time, $4l$, it has been transformed into two inward- or outward-moving fronts on the string. This is, of course, also the way any excitation on an ordinary rubber band will perform (try it on your kitchen table, which hopefully will have little friction, with a real rubber band!).

In order to exhibit the Lorentz covariance of this picture we describe in Fig. 15.2 how the motion will appear in a different frame, in this case the cms of the system. It is again perfectly feasible to trace the motion using the same simple rules of local energy-momentum conservation as we have used repeatedly. The reader is encouraged to carry through the calculations in order to see the details.

2 The space-time picture

We will next provide a picture of the Lund gluon model in space-time. We use the notion of a *light ray* to describe a lightlike direction in space-time, e.g. the direction of the energy-momentum of one of the partons. We will also use *lightcone distance* to refer to the distance such a massless parton will move before it changes direction.

In Fig. 15.3 we show at the top the situation at the time of meeting of the three particles. This corresponds to a quarter-period after the start, in the description of Fig. 15.1. The subsequent motion is then considered in the cms and is therefore a space-time version of Fig. 15.2. The two endpoints move outwards along their lightcones and the string at first consists of two segments moving between the light rays of qg

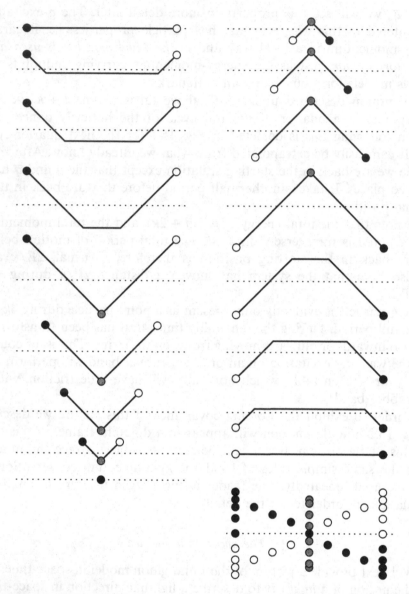

Fig. 15.2. The butterfly mode in the cms, using the same conditions and notation as in Fig. 15.1. The actual orbits of the quark (open circles), the antiquark (solid circles) and the gluon (grey circles), whenever it appears, are shown in the last combined picture.

and $g\bar{q}$, respectively. In this way the string is spanned via the g's light ray.

When the gluon has disappeared, the two segment fronts continue and there is a straight (although, in this frame, moving) string part connecting

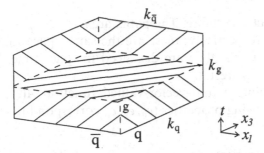

Fig. 15.3. The butterfly mode in space-time. The sets of parallel solid lines show the string at different times, the energy-momentum-carrying excitations move along the outer solid lines and the bends along the broken lines.

the fronts (see Fig. 15.2). Note that the bends between the flat part and the moving fronts of the string move along light rays parallel to the original directions of the q or $\bar{q}$.

Halfway through this (half-)period the string is totally straight. The two endpoints continue along the original direction of the g and then again two new string wave fronts are produced, which this time move inwards. The bends between the fronts and the remaining straight string again move along the lightcone directions determined by the original light rays of the q and $\bar{q}$. When the two fronts meet, the gluon reappears and the three particles approach each other, meet and separate again. During the motion the string evidently spans a surface in space-time and we will next consider some of the properties of this region.

We firstly note that all its characteristics are determined by the three lightcone distances contained in the three original excitations. In particular the right-hand, i.e. q-side, boundary line is obtained by adding in turn the energy-momenta k_q, k_g and $k_{\bar{q}}$ of the particles q, g and $\bar{q}$ (divided by κ, of course, but for the moment we will put $\kappa = 1$). Remember that the g loses twice the amount of energy-momentum per unit time to the adjoining string compared to the q and $\bar{q}$. Therefore the original lightcone distance indicated in Fig. 15.3 by g actually corresponds to half its energy-momentum, k_g. The lightcone distances that the q and $\bar{q}$ move, after using up their original energy-momentum, correspond to the true size of g's energy-momentum.

Thus the first conclusion is that the boundary line, which corresponds to the motion of the q, is given by $k_q + k_g + k_{\bar{q}}$, while the corresponding boundary line for the $\bar{q}$ is $k_{\bar{q}} + k_g + k_q$. We will soon find that there is a direct generalisation of this property to more complex string motions. In particular, everything q does is done in the opposite order by $\bar{q}$.

The corresponding conclusion for the g is that we may describe it either in terms of the motion of the right front's bend (i.e. that closer to q) or the left front's bend; cf. the figure. The paths are $k_g/2 + k_q + k_{\bar{q}} + k_g/2$ or $k_g/2 + k_{\bar{q}} + k_q + k_g/2$ respectively; these are the same space-time orbits as for the $\bar{q}$ and q except that they are displaced in space-time. It is again possible to generalise this result.

To that end we define a four-vector-valued function $A(\xi)$, called the *directrix*, with the following properties ($e_q, e_g, e_{\bar{q}}$ correspond to the energies of the three particles)

I   If $0 < \xi \le e_q$, then $A(\xi) = k_q(\xi/e_q)$.

II If $e_q < \xi \le e_q + e_g$, then $A(\xi) = k_q + k_g(\xi - e_q)/e_g$.

III If $e_q + e_g < \xi \le e_q + e_g + e_{\bar{q}}$, then $A(\xi) = k_q + k_g + k_{\bar{q}}(\xi - e_q - e_g)/e_{\bar{q}}$.

IV $A(\xi) = -A(-\xi)$.

V $A(2E + \xi) = A(\xi) + 2(k_q + k_g + k_{\bar{q}})$, with $E = e_q + e_g + e_{\bar{q}}$.

The orbit of the q, i.e. what we have referred to as the right-hand boundary line, is then $A(t)$ and the corresponding orbit for the $\bar{q}$ is $[A(t + E) + A(t - E)]/2$ (this should be checked by the reader). It is less easy to convince oneself that the orbit of the right-hand bend discussed above is given by $[A(t + e_q) + A(t - e_q)]/2$ and that of the corresponding left-hand bend is $[A(t + E - e_{\bar{q}}) + A(t - (E - e_{\bar{q}}))]/2$. But it is worth doing because this is the general behaviour of any point on the string. We learned at the very beginning that the string does not conserve its length (nor does any rubber band on your kitchen table!). Therefore the points on the string cannot be characterised in terms of their space position only. But it is possible to characterise a point fully by means of the amount of energy possessed by the string to its right (i.e. towards the q-side) or equivalently to its left and we will now proceed to give a description of general string motion using this approach.

The result of this exercise also explains why the motion is periodically simple and in particular why a combined translation $2E/\kappa$ in time and an accompanying $2\mathbf{P}/\kappa$ in space always brings the string system back to the same situation.

15.3   The general description of string motion

1   *The equations and their solutions*

As we saw above, a point on a string will be characterised by means of the amount of energy available between the point and the q-endpoint. We call this parameter σ and we write $\mathbf{X}(\sigma, t)$ for the space position of the point and $e_q(t)$ for the q-particle energy at time t.

A formula for σ is given by the integral along the string of the energy

$$\sigma = \int_{\mathbf{X}_q(t)}^{\mathbf{X}(\sigma,t)} \kappa dl \gamma(v_\perp) + e_q(t) \tag{15.1}$$

where $\gamma(v_\perp)$ as usual is $1/\sqrt{1 - v_\perp^2}$.

The transverse velocity is denoted $\mathbf{v}_\perp$ and the string tension $\mathbf{T}$. The tension is everywhere directed along the string tangent $\partial\mathbf{X}/\partial\sigma$ and it should, when the string piece considered is at rest, have the size κ. From this it is evident that we must have

$$\frac{\partial\mathbf{X}}{\partial t} = \mathbf{v}_\perp, \quad \kappa^2 \frac{\partial\mathbf{X}}{\partial\sigma} = \mathbf{T} \tag{15.2}$$

There will be two extra conditions stemming from our choice of parametrisation, and from the fact that the string has no longitudinal degrees of freedom so that the velocity and the tension must be orthogonal. From a variation for fixed t of Eq. (15.1) we obtain

$$d\sigma = \frac{\kappa|d\mathbf{X}|}{\sqrt{1 - v_\perp^2}} \tag{15.3}$$

This result contains an evident connection between the length of the tension vector $|\mathbf{T}|$ and the velocity. We have actually encountered and discussed this condition before. It can be expressed as in Chapter 6 as the result of time dilation. In this way we obtain the two conditions

$$\mathbf{T} \cdot \mathbf{v}_\perp = 0, \quad \frac{\mathbf{T}^2}{\kappa^2} + v_\perp^2 = 1 \tag{15.4}$$

Next we note that the momentum carried by a small energy 'grain' $d\sigma$ is $d\mathbf{p} = d\sigma\mathbf{v}_\perp$ (remember that $d\mathbf{p}/de = \mathbf{v}$ for an on-shell particle). Therefore the change in momentum with time for this energy grain is given by

$$\frac{d(d\mathbf{p})}{dt} = \mathbf{T}(\sigma + d\sigma) - \mathbf{T}(\sigma) \quad \Rightarrow \quad d\sigma\frac{\partial^2\mathbf{X}}{\partial t^2} = d\sigma\kappa^2\frac{\partial^2\mathbf{X}}{\partial\sigma^2} \tag{15.5}$$

Thus we obtain (after division by $d\sigma$) the usual wave equation for the motion of the points on a string (i.e. in the limit when the energy-momentum grains referred to above become infinitesimal). We therefore conclude that the general solution must be

$$\mathbf{X}(\sigma,t) = \tfrac{1}{2}[\mathbf{B}(t + \sigma/\kappa) + \mathbf{C}(t - \sigma/\kappa)] \tag{15.6}$$

where $\mathbf{B}$ and $\mathbf{C}$ are two arbitrary vector-valued functions. This solution corresponds to two moving fronts and we will now consider suitable boundary conditions to determine them. These conditions are simple in this case because we note that *for an open string with endpoints the tension of the string must vanish at the endpoints*. Therefore for $\sigma = 0$ and $\sigma = E$,

E being the total string system energy, we must have

$$\kappa^2 \frac{\partial \mathbf{X}}{\partial \sigma} = 0 \tag{15.7}$$

Expressed in terms of $\mathbf{B}, \mathbf{C}$ this means

$$\dot{\mathbf{B}}(t) = \dot{\mathbf{C}}(t), \quad \dot{\mathbf{B}}(t + E/\kappa) = \dot{\mathbf{C}}(t - E/\kappa) \tag{15.8}$$

where the dots indicate derivatives. The two equations can be readily integrated and we may write

$$\mathbf{B} = \mathbf{C}, \quad \mathbf{B}(t + 2E/\kappa) = \mathbf{B}(t) + 2\mathbf{P}/\kappa \tag{15.9}$$

Therefore the general solution can be expressed in terms of a single function $\mathbf{B}$:

$$\mathbf{X}(\sigma, t) = \tfrac{1}{2}[\mathbf{B}(t + \sigma/\kappa) + \mathbf{B}(t - \sigma/\kappa)] \tag{15.10}$$

with the requirement that $\mathbf{B}$ should be periodic, with a constant translation $2\mathbf{P}/\kappa$ over the period $2E/\kappa$. In particular we note that the q-endpoint moves along the function $\mathbf{X}(0, t) = \mathbf{B}(t)$ and the $\bar{q}$ along $\mathbf{X}(E, t) = [\mathbf{B}(t + E/\kappa) + \mathbf{B}(t - E/\kappa)]/2$.

The conditions in Eq. (15.4) mean that

$$\dot{\mathbf{B}}^2 = 1 \tag{15.11}$$

i.e. the endpoints always move with the velocity of light.

We have in this way obtained a complete description of any string with endpoints. In particular we find that the results for the simple $qg\bar{q}$-state described in the earlier section is true for all points on the string. The directrix function A defined there evidently coincides with the four-vector $(\xi, \mathbf{B}(\xi))$. (The result is easily generalisable to a string with many gluons and this is a useful exercise. We will later discuss an example with two gluons.) The condition IV on the directrix A is, however, peculiar to a string which passes through a single space-time point, i.e. the point where the three particles start out.

We will end this section with a few considerations on the energy-momentum content of a string region. We note the relation used above for the energy grains, $d\mathbf{p} = d\sigma \mathbf{v}_\perp$. From this we may by introducing our solution calculate the total momentum flowing across a spacelike surface in the region ABCD, depicted in space-time in Fig. 15.4.

The region is limited by the two sets of curves $t - \sigma$ and $t + \sigma$ equal to constants ($\kappa = 1$ again for simplicity). The parameter values are for A (t, σ_1), for B (t, σ_2) and for the pair CD the 'earlier' and 'later' crossing points.

Then we obtain by integrating out the energy-momentum content in

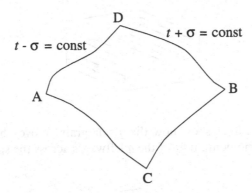

Fig. 15.4. The points A and B are at a spacelike distance (we choose equal times) and there is a region spanned by them and the points C (earlier) and D (later), which is the causal dependence region.

between A and B at the fixed time t,

$$
\begin{aligned}
\mathbf{P}_{AB} &= \int_A^B d\sigma \frac{\partial \mathbf{X}}{\partial t} \\
&= \tfrac{1}{2}[\mathbf{X}(t+\sigma_2) + \mathbf{X}(t-\sigma_1) - \mathbf{X}(t-\sigma_2) - \mathbf{X}(t+\sigma_1)] \\
&= \mathbf{X}_D - \mathbf{X}_C
\end{aligned} \tag{15.12}
$$

Thus the difference vector between two points on the string surface is directly given by the *energy-momentum that flows inside the causal dependence region*. This result was freely used in the $(1+1)$-dimensional model, where the lightcone directions coincided with the parameter curves $t \pm \sigma$ (a useful exercise is to find the directrix for the description of a straight $q\bar{q}$-string). Equation (15.12) means that this result is also true for a general string surface if we use the causal dependence region.

It is in the same way possible to calculate the energy-momentum that will flow across a timelike curve between the points C, D (note that these points have the same value of σ):

$$
\int_C^D dt\mathbf{T} = \mathbf{X}_B - \mathbf{X}_A \tag{15.13}
$$

This is the result first pointed out by Artru, [25], cf. Chapter 9: the momentum transfer between a group of particles moving to the left with respect to a particular breakup vertex and one moving to the right is given by the variable Γ. This variable also corresponds directly to the proper time, in this case between C and D.

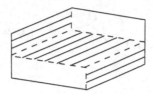

Fig. 15.5.   The solid lines show how the q-side grains move, thereby translating the original energy-momentum k_q of the q leftwards across the space-time surface.

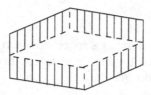

Fig. 15.6.   The two sets of g-grains moving apart, thereby translating half of the original energy-momentum k_g in each direction across the space-time surface.

2   *The space-time surface of a $qg\bar{q}$-state*

We now return to the butterfly-state motion to consider the space-time surface in the light of what we learned from the general behaviour of the MRS in the last subsection. The two wave fronts described by the directrix can be thought of as a (continuous) stream of energy grains moving to the left (from the q-side) and to the right (from the $\bar{q}$-side). They move throughout with the velocity of light and can be thought of as having been emitted by the excitation particles.

The conditions in Eq. (15.7) means that the grains coming in towards the left bounce out towards the right and vice versa. Note, however, that the grains often stay at the endpoint (and the g-excitation) positions for some time. We show in Figs. 15.5, 15.6 and 15.7 the way the energy-momentum vectors 'march' across the surface, thereby spannning it.

If we follow the q-side boundary line the first part can be thought of as corresponding to emission of the k_q-grains and the next part as corresponding to absorption of (half) the k_g-grains (i.e. those sent out by the g in that direction). The part after that corresponds to re-emission of these k_g-grains and is followed by absorption of the $k_{\bar{q}}$-grains by the q.

Thus we find that the reason why the string does not keep its size is that grains may be gathered up at certain space-time positions during the cycle. These positions correspond to the excitation particles either at the

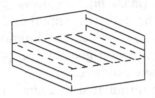

Fig. 15.7. The $\bar{q}$-side grains moving across the space-time surface thereby translating the original energy-momentum $k_{\bar{q}}$ of the $\bar{q}$.

endpoints or in the centre and they have the property to absorb or emit
grains at a constant rate in space-time.

The reason that the g interacts at twice the rate of the q or the $\bar{q}$ is that
in this case there are grains coming from or going towards both sides. A
bend corresponds to the situation when the grains come in and go out
again at the same rate.

In the same way we could describe the emergence of the straight string
piece connecting the two fronts as a combination of the k_q-grains coming
from the right and the $k_{\bar{q}}$-grains coming from the left, while the right (left)
wave front region corresponds to the combination of half the k_g-grains
with the k_q-grains ($k_{\bar{q}}$-grains).

Each region therefore corresponds to a lightcone diamond spanned by
two lightcone directions, each with a length corresponding to one of the
characteristic original particle energy-momenta (half for the g, however,
each time).

From the results of this discussion it is easy to calculate the area of the
surface for the half-period discussed. We find that $k_q k_{\bar{q}} + k_g k_q + k_g k_{\bar{q}} =
(k_q + k_g + k_{\bar{q}})^2/2 = M^2/2$. This is again in accordance with our earlier
result that the space-time surface area for the full period is given by the
squared system mass.

At this point we would like to make a few historical remarks. At the
basis of all advanced dynamics situations there can be formulated an
action principle. Thus, according to Hamilton's principle, the motion of a
system from time t_1 to t_2 is such that the line integral of the Lagrangian
L,

$$I = \int_{t_1}^{t_2} dt L(x(t), \dot{x}(t)) \qquad (15.14)$$

has an extremum along the path $x(t)$. Here L is expressed in terms of
(possibly many) coordinates x and velocities $\dot{x}$.

This statement is closely related to the behaviour of geodesics on
surfaces defined by a differential geometry and a metric. Thus a free

relativistic point particle (mass m) will move in such a way that the (invariant) length along the path is minimal and one can choose $Ldt = -m\sqrt{(dt)^2 - (d\mathbf{x})^2} = -mdt\sqrt{1 - \mathbf{v}^2}$ as the Lagrangian. The inclusion of electromagnetic fields introduces a geometry in phase space and modifies the particle motion to a new geodesic.

String dynamics can be formulated in a similar way, [62], by requiring that *a surface area should be minimal*. This can be expressed in very general forms (and in the reviews on the subject, [62], you will find very learned discussions). For the situation at hand we may formulate this surface area as a two-dimensional integral with integration element

$$d\Sigma = -\kappa dl dt \sqrt{1 - \mathbf{v}_\perp^2} = -\kappa dt d\sigma \left|\frac{\partial \mathbf{X}}{\partial \sigma}\right| \sqrt{1 - \left(\frac{\partial \mathbf{X}}{\partial t}\right)^2}$$

$$= -\kappa dt d\sigma \sqrt{\left(\frac{\partial X}{\partial \sigma}\right)^2 \left(\frac{\partial X}{\partial t}\right)^2 - \left(\frac{\partial X}{\partial \sigma} \frac{\partial X}{\partial t}\right)^2} \tag{15.15}$$

In the last line we have extended $\mathbf{X}$ into a four-vector $X = (t, \mathbf{X})$.

Use of Euler's variational calculus on such a two-dimensional integral leads to the wave equation considered in Eq. (15.5). *The main point is, however, that the string surface always is a minimal surface.* That is the reason why its behaviour is directly describable by means of the boundary curve, i.e. the directrix. Every young person who ever twisted a wire into some closed shape and dipped it into soapy water has seen the beauty of the shimmering thin surface emerging and probably also noted that this minimal surface is directly related to the bends and the twists on the wire. These features correspond in the MRS to the elementary excitations on the string. This illustrates why we can describe the string surface in terms of only the endpoint $q\bar{q}$- and the internal g-excitation paths.

15.4   Multigluon states and some complications

1   On the color-flow connections

We will not study the most general multigluon scenario that is possible within the Lund model but will be content to consider a state with two gluons having a general appearance similar to the earlier one-gluon case (Fig. 15.8).

We immediately encounter the problem that there are two ways of drawing the Lund string between the excitations in this case. These ways are shown in Fig. 15.8. The two cases correspond to different *color-flow directions* around the gluon corners. Classically they are mutually exclusive but it is a complex question whether quantum mechanics will allow interference between the two color-flow states.

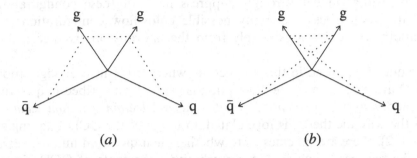

Fig. 15.8. The initial situation for a symmetrical two-gluon state with the momentum vectors of the four partons indicated. The broken and dotted lines show the string for the two possible color flows in the situation.

We will come back to the problem in Chapter 17 when we discuss multigluon bremsstrahlung emission. We note that *the question is basically whether it is sufficient to know the charges in order to obtain the fields*. This is always the case for abelian fields like those of electromagnetism. Besides very small quantum corrections due to photon-photon scattering it is always possible to describe the emerging electromagnetic fields as a superposition of the fields due to the separate charges.

One basic assumption in the Lund model is that the color electric fields do have a meaning *per se*, because all the final-state particles stem from the breakup of these fields. In a totally perturbative QCD scenario only the charges, i.e. the q, $\bar{q}$ and the g's, appear in the final state and one is, in general, summing over all their connecting color indices. Then the color-flow connection needed for the Lund model string fragmentation is not obvious. It turns out that both of the color-flow configurations described above will occur at the matrix element level as distinct contributions (cf. [71]). At the cross section level (i.e. after squaring the matrix elements) there will be interference between the color-flow configurations, however.

The interference terms are in general smaller by a factor of $1/N_c^2$ (with N_c the number of colors) compared with the terms corresponding to definite color-flow directions. Therefore one may hope that they should not be very noticeable for the observables in an actual experiment. From the point of principle they are, however, of great interest. There are some possible cases for which these corrections can be studied, [103], although there is at present no convincing experimental proof of their existence. The problem is that there are $n!$ possible color flows obtainable in connection with the general n-gluon state. Although, as we will see in the following chapters, the coherence properties of gluon radiation will strongly suppress most of these configurations there are nevertheless too many possible color-flow configurations left to pinpoint the differences simply from the hadrons observed in a final state.

The question, raised in this subsection, whether the field configurations in QCD are part of the state description is generally described in quantum field theory as *the problem of possible super-selection quantum numbers*. From the way the theory is formulated, in terms of the QCD Lagrangian, cf. e.g. [52], there are no clues as to whether such quantum numbers exist. Only the charges occur, in the perturbative treatments of QCD. Super-selection quantum numbers are therefore not observable unless one sums all the perturbative contributions.

2   A two-gluon state

Leaving aside this question we proceed to study one of the color-flows in the two-gluon state shown in Fig. 15.8. This state is the most 'natural' one in the sense that the string does not contain sharp bends. It is also the one with the largest probability of occurring owing to the above-mentioned coherence properties of gluon emission.

In Fig. 15.9 we show the space-time behaviour of this string state. It is easily understood as soon as we provide the directrix, which, this time, corresponds to the ordered curve between $k_q, k_{g1}, k_{g2}, k_{\bar{q}}$. It is obviously possible to expand the definition of the directrix in subsection 2 of section 15.2 to any number of color-connected gluons along the same lines.

We note again, in particular, how the grains transport the vectors of the elementary excitations diagonally across the surface. The initial region between the two gluons is spanned by $k_{g1}/2$ and $k_{g2}/2$ with the grains of the first coming from the left and those of the second from the right. This piece of surface appears four times during this half-cycle of the string motion, first between g_1 and g_2, next between the q and one bend, then between another bend and the $\bar{q}$ and finally in the rebuilding of the two gluon excitations.

It is also of interest to compare the situation for a single gluon in Fig. 15.3 with the one described by Fig. 15.9: note that on the surface of the butterfly-dance mode the single gluon 'ridge' along the lightcone has split up into a diamond between the two gluons. Evidently if the two gluons are close together then this diamond will approach the original single gluon ridge. In this case the two bends on the wave fronts denoted b_1 and b_2 in the figure will merge and re-form a single gluon.

This means that *the interpretation of the surface in the Lund model is infrared stable*, i.e. whether two collinear gluons are described as a single entity or as two distinct parts the surface will look the same. This property

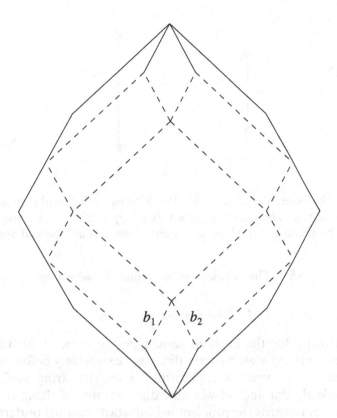

Fig. 15.9. The space-time surface spanned by a symmetrical two-gluon state.

is of the utmost importance for the success of the Lund model. The same
thing is the case if one, or both, of the gluons is collinear with the
q and/or the $\bar{q}$. Again some parts of the surface will become thinner
until they finally coincide with the lightcone motion of the q and/or the
$\bar{q}$.

If the gluon is soft and central, i.e. it contains little energy, then it will
quickly become two wave fronts. In this case the wave fronts are very
tiny disturbances on a basically straight string. This is again an expression
of the infrared stability of the Lund string fragmentation. A soft gluon
does not influence the fragmentation process; the final state will look very
much as if the gluon never had been there!

In making the comparison between the two figures it is interesting that
the appearance of more gluons in a certain sense 'smooths off' the string
surface. Evidently a general string surface can be described in terms of
lots of soft and collinear gluons crawling along 'eating' and 'spitting out'
the energy grains to which we have repeatedly referred.

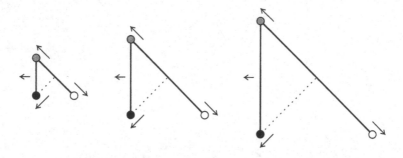

Fig. 15.10. The butterfly-dance mode after a boost such that the qg-segment of the string is at rest. The quark is shown as an open circle, the antiquark as a solid circle, the gluon grey and the directions of motion are marked with arrows.

15.5   The breakup of a gluonic Lund string

1   The possible problems

The general rules for the breakup should be the same as for the simple $(1+1)$-dimensional $q\bar{q}$-state we have discussed extensively before (Chapters 7–10). There are, however, complications when the string surface is no longer completely flat and we will now discuss some of them. In order to orient ourselves towards the problem we will start with the butterfly-dance mode again. This time we perform a boost transversely to the segment qg with a velocity $v = 1/\sqrt{2}$ (cf. the situation discussed in connection with Fig. 15.1).

In Fig. 15.10 we show the appearance of the $qg\bar{q}$-state after this boost. The q-particle is now moving outwards along a straight string at rest and the g is going in the other direction. At the q-end there is no reason to expect any difference from the $(1 + 1)$-dimensional model. The other segment between the g and $\bar{q}$ is of course moving away in a different direction. In the figure we show by a dotted line the path of the $\bar{q}$ in this frame. It is useful to carry through the calculations necessary to prove that the motion indicated in Fig. 15.10 really describes the situation!

We therefore assume that this part of the string segment will break up in its own rest frame, as before, and the same evidently goes for the region around the $\bar{q}$ (although that rest system is different). The difficulty occurs only in traversing the gluon corner from the straight string segment on the one side to the segment on the other (note that the segments are in general moving apart in different directions!).

There have been different suggestions in different models of how to handle a gluon corner in fragmentation. It is possible, for instance, to assume that the gluon is split up into a $q\bar{q}$-pair according to one or another rule. Then one would be able to handle the breakup by considering the two new strings obtained. *In the Lund model we will keep to a connected-string situation*, however.

Evidently, some difficulties may arise owing to the fact that the gluon in the Lund model is not always a pointlike particle: according to our findings, it dissolves into two wave fronts moving apart with a straight segment in between the two bends 'left over' from the g. *It turns out that there will be problems with respect to the time sequences of the breakups.* The main problem is whether it is possible to produce a scheme that has the same fragmentation results if we start the process along the straight segment on one side of the gluon corner if we start it on the other side and fragment in the opposite direction. *This turns out to be impossible* if we allow the string states to move independently in space-time. It is only for the simple $(1 + 1)$-dimensional model that the breakup of a string always produces two dynamical systems that are identical apart from their size.

To see the difficulty assume that we break the string around one of the wave front bends. Suppose we produce a $q_1\bar{q}_1$-pair in the string segment ending on the $q \equiv q_0$. Then we obtain a straight segment starting on q_0 and ending on $\bar{q}_1$ which behaves like an independent string state (and may be fragmented further in the usual Lund way). Besides the momentum transfer at the breakup this part will also continue to move as before, i.e. its space-time surface is (part of) the original string surface. The 'leftover' state with q_1 at the end and the wave front bend approaching also forms an independent string system *but this system will no longer move in space-time as before when it was connected to q_0.*

The new state will trace out a different space-time surface, which is not part of the original one. It is rather easy to convince oneself that if we trace it backwards in time then the g-excitation (from which the wave front bend stems) will be different (or even non-existent!). Therefore we obtain by this breakup two systems and one of them is not dynamically equivalent to a part of the original system.

If we instead consider a string breakup on the other side of the wave front bend, producing a $q_2\bar{q}_2$-pair, then neither the system composed of $\bar{q}_2$, the wave front bend and the straight segment to q_0 nor the remaining system ending on q_2 will behave in a simple way. They will trace out in space-time string surfaces without any simple resemblance to the original one; see for example [104]. What is even worse is that the behaviour of the subsystems depends upon the order in which we produce the $q_1\bar{q}_1$- or the $q_2\bar{q}_2$-breakups. The two production points are always at a spacelike distance (this is of course always the case for the production vertices along a Lund string!). We are, however, used to being able to introduce e.g. a proper time ordering with respect to the starting vertex. But even this

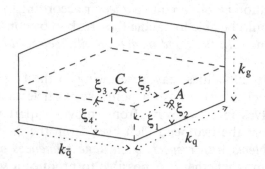

Fig. 15.11. The coordinates of some points on the string surface of the butterfly-dance mode.

proper time ordering may be different due to the fact that the produced string systems will move differently according to whether we break the string first at the $q_1\bar{q}_1$-vertex or at the $q_2\bar{q}_2$-vertex.

The conclusion of Sjöstrand's paper [104] is that *it is necessary to implement the string breakup as a process on the original string surface.* This means that we consider the string surface to be given as a 'frozen' geometrical object. Under these circumstances it is perfectly feasible to implement the symmetrical Lund model fragmentation process.

2   *The gluon fragmentation model of Sjöstrand*

We will make the following basic assumption.

- A string piece, if it fulfils the mass-shell condition, can be projected onto a hadronic state with the same probability irrespective of whether it contains at the semi-classical level internal excitations, bends etc.

Sjöstrand [104] has produced one version of a possible gluon fragmentation scheme based upon this assumption. It is incorporated into the Monte Carlo program JETSET, [105]. One of his findings in [104] is that there are only small differences between various possible schemes from the point of view of observables.

In order to describe his scheme we note that each point on the string surface can be given a 'proper time' with respect to the starting point of the original particles. In order to understand this we consider again the surface of the butterfly-dance mode (see Fig. 15.11).

The regions between the q and the g, and between the g and the $\bar{q}$, are of the same kind as we met in the $(1 + 1)$-dimensional model, i.e they are simply two lightcone regions. Consider an arbitrary point in the figure such as A and note that it can be described by means of two coordinates:

$$A = \xi_1 k_q + \xi_2 k_g \tag{15.16}$$

Therefore the squared proper time is given by $\Gamma_A = A^2 = 2\xi_1\xi_2 k_q k_g = \xi_1\xi_2(k_q + k_g)^2 = \xi_1\xi_2 M_{q,g}^2$. Its relation to an area can be inferred from the figure (we have again used dimensions such that $\kappa = 1$). The same goes for all parts of the qg and $g\bar{q}$ regions.

A more complex point is C, also indicated in the figure. It can be described as follows:

$$C = \xi_3 k_q + \xi_4 k_g + \xi_5 k_{\bar{q}} \tag{15.17}$$

where $\xi_4 = 1/2$. As for A we may identify $\Gamma_C = C^2$ and express it, this time, in terms of three coordinates ξ_j and the squared masses between the original partons. This is again an area and it is useful to construct it on the figure! There is no difficulty in convincing oneself that this procedure can be extended to any point on the surface.

It is also possible to define steps similar to the production steps in the $(1 + 1)$-dimensional model. If we imagine ourselves at the point A and would like to pick up a particular energy-momentum from the string by a step to, e.g. the point B, then if B is in the same segment as A there is again no difference from the $(1 + 1)$-dimensional case.

If B and A are in different string regions (for $B \equiv C$ we have such a case in the figure; C is on the flat string region between the two outward-moving fronts) then it is again possible to define a difference vector P_{AC} between A and C in terms of the original parton energy-momenta:

$$P_{AC} = \rho_1 k_q + \rho_2 k_g + \rho_3 k_{\bar{q}} \tag{15.18}$$

(Note that ρ_2 is determined by the starting position A and that there is a relation between ρ_1, ξ_3 and A's position, $\xi_1 = \rho_1 + \xi_3$, and one further condition, $\rho_3 = \xi_5$).

The requirement that P_{AC} should be on the mass shell then provides a condition among the coordinates ρ_j. The mass square can again be described in terms of certain areas on the surface. The main point is, however, that if we know the position of A, the size of Γ_C and the squared mass P_{AC}^2 then the position of C is also determined if it is on the string surface. (Convince yourself of that!)

The way JETSET implements the fragmentation is then step by step:

J1 With a knowledge of the original flavor (or antiflavor) a new $q\bar{q}$-pair is chosen with the probabilities described before.

J2 A meson with mass m is produced and a value of the fragmentation variable z is chosen from the symmetric fragmentation function.

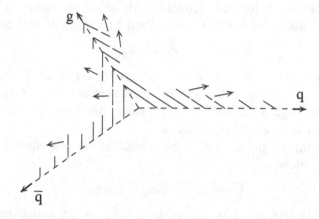

Fig. 15.12. The space-time development of a $qg\bar{q}$-state; the original directions of the partons are shown as broken lines, the string positions at different times are shown as solid lines and the momentum vectors of the emerging yoyo-hadrons are shown by the thinner arrows.

J3 The next $\Gamma_C = \Gamma_f$ is calculated from the earlier one, $\Gamma_A = \Gamma_i$, by $\Gamma_f = (1 - z)(\Gamma_i + m^2/z)$. This relation is exact in any of the 'simple' regions defined by two lightcone directions.

J4 The new breakup point is chosen as the point which has the value Γ_f and the step vector $P_{AC}^2 = m^2$. This is a unique prescription and determines the point C.

We have left out the transverse momentum generation, which is done in the same way as before, i.e. with a gaussian distribution. There are some complications about the directions that should count as transverse to the string direction in the relevant region, cf. [104]). The final mass, m, is then the transverse mass.

Some further technical problems are discussed in Sjöstrand's work, [104], but there is no need to delve into them here. We will instead turn to the experimentally observable consequences of the Lund gluon model.

15.6 The final-state particles in the breakup of a $qg\bar{q}$-state

1 General properties and the string effect

In Fig. 15.12 we illustrate the appearance of the final-state breakup in space-time for a one-gluon state. The three original excitations are moving out along the directions shown in the figure. The string is spanned via the g from q to $\bar{q}$ and a set of small final-state yoyo strings is depicted (for simplicity, at the moment of their emergence as independent entities) together with their space sizes and their momentum vectors.

The most noticeable thing is that most of the final-state yoyo particles move out along the three original parton directions with varying energies. The reason for this is that a moving string is Lorentz-contracted, as we have seen before. Therefore the size of one of the yoyo string pieces that moves quickly along e.g. the q-direction may appear very much smaller than one of the yoyo pieces produced at the centre. Nevertheless in its own rest system it is, of course, the same size. There will then be many more yoyo-hadrons from the seemingly small string pieces close to the trajectories of the three partons.

Quantitatively we may make the following estimates. Suppose that we consider a Lorentz frame in which the gluon goes out at an angle $\pi/2$ with respect to the q-direction. Then the longitudinal size l (i.e. the size along the q-direction) of a hadron with mass m and energy E is proportional to m/E. Such a string piece will contain an amount of gluon momentum $k_g \propto l$, i.e. $\propto 1/\cosh y$, with the rapidity y along the q-direction being calculated in this frame. Therefore we conclude that a gluonic disturbance is in general only noticeable within a small rapidity region (of order $\delta y \sim 1$) around the gluon direction (remember that angle and rapidity are connected). It will fall off as $\exp(-|y - y_g|)$ for larger rapidity differences.

There are some corrections to this, stemming from (almost) collinear gluon emission along the original gluon direction. Such emissions will tend to broaden the angular region affected around a hard parton but most of the parton energy still remains within a tiny angle even after a gluonic cascade and fragmentation; see the discussion in the following chapters.

Thus there will be three jets of particles basically along the three original directions (although there are some interesting differences between the jet directions and the original directions, to which we return).

From this picture we also conclude that the slower particles at the centre in general emerge earlier in time than the faster ones. This effect has been noted earlier and discussed in Chapter 7 in connection with the notion of the formation time.

The next experimentally observable result is that *there will be a few particles produced in the angular sectors between the q and the g and between the g and the q̄ but there are none produced between the q and the q̄ because there is no string spanned over this sector.* This is the nowadays well-known *string effect*, which was predicted (see [18]) before it was observed by the JADE group at PETRA.

There are several problems in disentangling this effect in an experiment, however. There are firstly the transverse momentum effects from the gaussian zero-point fluctuations. This means that the particles, which in the mean will emerge along two hyperbolas in momentum space, as

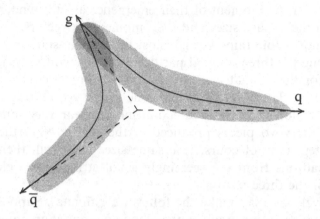

Fig. 15.13. The momentum-space picture of the final-state particles, which in the mean emerge along two hyperbolas. Due to transverse momentum fluctuations during the fragmentation the particles are diffused over the shaded regions.

shown in Fig. 15.13, are in reality diffused over the shaded areas. The typical distance of the hyperbola from the origin is of the order of 300 MeV/c, which is also the size of the transverse momentum fluctuations.

The second problem is to know which of the three jets is the gluon jet. In general the gluon jets contain less energy than the q or $\bar{q}$ jets do, but there are large variations according to the QCD emission probabilities. Nowadays this problem has diminished owing to the very large statistics produced in the LEP experiments. In these experiments it is even possible to tag one or both of the q and $\bar{q}$ jets by observing semi-leptonic heavy quark decays.

We may conclude that the string effect which was already quite noticeable in the JADE data nowadays provides strong confirmation of the existence of color-flow asymmetries in connection with gluon emission. There is, within perturbative QCD [27] also, such an effect, which is related to the coherence properties of gluon bremsstrahlung (cf. Chapters 16, 17).

It is interesting to note that if one only considers particles which have a large transverse momentum out of the production plane (the $qg\bar{q}$-state evidently defines a single plane in momentum space) then the string effect is even larger. The same applies if one considers only heavy particles, such as kaons and baryons. The reason, within the string scenario, is that the production of large (transverse) masses will use up larger pieces of the string and therefore such particles will feel more of a push from the string motion.

2 The jet-axes problem

A general problem in e^+e^- annihilation experiments is to determine the 'true' jet-axes' directions. In this case, as compared with e.g. hadronic interactions or inelastic lepto-production, there is no obvious initial direction along which the main dynamics proceeds.

The initial annihilation current of the e^+e^--pair is in the cms directed in the plane transverse to the momentum direction of the pair, if we neglect the rest masses. This was discussed for the current matrix elements in Chapter 4. The same also goes for the current of the $q\bar{q}$-pair produced in the annihilation, with respect to their momentum directions.

Therefore there is a correlation between the initial e^+e^--direction and the $q\bar{q}$-direction, corresponding to the current overlap $|\mathbf{j}_e \cdot \mathbf{j}_q|^2 \propto (1 + \cos^2\theta)$, θ being the angle between the two directions. This is a rather soft correlation, varying only between 1 and 2.

In order to analyse the final state in an event it is therefore necessary to define some directions by means of the observed particles. Several such methods are currently in use for doing jet analysis but we will not go into many details. We would like to point out, however, that the description of the events in terms of directions defined from the events themselves almost necessarily leads to some bias.

One rather obvious possibility is to consider a tensor $I_{\alpha\beta}$ constructed from the final-state momentum vectors $\mathbf{p}_j = \sum_\alpha p_{j\alpha}\mathbf{e}_\alpha$ of the N observed particles in an event:

$$I_{\alpha\beta} = \sum_{j=1}^{N}(\mathbf{p}_j^2\delta_{\alpha\beta} - p_{j\alpha}p_{j\beta}) \tag{15.19}$$

This tensor plays a role similar to the inertia tensor in the theory of solid bodies. Taken as a matrix it is possible to diagonalise it and to construct the eigenvalues λ_α as well as the (unit) eigenvectors $\mathbf{e}_\alpha$, $\alpha = 1, 2, 3$. From its construction we conclude that there will be a smallest eigenvalue, conventionally λ_3, and one defines the corresponding eigenvector $\mathbf{e}_3$ as the *sphericity axis* and the *sphericity S* as

$$S = \frac{3\lambda_3}{\sum\lambda_j} = \min_{\mathbf{e}}\left[\frac{3\sum_{j=1}^{N}(\mathbf{e} \times \mathbf{p}_j)^2}{2\sum_{j=1}^{N}\mathbf{p}_j^2}\right] \tag{15.20}$$

where the minimum corresponds to $\mathbf{e} = \mathbf{e}_3$. In this way one finds the *axis along which the sum of the (squared) transverse momenta is minimal.*

There is another way, already mentioned in Chapter 13, to find the *direction along which the sum of the longitudinal momenta is maximal, the thrust axis* $\mathbf{n}_T$ *such that the thrust* T *is maximal*:

$$T = \max_{\mathbf{n}} \left[\frac{2 \sum_{j=1}^{N} \Theta(\mathbf{n} \cdot \mathbf{p}_j) \mathbf{n} \cdot \mathbf{p}_j}{\sum_{j=1}^{N} |\mathbf{p}_j|} \right] \tag{15.21}$$

For events in which the observable momentum is conserved (meaning that no particle has evaded the detectors) we can change the thrust definition to

$$T = \max_{\mathbf{n}} \left[\frac{\sum_{j=1}^{N} |\mathbf{p}_j \cdot \mathbf{n}|}{\sum_{j=1}^{N} |\mathbf{p}_j|} \right] \tag{15.22}$$

There are no perfect detectors so the first definition is often the safer one.

It is evident that from an analytical point of view the sphericity measure is more regular. But at the same time it will give a quadratic weight to the momenta. Therefore a single particle with a large momentum will in general provide a larger contribution than a group of particles which together have this momentum.

This is particularly inconvenient if we consider an event before or after the decay of some of the particles. The thrust definition is less sensitive to these features. However, although thrust is not easy to work with analytically, it is very simple in general to generate a computer routine to find out where the thrust axis is for the observed particles in a given event. A general feature is that the thrust axis connects the two groups of particles which together have the largest (and oppositely directed) momenta. Therefore thrust directly serves as a 'handle' on the way the event looks.

Both the sphericity and the thrust variables provide a means to assess quantitatively the amount of gluon emission. For a large-energy single $q\bar{q}$-event the thrust $T \simeq 1$ and the sphericity $S \simeq 0$. They will deviate noticeably from these values for events containing one or more hard gluons because in that case a large amount of energy is moving transversely.

The string effect in the Lund model fragmentation actually produces some (minor) distortions in the particle distributions due to the way thrust and sphericity are defined. Suppose that there is a gluon emitted at a finite angle with respect to either the q- or the $\bar{q}$-direction and suppose that the mass of the two partons is not small. From this configuration we expect that some particles will be produced in the angular region between the q ($\bar{q}$) and the g.

The thrust and sphericity axes will both be tilted towards the most energetic of the q or $\bar{q}$ and the g but the particles in between will also influence the determination of the axes. In particular any jet-finding algorithm [2] would tend to create jets such that there is a slightly smaller angle between the observed directions of the qg- and $g\bar{q}$-jets than between

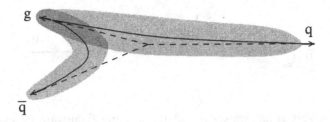

Fig. 15.14. A collinear configuration of a $qg\bar{q}$-state, and the ensuing final-state hadrons, described in momentum space.

the original parton directions, in order to accommodate the extra particles in between. The field of jet-searching algorithms is, however, still under intense development and we refer to the discussions in e.g. [2] for those readers with a technical interest in them.

3 Infrared stability

We have already referred to this notion. In the next chapter we will show that the cross sections for gluon emission are divergent for soft and collinear emissions. Therefore the number of gluons is not a well-defined notion but *the effect of the gluon emission is observable*. It is an essential point that if soft or collinear gluons are emitted in a bremsstrahlung process, it is their combined activity that will play a role for the fragmentation. We have already seen that the surface of the MRS is infrared stable, in the way the concept is employed in the Lund model. A soft or collinear gluon only has small and, in general, local effects.

In Fig. 15.14 we exhibit the result in momentum space for the fragmentation of a $qg\bar{q}$-state in the case where the gluon is close in angle to the $\bar{q}$. Again, the shaded area is the one inside which the final-state particles emerge. It is then noticeable that, as the mass of the $g\bar{q}$-pair diminishes towards the mass of the final-state hadrons, no hadrons are produced in between the two partons. Instead final-state particles may occur having a larger energy than any of the partons!

If we instead consider the emission of a 'soft' gluon in the centre of the event, i.e. a gluonic disturbance containing small cms energy then there are only very small effects even in the neighborhood of the gluon rapidity. In Fig. 15.15 we show again the shaded area in momentum space where the final-state particles emerge for such a soft gluon emission. The gluon is only noticeable as a small localised transverse momentum 'bump' in the distribution.

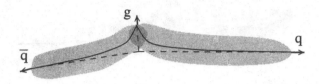

Fig. 15.15. A soft gluon emission and the ensuing final-state particles in momentum space.

As a general rule of thumb, the effect of a gluon excitation is hardly noticeable when the transverse momentum of the gluon or the gq- or $g\bar{q}$-pair's mass is smaller than 2 GeV. Actually there is a moving interface between the fragmentation and the gluon emission processes according to the Lund model. One can stop the emission of gluons basically anywhere between a cutoff at $k_\perp = 7$ GeV down to a few hundred MeV and still obtain the same distribution of final-state hadrons. One needs different fragmentation parameters, however, and we will present an interesting systematics for this phenomenon in Chapter 17.

4 *The decay of heavy quarkonia*

One of the true revolutions in high-energy physics occurred when the very long-lived resonance state, the J/Ψ, was found in October 1975. It was amazing to disentangle a state which is so massive. The J/Ψ-mass is around 3.1 GeV, i.e. about four times the ρ- and ω- masses and three times the proton and neutron masses. A more important fact was that the J/Ψ is so long lived. This meant that there must be new physics involved.

After the first few months of frantic discussions and investigations the high-energy physics world settled for the fact that there was a c-flavor quark and that the J/Ψ was a bound state, of vector character, of a $c\bar{c}$-pair. The other vector mesons, the ρ, ω and ϕ, are all built from the light quarks and all decay rather quickly; it is only necessary to produce one or two new light $q\bar{q}$-pair(s) to make them decay into a mixture of light pseudoscalar meson states.

We note that flavor is a conserved quantity in QCD-initiated reactions. For the J/Ψ (and also in connection with the later-discovered Υ, a bound state of a $b\bar{b}$-pair with mass around 9.5 GeV) the corresponding mesons, the D-states and the B-states, contain the c ($\bar{c}$) and b ($\bar{b}$) together with a light $\bar{q}$ (q). It turns out, however, that even the lightest $D\bar{D}$ ($B\bar{B}$) has too large a rest mass to allow the decay $J/\Psi \rightarrow D\bar{D}$ ($\Upsilon \rightarrow B\bar{B}$). It is then necessary either that one of the c (b) or $\bar{c}$ ($\bar{b}$) decays semileptonically (which we will not treat in this book; owing to the small

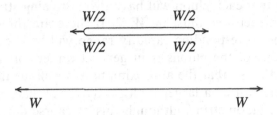

Fig. 15.16.   A state in which two gluons separate, spanning two string regions each carrying a mass $s/4 = (W/2)^2$, compared with a single string spanned by a $q\bar{q}$-state of mass $s = W^2$.

coupling constants these are rather suppressed reactions) or that there is an interaction channel allowing for the annihilation of the heavy flavor and antiflavor into multigluon states.

This latter alternative requires the c and $\bar{c}$ (b and $\bar{b}$) to meet, i.e. the decay is governed by the wave function at the origin of the relative coordinates, $|\psi(0)|^2$, which serves as a form factor suppressing the decay. The possible decay channels are governed by the internal quantum numbers of the J/Ψ or the Υ. The simplest such state is a three-gluon state (but multigluon states would also be allowed).

In this case *a closed string will emerge spanned by the three gluon corners*. We have not treated this situation in the general description of string motion above, mainly because we do not need the details in a general description.

It is rather easy to imagine how the closed string is stretched and the only feature of interest for this discussion is that *it has no endpoints*. We have seen before that the existence of a string endpoint means necessarily that there is also a fragmentation region governed by the flavor at the endpoint. In practice this means a lower yield of final-state particles within 1–2 units in rapidity. For a closed string (which is the same all over) these suppressions are not available and this is particularly noticeable for baryon-antibaryon production. The gluon is flavorless. There is also the fact, to be further discussed in the next section, that the appearance of gluons increases the phase space for particle production.

As a minimum size for the increase in phase space we may imagine that one of the three gluons is very soft, so that we obtain a situation in which two gluons move out in opposite directions, as in Fig. 15.16. In the figure we also show a state with the same mass $s = W^2$, but containing only a $q\bar{q}$-pair. We note that if the total rapidity for the $q\bar{q}$-state is $(\Delta y)_q = \log(s/s_q)$, then the total rapidity within which we can produce particles will be $(\Delta y)_g = 2\log(s/4s_g)$ in the gg-state.

The reason is that each gluon will have two adjoining string pieces and that the mass in each one must be $W/2$. If there are three hard gluons on the string the corresponding rapidity range will be even larger but it turns out that one of the gluons is in general rather soft in the process under consideration so that the approximation is well justified.

Therefore there will be a larger multiplicity for all hadrons in the case of a closed three-gluon string, although this of course depends upon the two scales s_q and s_g which govern the fragmentation. If we set $s_q = s_g = 2$ GeV^2 we obtain for Υ

$$(\Delta y)_q \simeq 3.8, \ (\Delta y)_g \simeq 4.8 \tag{15.23}$$

This implies that there should be around 1.3 times as many mesons on the Υ-resonance with the three-gluon decay as in the continuum surrounding it (which seems to be a reasonable approximation). *But there will be a noticeable enhancement of baryons as compared with an open string.*

According to the simple baryon-antibaryon production model discussed in Chapter 13 there is a region of around 1.5 units in rapidity, close to the endpoints, which is lost for baryon-antibaryon production. For the gluon string there is no 'flavor direction' and consequently this suppression is not available. Therefore the enhancement of baryons should in this case be very noticeable for a gg-state with the same mass as a $q\bar{q}$-state in the continuum. From the numbers above we would expect that the ratio of the number of baryons would be

$$\frac{(B\bar{B})_\Upsilon}{(B\bar{B})_{cont}} \simeq \frac{4.8}{3.8 - 1.5} = 2.1 \tag{15.24}$$

The estimates presented above are not far off the experimental results from ARGUS, although we have certainly used a very simple model!

15.7   A measure of multigluon activity, the generalised phase-space rapidity

Based upon the ideas presented in [48] we will in this section introduce a useful new variable, the total generalised rapidity λ. We will be content to consider a single hard gluon emission and extend the definition of λ to multigluon situations in section 17.4.

We have already seen that the appearance of gluonic excitations in a string state produces certain regions, close to a hard gluon emission region, where more particles will emerge. Therefore for such events there will no longer be an essentially constant rapidity plateau, which was characteristic for the simple $(1 + 1)$-dimensional $q\bar{q}$-model. (This result is independent of the axes chosen to define the rapidity variable).

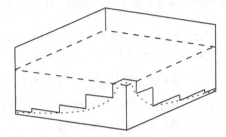

Fig. 15.17. The particles are produced along hyperbolas corresponding to fixed values of the squared proper time Γ.

It is useful to introduce some variable which follows the production region, as does the ordinary rapidity for a straight $q\bar{q}$-string. We would like such a variable to have the following properties.

- The measure λ should be well defined for each event.

- The mean value of λ should be proportional to the corresponding mean value of the multiplicity of the events.

- The distribution in multiplicity for events with a given value of λ should be almost Poissonian (although slightly narrower, as we found for the Lund model properties in Chapter 11).

We can rather easily obtain such a variable if we generalise the mean hyperbola decay picture we used in Chapter 9. There we found that the breakup vertices of the string are on the average distributed along a curve with a constant value of Γ, the squared proper time. In energy-momentum-space language the squared proper time corresponds to the squared momentum transfer between the particles produced to the left and to the right of the production vertex. From this dual relationship (cf. Figs. 9.4 and 9.5) the hyperbola decay corresponds to ladder diagram chains for which the momentum transfers are all the same.

If we consider a typical breakup, such as the one shown in Fig. 15.12, in space-time we obtain a picture like that in Fig. 15.17. We again notice the two hyperbolas in the regions between the q and the g and between the g and the $\bar{q}$ together with a few particles produced near the gluon corner.

In order to describe the situation we introduce the following notation. The total energy-momentum of the event is P_{tot} with $s = P_{tot}^2$ and the three energy-momenta of the partons are k_j, where

$$\sum_{j=1}^{3} k_j = P_{tot}, \quad s_{ij} = 2k_i k_j = (k_i + k_j)^2, \quad s = s_{12} + s_{23} + s_{13} \quad (15.25)$$

(indices 1 and 3 represent the q and $\bar{q}$, respectively, and index 2 represents the g).

Thus we obtain a generalisation of the total rapidity range from the case where the event is of the $q\bar{q}$-type to the case where it is of the $qg\bar{q}$-type:

$$
\Delta y = 2\log(\sqrt{s}/W_q) = \log(s/s_q)
$$
$$
(\Delta y)_{gen} \equiv \lambda = \log(s_{12}/W_g W_q) + \log(s_{23}/W_g W_q) \tag{15.26}
$$
$$
= \Delta y + \log(s_{12}s_{23}/ss_g)
$$

The two terms in the definition of λ are the lengths of the two hyperbolas in the qg- and the $g\bar{q}$-sectors.

We now assume that there are regions close to the q- and $\bar{q}$- ends, respectively, that correspond to fragmentation regions, in which there is a lower density of particles. Thus we 'lose' $\log W_q = (1/2)\log s_q$ in each q- and $\bar{q}$-region. Similarly we assume that on both sides of the gluon corner there is a corresponding loss governed by $\log W_g = (1/2)\log s_g$.

We may then conclude that the rapidity region has increased owing to the emission of the gluon, and that the quantity

$$
\log(s_{12}s_{23}/ss_g) \equiv \log(k_\perp^2/s_g) \tag{15.27}
$$

is a measure of the increase. We will next show that the quantity $k_\perp$ occurring in Eq. (15.27) in fact corresponds to the transverse momentum of the gluon.

To see this we consider the event again in a frame where the q and $\bar{q}$ separate in opposite directions with energies e_1 and e_3, respectively. The g will move away transversely with energy e_2. Then we obtain by direct calculation that

$$
s_{12} = (e_1 + e_2)^2 - e_1^2 - e_2^2 = 2e_1e_2
$$
$$
s_{23} = (e_2 + e_3)^2 - e_2^2 - e_3^2 = 2e_2e_3
$$
$$
s_{13} = (e_1 + e_3)^2 - (e_1 - e_3)^2 = 4e_1e_3 \tag{15.28}
$$
$$
k_\perp^2 = \frac{s_{12}s_{23}}{s} = \frac{e_2^2}{1 + (e_2/2)(1/e_1 + 1/e_3)} \simeq e_2^2
$$

the approximation being valid unless the g's energy is of the same order as the energies of the q and $\bar{q}$. Another way to obtain the result is to note that there is a direction in the cms (approximating the directions of the q and $\bar{q}$) along which $k_\perp^2$, as defined above, is identical to the gluon's transverse momentum. We will show this in Chapter 17 after we have introduced a few more kinematical notions.

The result for the phase-space extension is clear. What happens is that the single hyperbola for the flat string is exchanged for two hyperbolas, the connecting point being 'pulled out' by the gluon corner. The tip formed in this way corresponds to an extension of phase space (not only for emission of final-state hadrons but also for further gluon emission, cf. Chapter 17), whose size is determined by the transverse momentum of the gluon.

The size of the extension is calculated in terms of a scale s_g characteristic of the particle production around the gluon corner. In the same way the original hyperbola is measured by means of a scale s_q characteristic for production at the q- and $\bar{q}$- endpoints of the string. In the section on Υ-decay we used the estimates $s_g = s_q = 2 \text{ GeV}^2$.

In section 17.4 we will consider the necessary steps for a generalisation of the λ-measure to multigluon situations. We note, however, that the present definition is only useful when the squared masses between the partons exceed the scales s_q and/or s_g and we will therefore in section 18.7 extend the definition to an infrared-stable λ-measure.

At the same time we will be able to introduce a 'local' value of the λ-measure. Up to now λ as it is defined evidently corresponds to the total available region for particle production (and, as we will later also find, for gluon emission in perturbative QCD). Therefore it is similar to Δy, the total available rapidity region for the decay of a straight $q\bar{q}$-string. Using the directrix function (generalised to multiparton situations) it is possible to introduce a value $\lambda(\sigma)$ that varies from e.g. $\lambda(\sigma = 0) = 0$ to $\lambda(\sigma = E/\kappa) = \lambda$ (for the variable σ see Eq. (15.10)), just as the cms-rapidity y varies from $(-1/2)\Delta y$ to $(+1/2)\Delta y$.

16
Gluon emission via the bremsstrahlung process

16.1 Introduction

Bremsstrahlung emission is an inherent property of all gauge field theories. It can be understood even within classical mechanics, at least for the soft part of the spectrum. Suppose that we consider a charge surrounded by its Coulomb field, which necessarily is extended in space outside the charge. Then suppose that there is a sudden change in the state of motion of the charge itself. The result will be that the outlying field will need some time to readjust to the new situation.

Therefore there will be, as in all other situations of sudden change in physics, a brief interlude of compressions and extensions in the field before it comes back to a stable state. *The ensuing radiation field, to be described below, is a bremsstrahlung field.* Its properties depend upon the way in which the charge distribution is changed. For a single charge with a sudden momentum transfer, or for the situation when a charge and anticharge suddenly emerge, the bremsstrahlung is essentially of a *dipole character*. This approximation means that the current contains a direction, the dipole axis, but the size of the interaction region is neglected. We will consider a 'classical' current with these properties.

Some warning is needed against taking the classical picture too far. We have shown in Chapter 2 how the method of virtual quanta describes the Coulomb field of a fast-moving charge. In particular we have shown that the virtual field quanta have a distribution in rapidity and transverse momentum. In this chapter we will meet this again, as the bremsstrahlung distribution.

If we make a measurement on the field that really interacts with one of the quanta then the field will change. This will in turn (i.e. causally) also affect the current-charge itself. Therefore the bremsstrahlung process is difficult to visualise in a classical scenario, i.e. it is not possible to say

302

whether the field quanta exist before a measurement is made on them or whether they come to existence because of the measurement. We will in this chapter consider the bremsstrahlung process in some detail. We derive the cross section from first principles and express it in different ways in order to stress different properties.

Dipole bremsstrahlung contains coherence conditions, i.e. inside some regions the waves stemming from the different parts of the current will interfere constructively and in other destructively. In order to take coherence into account it is necessary to carefully preserve gauge invariance. If the interference diagrams are all taken into account then it is possible to use any gauge to evaluate the result. It turns out that in the emission of coherent dipole bremsstrahlung there is a close connection between the regions with positive interference and the regions allowed by energy-momentum conservation. As in many other cases the laws of nature ensure consistency. In this case one is evidently not allowed to emit more radiation energy than the energy carried by the charge!

These conditions imply that bremsstrahlung emission may only occur inside a chacteristic emission region, which can be expressed in terms of the transverse momenta and rapidities of the emitted quanta. In order that the conditions should be valid in any Lorentz frame these variables are most conveniently expressed in terms of Lorentz invariants. The bremsstrahlung spectrum from unpolarised charges must be independent of the azimuthal angle around the dipole axis in the rest frame. Together with the requirements on rapidity and transverse momentum this requirement translates into certain allowed conar emission regions in a moving frame. As long as one considers the emission from the full dipole these regions are easily traced.

The total bremsstrahlung from the dipole is in many models, e.g. HERWIG [94] and JETSET [105], subdivided into contributions from the individual charges. This is, of course, an allowed operation as long as one avoids double counting, i.e. the total coherence conditions are invoked. We will derive a condition referred to as the *strong angular ordering condition*, [59], in this connection.

We will also indicate that a too-literal application of strong angular ordering means that some, usually soft, emission will be displaced in phase space. Clever model builders, like the authors of the two Monte Carlo models mentioned above, have taken some precautions in this respect.

16.2 The matrix element for dipole emission

We will use a semi-classical picture and assume that the electromagnetic current **j** is suddenly changed, e.g. by an external agent. As a simple model for such a current distribution we assume the shape

$$\mathbf{j}(\mathbf{x}, t) = g\mathbf{v}(t)\delta(\mathbf{x} - \mathbf{x}(t)), \quad \mathbf{v}(t) = \frac{d\mathbf{x}}{dt}(t) \tag{16.1}$$

We also assume that $\mathbf{v}(t)$ suddenly changes at $t = 0$ from $\mathbf{v}(-\delta) = \mathbf{v}_-$ to $\mathbf{v}(+\delta) = \mathbf{v}_+$ so that we are in effect considering the case where a charged particle (charge g) moves in a pointlike way along some straight line $\mathbf{x}(t)$ (with velocity $\mathbf{v}(t) = d\mathbf{x}/dt$) and suddenly during a very short time interval $-\delta < t < +\delta$ changes to another straight-line orbit.

The number of quanta, i.e. photons, emitted with energy-momentum vector k is as usual given by Fermi's Golden Rule. By means of the methods we have used several times before we obtain, cf. Eq. (3.104)

$$dn_\gamma = \sum_{\mathbf{k}} \frac{w}{\Delta t} = \frac{|\mathcal{M}|^2}{2V\omega} \frac{V d^3k}{(2\pi)^3} = \frac{dk}{(2\pi)^3} \delta(k^2) |\mathcal{M}|^2 \tag{16.2}$$

The transition matrix element $\mathcal{M}$ is given by

$$\mathcal{M} = \int dt d^3x \, \mathbf{j}(\mathbf{x}, t) \cdot \mathbf{A}(\mathbf{x}, t) \tag{16.3}$$

The vector potential $\mathbf{A}$ describes a free photon with polarisation vector ϵ, i.e. it corresponds to a transverse wave

$$\mathbf{A} = \epsilon \exp(ikx), \quad k = (\omega, \omega\mathbf{n}), \quad \epsilon \cdot \mathbf{n} = 0 \tag{16.4}$$

Note that the normalisation factor $1/\sqrt{2V\omega}$ already has been used in connection with Eq. (16.2). We will sometimes write the polarisation ϵ as a four-vector.

Under these assumptions we can immediately obtain a result for $\mathcal{M}$, by means of an integration over time:

$$\begin{aligned}
\mathcal{M} &= \int dt g\mathbf{v} \cdot \epsilon \exp[i\omega(t - \boldsymbol{n} \cdot \mathbf{x}(t))] \\
&= \int \frac{g\mathbf{v} \cdot \epsilon}{i\omega(1 - \boldsymbol{n} \cdot \mathbf{v})} id[\omega(t - \boldsymbol{n} \cdot \mathbf{x}(t))] \exp[i\omega(t - \boldsymbol{n} \cdot \mathbf{x}(t))] \\
&= \int dt ig \frac{dX}{dt}(t) \exp[i\omega(t - \boldsymbol{n} \cdot \mathbf{x}(t))]
\end{aligned} \tag{16.5}$$

where we have neglected a surface term in the integral corresponding to times well before or well after the emission and have written

$$X(t) = \frac{\epsilon \cdot \mathbf{v}(t)}{\omega(1 - \boldsymbol{n} \cdot \mathbf{v}(t))} \tag{16.6}$$

In the second line of Eq. (16.5) we have changed the integration variable in an obvious way. The dipole approximation corresponds to the assumption that the quantity X changes much faster than the exponential in the last line of Eq. (16.5) so that we may take the exponential phase factor outside

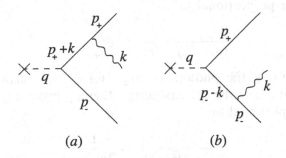

Fig. 16.1.   The emission of bremsstrahlung either before or after the encounter with an external-momentum-transfer producing agent at time $t = 0$.

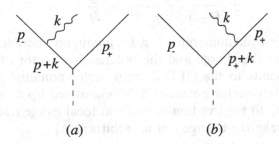

Fig. 16.2.   The production of a pair at $t = 0$ and the ensuing bremsstrahlung from each of the charges.

the integral and write

$$\mathcal{M} = \exp(i\Phi)[X(+\delta) - X(-\delta)] \tag{16.7}$$

We will from now on ignore the unobservable phase factor $\exp(i\Phi)$. The two terms in Eq. (16.7) can be written, incorporating the assumptions on $\mathbf{v}$, as

$$X(\pm\delta) \equiv X^{\pm} = \frac{g\epsilon p(\pm)}{kp(\pm)}, \quad p(\pm) = (E_{\pm}, \, E_{\pm}\mathbf{v}(\pm)) \tag{16.8}$$

In order to obtain this formula we have multiplied by the energies E_- and E_+ of the current particle before and after the emission in the numerators and denominators of the two terms, respectively.

The result is rather easy to interpret in terms of Feynman graphs (see Figs. 16.1 and 16.2). In Fig. 16.1(a) we see a particle coming in on-shell with energy-momentum p_-, suddenly changing during the encounter with the external agent to a virtual particle with momentum $p' = p_+ + k$, and with propagator proportional to

$$\frac{1}{(p_+ + k)^2 - m^2} = \frac{1}{2p_+k} \tag{16.9}$$

and finally emitting the photon. In Fig. 16.1(b) the particle emits the photon before it meets the external agent, thereby becoming virtual with propagator proportional to

$$\frac{1}{(p_- - k)^2 - m^2} = \frac{-1}{2p_-k} \tag{16.10}$$

Similarly we may interpret Figs. 16.2 as an emission from a produced pair with charges $\pm g$ and with energy-momenta $p_\pm$,

$$\pm g \frac{\epsilon p_\pm}{(p_\pm + k)^2 - m^2} = \frac{\pm g \epsilon p_\pm}{2p_\pm k} \tag{16.11}$$

The appearance of the numerator $g\epsilon p$, i.e. a coupling between the particle momentum times the charge and the polarisation vector of the radiated photon, corresponds to the QED current-vector-potential interaction in Eq (16.3). The four-vector potential A is determined up to a gradient (cf. Chapter 2) owing to the freedom to perform local gauge transformations. Thus we may make the change (for an arbitrary $\tilde{\Lambda}$)

$$\epsilon \to \epsilon + \tilde{\Lambda}(k)k \tag{16.12}$$

It is essential to have a difference between X^+ and X^- because then each term will obtain the same contribution $\tilde{\Lambda}(k)$, which vanishes in the difference. Thus *in order that the matrix element $\mathcal{M}$ should be gauge-invariant the contributions must occur with a relative minus sign.*

We have evidently obtained the same gauge-invariant result whether we imagine a sudden change in the equations of motion of a single particle with charge g, or the equally sudden production of a pair of particles with charges $\pm g$. In both cases the Coulomb field changes, in the first by rebuilding and in the second by starting up. We will come back to this picture again in Chapter 20. For now we note that the result that the matrix elements of the two processes are the same is a general one in relativistic quantum field theory, corresponding to the property of *crossing symmetry* for the S-matrix.

A final remark at this point is that it is often dangerous to make too-literal interpretations of Feynman graph techniques. According to the discussion in Chapter 3 the Feynman propagator is completely satisfactory with respect to the requirements of Lorentz covariance, causality and the quantum conditions put up by the Heisenberg indeterminacy relations. In that case, however, the interpretations are well defined.

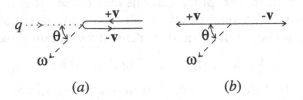

Fig. 16.3. The two different situations when bremsstrahlung γ's are emitted either (a) from a single charge bouncing back (the Breit frame) or (b) in connection with the production of a pair in the cms.

16.3 The dipole cross section

1 The dependence on energy and rapidity

We will provide several forms for the cross section in Eq. (16.2) in order to stress different properties of the process. We will start with a description in terms of the photon's energy ω and rapidity y:

$$y = \frac{1}{2} \log\left(\frac{\omega + k_3}{\omega - k_3}\right) = \frac{1}{2} \log\left(\frac{1 + \cos\theta}{1 - \cos\theta}\right) = \log\cot(\theta/2) \quad (16.13)$$

where θ is the angle between the dipole (3-)axis and the photon direction **n**. We will for now work in a Lorentz frame in which the two particle momenta, $\mathbf{p}(\pm)$, are along the 3-axis, equal in size and oppositely directed.

This means that when there is a sudden momentum transfer to bring the incoming state particle with $\mathbf{p}(+)$ to the final state $\mathbf{p}(-)$ this momentum transfer must be directed oppositely to the particle's momentum so that it comes in and bounces back again (in the Breit frame). For the case when a pair, $(\mathbf{p}(+), \mathbf{p}(-))$, is produced the two particles move in opposite directions along the 3-axis (see Fig. 16.3). The rapidities of the two particles will be called $\pm y_0$.

We next choose two independent directions to describe the polarisation vector ϵ. For simplicity we choose one of them to be in the plane of **n** and the 3-axis and the other out of this plane. Therefore we will only obtain a contribution to the matrix element from the one in the plane. That contribution is given by

$$\epsilon \cdot \mathbf{v}(\pm) = \pm \tanh y_0 \frac{1}{\cosh y} \quad (16.14)$$

where we have used the formula

$$\sin\theta = \frac{2\sin(\theta/2)\cos(\theta/2)}{\sin^2(\theta/2) + \cos^2(\theta/2)} = \frac{1}{\cosh y} \quad (16.15)$$

(An exercise for the reader: prove that one may chose the two polarisation directions in any orthogonal way and still obtain the same result!).

For the denominators in the matrix element we obtain directly

$$\omega(1 - \mathbf{n} \cdot \mathbf{v}(_\pm)) = \omega(1 \mp \tanh y_0 \tanh y) \qquad (16.16)$$

using a similar trick to express $\cos\theta$ in terms of the rapidity variable y. Putting it together we obtain the result of summing over the polarisation directions:

$$|\mathcal{M}|^2 = \frac{2\cosh 2y_0 \cosh^2 y}{\omega^2 \cosh(y + y_0)\cosh(y - y_0)} \qquad (16.17)$$

The size of a phase-space element is

$$dk\delta(k^2) = \tfrac{1}{2}\omega d\omega d\phi \sin\theta d\theta \qquad (16.18)$$

where ϕ is the azimuthal angle (over which we can evidently integrate to give 2π) and where the θ-dependence easily transforms to a rapidity dependence:

$$\sin\theta d\theta = \frac{dy}{\cosh^2 y} \qquad (16.19)$$

Therefore the number of γ's per unit energy and unit rapidity is

$$dn_\gamma = \left(\frac{g^2}{4\pi^2}\right)\frac{d\omega}{\omega}dy\frac{\cosh(y_0 + y_0)}{\cosh(y + y_0)\cosh(y - y_0)} \qquad (16.20)$$

The somewhat fancy way we have used to write the arguments in the hyperbolic sine and cosine functions is made in order to exhibit the Lorentz invariance of the formula: it only depends upon the rapidity differences $y_0 - (-y_0)$ and $y - (\pm y_0)$. Therefore it is the same in any Lorentz frame obtained by boosting along the dipole axis.

A closer examination of the rapidity-dependent factor also reveals that it is basically a constant for rapidities

$$|y| < |y_0| \qquad (16.21)$$

and that it falls off exponentially fast outside this region. Therefore the spectrum is, to a good approximation

$$dn_\gamma = \left(\frac{2\alpha}{\pi}\right)\frac{d\omega}{\omega}dy \qquad (16.22)$$

where the requirement in Eq. (16.21) must be incorporated. We have here as usual introduced the fine structure constant $\alpha = g^2/4\pi$ under the assumption that we are dealing with electrons and positrons. We will shortly come back to the difference when we consider color-charged q- and $\bar{q}$-particles.

If we rewrite the energy dependence in terms of a transverse momentum dependence for fixed rapidity,

$$k_\perp \equiv \omega \sin \theta = \frac{\omega}{\cosh y} \tag{16.23}$$

we obtain the formula found in connection with the method of virtual quanta in Chapter 2:

$$dn_\gamma = \frac{\alpha}{\pi} \frac{dk_\perp^2}{k_\perp^2} dy \tag{16.24}$$

(we may of course also make the change $dy \to dx/x$ in the same way).

Consequently, *the bremsstrahlung spectrum arising from a change in the current distribution is equivalent to the flux of virtual quanta which can be used to describe the electromagnetic field around a fast-moving charge.* Quantum mechanics does not tell you before you measure what you may find in your detector!

2   The invariant cross section for dipole emission

If we neglect the particle masses we may write for the two denominators in Eq. (16.8), using the conventional notation and so calling the positive charged particle's energy-momentum $p(+) \equiv p_1$, that of the negative charge $p(-) \equiv p_3$ and that of the emitted photon $k \equiv p_2$:

$$p(+)k \equiv \frac{s_{12}}{2}, \quad p(-)k \equiv \frac{s_{23}}{2} \tag{16.25}$$

where we have introduced the squared masses of the particle pairs.

Squaring the matrix element and summing over the polarisation directions we obtain for the polarisation sum (cf. Eq. (4.40))

$$\sum_{polarisation} \epsilon_j \epsilon_l = \delta_{jl} - \frac{k_j k_l}{\mathbf{k}^2} \tag{16.26}$$

There will then be three terms in the squared matrix element. The first one can be written as (using $\sqrt{\mathbf{k}^2} \equiv \omega$)

$$\frac{4 \left[(\mathbf{p}_1)^2 - (\mathbf{p}_1 \cdot \mathbf{k})^2 / \mathbf{k}^2 \right]}{s_{12}^2} = -\frac{1}{\omega^2} + \frac{4E_1}{\omega s_{12}} \tag{16.27}$$

The second term is the same but with the obvious exchange of index $1 \to 3$. The third becomes

$$\frac{2}{\omega^2} + \frac{8p_1 p_3}{s_{12} s_{23}} - \frac{4E_1}{\omega s_{12}} - \frac{4E_3}{\omega s_{23}} \tag{16.28}$$

so that the total result is

$$\sum_{polarisation} |\mathscr{M}|^2 = \frac{4s_{13}}{s_{12}s_{23}} \qquad (16.29)$$

For the phase-space factors we obtain, by fixing the two squared masses s_{12} and s_{23},

$$\int dk\delta(k^2)\delta(2kp_1 - s_{12})\delta(2kp_3 - s_{23}) = \frac{\pi}{2s_{13}} \qquad (16.30)$$

Then the total γ-multiplicity is given by

$$dn_\gamma = \left(\frac{\alpha}{\pi}\right)\frac{ds_{12}ds_{23}}{s_{12}s_{23}} \qquad (16.31)$$

Although the result in Eq. (16.31) is derived by semi-classical methods it agrees in detail with the results of a complete quantum mechanical calculation *for soft γ-radiation*. But when it comes to hard bremsstrahlung, i.e. when $s_{13} \simeq s_{12}$ and/or s_{23}, there are corrections. The formula used for the current, cf. Eq. (16.1), in the calculation of the matrix element does not account for the fact that the electrons and positrons are spin 1/2 particles. There are then, just as in connection with the Rutherford scattering matrix elements in section 5.5, also contributions from the spin structure. Further, the treatment of the phase space in Eq. (16.2) leading to Eq. (16.31) is also too simple. As subsequently we will need a formula also for the hard radiation we will briefly exhibit the steps necessary to obtain a more precise formula.

Firstly, the current in Eq. (16.1) should be changed as follows:

$$\frac{g}{E}\mathbf{p}\delta(\mathbf{x} - \mathbf{x}(t)) \rightarrow \frac{g}{E}(\mathbf{p} + \boldsymbol{\sigma} \times \nabla)\delta(\mathbf{x} - \mathbf{x}(t)) \qquad (16.32)$$

(we have for simplicity written $\mathbf{v} \equiv d\mathbf{x}(t)/dt = \mathbf{p}/E$) with $\boldsymbol{\sigma}$ describing the spin (cf. section 4.4) of the fermions. It is an axial vector, which means that the term $\boldsymbol{\sigma} \times \nabla$ corresponds to a proper vector, as is $\mathbf{p}$, and therefore it is an 'allowed' contribution to the current in a parity conserving theory. Further for a massless fermion the helicity can only take on two values ($\pm 1/2$) corresponding to spin 'along' and 'opposite to' the direction of motion and we must sum over the two values in the final squared matrix element if we have unpolarised fermions.

It is evident that this extra contribution will change the result in Eq. (16.8) into

$$X = \frac{\boldsymbol{\epsilon} \cdot (\mathbf{p} + i\boldsymbol{\sigma} \times \mathbf{k})}{pk} \qquad (16.33)$$

(with appropriate indices). When we square the matrix element using this expression for the X-factors we obtain extra terms as compared to Eq.

(16.29), which, after summing over the photon polarisations according to Eq. (16.26), will be

$$\mathscr{D}_1(\pm) = \frac{(\boldsymbol{\sigma}(\pm) \times \mathbf{k})^2}{(p(\pm)k)^2}$$

$$\mathscr{D}_2 = -2\frac{[\boldsymbol{\sigma}(+) \times \mathbf{k}] \cdot [\boldsymbol{\sigma}(-) \times \mathbf{k}]}{(p(+)k)(p(-)k)} \tag{16.34}$$

(the remaining interference terms, as e.g. those proportional to

$$\mathbf{p}(\pm) \cdot [\boldsymbol{\sigma}(\pm) \times \mathbf{k}]$$

vanish because $\mathbf{p}(\pm) \times \boldsymbol{\sigma}(\pm) = 0$ as we mentioned above). The result in Eq. (16.34) should then be summed over the possible values of $\boldsymbol{\sigma}(\pm)$; only the two quantities $\sum_{spins} \mathscr{D}_1(\pm) \to k_\perp^2/[p(\pm)k]^2$ are nonvanishing (with $k_\perp^2$ defined in Eq. (16.39) below). Therefore the result in Eq. (16.29) is changed as follows:

$$\frac{4s_{13}}{s_{12}s_{23}} \to \frac{4s_{13}}{s_{12}s_{23}} + \frac{2s_{12}}{ss_{23}} + \frac{2s_{23}}{ss_{12}} \tag{16.35}$$

For soft radiation the last two terms are negligible compared to the first term.

Secondly, the phase-space factor in Eq. (16.2) should be exchanged for the three-particle phase space we obtained in Eq. (4.14) (with the modification that we have defined this phase space with a factor $(2\pi)^3$ too large according to Eq. (4.4)). Putting it all together (with the right numerical factors) and introducing the squared pair-masses in terms of the x_j-variables:

$$x_j = \frac{2E_j}{\sqrt{s}}, \quad \sum_{j=1}^{3} x_j = 2 \tag{16.36}$$

For example, we have for s_{12}

$$s_{12} = (p_1 + k)^2 = (P_{tot} - p_3)^2 = s - 2P_{tot}p_3 = s(1 - x_3) \tag{16.37}$$

We obtain after straightforward algebra (note that $2(1 - x_2) + (1 - x_1)^2 + (1 - x_3)^2 = x_1^2 + x_3^2$)

$$dn_\gamma = \frac{\alpha}{2\pi} \frac{x_1^2 + x_3^2}{(1 - x_1)(1 - x_3)}dx_1dx_3 \tag{16.38}$$

The introduction of the fermion spin means that we exchange 1 for $(x_1^2 + x_3^2)/2$ but the new factor is in general close to unity because of the two pole factors in Eq. (16.38). In section 17.7, when we consider collinear bremsstrahlung, we will discuss the results of this modification.

We end this subsection with a few comments. We firstly note that, while the *spin* (for massless particles) is along (or opposite to) the direction of motion, the *polarisation* of the current, i.e. the added cross-product in Eq. (16.33), is transverse to this direction. This is the same behaviour as for the electromagnetic fields.

A vector product $\mathbf{a} \times \mathbf{b}$ is not a true vector but instead describes the components of an antisymmetric tensor (which has the same transformation properties with respect to rotations as a vector but is different with respect to space reflections). There is a single 3-tensor $\epsilon_{jlm} = \pm 1$, depending upon whether the permutation jlm among the numbers 123 is even or odd (e.g. 231 is even and 213 is odd); the axial vector $(\mathbf{a} \times \mathbf{b})_j$ can be written as $\epsilon_{jlm}a_lb_m$ with a sum over repeated indices. Actually this latter quantity can also be described as the $\mu = 0$ (the 'time') component of the antisymmetric 4-tensor $\epsilon_{\mu\nu\sigma\lambda}$, which is defined in the same way in terms of the four indices 0123. It is interesting to note that the relationship to the electromagnetic fields can in this way be taken even further because the polarisation term of the current is then $\epsilon_{0jlm}\sigma_l\nabla_m$. The polarisation of the electromagnetic field is conventionally taken along the electric field $\mathcal{E}_j$; this is likewise the $0j$-component of the field tensor.

The use of an axial vector to describe the polarisation also means some loss of gauge and Lorentz invariance (although these symmetries may be restored by a more elaborate formalism). But while the current term based upon the true vector $(\mathbf{p}, E) \propto (d\mathbf{x}(t)/dt, 1)\delta(\mathbf{x} - \mathbf{x}(t))$ may easily be seen to fulfil a Lorentz-invariant current conservation requirement, $\nabla\mathbf{j} + \partial j_0/\partial t = 0$, the added axial vector term obeys only space-current conservation $\nabla\mathbf{j} = 0$ as well as the corresponding invariance under 'transverse' gauge transforms $\epsilon \cdot \mathbf{j} \equiv (\epsilon + i\mathbf{k}\Lambda) \cdot \mathbf{j}$. Nevertheless we may use the shape of the current we have introduced above to derive the tensors $(T_1 + T_2)_{\mu\nu} \propto \sum_{spins} \langle 0| j_\mu |k_1, k_2\rangle \langle k_1, k_2| j_\nu |0\rangle$, which we discussed in section 4.4. (What are the necessary normalisation factors?)

3   *The invariant transverse momentum, the rapidity and phase space*

It is useful to introduce the invariant transverse momentum and rapidity for the photon,

$$k_\perp^2 = \frac{s_{12}s_{23}}{s} \equiv s(1 - x_1)(1 - x_3)$$

$$y = \frac{1}{2} \log \left(\frac{1 - x_1}{1 - x_3}\right) \tag{16.39}$$

in terms of which we may obtain the inclusive photon multiplicity distribution from Eq. (16.31),

$$dn_\gamma = \frac{\alpha}{\pi} \frac{dk_\perp^2}{k_\perp^2} dy \tag{16.40}$$

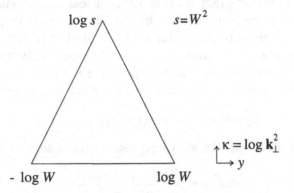

Fig. 16.4. The phase space for photon emission in terms of the logarithmic variables κ and y described in the text.

i.e. the same result as in Eq. (16.24). This time it is, however, expressed in terms of invariants.

The total phase space is, in terms of the invariants $k_\perp$ and y (we will from now on only use the variables in that sense so we drop the word invariant),

$$s \geq s_{12} + s_{23} = 2k_\perp \sqrt{s} \cosh y \qquad (16.41)$$

which can be conveniently approximated by

$$|y| \leq (L - \kappa)/2 \qquad (16.42)$$

with $\kappa = \log(k_\perp^2/s_0)$ and $L \equiv \kappa(k_\perp^2 = s)$. Here s_0 is some scale which is not determined by our present considerations. We note that the phase space has in this way changed to the interior of a triangle (Fig. 16.4). The meaning of the cross section is evidently now that there is a density of photons given by α/π inside the triangular phase space because the cross section is $dn_\gamma = \alpha d\kappa dy/\pi$.

We will use this picture extensively in the following. The energy-momentum conservation requirement in Eqs. (16.41) and (16.42) is evidently very similar to the results from the coherence calculations in Eq. (16.21). In that case we found that radiation is only allowed inside a certain (pseudo)rapidity region determined by the rapidity of the emitters.

The result in Eq. (16.41) is valid for massless emitters. For massive ones, there should be a region, close to the rapidity endpoints for the massless case, where there is suppression for photon emission. Although we will not consider this situation we note that dipole emission is only allowed within an angular cone which is characteristic for the dipole.

It is worthwhile to note that the $k_\perp$-variable in Eq. (16.39), although defined in a rather abstract way, is nevertheless a reasonable measure of the 'true' transverse momentum of the photon with respect to some dynamical axis. This is in particular so if the photon is soft. We will now show that there is always a direction $\mathbf{e}$ such that the transverse component of the cms momentum of the γ with respect to $\mathbf{e}$ is equal to $k_\perp$. If the angle between the γ's momentum direction and $\mathbf{e}$ is θ we obtain the requirement

$$E_2^2 \sin^2\theta = k_\perp^2 = s(1-x_1)(1-x_3) \tag{16.43}$$

Using the relation in Eq. (16.36) to express x_2 we obtain

$$\tan^2\theta = \frac{4(1-x_1)(1-x_3)}{(x_1-x_3)^2} \tag{16.44}$$

This means that e.g. when the electron and positron afterwards have the same energies then the direction of $\mathbf{e}$ is at an angle $\pi/2$ to the γ's direction and the whole γ-momentum is transverse to $\mathbf{e}$. Note, however, that in order to conserve momentum the charged emitters will afterwards also move at an angle to the $\mathbf{e}$-axis (for the recoil problems in the emissions, cf. section 17.8).

We have in Chapter 4 described the changes necessary when we go from QED to QCD. The number of color configurations which contribute for a color-$(3, \bar{3})$ dipole is $N_c - 1/N_c$ with $N_c = 3$ the number of colors. There is also the unfortunate definition of the QCD charge to take into account so that we should change α_{QED} to $(N_c - 1/N_c)\alpha_s/2$; all in all this leads to

$$\alpha_{QED} \rightarrow \alpha_{QCD,effective}(q\bar{q} \rightarrow qg\bar{q}) = \frac{4\alpha_s}{3}$$

$$dn_g = \left(\frac{2\alpha_s}{3\pi}\right) \frac{x_1^2 + x_3^2}{(1-x_1)(1-x_3)} dx_1 dx_3 \tag{16.45}$$

16.4   The antenna pattern of dipole emission

In this section we will describe the physics corresponding to the *strong angular ordering condition* [59]. We assume that the dipole is boosted transversely to its axis as described by Figs. 16.5(*b*) and (*c*). This means that the angle between the directions of motion of the charges is no longer π as in the rest frame but 2ψ, with $v = \cos\psi$ as the relative velocity of the frames.

From Eqs. (16.2) and (16.29) we may obtain an angular emission pattern, which is called the *antenna pattern* in [27]. When this is expressed in angular variables (or rather in the scalar products betwen unit vectors) we obtain, using e.g. $s_{12} = 2E_1 k(1 - \mathbf{n}_1 \cdot \mathbf{n}_2)$, the following angular dependence

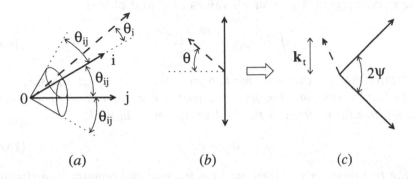

Fig. 16.5. (a) The angular emittance cones around the partons i and j and the region O with no emission according to the strong angular condition. (b) A gluon with a certain k_t is emitted at an angle $\pi/2 - \theta$ from a dipole in its cms. (c) The system is boosted to a frame with velocity $v = \cos\psi$, for the notation see the text.

on the gluon direction (note $dk\delta^+(k^2) = kdk\sin\theta\,d\theta d\phi/2$):

$$W_{1,3}(\mathbf{n}) = \frac{a_{13}}{a_1 a_2}$$

$$a_{ij} = 1 - \mathbf{n}_i \cdot \mathbf{n}_j, \quad a_{i2} \equiv a_i, \quad i = 1,3, \tag{16.46}$$

The angular distribution $W_{1,3}$ contains a dependence both on the relative angle between the two emitters 1 and 3, to be called θ_{13}, and on the polar and azimuthal angles θ and ϕ of the emitted g, see Fig. 16.5(a). It can be written as a sum of two terms:

$$W_{1,3} = U_{1,3} + U_{3,1}, \quad U_{i,j} = \frac{1}{2}\left(\frac{1}{a_i} + \frac{a_{ij} - a_i}{a_i a_j}\right) \tag{16.47}$$

We can calculate the polar angle θ with respect to either the i-direction or the j-direction; an index on θ will indicate which one we are using. For the expression $U_{i,j}$ we note that if we fix θ_i and θ_{ij} the only azimuthal angular dependence is that of the second term in the bracket.

The numerator of the second term is $2[\sin^2(\theta_{ij}/2) - \sin^2(\theta_i/2)]$. This provides a positive or negative contribution depending upon whether θ_i is smaller or larger than θ_{ij}. It is useful for the reader to check for himself/herself that the partitioning is done in such a way that this numerator will have no pole in $U_{i,j}$ if $\theta_j = 0$.

The expression for $U_{i,j}$ is therefore only large when the emitted g is close in angle to the parton i. The same is evidently also valid for the corresponding term $U_{j,i}$ with respect to the parton j.

One may integrate $U_{i,j}$ over all values of ϕ and obtain

$$\int \frac{d\phi}{2\pi} U_{i,j} = \frac{\Theta(\theta_{ij} - \theta_i)}{a_i} \tag{16.48}$$

This means that *the average emission from the term $U_{i,j}$ is the same as if there had only been emission from the parton i inside a cone such that the following angular relation in the Θ-distribution is fulfilled*:

$$\theta_{ij} \geq \theta_i \tag{16.49}$$

Thus the two terms in $U_{i,j}$ turn out to give equal and opposite contributions outside this 'mother' cone. This makes it possible to interpret the dipole emission formula as terms of independent emission from either the i- or the j-parton.

We will now investigate the way in which such an angular condition works. We consider Fig. 16.5 and first concentrate on the condition for emission from the parton i. The condition in Eq. (16.49) then means that i can emit inside the upper angular cone around the direction i. Similarly j can emit inside the lower angular cone around j.

Therefore both can emit in the region between them and neither can emit in the region indexed O. Due to the partitioning above we also know that the amount which is lost inside the region O is gained in between the partons. Thus a literal use of the angular condition means that some, in general soft, gluon radiation is 'misplaced' in phase space.

In order to inform ourselves about the size of this problem we will make the following calculation. We assume that in its rest frame a dipole emits a gluon at an angle $\pi/2 - \theta$ to the dipole axis; see Fig. 16.5(b). Then in a coordinate system in which the dipole moves with velocity $v = \cos\psi$ transverse to its axis and at an azimuthal angle π with respect to the gluon, (see Fig. 16.5(c)) all those gluons with $\theta \leq \theta_{max}$, where

$$\cos\theta_{max} = \frac{\cos\psi - \cos 3\psi}{1 - \cos\psi \cos 3\psi} \tag{16.50}$$

will be forbidden by the strong angular condition.

This means that when $v \to 1$ then $\theta_{max} \to 0.64$ while $\theta_{max} = 0$ for $v = 0.5$. *The strong angular condition is an inclusive statement* in the sense that if all possible gluon emissions are allowed then the errors compensate. In a Monte Carlo simulation of single events the errors can be appreciable event for event, however.

A clever model builder can to some extent compensate for the error. In particular the most popular Monte Carlo models on the market, JETSET [105], HERWIG [94] and ARIADNE [92] implement the full dipole matrix element in the first emission (in different ways). While in ARIADNE emission continues according to the dipole formula, in the other programs the conar conditions are applied later on in the cascades. If the conar conditions are neglected, however, then there will be considerable double-counting and far too many gluons emitted.

17

Multigluon emission, the
dipole cascade model and
other coherent cascade models

17.1   Introduction

In the last chapter we considered the bremsstrahlung cross section for dipole radiation. This cross section is valid even for multiple QED bremsstrahlung. But there is a major difference between an abelian and a nonabelian gauge field theory in connection with multiquantum radiation.

For an abelian gauge theory, like QED, the emitted quanta are chargeless. Therefore the current is the same before and after the radiation (besides the recoils, which pose particular problems in all theories).

For QCD, the emission of (the color-8 charged) bremsstrahlung gluons may actually be disastrous. The original current in e.g. an e^+e^- annihilation event consists of a color-3 and a color-$\bar{3}$ charge (the original $q\bar{q}$-pair), forming the primary dipole. But after the emission of a gluon the current consists of a state with color-$(3, 8, \bar{3})$ charges moving in different directions. It is a great simplification that *the three charges to a very good approximation can be tretaed as two independent dipoles* [27].

We will start by presenting this result and then continue the discussion in terms of the *Lund dipole cascade model*, the DCM [75]. In this model the production of new gluons stems at every step from the formation of dipoles by pairs of previously emitted partons, and the process leads to new and smaller dipoles. The coherence conditions can in a simple way be realised in the DCM. The process is implemented in a Monte Carlo program, ARIADNE [92]. Within the DCM it is also easy to clarify the way the directrix of the final-state string emerges. We will show that the coherence conditions of multigluon emission in this model tend to bend the directrix in a characteristic way towards an ever smoother curve.

After that we will turn to the description of models in which the subsequent gluon radiation is related to a single one of the already existing partons. It is then necessary to partition the dipole cross section

318

in a consistent way into the contributions from the two charges at the endpoints. It is necessary to take coherence into account by means of the strong angular ordering condition derived in Chapter 16.

We will also be more precise with respect to the polarisation correlations. We will exhibit the so-called splitting functions, which correspond to approximations to the dipole emission formulas valid when the radiated gluon is collinear to one of the charges.

We briefly describe the procedures in two Monte Carlo models of this kind, HERWIG, [94], and JETSET, [105]. At the same time we consider some features of the Webber fragmentation model. We will also discuss the *gluon splitting process*. In particular we exhibit *the results of a competition between different stochastic processes*, in this case gluon emission, $g \to gg$, and gluon splitting, $g \to q\bar{q}$.

We do not know the higher-order perturbative results for the cross sections. Therefore there is a problem in connection with how to partition the recoils in the emissions. We will show that within the DCM the results are stable and consistent.

There is actually a particularly nice relationship between the DCM and the Lund fragmentation model. It turns out that the dipoles of the DCM occur just in the regions where the Lund model would span a string. *Consequently all 'new' gluon emissions, with ensuing activity, occur where the Lund model already provides for particle production.*

In other words the DCM (and models containing a correct treatment of the coherence conditions) provide for gluon production in accordance with the string effect, discussed in Chapter 15. The softer gluon radiation (softer because the corresponding dipole masses in general are smaller) in the string regions only serves to provide smaller gluon excitations on an already existing string.

The result is that *there is a moving interface between the radiation of more and softer gluons and the fragmentation process of the Lund model.* We will exhibit this property and discuss the consequences in some detail.

The reason that the Lund model results and the results of the Webber-Marchesini model [94] agree so well, despite the large conceptual differences in the models, is that both models implement the bremsstrahlung coherence conditions. In other words *both models contain* (in a statistical sense) *activity inside the same regions of phase space.*

17.2 The consequences of the second-order matrix element

The main difference between QED emission of photons and QCD emission of gluons is the final-state current distribution. In the QED case the γ's are chargeless and apart from the recoil problems the original current is still the same. In QCD, however, we start with a color-$3\bar{3}$ dipole and afterwards end up with a $(3, 8, \bar{3})$-charge situation.

There is, however, one simplification. We started out with a color singlet composed of the 3- and $3\bar{3}$-charges and it is evident that the three charges $(3, 8, \bar{3})$ must also together form a color singlet. Therefore in particular the combined qg-charge must compensate the $\bar{q}$-charge and similarly the combined $g\bar{q}$-charge must compensate the q-charge. Such a charge situation may have implied the occurrence of higher multipole charge distributions. However, *up to a small correction we only obtain two new dipole emitters*. We will now treat the radiation of one more gluon along these lines.

The cross section for the process $e^+e^- \rightarrow qg_1g_2\bar{q}$ has been calculated in great detail and the full expression, [56], is very long and rather complicated. If we assume that the cms energies of the particles are strongly ordered, i.e. $E_2 \ll E_1 \ll E_q, E_{\bar{q}}$ then it is much simplified. The total angular distribution is then a product of two expressions, [27], where we make use of the antenna pattern distribution W defined in Eq. (16.46). The factor N_c is the number of colors and is proportional to

$$W_{q,\bar{q}}(\mathbf{n}_1) \times \left[W_{q,1}(\mathbf{n}_2) + W_{1,\bar{q}}(\mathbf{n}_2) - \frac{W_{q,\bar{q}}(\mathbf{n}_2)}{N_c^2} \right] \qquad (17.1)$$

Therefore in this limit we can regard the process as if

- there is a first emission of g_1 from the original $q\bar{q}$-dipole,

- then there is a second emission of g_2 either from the qg_1-dipole or from the $g_1\bar{q}$-dipole. *The two dipoles in this way work independently.*

- the third term in the brackets is small and may be neglected

The dipole cascade model

In this model, [75], the pattern exhibited in Eq. (17.1) is taken all the way. Thus, the radiation of two gluons produces three dipoles and then these new dipoles are allowed again to decay independently. At each step there is a new gluon emitted and the corresponding dipole is then subdivided into two. After n gluon emissions there are then $n + 1$ dipoles.

The cross sections used in the Monte Carlo simulation program ARIADNE [92], which implements the model, are

$$dn_{q\bar{q} \rightarrow qg\bar{q}} = \frac{2\alpha_s}{3\pi} \frac{(x_1^2 + x_3^2)dx_1dx_3}{(1-x_1)(1-x_3)}$$

$$dn_{qg \rightarrow qgg} = \frac{3\alpha_s}{4\pi} \frac{(x_1^3 + x_3^2)dx_1dx_3}{(1-x_1)(1-x_3)} \qquad (17.2)$$

$$dn_{gg \rightarrow ggg} = \frac{3\alpha_s}{4\pi} \frac{(x_1^3 + x_3^3)dx_1dx_3}{(1-x_1)(1-x_3)}$$

We will come back later to the powers in the polarisation sum in the numerators. Note that if a g is the emitter then there is a power 3 for the corresponding variable. For a q- or $\bar{q}$-emitter there is a power 2.

The color factors in the cross sections are also different. There is no suppression if a gluon splits up into two gluons. If a q and a $\bar{q}$ in a pair are the emitters there is (cf. Chapter 4) one chance in nine of obtaining a color singlet combination. Therefore only 8/9 of the color combinations are gluons, which is just the ratio between the color factors in Eq. (17.2).

We will from now on use the variables $k_\perp, y$ defined in Eq. (16.39) of Chapter 16. In particular the stochastic process of multigluon radiation in ARIADNE uses the transverse momentum variable $k_\perp$ as the *ordering variable*. We will start by clarifying the meaning of this notion.

17.3 An aside on ordering and the Sudakov form factors

A stochastic process is always defined by means of a direction. In order not to double-count the contributions they must be organised according to some system, so that we have the first step, the second etc. In the Lund fragmentation model the process is e.g. ordered along the lightcone(s) (the equivalence of the orderings actually provides a unique process, cf. Chapters 7–9). For multigluon emission processes the model builders use different ordering variables but the choices are in general made so as to account for the coherence conditions of the radiation.

If the available phase-space cells are subdivided and numbered according to the prescribed ordering variable, each with a given probability a_k that an event should happen, *then the very first event is defined by the requirement that it happens in the cell j with probability $P^{(1)}(j)$, where*

$$P^{(1)}(j) = a_j \prod_{k=1}^{j-1}(1 - a_k) \qquad (17.3)$$

The product corresponds to the requirement that nothing has happened in the first $j - 1$ cells.

In the limit when the subdivision of phase space becomes more and more fine-grained and the number of cells grows correspondingly we obtain

$$dP(A) = dn_g(A)\exp\left(-\int_\Omega dn_g\right) \qquad (17.4)$$

This is the probability that no emission has occurred in the region Ω and one emission occurs at the boundary point A. The exponential factor in Eq. (17.4) is usually referred to as a *Sudakov factor* [107]. It is of course also the factor occurring in decay formulas generally (cf. the description of the Artru-Menessier-Bowler model in Chapter 8 for a different context).

We have in this way made use of general probability concepts. The properties of the Sudakov factor are governed by the density dn_g. It may happen that dn_g is not a local quantity independent of the prehistory, i.e. the path from the starting point to A. Then the integral must be correspondingly 'path-ordered'. In the DCM where the gluon emissions are $k_\perp$-ordered the integral is over the values of $k_\perp \geq k_{\perp A}$ and dn_g is given by the relevant formula in Eq. (17.2). As we will see later on there are also other ways to order a QCD perturbative parton cascade, still keeping to the coherence conditions. The coherence conditions in QED photon emission are simpler. One may generally choose any ordering in the phase space which is defined by the properties of the charged current.

Actually it was for multiple photon emissions in QED that Sudakov first constructed the form factor. His arguments were, however, not based upon probabilities. He pointed out that if we consider the emission of a fixed (*exclusive*) multiphoton state then there are many Feynman diagrams contributing to the matrix element. In particular at every new order in the coupling constant there are virtual corrections, corresponding to emission and reabsorption of photons in the available phase space. He was able to calculate the leading contributions from this series, i.e. those with the largest energy dependence, and *it turned out that the resulting sum 'exponentiated' into just the form factor in Eq. (17.4)*.

In this way the probability of emitting nothing is directly in perturbative field theory related to the 'virtual corrections' from emitting and absorbing anything else besides the exclusive state. The 'real' emission density dn_g is then the same as the virtual emission-reabsorption density!

This result is also reasonable from Eq. (17.4). We may imagine that there is a small region $\delta\Omega$ in Ω around the boundary point A. If we allow for emission of anything inside $\delta\Omega$ but neglect to observe the results then we go over to an inclusive distribution: we observe that there is nothing in $\Omega - \delta\Omega$, there is something at the boundary point and there may be anything in $\delta\Omega$. The probability for this is given by Eq. (17.4) with $\Omega \to \Omega - \delta\Omega$. *Therefore summing up contributions in a region means that the region vanishes from the Sudakov factor*!

We will meet this situation again in a somewhat different context in connection with deep inelastic scattering (DIS) in Chapters 19–20. We would then like to subdivide the total radiation in a state into two sets. One of these corresponds to the production of a state with a well-defined set of gluons, usually referred to as the initial-state bremsstrahlung (ISB). In the linked dipole chain model these gluons correspond to a set of connected dipoles. The other set, the final-state bremsstrahlung, (FSB) corresponds to the available radiation from these dipoles. Then we may sum up the FSB contributions to the Sudakov factor when we calculate the contribution from the ISB to the cross section.

But for every exclusive state there is, of course, a Sudakov factor for that particular FSB emission from the ISB dipoles. Taking all the states into consideration, however, they do sum up to a factor 1 in the cross section.

17.4 The generalisation of the λ-measure to multigluon situations

In section 15.7 we have introduced a generalised rapidity variable called λ that has the same properties for an event containing a hard gluon emission as the ordinary rapidity variable has for two-jet events, i.e. those events that correspond to the fragmentation of a straight $q\bar{q}$-string.

Based upon the properties of multigluon emission as it is described above in the dipole cascade model we will now generalise the definition of the λ-measure to multigluon emission. There are two major properties of the cascade and the fragmentation processes that we need.

Firstly we note that according to the perturbative dipole cascade model each dipole may during the cascade be subdivided into two new dipoles by the emission of a gluon. While the original dipole moves as a straight string segment between the two light rays of the endpoint partons the two 'new' dipoles will move apart as string segments spanned between each of the original endpoints and the gluon. In this way the dipoles correspond to color-field *links* between the *corners* defined by the partons. The new dipoles will move apart so that the fields are stretched over the gluon light ray (we use the same notions as in the description of the string surface in space-time in Chapter 15). An example is given by Fig. 15.9 in which a state containing a q, two g's and a $\bar{q}$ (they will be indexed $1, 2, 3, 4$, respectively) is shown as it develops in space-time.

Secondly, after the emission of the two gluons there are three lightcone regions spanned between the q_1 and the g_2, between the g_2 and the g_3 and between the g_3 and the $\bar{q}_4$. Inside each lightcone region there will be a typical hyperbola decay during the fragmentation process as we have described before (see Chapters 9 and 15). Thus the final-state particles are on the average produced along a hyperbola with a fixed proper time with respect to the origin. There may also be a few particles produced around the gluon corners, i.e inside the gluon fragmentation regions (see Fig. 17.1).

It is then evident how to generalise the λ-measure. The result in Eq. (15.26) that the total available region for particle emission is changed from the straight, string fragmentation result $\Delta y = \log(s/s_0)$ to $\lambda =$

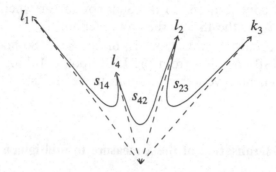

Fig. 17.1. The situation after the appearance of a second gluon. Note the two 'tips' dragging out hyperbolas around the gluon corners, which each contain a number of particles proportional to the value of $\log k_\perp^2$ for the gluon in question.

$\log(s/s_0) + \log(k_\perp^2/s_0)$, will be further changed to

$$\lambda = \log(s_{12}/s_0) + \log(s_{23}/s_0) + \log(s_{34}/s) \tag{17.5}$$

i.e. into the hyperbolic size of each of the three lightcone sections. In Eq. (17.5) we have used the same scale variable s_0 whether it is a gluon or a quark fragmentation region (cf. the discussion in subsection 15.6.4).

According to the dipole cascade model we may consider the size of λ as emerging in a step-by-step process in which we first have the original dipole $q\bar{q}$ (total energy-momentum P_{tot}) changing into e.g. $q_1 g_2' \bar{q}_4'$ (with energy-momenta k_1, k_2', k_4') and finally into $q_1 g_2 g_3 \bar{q}_4$ (with energy-momenta l_1, l_2, l_3, l_4). (The primed variables are introduced to indicate that in the last emission, which in this case according to the dipole cascade model corresponds to emitting the g_3 from the dipole $g_2' \bar{q}_4'$, there will be recoil changes in the emitters' energy-momenta). Energy-momentum conservation implies

$$P_{tot} = k_1 + k_2' + k_4' = l_1 + l_2 + l_3 + l_4 \tag{17.6}$$

and the independence of the emission according to the dipole cascade model implies that $k_1 = l_1$. Therefore $k_2' + k_4' = l_2 + l_3 + l_4$. Consequently the first gluon is emitted at an invariant $k_{\perp 2}^2 = s_{12}' s_{24}'/s$ and the second at an invariant $k_{\perp 3}^2 = s_{23} s_{34}/s_{234}$ where $s_{234} \equiv s_{24}'$ in easily understood notation.

The result in Eq. (17.5) can then be reformulated in the following way:

$$\lambda \simeq \log s + \log(s_{12} s_{234}/s) + \log(s_{23} s_{34}/s_{234}) \equiv \log s + \sum_{j=2}^{3} \log k_{\perp j}^2 \tag{17.7}$$

where we have approximated s_{12} by s_{12}', thereby neglecting the recoil of the gluon emitter 2 in the second emission.

Equations (17.5) and (17.7) can evidently be generalised into multigluon situations and are well defined as long as all the squared masses $s_{i,i+1}$ of two color-connected gluons are larger than the scale s_0. We will later extend the definition in an infrared-stable way.

Finally it is also obvious that the measure λ defined above corresponds to the total generalised rapidity region available for final-state hadron emission, i.e. it could have been called $\Delta\lambda$ in the same way as we introduced $\Delta y = \log(s/s_0)$ for the two-jet events. When we extend the definition of λ in section 18.7 it is possible to define a local value $\lambda(\sigma)$ where σ parametrises the points along the directrix of the state (cf. the general description of string motion in terms of the directrix in Chapter 15). Every point along the hyperbolas spanned between the color-connected gluons will then have a well-defined value of $\lambda(\sigma)$ between $\lambda(0) = 0$ and $\lambda(E/\kappa) = \lambda$.

17.5   The phase-space triangles of DCM

We will next discuss the available phase space. The subdivision of the very first dipole by the radiation of a gluon with $k_{\perp 1}$ and y_1 will in accordance with Eqs. (16.39) lead to a splitting of the original mass-square s into the two dipole mass-squares

$$s_{12} = k_{\perp 1}\sqrt{s}\exp y_1, \quad s_{23} = k_{\perp 1}\sqrt{s}\exp(-y_1) \tag{17.8}$$

Thus, in the phase space approximately described by the triangular region in Fig. 16.4 there is one point shown in Fig. 17.2 corresponding to the $(k_{\perp 1}, y_1)$-variables of the first emission. The region above this point, i.e. the triangle above the hatched line, does not contain any gluons. If this is the first emission the region Ω in Eq. (17.4) is just this triangle according to the ordering in ARIADNE.

We next construct the phase space for the 2nd emission. The logarithm of the squared masses in Eq. (17.8) can be described in size according to Fig. 17.2. Thus

$$L_{12} \equiv \log(s_{12}/s_0) = (\kappa_1 + L)/2 + y_1$$
$$L_{23} \equiv \log(s_{23}/s_0) = (\kappa_1 + L)/2 - y_1 \tag{17.9}$$

where we have introduced the variables from Eq. (16.42). In the figure, starting from the emission point, we have drawn out a triangular fold, which sticks out of the earlier triangle. The length of each triangular side of the fold baseline is $\kappa_1/2$ and the triangular height is of course κ_1. Therefore the distances from the lower left and lower right corners of the original triangle along its baseline to the tip of the fold are

$$\frac{L}{2} + y_1 + \frac{\kappa_1}{2} = L_{12}, \quad L_{23} = \frac{L}{2} - y_1 + \frac{\kappa_1}{2} \tag{17.10}$$

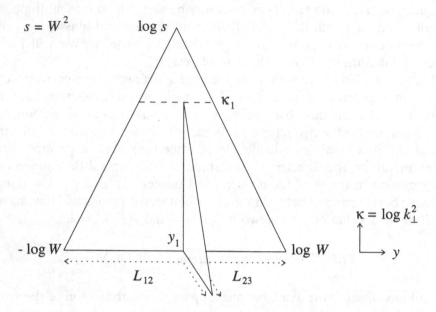

Fig. 17.2. The production of two new dipoles, by the emission of a gluon with $k_{\perp 1}, y_1$ described in the logarithmic phase-space variables κ, y. The (logarithmic) size of the dipoles is indicated along the baseline and out to the tips of the added triangular fold.

respectively. The folded triangle is the increase in phase space in QCD for further gluon emissions, given the variables of the first emission. It was introduced as the generalised rapidity variable λ in Chapter 15. The length of the baseline (including the triangular fold) is evidently $\lambda = L + \kappa_1$, i.e. that obtained in Eq. (15.27).

At this place we make the following further comment:

- if we use another scaling variable s_1 instead of s_0 in the definition of κ then each triangular construction is lengthened or shortened by the factor $\log(s_1/s_0)$. Each dipole size is changed and therefore the sum of the two dipole lengths is changed correspondingly.

In particular if we use the scale $s_1 = k_{\perp 1}^2$ then the baseline changes to the size $L - \kappa_1$. This is the rapidity region available for the first radiation. In Chapter 18 and also in Chapter 20 we will consider κ_1 as the (logarithmic) 'virtuality' of the dipole and then $L - \kappa_1$ is the size of the dipole just before it decays.

The phase-space triangle (before the second emission) has been changed into two cutoff triangles, corresponding to the two new dipoles and one

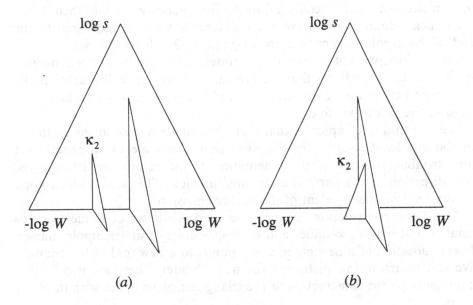

Fig. 17.3. The second emission in the cascade corresponds to one more projecting fold. The two cases corresponding to (*a*) and (*b*) are discussed in the text.

empty part (note that the next emission must be below κ_1), as follows. We have

1 one triangle above κ_1 in which there are no gluons;

2 one left (cutoff) triangle corresponding to the new dipole qg with mass s_{12} (to see this triangle imagine that you bring the fold towards the right in Fig. 17.2 into the old triangle);

3 one (also cutoff) triangle corresponding to the new dipole $g\bar{q}$ with mass s_{23} (bring the fold towards the left in Fig. 17.2 into the old triangle).

We may now repeat the whole process, moving downwards towards smaller κ in each of the two triangles, until somewhere we find the next emission, (κ_2, y_2). This new radiation may then occur in any one of the two independent dipoles.

In Fig. 17.3(*a*) we have divided the dipole s_{12} and added a new fold which at the baseline has the (double-sided) size κ_2. In Fig. 17.3(*b*) the corresponding division occurs in the folded triangle stemming from the first emission. This construction warrants further comments. In principle we could say that the situation described in Fig. 17.3(*b*) corresponds to the emission of a gluon 'collinear' to the first gluon. We would then define a collinear gluon as one having a y-difference variable measured to the end of the dipole smaller than the original $\kappa_1/2 = \log(k_{\perp 1}/\sqrt{s_0})$.

But for independent dipoles such a statement would be rather arbitrary. A little reflection tells us that the crucial variable for 'collinearity' is the difference between the emission point and the point closest to it along the edge of the triangular fold.

We will find in Chapter 18 that there is actually a scale at the endpoint of the dipoles. Closer to the endpoint than this scale we in general get into trouble, because of the kinematics of the recoils, in defining the real direction of the partons after any emission. This scale is to a good approximation independent of the earlier emissions.

The same construction can now be used when we continue towards smaller and smaller κ-values and correspondingly smaller dipole masses. Every radiation of a new gluon corresponds to a new fold in the triangle. We will return to this picture in the next chapter. There are two further comments on the construction of the triangular phase space with its many folds.

Firstly in Chapter 15 we have shown the string space-time surface when there are several gluons radiated. The measure λ was then described as the length of a set of hyperbolas spanned between the gluon peaks. These hyperbolas correspond in the phase-space triangle to the straight lines between the peaks of the projecting triangles and the folds themselves correspond to the gluons.

Secondly the variables used in the triangular phase space are the invariants $k_\perp, y$. It may be dangerous to associate these variables with the 'true' transverse momenta and rapidities with respect to a physical axis in the event. The invariant variables have a meaning in each dipole's rest frame. This means that the size and the place of the phase-space folds depend upon the earlier emissions.

17.6 The description of multigluon emission as a process on the directrix

1 The process

We will in this section demonstrate the way the directrix is gradually changing during the radiation cascade of gluons. The directrix is in Chapter 15 defined as the connected curve obtained when the parton energy-momentum vectors are laid out in color order.

We will only consider the space parts of the directrix in this section and only the first half-cycle of the curve. The second half is the same but with the parton energy-momenta laid out in the opposite order, according to the prescriptions for the directrix of a string starting at a single point.

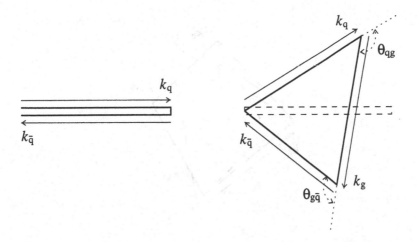

Fig. 17.4. The $q\bar{q}$-state directrix and the triangular directrix after the first gluon emission.

The directrix corresponding to the original $q\bar{q}$-state is in the cms just a double line, each part having length $W/2$ with W the cms energy of the state (see Fig. 17.4). The first gluon emission changes the double line into a triangle with the sides corresponding in turn to the q-momentum, $\mathbf{k}_q$, the emitted g-momentum, $\mathbf{k}_{g1}$, and the $\bar{q}$-momentum after the emission, $\mathbf{k}_{\bar{q}}$.

This configuration is also shown in Fig. 17.4. The momentum recoils are in this example quite noticeable. In particular they are represented by the angles between the consecutive vectors. *The strong angular condition will require that any new emission must provide a smaller angle than the one characteristic of the emitting dipole.* We now assume that the second gluon is emitted from the dipole between the q and the g_1, see Fig. 17.5.

The directrix is then changed from a triangle to a quadrilateral curve. The strong angular condition corresponds to the coherence requirements on the radiation. The condition will in this case require the second gluon, g_2, to cut off the corner between the q- and the g_1-momentum vectors. We note that these two will recoil (the perimeter of the polygon still corresponds to W!) and the result will evidently lead to a shape with less violent bends.

It is of some interest at this point to consider the other possible color-ordering, i.e. the one which would occur if the same g_2 had been emitted from the dipole between the g_1 and the $\bar{q}$. The corresponding directrix is shown by the heavy broken line in Fig. 17.5. The most noticeable thing is that this color-order produces a directrix which contains sharper corners.

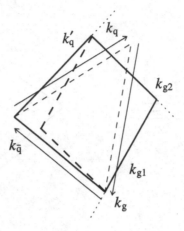

Fig. 17.5. The second gluon emission in the dipole between the q and the g_1 changes the directrix into a quadrilateral.

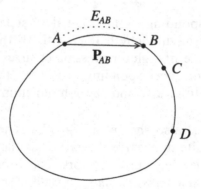

Fig. 17.6. A smooth directrix curve stemming from multigluon emission.

This possibility is not necessarily excluded, because we are working in a three-dimensional space during this emission and the angles are more complex than in a plane. The configuration is, however, in general strongly suppressed as compared to the one with smoother corners.

The characteristic conar angles for the three dipoles are also shown in Fig. 17.5 and they are in general smaller than those obtained after the first emission. It does not take much imagination to understand that *in every new step the strong angular condition will drive some of the new angles towards smaller values, i.e. the directrix curve becomes more and more smoothly bent* as shown in the example in Fig. 17.6. In this case we have gone over to a continuous curve but it can be considered as emerging from continually smaller gluon emissions as we will show below.

2 A self-similar string directrix

One immediate question is the general structure of the directrix curves we obtain in a cascade containing the coherence conditions of QCD bremsstrahlung. In order to answer this we must be able to make comparisons between different parts of the directrix curves. Such parts correspond to sets of color-connected gluons. The relevant question is whether a group of gluons in one part of the state will look 'similar' to another group somewhere else. To understand the problem we will in particular focus on the regions between the points indicated in Fig. 17.6.

At first sight the two parts AB and CD evidently do not have the same appearance. According to the definition of the directrix the distance along the curve between A and B corresponds to the energy and the vector $\mathbf{AB} \equiv \mathbf{P}_{AB}$ corresponds to the momentum of that part of the directrix. Therefore it is perfectly feasible to go to the rest frame of the region AB and consider the result.

This is the only relevant way to compare different parts of the directrix. In a theory which is Lorentz-covariant we are not interested in differences corresponding to the use of different Lorentz frames nor differences due to rotations of the state within the frames. We will now exhibit some rather puzzling and interesting features of the particular curve drawn in Fig. 17.6, [47].

We can do the same procedure for CD as we have performed for AB. The interesting thing is that the two parts will look exactly alike in their rest frames! Not only that: the original curve is chosen in such a way that the curve itself, the parts AB, CD and any connected part of the curve have the same shape if we consider each in its own rest frame. All parts are *self-similar*. This means that apart from a scale factor and a possible rotation in space they can all be described in the same way:

$$
\begin{aligned}
A_0(\xi) &= R(2\xi - 3\xi^2 + 2\xi^3) \\
A_1(\xi) &= R(-\xi + 3\xi^2 - 2\xi^3) \\
A_2(\xi) &= R\sqrt{3}(-\xi + \xi^2) \\
A_3(\xi) &= 0
\end{aligned}
\tag{17.11}
$$

Here, the space part is chosen in the 12-plane and is required to pass through the origin for $\xi = 0$ and $(R, 0)$ for $\xi = 1$. The parameter R corresponds to the mass of the state and is the only available parameter for the curve.

The curve in Eq. (17.11) is unique (besides an arbitrary Lorentz transformation) and has the particular property that

$$\dot{A}(\xi_1)\dot{A}(\xi_2) = 6R^2(\xi_1 - \xi_2)^2 \qquad (17.12)$$

The reader may convince himself/herself that the results described above are correct by choosing different connected parts, calculating the corresponding boost parameters and considering the emerging function in the rest frame!

The completely self-similar curve in Eq. (17.11) is a limiting case of the following situation. Suppose that we emit a set of gluons in such a way that there is one gluon in the centre at the rapidity $y = 0$ and then symmetrically placed gluons at $\pm n(\Delta y)$ for $n = 1, \ldots, N$, all with the same transverse momentum $k_\perp$. Both the rapidity and the transverse momenta are 'real' variables in the sense that they are measured with respect to a particular axis in the Lorentz frame we are considering.

The color connection will be the obvious one, i.e. from the outermost parton at $N\Delta y$ to $(N - 1)\Delta y$ etc., ending on the gluon at $-N\Delta y$. We will assume that the two at the endpoints are the q and the $\bar{q}$ for an open string.

This means that the nth gluon (we will allow n to take both positive and negative values) will have energy and momentum equal to

$$\begin{aligned} e_n &= k_t \cosh(n\Delta y) \\ k_{1n} &= k_t \\ k_{2n} &= k_t \sinh(n\Delta y) \end{aligned} \qquad (17.13)$$

We are obviously not describing the state in the cms. Nevertheless, this particular state will on a local level look exactly the same (almost) everywhere because the mass of a neighboring group of gluons is always

$$\begin{aligned} M_2^2 &= 2k_t^2[\cosh(\Delta y) - 1] \\ M_3^2 &= 2M_2^2 + 2k_t^2[\cosh(2\Delta y) - 1] \end{aligned} \qquad (17.14)$$

etc. with the lower index corresponding to the number of neighbors counted.

The total energy of the state is easy to sum up:

$$E = k_t \frac{\sinh[\Delta y(2N + 1)/2]}{\sinh(\Delta y/2)} \qquad (17.15)$$

and correspondingly the momentum is $P_1 = (2N + 1)k_t$ along the 1-axis. We may then boost the state by the velocity P_1/E to the cms.

We now consider the case when $N \to \infty$ and $\Delta y \to 0$ in such a way that the product $N\Delta y \to \beta$. Then we obtain for the derivative of the directrix function with respect to ξ, where $\xi = (N - n)/(2N)$,

$$\dot{A}_0 = k_\perp \frac{\sinh \beta \cosh[(1 - 2\xi)\beta] - \beta}{\sqrt{\sinh^2 \beta - \beta^2}}$$

$$\dot{A}_1 = k_\perp \frac{\sinh \beta - \beta \cosh[(1 - 2\xi)\beta]}{\sqrt{\sinh^2 \beta - \beta^2}} \tag{17.16}$$

$$\dot{A}_2 = k_\perp \sinh[(1 - 2\xi)\beta]$$

We also note that ξ, defined in this way, always fulfils $0 \leq \xi \leq 1$. In this way we obtain a set of possible directrix functions containing two shape parameters, k_t and β. It is easy to integrate Eqs. (17.16) but we will leave that for the interested reader.

Instead we consider the limit when $\beta \to 0$ and $k_t \to \infty$ so that $k_t\beta \to R\sqrt{3}$. Then it is easy to prove that the vector in Eq. (17.16) will approach the derivative $\dot{A}$ of Eq. (17.11) at every point ξ.

The states we have introduced in this way seem to be unusual in the sense that the gluons all have the same local properties, in particular they all have the same value of the transverse momentum variable k_t. It is instructive to notice that this value of the transverse momentum is not equal to the invariant transverse momentum we would use to order the emissions in the dipole cascade model.

Thus we obtain for the invariant transverse momentum (Eqs. (16.39) and (17.14)):

$$k_\perp^2 = \frac{M_2^4}{M_3^2} = \frac{2k_t^2 \sinh^2(\Delta y/2)}{1 + 2\cosh^2(\Delta y/2)} \tag{17.17}$$

In order to understand that such a state actually can come out of a cascade we note that we may start with a situation where the original q and $\bar{q}$ are in some frame going out with rapidities $\pm N\Delta y$ and the same transverse momentum k_{t1}. This means that the total squared mass of the system is $s = 2k_{t1}^2[\cosh(2N\Delta y) - 1]$. Next we emit a gluon with the invariant $k_{\perp 1}$ at the invariant rapidity $y = 0$. The rapidity condition means that the two emerging dipoles are exactly equal. This means that we can find a new frame with the q, g_1 and $\bar{q}$ moving at rapidities $\pm N\Delta y$ and 0 in that frame. We define $k_{\perp 1}$ so that all the three partons have the same transverse momentum, k_{t2}, in that new frame.

To determine the variables it is only necessary, due to the symmetry, to conserve the total mass:

$$s = 2k_{t1}^2[\cosh(2N\Delta y) - 1]$$
$$= 2k_{t2}^2[(\cosh(2N\Delta y) - 1] + 2[\cosh(N\Delta y) - 1)] \tag{17.18}$$

This provides an equation that determines k_{t2} in terms of Δy and k_{t1}. The invariant $k_{\perp 1}$ is given by

$$k_{\perp 1}^2 = \frac{s_{qg_1} s_{g_1 \bar{q}}}{s} = k_{t2}^2 \frac{2 \sinh^2(N \Delta y / 2)}{2 \cosh^2(N \Delta y / 2) + 1} \qquad (17.19)$$

and we recognise this for $N = 1$ from Eq. (17.17).

In the next step we divide each of the dipoles qg_1 and $g_1 \bar{q}$ in the middle and again go to a new frame in which the q and $\bar{q}$ have rapidities $\pm N \Delta y$, the two new gluons have rapidities $\pm N/2 \Delta y$ and the original gluon still goes out at $y = 0$, all with transverse momenta k_{t3}. Again it is, due to symmetry, only necessary to fulfil the total squared mass condition:

$$s = 2k_{t3}^2 \{ [\cosh(2N \Delta y_3) - 1] + 2[\cosh(3N/2 \Delta y_3) - 1]$$
$$+ 3[\cosh(N \Delta y_2) - 1] + 4[\cosh(N/2 \Delta y_2) - 1] \} \qquad (17.20)$$

This condition will fix k_{t3} in terms of the earlier k_t-variables. We can evidently continue this process every time choosing the Δy-variables to be the same at each level and filling in more and more gluons. The k_t- and the invariant $k_\perp$-variables will quickly decrease because of the number of terms in the mass condition.

The ordinary cascade states contain many more irregularities. It is, however, a fact that most of the fluctuations in the cascade states are connected to the first two gluon emissions. We will discuss this result, [12], in Chapter 18 after we have developed more analytical tools.

17.7 Single-parton emission compared to the DCM procedure

1 *The splitting formulas*

We start with a partitioning of the dipole radiation formulas, Eq. (17.2), into two parts, each corresponding to the contribution from one of the charges. In section 16.4 we have derived the strong angular ordering condition, which is one way to include the coherence conditions in QCD bremsstrahlung. Then the dipole emission region is divided into two parts, according to the angle with respect to the existing partons. Although the formula in Eq. (16.47) is a good approximation to the cross section for soft gluon emission it is necessary, for hard and collinear gluon emissions, to account for the contributions from the polarisation correlations in the numerator of the cross section. It is therefore necessary to provide more precise expressions.

We assume that the gluon (index 2) is, in connection with the first formula in Eq. (17.2) emitted close to the q-particle (index 1). Then the squared mass $s(1 - x_3) = s_{12} \equiv Q^2$ is small and Q is usually referred to as the virtual mass of the qg-pair. The reason is that intuitively we may consider the process as being in two steps: firstly there is the production of an off-shell q (with positive mass Q corresponding to a time like vector); secondly this then decays into the on-shell qg-pair, see Fig. 16.2 (note the discussion in the corresponding text that such a statement is not gauge-independent and that therefore one must be careful).

Also, $x_3 \to 1$, i.e. the $\bar{q}$ takes energy $\simeq \sqrt{s}/2$. The q and g will together take $x_1 + x_2 \simeq 1$ of the remaining energy $\sqrt{s}/2$. We may then write the bremsstrahlung formula in the first line of Eq. (17.2) as the product of a factor corresponding to the production of Q^2 and another factor, *the splitting function $\mathscr{P}_g^q(z)$ for the virtual q to emit the g* with g-fraction $z = x_2$:

$$dn_g = \frac{\alpha_s}{4\pi} \frac{dQ^2}{Q^2} \mathscr{P}_g^q$$

$$\mathscr{P}_g^q = \frac{4}{3} \frac{[1 + (1-z)^2]}{z}$$

(17.21)

The corresponding splitting function for the (virtual) q to emit the on-shell q with fraction $z = x_1$, $\mathscr{P}_q^q(z)$, evidently becomes

$$\mathscr{P}_q^q = \frac{4}{3} \left(\frac{1+z^2}{1-z} \right)$$

(17.22)

In order to exhibit the polarisation contributions in the numerator of the formulas in Eqs. (17.2) and (17.21), (17.22) we will derive the results from the Rutherford scattering formula we obtained in Eq. (5.40). We consider the scattering of two (massless) spin $1/2$ particles in the cms, described initially by the vectors $p_{1,2} = (W/2, \pm W/2, \mathbf{0}_\perp)$, with momentum transfer q to final states $p_{3,4} = (W/2, \pm p_\ell, \pm \mathbf{p}_\perp)$. We will then have the kinematical relations (in terms of the scattering angle θ)

$$p_\ell = \frac{W \cos\theta}{2}, \quad p_\perp^2 = \frac{W^2 \sin^2\theta}{4}$$

$$q = \left(0, \frac{W(1 - \cos\theta)}{2}, -\mathbf{p}_\perp \right)$$

(17.23)

If we define the splitting variable z at the vertex $p_1 \to qp_3$ by the lightcone fraction $q_0 + q_\ell = z(p_{01} + p_{\ell 1})$ (note that this means that

$$1 - z = (p_{03} + p_{\ell 3})/(p_{01} + p_{\ell 1}) = (1 + \cos\theta)/2)$$

then we obtain for the variables in Eq. (5.40)

$$s = \hat{s} = W^2, \quad p_\perp^2 = z(1-z)s, \quad Q^2 = -q^2 = zs$$

(17.24)

We then obtain from the Rutherford scattering cross section (remember that the fine structure constant α becomes in QCD $C\alpha_s/2$, with C a color factor and α_s the coupling):

$$d\sigma = 2\pi \left(\frac{4\pi\alpha_s}{3s} \right) \frac{\alpha_s dQ^2}{4\pi Q^2} \mathscr{P}_g^q \qquad (17.25)$$

i.e. we obtain the two factors mentioned above corresponding to the production of the (virtual) Q^2 multiplied by the splitting $q \rightarrow q, g$. We note that in this case we start with an on-shell (massless) q, which splits into another massless q and a spacelike gluon propagator $Q^2 = -q^2$. Any splitting process must bring (at least one of) the final-state particles to a smaller (or spacelike) mass and we will come back to this in the treatment of deep inelastic scattering in Chapters 19 and 20.

The first factor in Eq. (17.25) stems from the azimuthal angular (in)dependence and the second has the intuitive meaning of the interaction cross section for two waves with (longitudinal) wavelengths $\simeq 1/\sqrt{s}$ and interaction constant $C\alpha_s$ (or, if we go back to the derivation of the Rutherford formula in Eq. (5.40), we find $1/s$ as the 'flux factor' from the incoming state).

We may consider along the same lines the process $g \rightarrow q\bar{q}$ from Eq. (5.41), which in field theoretical language corresponds to the crossing-symmetric result of Rutherford scattering. It is straightforward to see that we obtain the splitting functions $\mathscr{P}_q^g$ and $\mathscr{P}_{\bar{q}}^g$ as

$$\mathscr{P}_q^g = \mathscr{P}_{\bar{q}}^g \quad \propto \quad z^2 + (1-z)^2 \qquad (17.26)$$

(note that, although the process must be symmetric, in this case there will be no z- or $(1-z)$-pole! We have discussed this in connection with the properties of the polarisation function in QCD, section 4.5. It stems from helicity conservation and we will return to the implications in the next subsection). The normalisation is discussed in Eq. (17.35).

Along the same lines one can derive, [5], the splitting function for a $g \rightarrow gg$ process:

$$\mathscr{P}_g^g(z) = 3 \left[\frac{1-z}{z} + \frac{z}{1-z} + z(1-z) \right] \qquad (17.27)$$

It is important to clarify the notion of virtual mass. Consider the decay situation described in Fig. 17.7. One particle with energy-momentum vector (in lightcone coordinates along its direction of motion) $Q = (W_+, W_-, \mathbf{0}_t)$ decays into two, which share the positive lightcone component in the fractions z and $1-z$ and have transverse momenta $\pm\mathbf{k}_t$. If we assume that the decay products are massless then we obtain immediately for their energy-momenta

$$\left(zW_+, \frac{\mathbf{k}_t^2}{zW_+}, \mathbf{k}_t \right) \quad \text{and} \quad \left((1-z)W_+, \frac{\mathbf{k}_t^2}{(1-z)W_+}, -\mathbf{k}_t \right) \qquad (17.28)$$

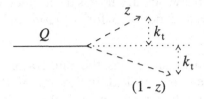

Fig. 17.7. The decay of a particle with mass Q into two particles with compensating transverse momenta and fractions z and $1-z$ of the lightcone component of the mother particle.

Energy-momentum conservation then means that

$$W_- = \frac{\mathbf{k}_t^2}{zW_+} + \frac{\mathbf{k}_t^2}{(1-z)W_+} \tag{17.29}$$

which implies that

$$Q^2 \equiv W_+ W_- = \frac{\mathbf{k}_t^2}{z(1-z)} \tag{17.30}$$

If the decay-product mass-squares are known to be Q_1^2, Q_2^2 then $\mathbf{k}_t^2 \rightarrow \mathbf{k}_t^2 + (1-z)Q_1^2 + zQ_2^2$ in Eq. (17.30).

The relationship in Eq. (17.30) means that the mass parameter Q is proportional to the transverse momentum. The proportionality factor depends upon the fractional partitioning in energy-momentum. Different authors have used different definitions of z and of k_t and this changes the relationship a bit. But it is anyhow evident that

- it is possible to calculate this 'virtual mass' of the decaying particle from the decay products, whether we know x_1 and x_3 (in the DCM and ARIADNE) or z and k_t, or equivalent variables (in the Webber-Marchesini HERWIG and Sjöstrand's JETSET);

- the relationship between the virtual mass and the transverse momentum means that in general both of them will diminish in a similar way as the cascade proceeds downwards and that they can both be used as ordering variable.

We will see in the next subsection that if the stochastic process contains two or more competing subprocesses then the choice of ordering variable does play a role and may lead to different physical results.

We end by describing the reasons for choosing the numerator polarisation sums in Eq. (17.2) with these particular powers for the q- and g-emissions.

There are two features which are necessary to understand in connection with these choices, i.e. the power 3 for a gluon and the power 2 for a quark (antiquark) emitter:

- outside the collinear situations the numerator is a slowly varying function and there is no clear indication of the way one should choose the interpolation between the pole-dominated regions.

- all formulas should be arranged so that there is no double-counting in the cross sections.

The formulas are derived in such a way that they will fulfil these two requirements, i.e. exhibit the right splitting function structure at the pole and correspond to a smooth interpolation away from it. To see this assume that we have an emission which is collinear to an already emitted gluon. This means that we consider the situation when $x_3 \to 1$ and so $1 - x_1 = z$, i.e. the splitting variable. Then the result for a gluon emission cross section in Eq. (17.2) is

$$\frac{1 + (1 - z)^3}{z} \tag{17.31}$$

Together with the corresponding factor from the adjoining dipole (obtained by putting z equal to $1 - z$) we then have

$$\frac{1 + (1 - z)^3}{z} + \frac{1 + z^3}{1 - z} \quad \propto \quad \left[\frac{z}{1 - z} + \frac{1 - z}{z} + z(1 - z) \right] \tag{17.32}$$

which is just the right Altarelli-Parisi splitting function (apart from a color factor), cf. Eq. (17.27).

2   *The gluon splitting process*

Up to now we have only been concerned with the emission of new gluons. There is in QCD also the process of gluon splitting, i.e. when a gluon decays into a quark-antiquark pair $g \to q\bar{q}$. This is a rather small correction but it is of large interest when the experimentalists are able to provide precise data on the appearance of heavy quarks in the centre of phase space. We will follow [10] in this description.

We have already mentioned before that e.g. charm and bottom flavors are so heavy that they cannot be produced in a soft fragmentation situation. They can, however, be produced 'immediately', i.e. as first pairs in an e^+e^- annihilation event when we have passed the mass threshold $2M_Q$ with M_Q the heavy quark mass. In that case they will for large energies end up in final-state particles with large rapidities. *They can, however, also be produced from the gluon splitting process and this is the only possible source for small rapidities.*

Just as in connection with the emission of a second gluon in the cascade in Eq. (17.1) there has been an explicit formula derived, [56], for the process $e^+e^- \to qQ\bar{Q}\bar{q}$, where we use the symbol $Q\bar{Q}$ for a second quark-antiquark pair. If we use the indices 1–4 for the particles as they are written in the production description then the exact (rather long and involved) formula simplifies as follows:

$$\frac{d^4\sigma}{dx_1 dx_4 ds_{23} dz} \sim \frac{x_1^2 + x_4^2}{(1-x_1)(1-x_4)} \frac{z^2 + (1-z)^2}{s_{23}} \tag{17.33}$$

This approximation is valid when the original pair has cms energy fractions x_1 and x_4 which are not too small and the mass square s_{23} of the extra $Q\bar{Q}$-pair is not large. Finally the variable z is the lightcone fraction of the energy-momentum of the $Q\bar{Q}$-pair which is carried by the particle indexed 2 and the expression is of course symmetric between z and $1-z$.

This expression is factorisable and can be understood as follows:

- There is a first emission of a gluon (23) from the original $q\bar{q}$-pair (at the end indexed 14) with the cross section in Eq. (17.2).

- There is after that a splitting of the gluon into the pair 2 and 3 with a cross section

$$\simeq \frac{n_f \alpha_s}{4\pi} \frac{dQ^2}{Q^2} [z^2 + (1-z)^2] dz \tag{17.34}$$

In Eq. (17.34) we have used the conventional variables, the virtuality Q^2 and the fractional energy-momentum-sharing variable z. We also note the appearance of a factor n_f for the number of flavors that can be produced. The expression

$$\mathscr{P}_q^g = n_f [z^2 + (1-z)^2] \tag{17.35}$$

is usually referred to as the splitting function for $g \to q\bar{q}$ and it occurs in that form in the QCD single-parton cascades.

It is useful to try to compensate for some of the approximations in the derivation of Eq. (17.33) by choosing the kinematical variables with care. We will not go into details; they can be found in the original paper, [10], for ARIADNE and there are corresponding choices for the JETSET and the HERWIG Monte Carlo simulation programs.

There are two interesting features about the gluon splitting process.

- There is only the virtuality pole in Eq. (17.34). If we compare with the splitting formulas we derived above in subsection 1 we note that both the processes $q \to qg$ and $g \to gg$ also have poles in the z-variable. This is the major reason why the gluon splitting process is much smaller than the other two.

- When we introduce the gluon splitting together with the gluon emission process there will be competition between the possibilities that a particular dipole will emit a new gluon or that one of the already produced gluons between two dipoles is split up into a $q\bar{q}$-pair.

In ARIADNE the choice has been to partition the cross section for gluon splitting equally between the two dipoles connected to a particular gluon (this is the reason why there is a factor of 2 in front in the expression in Eq. (17.34)). There is nothing fundamental about this but we have checked that it is a good approximation to the precise matrix element almost everywhere in the phase space.

The competition between the two processes means that now the Sudakov factor in Eq. (17.4) will contain two (three) contributions for each qg- or $g\bar{q}$- (gg-) dipole, one for the emission of a further gluon and one for the splitting up of the gluon(s) at the corners. *And now the ordering variable becomes interesting.* We have already seen that the virtuality, Q^2, is related to the squared transverse momentum by Eq. (17.30).

But, due to the occurrence of the denominator $z(1-z)$ in Eq. (17.30), a given (large) Q^2 can be obtained either from the situation when we have a large $k_{\perp}^2$ and a value of $z \sim 1/2$ or from a small value of $k_{\perp}^2$ and a $z \sim 1$ or $z \sim 0$. Thus if we order in Q^2 we would be comparing soft and/or collinear gluon emission to hard, i.e. large $k_{\perp}^2$, emission of a $Q\bar{Q}$-pair.

The major result of [10] is that if we want to use Q^2 as the ordering variable (which is done in the JETSET cascade) rather than $k_{\perp}^2$ as ordering variable (for the precise definition of this variable consider [10]) one obtains at least a factor of 2, sometimes 3–5, fewer gluon splitting pairs.

We are, however, talking about a change from about 5%–8% ($k_{\perp}$-ordering) to about 2% (Q^2-ordering) when comparing gluon splitting into $q\bar{q}$-pairs to gluon emission into 'new' gluons. But it is of large interest as it is the only known mechanism that provides heavy flavors in the centre of phase space. In the case of gluon splitting into a $Q\bar{Q}$-pair, then the final state is treated in the Lund model as two independent strings, one from a 'forward' q to the $\bar{Q}$ and one from the Q to a 'backward' $\bar{q}$. This means that after a splitting the remaining string mass(es) may be very much reduced, implying effectively that further radiation is correspondingly reduced.

3 Single-parton coherent cascades

The splitting functions correspond to *collinear approximations to the radiation cross sections.* They are correct when the emitted parton is sufficiently close to the original parton. However they can be more or less good approximations when the gluon is further away. The results depend upon the definitions of the kinematical variables used.

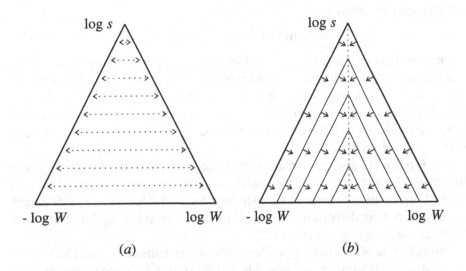

Fig. 17.8.   The triangular phase space for gluon emission with arrows showing
the ways (*a*) HERWIG and (*b*) JETSET search it to find the possible parton
emissions. HERWIG goes sideways in an angle or rapidity variable, all the time
looking upwards and downwards at all $k_\perp$ allowed for that rapidity. JETSET
orders in Q^2: the thin solid lines corresponds to fixed Q^2, and are investigating
different values of z (left-hand side) or $1 - z$ (right-hand side).

We will in this subsection consider two different ways to implement
the concept of a single-parton cascade, i.e. when each already emitted
parton is independently allowed to continue to radiate new partons in a
branching process. We have shown in Chapter 16 that in the mean, i.e. for
inclusive distributions, one may, by imposing the strong angular ordering
condition, obtain correct results at least in the collinear limit.

It is evident that, whichever model one uses the same phase space is
around with respect to further emissions. In this book we have used a
phase-space description relevant to the DCM in terms of a triangle in
$\kappa = \log k_\perp^2$ and y. We show it again in Fig. 17.8 this time in order to
exhibit the ways in which HERWIG, [94], and JETSET, [105], include the
phase-space restrictions.

We have already said that in ARIADNE and the DCM the ordering
variable is κ, i.e. one starts at the largest available $k_\perp$ and then proceeds
downwards in the triangle all the time looking, via the Sudakov factor, in
the allowed rapidity regions for new emissions.

A HERWIG cascade can be easily traced in the DCM triangle. In
HERWIG the authors use as ordering variable the angle θ with respect
to some axis, chosen at random. Remembering the relation between angle

and (pseudo)rapidity, y,

$$y = \log[\cot(\theta/2)] \simeq -\log(\theta/2) \qquad (17.36)$$

this means that they are not going downwards in phase space, they are instead going sideways in the original triangle, *thereby obviously fulfilling the strong angular ordering condition for coherence in the QCD bremsstrahlung.* They further use the variable $\xi = E\theta \sim k_\perp$, which in general is a good approximation to the relation between the energy and transverse momentum of the emitted parton.

The HERWIG authors arrange it in the same way but in the opposite direction, i.e. towards the $\bar{q}$-end with the obvious exchange $\theta \to \pi - \theta$. Again they continue towards smaller angular variables, i.e. towards larger rapidities in that direction. The argument in the splitting functions and the running coupling constant is ξ.

Though it is obviously opposite to the arrangement in the DCM, one can just as well sample the possible jets by going sideways and looking up and down as going downwards and looking right and left. Any time a jet is found for some value of (θ, ξ) it is implemented as a fold like those drawn out of the original triangle in Chapter 15. Then this jet region, i.e. the triangular fold, is searched through in the same way and subtriangles, i.e. subjets, are noted and followed up, etc.

At the end all the emitted partons have been sampled and all the precise kinematical variables have been calculated. Although the choice of energies and angles does not give us bona fide Lorentz-covariant variables the process works very well and seems to give results very close to those of the DCM. We will come back shortly to the Webber cluster fragmentation model, which is used at the end to provide final-state hadrons.

Before that we will briefly consider the way the cascade is implemented in Sjöstrand's JETSET. There the variables Q^2 and z are defined by means of Lorentz invariants. For the precise choices we refer to the original papers, [105]. This means, however, a different way of searching through the triangular region. The process corresponds to passing inwards towards the centre of the triangle from both sides; one of the sides corresponds essentially to using z as the positive lightcone energy-momentum fraction and the other to the negative lightcone fraction. From the relationship in Eq. (17.30) it is evident that if we identify the triangle variables $\kappa \sim \log Q^2 + y_z$ and $y \sim y_z$ then using on one side $y_z = \log(z/2)$ and on the other side $y_z = \log(1-z)/2$ provides a mapping.

It is necessary to calculate the available phase space in z for a given Q^2 etc., but all this is done in a very effective way in the Monte Carlo routines. Once again the background triangle is searched through and every time a parton emission occurs then it is accepted as long as the correct angular ordering condition is fulfilled. Then afterwards this new fold is searched through along the same lines to obtain subfolds, which again are searched through, etc.

There is some possible bias in this procedure of checking the angular ordering afterwards, [70]. There are, however, to our present knowledge no observable consequences of such a bias. It should also be remarked that as the final fragmentation is done by means of the Lund model string scenario the suggested process is not only infrared stable but the possible errors may become hidden. It is known that some earlier, even rather gross, violations of the strong angular conditions in the partonic cascade can be overcome by imposing string fragmentation.

4   The Webber cluster decay model for fragmentation

We will end this section with a discussion of the Webber fragmentation model, [110]. The basic idea is to continue the cascade to a certain level and then to let all remaining gluons decay into $q\bar{q}$-pairs. The final state is then sampled to ensure that it will be composed from (color-singlet) clusters stemming from a q from one gluon and a $\bar{q}$ from a color-adjacent gluon.

The way to implement this is to provide the gluons with a fictitious mass m_g so that below a certain virtuality there is no longer any possibility left of emitting more gluons. All the available gluons should then split up into lower-mass $q\bar{q}$-pairs. In practice it is sometimes necessary to force this breakup. In this way there will be a set of clusters, containing the energy-momentum of the (color-)adjacently produced q and $\bar{q}$.

This is similar to the Lund model prescription but there is no requirement that the clusters should have a fixed mass. Instead there will be, as in the Artru-Menessier-Bowler model, a continuous mass spectrum. This time there is a lower cutoff but it does happen frequently that some of the clusters will attain large masses.

The next step is to let the clusters decay into two-particle states, conserving all quantum numbers and only using phase space and The Particle Data Group tables. In practice it is necessary to work hard, just as Sjöstrand has done in connection with the JETSET fragmentation routines, to decide upon the branching ratios that are relevant for different decay channels. There are three practical problems in this program (even after any amount of hard work), W1–W3 as follows.

W1 The large-mass clusters cannot be allowed to decay isotropically into two-particle states, because there will then be much too much transverse momentum generated. This is solved by using a string-breaking routine (which works to cut up the clusters longitudinally, i.e. along the color-connected gluon directions) so that the large masses are brought down below a certain level to where the method works.

The reason to choose a string-breaking routine is that in this way the final states are not distorted from the cascade distributions, which fulfil the strong angular conditions. The large-mass states typically stem from situations in which there are one or more collinear partons going along the original q-direction, and likewise a few along the $\bar{q}$-direction with little gluonic activity in between. These are the states, usually called two-jet states, which in the Lund model would be similar to the original $(1 + 1)$-dimensional string breakup situation. If all the transverse momentum (with respect to some chosen observed axis) stems from the cluster decays then it is necessary to 'tune' the cluster sizes in order to be able to account for the 'gaussian' fluctuations which are introduced in the Lund model.

W2 The requirement that the clusters should decay into just two final-state hadrons means that there are hardly any hadrons with a value of the fragmentation variable $z \sim 1$. The energy sharing will effectively damp large rapidity values for the decay products.

This is solved by introducing a certain number of single-particle clusters, together with a procedure to rearrange the corresponding cluster energies 'backwards', i.e. towards the neighboring 'ordinary' clusters.

W3 The straightforward application of a cluster decay to baryon-antibaryon production means that the B and $\bar{B}$ in a pair stem from the same cluster. The property that the observed $B\bar{B}$-pairs seem to be dragged apart longitudinally, i.e. in the Lund model along the string direction, and also the (lack of) correlations in the transverse momentum of the pair mean that this cannot be the major source of such production.

This is solved by allowing some of the gluons to split up into diquark-antidiquark pairs thereby producing clusters with baryon and antibaryon quantum numbers. By a reasonable choice of the number and kinematics of such breakups one obtains a good description of the observed baryon and antibaryon distributions. But there seems to be a set of similar problems as in the Lund model in describing the baryon resonances, in particular the baryons with strangeness.

The final result of the fragmentation routines is in most cases indistinguishable from the results of the Lund model and both models certainly are well in agreement with most parts of the present experimental data.

17.8 Some further comments

1 *The recoil and color-interference problems*

In this subsection we will treat some particular problems connected with the approximations which are made in the partonic cascades. We will be satisfied to investigate these problems within the DCM, where there has been an extensive investigation. There are problems of three different kinds:

1. the *recoil problems* along the cascade;

2. the relationship between the exact result from the second-order perturbative QCD results (which we will henceforth refer to as PQCD2) and the cascade results;

3. the quantum mechanical color-interference effects, which occur in the PQCD2, but are neglected in the cascades.

We start with the recoil problems. There are two kinds of recoil problem. There is firstly the loss of energy of the emitters and there is secondly the necessary momentum compensation. The energy loss corresponds to the obvious requirement that all the fractional cms energies after the emission fulfil $x_j < 1$. Further the relative angles between the partons after the emission are also defined by the x_j's. To see this we note that

$$s_{12} = s(1 - x_3) = 2E_1 E_2 (1 - \cos \theta_{12}) \quad \Rightarrow \quad \sin^2 \left(\frac{\theta_{12}}{2} \right) = \frac{1 - x_3}{x_1 x_2} \quad (17.37)$$

But *the relative angles between the original dipole and the final state* are not defined in this way. We have already mentioned that according to the transition matrix element they are given by the overlap of the original and final currents. This results in an angular correlation factor $1 + \cos^2 \theta$ with θ the angle between the directions of the initial-state $e^+ e^-$ annihilation current and the produced $q\bar{q}$-current.

The current in a $qg\bar{q}$-state is more complex; there is nevertheless a preferred direction in the final state, which can be most easily described as the axis with a minimum for the transverse momentum combination $k_{\perp 1}^2 + k_{\perp 3}^2$ with respect to the dipole axis. This means that the q- and $\bar{q}$-charges try to keep as much as possible to their original directions.

There is a prescription given by Kleiss, [85], of how to implement this correlation in a Monte Carlo generation and we refer to this original paper for the details. There is, however, no known prescription for the alignment between the final state and the original dipole direction when the emitters are gg-, qg- or $g\bar{q}$- pairs.

Some guidance can be obtained from the PQCD2 results for the last two cases. In [103] a comparison is made between the full PQCD2 and the Monte Carlo implementation of the DCM in ARIADNE.

The results are that ARIADNE with the ordering variable $k_\perp$ works very well over the whole phase space even when the transverse momentum $k_{\perp 1} \simeq k_{\perp 2}$, i.e. when the first and the second gluons have almost the same hardness. In that paper several different recoil strategies are investigated. Owing to the large amount of 'noise' from the multiparton distributions in the final state and the subsequent fragmentation into hadrons it is not possible to discern the differences between the different recoil schemes.

In the same paper the color-interference term between different color-flow situations has also been investigated. We have already mentioned this problem (see Chapter 15). It is by no means clear that a theory which only contains the production mechanism for the charges, as perturbative QCD does, is not neglecting some structure related to the fields themselves. The color-interference term turns out to be negative and therefore it is not easily introduced into any probabilistic scheme like a Monte Carlo simulation program. It is, however, possible to disentangle the effect so that the PQCD2 formula can be subdivided into two gauge invariant terms:

$$d\sigma \propto B_1 - \frac{B_2}{N_c^2} \tag{17.38}$$

where B_1 and B_2 are positive and N_c is the number of colors.

Intuitively B_2 corresponds, just as in Eq. (17.1), to the contribution from the $q\bar{q}$-dipole, which is recoiling with respect to the gluon that is firstly emitted. Therefore the term B_2 is large when the second gluon is oppositely directed to the first, while it is small when the two gluons go in the same hemisphere. Nevertheless the color-interference term is only of the order of 10% compared to the other contribution.

The possibility of measuring the existence of such a color-interference factor is of obvious interest. In [103] the method is to use ARIADNE to generate multigluon events and to stop the generation after the emission of two gluons. Then the result is corrected by a weighting factor between the full second-order matrix element and the ARIADNE probability, with and without taking the color-interference term into account.

Then the cascade is continued, the final state fragmented into hadrons and different configurations investigated. It is found that there is an effect of the order of 10% between just the two configurations mentioned above, i.e. when there are two jets in the same hemisphere and two jets in opposite hemispheres with respect to the thrust axis. For the necessary experimental cuts and the necessary statistics to find the effect we refer to the original paper.

2 A moving interface between the dipole cascade and the Lund fragmentation

We have noted that the dipoles in the dipole cascade model are spanned between the gluons in the same way as the segments of a Lund string (with the Lund interpretation of a gluon as an internal excitation on the string). Therefore in the Lund model further gluon emission in the DCM corresponds to excitations on an already existing string segment.

In accordance with our considerations in Chapter 15 this implies that a straight string segment will be bent and a region of the string surface which originally was spanned between two lightcone directions will after the new emission still be spanned between them but now via a third lightcone direction. The new excitation is, however, in general much smaller than the earlier ones. We will show that by explicit calculations on the mean cascade development in Chapter 18.

From the investigation of how gluon emissions will affect the directrix, section (17.6), we concluded that, for a given directrix, subsequent gluon emission tends to smooth out the sharp corners stemming from the first few excitations. The pole structure of the emission cross section means that most gluons will be collinear or soft as compared with the emitters.

Therefore these further emissions do not really change the general shape very much, although they may correspond to some increase in the total generalised phase space, i.e. the λ-measure we have introduced before. A natural question is to what extent it is possible to differentiate between the results of such multigluon emission and the final-state fragmentation process. In other words, is there some particular scale where the effects of the fragmentation takes over from the the dipole cascade?

We will use the results of [12] to answer this question. The softer gluons turn out to correspond to a noise of the same kind as the fragmentation process. They increase the multiplicity and the transverse momentum fluctuations. But it is possible to compensate these effects by changing the fragmentation parameters in the Lund fragmentation model in such a way that all inclusive event observables are the same independently of where we stop the cascade.

This is true at least if we use a cascade stop at $k_{\perp,c}$ with $k_{\perp,c} = 7$ GeV or any number below that (but above Λ_{QCD}). In this way we obtain a functional dependence on $k_{\perp,c}$ of the main fragmentation parameters, a and b in the fragmentation function and σ corresponding to the width in the zero-point transverse momentum fluctuations.

The result is shown in Fig. 17.9 and we conclude that while a and σ need small adjustments it is necessary to change b appreciably with $k_{\perp c}$. It is particularly interesting to see that the value of a deduced in this way tends to be stable a bit above 0.5, i.e. it is close to the value of the ρ-Regge

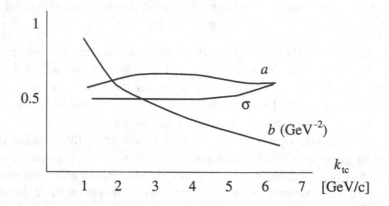

Fig. 17.9. The dependence of the parameters a, b and σ on the cutoff in $k_{\perp c}$ for the cascade.

intercept. This is in accordance with the interpretation we obtained from the discussion in Chapter 10. The parameter σ governing the transverse momentum fluctuations increases by around 25% from small cutoff values to the larger ones.

We have related the parameter b to the transverse width of the string in Chapter 11 and also, via the relation $b\kappa \sim \alpha_s/12$, to the coupling α_s. We find that b decreases from around 0.85 at $k_{\perp,c} = 1$ GeV to 0.15 at $k_{\perp,c} = 7$ GeV. This is just what we would expect from the interpretation that b is proportional to the inverse logarithm of $k_{\perp c}^2$, i.e. to the running QCD coupling.

If we continue downwards in the cascade we resolve the string better and better in transverse momentum. But this means that the string is less and less well resolved in the canonically conjugate space, i.e. in impact parameter space. Therefore we should expect that the transverse width of the string becomes larger and thus also the value of b. Using the value of the running coupling constant derived in Chapter 4 as a function of $k_{\perp,c}$ we obtain a reasonable agreement with the formula above for $\kappa \simeq 0.2$ GeV^2 although we need a somewhat large value of $\Lambda_{QCD} \sim 0.4$–0.5 GeV.

The main point is that in this way there is a moving interface between the fragmentation process and the dipole cascade in the Lund model. The model is very stable, in particular infrared stable.

18

The λ-measure in the leading-log and modified leading-log approximations of perturbative QCD

18.1   Introduction

We will in this chapter present the Lund model in a somewhat different and also more mathematical manner. We have already described the emergence of a string with the q and $\bar{q}$ at the endpoints and with a set of color-connected gluonic excitations in the interior. This corresponds to a more or less complex string space-time surface, which via the directrix may be traced back to all these original excitations. The string breakup is then in Sjöstrand's version a process on the surface leading to a set of final-state (mostly) yoyo-like small hadron strings. All this is in accordance with the Lund model.

Since we have seen in the previous chapters how the string surface is obtained and how the string breaks up along the surface the reader should at this point learn how to use Lönnblad's program ARIADNE, which provides a kinematically precise implementation of the dipole cascade model, and then compare with the results of Webber's HERWIG or Sjöstrand's JETSET Monte Carlo simulation programs. In each case the reader will be able to produce an ensemble of partonic cascade excitations, corresponding to the chosen treatment of the way the color force field will be stretched.

After that either JETSET's or HERWIG's routines can be used (but when doing experimental analysis the reader is advised to use all available methods) to obtain the final-state fragmentation distributions stemming from the chosen ensemble. It is instructive to find for oneself that these seemingly very different approaches will in the end lead to very similar predictions for most inclusive distributions. It is a challenge to find particular differences which are amenable to experimental analysis and also to be able to trace these differences to the dynamical input in the models.

A rather different approach is also possible. Many people, in particular theorists, want something more than just a Monte Carlo 'black box'. It is the intention of this chapter to show that it is possible to calculate many observables in analytical approximations.

Although it is very satisfying to produce an analytical expression for an observable we should be aware that the approach is more approximate than the Monte Carlo methods because it is only partly possible to take the different kinematical constraints into account. To be frank the approximations which are necessary to obtain solvable analytical equations very often have the unfortunate property that the results may be misleading with respect to the dynamics.

But the fact that we generally deal with a multiparticle situation does imply some simplifications. As long as we only consider *inclusive* distributions then many of the approximations actually do not show up because they will drown in the general 'noise' in accordance with the laws of statistics and in particular the law of large numbers.

We will in general concentrate on the λ-measure, which was introduced in [48] to describe the multiplicity in complex multigluon situations. It constitutes a generalised rapidity variable and we described it in Chapter 15 in connection with the triangular phase space with its extended folds, which is typical of (multi)dipole emission. We will also discuss the distributions in the dipole multiplicity (which is related to the gluon multiplicity) but as expected this multiplicity is not, in contrast to the λ-measure, an infrared-stable quantity.

The intention is to introduce methods to calculate analytically *inclusive λ-distributions*. This means that we consider an ensemble of states, e.g. produced from a particular partonic cascade, and calculate the distribution in the variable λ over this ensemble. We will use two different methods, which we will refer to as the L-method and the κ-method.

In the L-method, [48] we consider the analytical equations which governs the change in the λ-distribution when we increase the energy, i.e. in particular increase the variable $L = \log(s/s_0)$. There are sudden and large changes in the distribution when we thus move upwards in virtuality. This is due to the fact that we then encounter very hard gluon radiation which means large changes in the states. The corresponding changes in the distribution have, however, a structure so that it is easy to describe the *Laplace-transformed distribution*. We derive a second-order differential equation for the Laplace-transformed distribution and we also show how to obtain the moments in λ directly from the differential equation.

This will lead us to the notion of KNO scaling [86] for the multiplicity distributions, although we will find that at this level of approximation we are rather far away, in the analytical formulas, from the observed distributions in e^+e^- annihilation events.

After that we turn to the κ-method, [6], where the idea is to consider the production of new dipoles when we go downwards in the transverse momentum variable $\kappa = \log(k_\perp^2/s_0)$ from a fixed maximum $\kappa \simeq L$. This leads to linear partial differential equations of a gain-loss character.

It is possible to investigate many local properties of the distribution in λ by means of this method. We will, however, concentrate more upon the complementarity of the two methods. We will in particular derive a kind of master equation for the combined distribution in λ and n_d, with n_d the number of dipoles for a given L and κ. From this equation it is easy to derive any kind of moment equation for the two variables.

Up to now we have used an approximation to perturbative QCD which is known as the *leading-log approximation*, the LLA. We will meet this approximation again in Chapter 19. It is rather easy to extend the approximation to the *modified leading-log approximation*, [52]. The basic point is to note that one loses a region close to the endpoints of every dipole, [72], in connection with the emission of particles. This is mostly due to recoil problems during the emission but is also related to the spin coupling between the emitters and the final-state partons. After we have taken these corrections into account we obtain quite good analytical approximations to the Monte Carlo simulated distributions in λ and n_d.

Owing to the similarity between the classical gain-loss equations and the Callan-Symanzik equations for the changes in the renormalisation point in a field theory we will at this point also speculate about the meaning of the running coupling constant in QCD.

We will after that present a very simple approximation scheme, called *discrete QCD* [15], which is based upon the properties of the coupling constant. It is possible in this scheme to exchange the continuous triangular phase space, which we have discussed repeatedly for the dipole emissions, for a lattice, where only a discrete set of emission points is available. Each emission point has a simple probability that there will be a gluon emission and a corresponding new triangular fold extended. Then the procedure of discretisation can again be extended to this triangle and later to the subtriangles etc. Therefore the whole structure corresponds to a 'tree' containing 'subtrees', 'branches' and 'twigs' etc. in accordance with simple prescriptions. The procedure leads to very good analytical approximations and also provides further insight into the structure of the perturbative QCD parton-branching processes.

Then we will consider the notion of *fractality* or rather *multi-fractality* in connection with QCD parton cascades. We start by presenting a method to visualise the average distributions of the final-state hadrons already from the partonic state by deriving an equation for a curve called the *x-curve* in [20] and [48].

The *x*-curve has an everywhere timelike tangent compared with the directrix curve, which is everywhere lightlike. Their relationship is that the

x-curve is stretched along the hyperbolas which have the directrix curve as their asymptotes. The (invariant) length of the x-curve is equal to the measure λ, with λ this time defined in an infrared-stable way.

If the x-curve is cut up into pieces, each of a length corresponding to the mass of the final-state hadrons, then we recover the average energy-momentum distribution of the hadrons in accordance with the Lund model fragmentation process. In this way we have on the one hand derived the *local parton-hadron duality* concept in the Lund model, [53], and on the other hand presented a further way to visualise the relationship between the λ-measure and the final-state hadron multiplicity.

One may take the x-curve and its properties as the starting point for a fragmentation scheme, [21], in the spirit of the Lund model but with methods conceptually different from those in the Sjöstrand fragmentation scheme in Chapter 15. The idea is to find the variations around the x-curve from the Lund model fragmentation formulas. The final result is nevertheless similar to Sjöstrand's distributions but in this way we will be able to identify the transverse momentum correlation length, that was introduced in connection with the Ornstein-Uhlenbeck process in Chapter 12.

The curves under consideration in general do not exhibit simple regularity. If we go back to the phase-space triangle used to describe the emission region of gluons in Chapters 15 and 17, this statement is rather obvious. When we have drawn out its many folds from the gluon emissions it looks less like a smooth ordinary curve than one of the fractal curves which have been under intense investigation in recent years.

We will therefore investigate the dimensions of the curves we have derived from the point of view of such fractals. We will exhibit in some detail, [48], the fact that what has for a long time been known as the *anomalous dimensions of QCD* actually can be described also as the *(multi)fractal dimensions of the curves describing the λ-measure.*

18.2   The L-method

1   *The differential equations*

We will in this section introduce a set of differential equations, [48], for the distribution in λ stemming from the dipole cascade model. In particular, we will investigate the changes in the λ-distribution with increasing energy. Then the phase-space triangle is increased in the upwards direction towards larger values of $L = \log(s/s_0)$ (see Chapters 16 and 17 and Fig. 18.1). We will call the distribution in λ for a fixed value of L, $P(\lambda, L)$.

We start by noticing that the size of the λ-measure obtains independent contributions from each particular y-region (usually called a y-bin). We

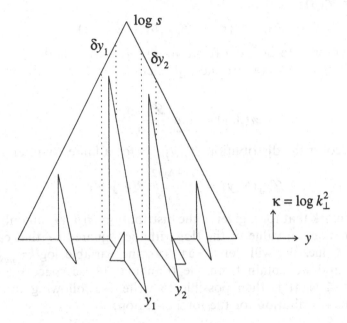

Fig. 18.1.   The triangular phase space with the folds corresponding to new gluon emissions, with two independent regions in rapidity, δy_1, δy_2, exhibited.

again stress the fact that neither $k_\perp$, and therefore κ, nor y necessarily correspond to directly observable momenta and rapidities. They are defined invariantly and recursively from the masses of the dipoles that arise.

The combined λ-distribution from the two regions, see Fig. 18.1, δy_1 around y_1 and δy_2 around y_2, is then using $P_j \equiv P_{\delta y_j}$ for the contributions from region j,

$$P(\lambda; \delta y_1, y_1; \delta y_2, y_2) = \int P_1(\lambda_1)P_2(\lambda_2)d\lambda_1 d\lambda_2 \delta(\lambda_1 + \lambda_2 - \lambda) \quad (18.1)$$

Thus the folds which occur in a certain region are in this approximation independent of the folds in a different region. Therefore it is natural to go over to a Laplace transform of the distribution. Then we obtain by the definition

$$\tilde{P}(\beta, L) = \int d\lambda \exp(-\beta\lambda)P(\lambda, L) \quad (18.2)$$

the following result for the Laplace transform of Eq. (18.1):

$$\tilde{P}_{\delta y_1 + \delta y_2}(\beta) = \tilde{P}_{\delta y_1}(\beta)\tilde{P}_{\delta y_2}(\beta) \quad (18.3)$$

This implies that the logarithm of the distribution is additive (using

$\log \tilde{P}(\beta) \equiv \mathscr{L}(\beta)$:

$$\mathscr{L}_{\delta y_1 + \delta y_2}(\beta) = \mathscr{L}_{\delta y_1}(\beta) + \mathscr{L}_{\delta y_2}(\beta) \tag{18.4}$$

It is then possible to define a function $\mathscr{R}(\beta, y)$ corresponding to the limit of a vanishing δy-interval around a particular rapidity y within the phase space:

$$\mathscr{R}(\beta, y) = \lim_{\delta y \to 0} \frac{\mathscr{L}_{\delta y}(\beta, y)}{\delta y} \tag{18.5}$$

We may recover the distribution $\mathscr{L}_{\Delta y}(\beta, y)$ for a finite Δy-interval by

$$\mathscr{L}_{\Delta y}(\beta, y) = \int_{y - \Delta y/2}^{y + \Delta y/2} \mathscr{R}(\beta, y') dy' \tag{18.6}$$

We next remark that for a given y the distribution $\mathscr{R}(\beta, y)$ can only depend upon the maximum value of the (logarithmic) squared $k_\perp$ that can occur for that y-value. We will denote as $\ell(y)$, the variable $\log(k_{\perp, max}^2 / s_0)$ for a given y and we obtain from the triangular phase space the equality $\ell(y) = L - 2|y|$. It is then possible to write the following formula for $\mathscr{L}(\beta, L)$, the distribution for the total L-region:

$$\mathscr{L}(\beta, L) = \int_{-L/2}^{L/2} dy \, \mathscr{R}(\beta, \ell = L - 2|y|) = \int_0^L d\ell \, \mathscr{R}(\beta, \ell) \tag{18.7}$$

We will now consider the change in $\mathscr{R}$ when $\ell \to \ell + \delta$ and concentrate upon a particular infinitesimal y-region Δ around y. There is the probability

$$\Delta \delta \frac{\alpha_0}{\ell} \tag{18.8}$$

of obtaining a new gluon inside the region which is shaded in Fig. 18.2. If there is such a gluon then the increase in λ is described by $P(\lambda, \ell(y))$, because with all its folds and subfolds it corresponds exactly to an isolated system with $L = \ell(y)$. In Eq. (18.8) we have introduced the runnning coupling constant of QCD, $3\alpha_s k_{\perp, max}^2 / (2\pi) \equiv \alpha_0 / l$. This means that we have identified the scale $s_0 = \Lambda_{QCD}^2$.

The remaining probability $1 - \Delta\delta\alpha_0/\ell$ corresponds to the case when there is no extra gluon and consequently λ is unchanged. We obtain

$$P_\Delta(\lambda, \ell + \delta) = (1 - \Delta\delta\frac{\alpha_0}{\ell}) P_\Delta(\lambda, \ell)$$

$$+ \Delta\delta\frac{\alpha_0}{\ell} \int d\lambda_1 d\lambda_2 P_\Delta(\lambda_1, \ell) P(\lambda_2, \ell(y)) \delta(\lambda_1 + \lambda_2 - \lambda) \tag{18.9}$$

We may now subtract $P_\Delta(\lambda, \ell)$ from both sides, take the Laplace transforms and go to the limit $\delta, \Delta \to 0$ to obtain the following result for $\mathscr{R}$:

$$\frac{d\mathscr{R}(\beta, \ell)}{d\ell} = \frac{\alpha_0}{\ell} [\exp \mathscr{L}(\beta, \ell) - 1] \tag{18.10}$$

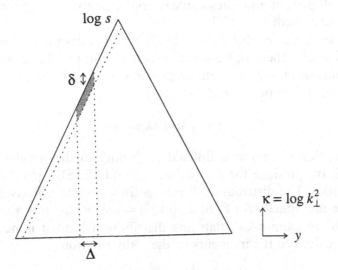

Fig. 18.2. The region for emitting a gluon obtained by increasing L and thereby also the local ℓ is shown as shaded.

Combining this equation with the result in Eq. (18.7) we obtain the second order differential equation

$$\frac{d^2\mathscr{L}(\beta,\ell)}{d\ell^2} = \frac{\alpha_0}{\ell}[\exp\mathscr{L}(\beta,\ell) - 1] \qquad (18.11)$$

In order to specify the function $\mathscr{L}(\beta,L)$ we must supplement Eq. (18.11) with the proper boundary conditions. For $L = 0$ we evidently have a δ-contribution, i.e. $P(\lambda,0) = \delta(\lambda)$, which implies that $\mathscr{L}(\beta,0) = 0$. Further for small values of L the contributions to λ from the gluons is of order L^2. This means that $d\mathscr{L}(\beta,0)/dL = -\beta$.

2   The moments in λ

Equation (18.11) with these boundary conditions has unfortunately no solution in terms of elementary functions. Numerical solutions of Eq. (18.11) indicate that a good approximation for the distribution $P(\lambda,L)$ is given by a Γ-distribution:

$$P(\lambda,L) \simeq \frac{\lambda^{v-1}\rho^v}{\Gamma(v)}\exp(-\rho\lambda) \qquad (18.12)$$

where v and ρ are slowly varying functions of $\alpha_0 L$. For large values of L the result in Eq. (18.29) implies that $v \to 3$. This estimate is, however, due to kinematical corrections rather bad and should be exchanged for

$v \sim 7$–8 at all present and foreseeable energies, although the general shape is reasonable according to [74].

In order to relate the distribution to the observables we assume that for a given value of λ there is, besides the distribution in λ for given s which we have considered above, a definite multiplicity distribution of final-state hadrons $P(n, \lambda)$, independent of the energy:

$$P_{tot}(n, s) = \int d\lambda P(\lambda, s) P(n, \lambda) \tag{18.13}$$

This assumption is very well fulfilled according to the simulations with ARIADNE (to produce the λ-distribution) and JETSET (to produce the fragmentation). The distribution $P(n, \lambda)$ is then close to a Poissonian with the average multiplicity for fixed λ , $\langle n \rangle (\lambda) = m\lambda$, where m is a constant.

Although we cannot calculate the distribution itself it is nevertheless possible to calculate the moments of the λ-distribution,

$$\langle \lambda^n \rangle = \int d\lambda \, \lambda^n P(\lambda, L) \tag{18.14}$$

directly from the differential equation.

Thus we have from the defining equation of the Laplace transform

$$\exp(\mathscr{L}(\beta, L) \equiv \tilde{P}(\beta, L) = \int d\lambda \exp(-\beta\lambda) P(\lambda, L)$$

$$= \sum_{j=0}^{\infty} \frac{(-\beta)^n}{(n)!} \int d\lambda \, \lambda^n P(\lambda, L) \tag{18.15}$$

It is straightforward to prove for the first two moments, the mean $\langle \lambda \rangle$ and the variance $V_\lambda = \langle \lambda^2 \rangle - \langle \lambda \rangle^2$, that if

$$\mathscr{L}(\beta, L) = -\beta G_1(L) + \beta^2 G_2(L) + O(\beta^3) \tag{18.16}$$

then

$$\langle \lambda \rangle = G_1(L)$$
$$V_\lambda \equiv \langle \lambda^2 \rangle - \langle \lambda \rangle^2 = \langle (\lambda - \langle \lambda \rangle)^2 \rangle = 2G_2(L) \tag{18.17}$$

The differential equation for G_1 is immediately obtained from the first-order expansion in β of Eq. (18.11):

$$\frac{d^2 G_1}{dL^2} = \frac{\alpha_0}{L} G_1 \tag{18.18}$$

(It is worthwhile to go through the calculations leading to Eqs. (18.15)–(18.18).)

The solutions to this equation are related to the modified Bessel functions of first rank, I_j and K_j. As we are going to use these solutions repeatedly we will introduce a special notation for them:

$$\mathscr{I}_1(\kappa) = \sqrt{2\kappa}I_1(2\sqrt{\alpha_0\kappa}), \quad \mathscr{I}_0(\kappa) = \sqrt{2\alpha_0}I_0(2\sqrt{\alpha_0\kappa}) \qquad (18.19)$$

with similar relations between $\mathscr{K}_j$ and K_j. These functions fulfil

$$\frac{d\mathscr{I}_1}{d\kappa}(\kappa) = \mathscr{I}_0(\kappa), \quad \frac{d\mathscr{I}_0}{d\kappa}(\kappa) = \frac{\alpha_0}{\kappa}\mathscr{I}_1(\kappa) \qquad (18.20)$$

with similar relations for $\mathscr{K}_j$ but with a negative sign. The functions are normalised so that

$$\mathscr{I}_1\mathscr{K}_0 + \mathscr{K}_1\mathscr{I}_0 = 1 \qquad (18.21)$$

We also note the limiting behaviour

$$\lim_{\kappa\to 0}\mathscr{K}_1(\kappa) = \frac{1}{\sqrt{2\alpha_0}}$$

$$\lim_{L\to\infty}\mathscr{I}_1(L) = \frac{L^{1/4}}{(2\pi)^{1/2}\alpha_0^{1/4}}\exp(2\sqrt{\alpha_0 L}) \qquad (18.22)$$

$$\lim_{L\to\infty}\mathscr{K}_1(L) = \frac{\pi^{1/2}L^{1/4}}{2^{1/2}\alpha_0^{1/4}}\exp(-2\sqrt{\alpha_0 L})$$

All these relations are easy to prove from any handbook on Bessel functions, e.g. [57]. It can also be seen from a combination of Eqs. (18.20) that the general solution to Eq. (18.18) is a linear combination of $\mathscr{I}_1$ and $\mathscr{K}_1$,

$$G_1 = A\mathscr{I}_1 + B\mathscr{K}_1 \qquad (18.23)$$

where the numbers A and B must be chosen so that the boundary conditions are fulfilled. Thus $G_1 = 0$ and $dG_1/dL = 1$ at the starting point, according to the boundary conditions for Eq. (18.11). We will choose the starting point to be a bit more general than before, $L = L_0$, and assume that L_0 may be different from 0.

It is then easy to prove by means of the relations in Eqs. (18.20), (18.21) and (18.22) that the general solution is

$$G_1(L, L_0) = \mathscr{I}_1(L)\mathscr{K}_1(L_0) - \mathscr{K}_1(L)\mathscr{I}_1(L_0)$$

$$\to \frac{(LL_0)^{1/4}}{2(\alpha_0)^{1/2}}\exp[2(\sqrt{\alpha_0 L} - \sqrt{\alpha_0 L_0})] \qquad (18.24)$$

$$G_1(L, 0) = \sqrt{\frac{L}{\alpha_0}}I_1(2\sqrt{\alpha_0 L}) \to \sqrt{\frac{L^{1/2}}{\alpha_0^{3/2}4\pi}}\exp(2\sqrt{\alpha_0 L})$$

where the limits correspond to $L \gg L_0 \gg 0$.

The corresponding differential equation for the variance in λ is obtained from the expansion coefficient proportional to β^2 in Eq. (18.11):

$$\frac{d^2 G_2}{dL^2} = \frac{\alpha_0}{L}\left(G_2 + \frac{G_1^2}{2}\right) \tag{18.25}$$

with the condition that $G_2 = dG_2/dL = 0$ at the starting point. The equation can then be easily solved by means of the Green's function method, i.e. we look for a solution to the equation

$$\frac{d^2 G}{dL^2} - \frac{\alpha_0}{L}G = \delta(L - L') \tag{18.26}$$

with the same boundary values as for G_2. This is obtained by the use of G_1 from Eq. (18.24) above:

$$G(L, L') = \Theta(L - L')\frac{G_1(L, L')}{2} \tag{18.27}$$

In this way we obtain the following result for G_2:

$$G_2(L, L_0) = \frac{1}{4}\int_{L_0}^{L}\frac{\alpha_0}{L'}G_1(L, L')dL'G_1^2(L', L_0) \tag{18.28}$$

We obtain using the asymptotic expression for $G_1 = \langle\lambda\rangle$

$$V_\lambda \to \frac{\langle\lambda\rangle^2}{3} \tag{18.29}$$

The reader is once again invited to carry through the necessary calculations to prove Eqs. (18.24)–(18.29).

3 The notion of KNO scaling and the fact that the dipole cascade is dominated by the first two gluon emissions

The result that the variance in the multiplicity of final-state particles obeys $V_n \simeq C_2\langle n\rangle^2$ has been known for a long time in high-energy physics. It was first known as the Wroblewski relation and later extended to the notion of KNO scaling, [86]. It was then applied to hadronic reactions but it is nowadays known that both e^+e^- annihilation and inelastic leptoproduction event multiplicities exhibit a similar structure.

The basic idea in KNO scaling is that the multiplicity distributions, $P(n, s)$, scale with the mean multiplicity. For a squared cms energy s the suggestion is that

$$P(n, s) = \frac{1}{\langle n\rangle(s)}F\left(\frac{n}{\langle n\rangle(s)}\right) \tag{18.30}$$

where $\langle n\rangle(s)$ is the mean multiplicity at the squared cms energy s and F a continuous function which depends upon the dynamics of the particular interaction but is independent of the energy. It is then evident that all moments $\langle n^j \rangle$, not only the second one, will be proportional to $\langle n \rangle^j$.

There are some problems to be faced in connection with the approximation of a discrete distribution in terms of a continuous function. Different investigators have therefore used somewhat different ways to define the relationship between F and P_n. One reason is that if we consider e.g. a proton-proton reaction and concentrate on the charged particles then charge conservation means that only particle pairs with one positive and one negative can be produced. Therefore it is only possible to observe an even number of charged particles. Further it is not known whether the particles carrying the original two positive charges should be included. One solution would be to define the variable x as $x = (n - n_0)/(\langle n \rangle - n_1)$, where n_0, n_1 are suitably chosen parameters for the argument in $F(x)$.

Nevertheless, Eq. (18.30) gives, if you allow for these uncertainties, a surprisingly good description of a large amount of multiplicity data from many different processes and energies. The function F is, however, process dependent.

There have been many speculations on the origin of the KNO scaling property. We will be satisfied with a few comments. Suppose that the basic particle-production mechanism corresponds to truly independent production as in the case of an external current acting on the different frequencies of a quantum field (the Schwinger model in Chapter 3). There are reasons for such a simple assumption. Multiparticle production generally leads to a flat central rapidity distribution without much correlation between the produced particles. Then it would be pure chance whether there is an observed particle in a small rapidity bin or whether it will be empty. This would, however, imply Poissonian statistics (just as we obtained for the external current). Therefore one should expect that the variance behaves as $V_n \propto \langle n \rangle$ instead of the KNO prediction. Consequently it is necessary to introduce more dynamical assumptions to obtain the wider KNO scaling distributions.

There is a set of necessary constraints on particle production in both the Lund model and other successful models. The total charge and energy-momentum must be conserved during the production process. Further there is resonance production, which will introduce local correlations in rapidity because the decay products are spread over 1–2 units in rapidity. All such phenomena can be accounted for within the simpler schemes used in the iterative cascade models, which were described in Chapters 7 and 9. In general these models predict, nevertheless, an essentially Poissonian statistics for the central production in rapidity. The breakup of a single Lund $q\bar{q}$-string also leads to something close to a Poissonian (although a somewhat narrower, cf. the treatment of the Feynman-Wilson gas in Chapter 11). It should, however, be understood that for small energies the fragmentation distributions are wider than Poissonian because then the phase space is dominated by the regions close to the string endpoints where the variations are larger.

If several different production processes *independently* contribute to the final hadronic state then the total multiplicity distribution will be a folding of the different distributions. In this case it is easy to prove that the variance will be simply additive, and there will not be much widening from such an assumption unless one distribution is very wide by itself,

$$P_{tot}(n) = \sum_{n_1+...+n_N=n} P_1(n_1) \cdots P_N(n_N)$$
$$\left\langle \sum n_j \right\rangle = \sum \langle n_j \rangle, \quad V\left(\sum n_j\right) = \sum V_{n_j}$$

(18.31)

It is, however, rather easy to obtain a broad distribution if there are several contributing processes *which exclude each other*. If we have two processes, $j = 1, 2$, with the probabilities α and $1 - \alpha$ occurring we have for the total multiplicity distribution and its moments

$$P_{tot}(n) = \alpha P_1(n) + (1 - \alpha)P_2(n)$$
$$\langle n \rangle_{tot} = \alpha \langle n \rangle_1 + (1 - \alpha) \langle n \rangle_2$$
$$V(n_{tot}) = \alpha V_1 + (1 - \alpha)V_2 + \alpha(1 - \alpha)(\langle n \rangle_1 - \langle n \rangle_2)^2$$

(18.32)

This is much broader than each of the distributions when the mean multiplicities are different in the two processes. In hadronic interaction models this is often used to explain KNO scaling because in this case there are, besides a general smooth central production, also *diffractive events* with smaller multiplicities concentrated around the incoming particle rapidities.

For hard QCD processes it is instead the bremsstrahlung of gluons which causes the broadening of the multiplicity distributions. This means that it is the folding in Eq. (18.13) which is responsible and there is a large width in the λ-distribution in accordance with what we have learned; this should be compared to the statement in Eq. (18.31).

The Lund model predicts, in good agreement with data, a KNO scaling result. This stems from a combination of the fragmentation process and the partonic cascade. Using the results from Eq. (18.13) we obtain

$$\langle n \rangle_{tot}(s) = m \langle \lambda \rangle(s) = \langle n \rangle(\lambda = \langle \lambda \rangle(s))$$
$$V_{tot} = m^2 V_\lambda(s) + V_n(\lambda = \langle \lambda \rangle(s)) \equiv V_{casc} + V_{frag}$$

(18.33)

In this way we have partitioned the variance into the cascade contribution, i.e. the variations in λ for fixed s, and the fragmentation contribution, the variations in multiplicity for a fixed generalised rapidity region λ. We can go further and obtain two independent contributions to the (squared)

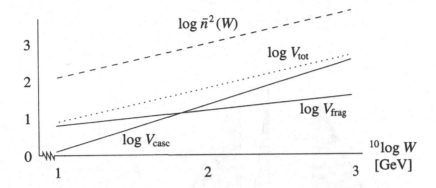

Fig. 18.3. The contributions to the multiplicity variance stemming from the dipole cascade model (ARIADNE for λ) and from the Lund fragmentation model (JETSET), together with the total variance (dotted line) and the square of the mean multiplicity (broken line). The variable W is the square root of s.

width over the mean:

$$\frac{V_{tot}}{\langle n \rangle^2_{tot}} = \frac{V_\lambda}{\langle \lambda \rangle^2} + \frac{V_n(\lambda = \langle \lambda \rangle (s))}{\langle n \rangle^2 (\lambda = \langle \lambda \rangle (s))} \tag{18.34}$$

The resulting distribution is shown in Fig. 18.3. While the contributions lead to a smooth, almost constant, ratio, one of them, the fragmentation contribution, dominates at small values of s and the other, the cascade contribution, dominates for large s. The crossing point is somewhere below the LEP energies.

We end this section with another result from the investigations of the dipole cascade model. *It is possible to show that almost all the variations in the λ-distribution stem from the emission of the first two gluons.*

To understand this result we will make use of the distributions $\mathscr{R}$ and $\tilde{P} = \exp \mathscr{L}$, which are defined in Section 1. They correspond to particular Laplace transforms of the λ-distribution.

We assume that the first gluon (i.e. the one with the largest $k_\perp$) is emitted at $\log(k^2_{\perp 1}/s_0) = \kappa_1$. From Fig. 18.4 we obtain that the total $\mathscr{L}$-distribution for these kind of events, $\mathscr{L}(\beta, L, \kappa_1)$, is given by

$$\mathscr{L}(\beta, L, \kappa_1) = 2\mathscr{L}(\beta, \kappa_1) + (L - \kappa_1)\mathscr{R}(\beta, \kappa_1) \tag{18.35}$$

The first contribution on the right-hand side stems from the two half-triangles at the outskirts and the folded triangle from the emission at κ_1. The rest stems from what is left of the background triangle, i.e. the rectangle with side length $L - \kappa_1$. Therefore the average and the variance in λ (the two lowest-order power contributions in β) are remembering that

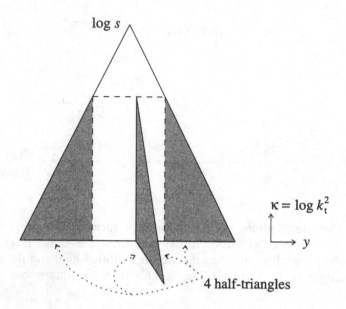

Fig. 18.4. The emission of the first and hardest gluon partitions the remainder of the λ-distribution for such states into one contribution from the $\mathscr{R}$, integrated over the centre, and the contributions from two triangular regions, one along the two boundaries and one from the gluon fold.

$\mathscr{R}$ is the derivative of $\mathscr{L}$,

$$\langle \lambda \rangle (L, \kappa_1) = 2 \langle \lambda \rangle (\kappa_1) + (L - \kappa_1) \frac{d}{d\kappa_1} \langle \lambda \rangle (\kappa_1)$$

$$V_\lambda(L, \kappa_1) = 2 V_\lambda(\kappa_1) + (L - \kappa_1) \frac{d}{d\kappa_1} V_\lambda(\kappa_1)$$

(18.36)

The total variance for a given L, $V_{tot}(L)$, stems from the variations in the first emission together with the variations of all the remaining gluons, averaged over the first emission. To calculate the variation in κ_1 we must as always introduce a Sudakov factor f, corresponding to no emission before κ_1,

$$V_{tot}(L) = \left\{ \int d\kappa_1 f(\kappa_1) \langle \lambda \rangle^2 (L, \kappa_1) - \left(\int d\kappa_1 f(\kappa_1) \langle \lambda \rangle (L, \kappa_1) \right)^2 \right\}$$
$$+ \int d\kappa_1 f(\kappa_1) V_\lambda(L, \kappa_1)$$

(18.37)

From the bremsstrahlung cross section we obtain, integrating over the available rapidity range with $3\alpha_s/2\pi = \alpha_0/\kappa$,

$$\frac{dn_g}{d\kappa_1} = \frac{\alpha_0}{\kappa_1}(L - \kappa_1) \qquad (18.38)$$

Then if $(L - \kappa_1)/L \ll 1$, f will asymptotically behave as a gaussian distribution:

$$f \simeq \alpha_0 \frac{L - \kappa_1}{L} \exp\left[-\frac{\alpha_0(L - \kappa_1)^2}{2L}\right] \qquad (18.39)$$

It is then rather easy to introduce the results for $V_\lambda(L, \kappa_1)$ and f to perform the second integral in Eq. (18.37). We obtain an error-function result which is equal to

$$\int d\kappa_1 f(\kappa_1) V_\lambda(L, \kappa_1) \simeq 0.53 V_{tot}(L) \qquad (18.40)$$

From this approximation we obtain that the variations from the first gluon emission in this asymptotic scenario contribute around 47% to the total variance.

It is possible to do the same calculation for the second, third etc. gluons and it is also possible to simulate the variations from ARIADNE in the way described in [12]. We find that over the PETRA-PEP energy range around 95% of the variance stems from the variations in the first gluon emission. Over the LEP-SLC region more than 87% of the variance stems from the first and in total more than 95% from the first two gluon emissions.

The reason for this result is that if we have a hard gluon emission $\kappa_1 \sim L$ we obtain a very different kind of event as compared to the case when $\kappa_1 \ll L$. In both cases there will be further gluon emissions but the variations in them will at the present energies give very small contributions to the variations in the final-state multiplicities.

We are basically invoking the *very slow development of the QCD cascades* and we will come back to this later on. But even the asymptotic result, i.e. that for all energies around half of the variations stem from the first gluon, is quite surprising!

18.3 The κ-method

1 Preliminaries

There is one inconvenient property of the L-method and that is that when we search the phase space for gluon emissions upwards, i.e. for increasing L, we obtain hard gluon contributions to the λ-distribution. This means that the L-method leads to discontinuous changes in λ. But the structure of independence in the y-variable means that the Laplace-transformed λ-distributions change in a smooth and differentiable way.

We will now introduce a different method, [6], in which for a fixed value of L we go downwards in κ in the phase space, looking for gluons with smaller and smaller $k_\perp$. In this way we obtain a smooth and differentiable distribution in the dipole productions and thereby also in λ.

The basic idea is to introduce the *notion of virtual dipole size at the level* κ, to be called μ for an individual dipole. We define the variable μ for a dipole of squared mass s_d as the rapidity size within which the dipole can emit gluons with transverse momentum $k_\perp$ (cf. the triangular phase space size)

$$\mu = \log(s_d/s_0) - \log(k_\perp^2/s_0) \equiv L_d - \kappa \qquad (18.41)$$

In order to see the general behaviour we will firstly study the (inclusive) distribution in the μ-variable for the dipole containing the original q-particle (the right-endpoint dipole), to be called $P_1(\mu, \kappa)$. For simplicity we suppress the dependence on the starting point of the evolution, κ_{max}.

We now show how to write gain-loss equations for the distribution P_1 when we change the *resolution scale* κ. We firstly note that the dipole size μ will obviously due to its definition increase to $\mu + \delta\kappa$ when we decrease κ to $\kappa - \delta\kappa$.

A dipole of size μ may, however, also split up into two dipoles, μ_1 and μ_2, when $\kappa \to \kappa - \delta\kappa$. The probability for this to happen is

$$\frac{\alpha_0}{\kappa}\delta\kappa d\mu_1 d\mu_2 \delta(\mu - \mu_1 - \mu_2) \qquad (18.42)$$

This is a rewriting of the ordinary cross section for gluon emission, once again with the effective running coupling $3\alpha_s/2\pi \equiv \alpha_0/\kappa$. We have assumed that there is in the dipole rest frame a rapidity value y such that $|y| \leq \mu/2$ and that a new gluon is emitted at (κ, y). Thus the dipole is split into two with virtual sizes $\mu/2 \pm y$; these are called μ_1 and μ_2 in Eq. (18.42).

After the splitting each new dipole will increase independently of the other one when we go downwards in the cascade, according to the basic assumption of the DCM. These dipole sizes correspond to the available emission hyperbolas from the endpoints of the μ-dipole, via the new gluon 'peak', according to the description in Chapter 15.

2 The distribution of the endpoint dipole

The distribution P_1 can thus change in three ways when we take the step $\kappa \to \kappa - \delta\kappa$:

1 The value of μ at $\kappa - \delta\kappa$ may correspond to the value $\mu - \delta\kappa$ at κ.

2 It is possible for a dipole of size μ to decay into smaller-size dipoles according to Eq. (18.42) (loss at the value μ).

3 It is possible that a larger dipole $\mu_1 > \mu$ will decay into μ according to Eq. (18.42) (gain at the value μ).

Gathering these contributions we obtain the following partial differential equation, using the effective coupling $\bar{\alpha} = \alpha_0/\kappa$,

$$\frac{\partial P_1}{\partial \kappa} - \frac{\partial P_1}{\partial \mu} \equiv \Delta_1 P_1 = \frac{\alpha_0}{\kappa} \left(\mu P_1 - \int_\mu d\mu_1 P_1(\mu_1, \kappa) \right) \quad (18.43)$$

The distribution P_1 should obviously be normalised to 1 for all κ, so that we have

$$N_{1j}(\mu, \kappa) = \int_\mu d\mu_1 (\mu)^{j-1} P_1(\mu_1, \kappa), \quad N_{11}(0, \kappa) = 1 \quad (18.44)$$

If we perform a partial integration in μ on the integro-differential equation in Eq. (18.43) we obtain a partial differential equation solely in N_{11}:

$$\Delta_1 N_{11} = \frac{\alpha_0}{\kappa} \mu N_{11} \quad (18.45)$$

We note the similarity to the Callan-Symanzik equation which was treated in Chapter 4. We will again solve it by integration along rays:

$$\frac{d\mu}{d\kappa} = -1 \quad \Rightarrow \quad C_1 = \mu + \kappa \quad (18.46)$$

where C_1 is a constant. We obtain including the right boundary condition

$$N_{11}(\mu, \kappa) = \Theta(L - \mu - \kappa) \exp \left(-\int_\kappa^{\mu+\kappa} dy (\mu + \kappa - y)\alpha_0/y \right) \quad (18.47)$$

The step function corresponds to the largest value the ray parameter C_1 can take on, i.e. the starting point of the cascade, L. The distribution P_1 obviously fulfils

$$P_1 = -\frac{\partial N_{11}}{\partial \mu} \quad (18.48)$$

and therefore contains a δ-distribution corresponding to the situation when there has been no gluon emission and we are left with the original dipole of size $\mu = L - \kappa$ at this virtuality. The coefficient in front of the δ-distribution is the Sudakov exponential factor, which was approximated by means of a gaussian in connection with Eq. (18.39):

$$f_{Sud}(L, \kappa) = \exp \left(-\int_\kappa^L dy (L - y)\alpha_0/y \right) \quad (18.49)$$

3 The general inclusive single-dipole distribution

In this subsection we will be concerned with the inclusive distribution $\mathscr{P}_1$ for all the available dipoles whether they are endpoint dipoles or central ones. There are then two differences compared to P_1. The first one is related to the normalisation. If we define

$$\mathscr{N}_{1j}(\mu, \kappa) = \int_\mu d\mu_1 \mu_1^{j-1} \mathscr{P}_1(\mu_1, \kappa) \tag{18.50}$$

then $\mathscr{N}_{11}(0, \kappa) \equiv M_1(\kappa)$ corresponds to the average number of dipoles at the virtuality κ. In the same way $\mathscr{N}_{12}(0, \kappa) \equiv M_2(\kappa)$ corresponds to the total length of these dipoles, which evidently must be identical to the average $\langle \lambda \rangle$, defined before.

The second difference is the fact that the gain factor from the decays of larger dipoles is 2 instead of 1 because each dipole of length $\mu_1 > \mu$ will decay into two dipoles and either one of them may have the length μ. Thus the integro-differential equation (18.43) becomes

$$\Delta_1 \mathscr{P}_1 = \frac{\alpha_0}{\kappa} \left[\mu \mathscr{P}_1 - 2 \int_\mu d\mu_1 \mathscr{P}_1(\mu_1, \kappa) \right] \tag{18.51}$$

These changes mean that if we again perform a partial integration in μ we obtain instead of a single equation a set of two coupled partial differential equations from Eq. (18.51) for the first two moments of $\mathscr{P}_1$:

$$\Delta_1 \mathscr{N}_{11} = -\frac{\alpha_0}{\kappa}(\mathscr{N}_{12} - 2\mu \mathscr{N}_{11}), \quad \Delta_1 \mathscr{N}_{12} = -\mathscr{N}_{11} + \frac{\alpha_0}{\kappa}\mu^2 \mathscr{N}_{11} \tag{18.52}$$

From these equations we immediately obtain for the quantities M_j, $j = 1, 2$, which are defined above,

$$\frac{dM_1}{d\kappa} = -\frac{\alpha_0}{\kappa}M_2, \quad \frac{dM_2}{d\kappa} = -M_1 \tag{18.53}$$

If we combine these two equations we obtain back the second-order equation already derived for $M_2 \equiv \langle \lambda \rangle$ before (Eq. (18.18)),

$$\frac{d^2 M_2}{d\kappa^2} = \frac{\alpha_0}{\kappa} M_2 \tag{18.54}$$

and the solution is the same (cf. Eq. (18.24)):

$$M_2(L, \kappa) = \mathscr{I}_1(L)\mathscr{K}_1(\kappa) - \mathscr{K}_1(L)\mathscr{I}_1(\kappa) \tag{18.55}$$

The second equation in (18.53) tells us that the mean number of dipoles is given by the (negative) κ-derivative of $\langle \lambda \rangle$:

$$\langle n_d \rangle = \mathscr{I}_1(L)\mathscr{K}_0(\kappa) + \mathscr{K}_1(L)\mathscr{I}_0(\kappa) \equiv M_1(L, \kappa) \tag{18.56}$$

From the properties of Eq. (18.21) we conclude that $M_1(L, L) = 1$ while $M_2(L, L) = 0$. It is a very general property of second-order differential equations that we can find another pair of functions M_3 and M_0 which also fulfils Eqs. (18.53) with $M_3 \to M_2$ and $M_0 \to M_1$:

$$M_3(L, \kappa) = \mathcal{I}_0(L)\mathcal{K}_1(\kappa) + \mathcal{K}_0(L)\mathcal{I}_1(\kappa)$$
$$M_0(L, \kappa) = \mathcal{I}_0(L)\mathcal{K}_0(\kappa) - \mathcal{K}_0(L)\mathcal{I}_0(\kappa) \tag{18.57}$$

This time the pair fulfils $M_3(L, L) = 1$ and $M_0(L, L) = 0$. Using the two pairs of solutions it is possible to write out the mean of the λ and the number of dipoles n_d in the general situation when we take the ensemble of states which starts at $L = \ell_0$ with $\lambda = \lambda_0$ and $n_d = n_{d0}$:

$$\langle \lambda \rangle (\ell_0, \kappa) = n_{d0}M_2(\ell_0, \kappa) + \lambda_0 M_3(\ell_0, \kappa)$$
$$\langle n_d \rangle (\ell_0, \kappa) = n_{d0}M_1(\ell_0, \kappa) + \lambda_0 M_0(\ell_0, \kappa) \tag{18.58}$$

We will meet all these functions later in a more general context but it is worthwhile to convince oneself that Eqs. (18.57) and (18.58) are correct.

While the mean value of λ (in both the cases discussed above) is finite also when $\kappa \to 0$ (cf. Eq. (18.24)) we find that the mean number of dipoles diverges logarithmically in that limit.

4 An aside on the rate of decay in QCD cascades

In this subsection we will give a simple explanation for the results in Eqs. (18.24) and (18.56) in order to understand the way the partonic cascades develop in QCD. We will start by analysing the probability that a dipole of mass $\sqrt{s}$, i.e. with logarithmic variable $L = \log(s/s_0)$, will decay into two dipoles by the emission of a (first) gluon at the value $\kappa = \log(k_\perp^2/s_0)$. From the Sudakov factor f_{Sud} in Eq. (18.49) and the bremsstrahlung cross section we obtain

$$dP = \frac{\alpha_0}{\kappa}d\kappa(L - \kappa) \exp\left[-\int_\kappa^L \frac{\alpha_0}{\kappa_1}d\kappa_1(L - \kappa_1)\right] \tag{18.59}$$

For the available rapidity range we have introduced the virtual size $\mu = L - \kappa$ of the dipole at κ. We note that this distribution is not normalised to unity but instead we obtain by integration over κ

$$\int_{\kappa_c}^L dP = 1 - f_{Sud}(L, \kappa_c) \tag{18.60}$$

where κ_c is a suitable cutoff. The interpretation is that the probability density dP is normalised to unity apart from the possibility that the dipole does not decay before we reach κ_c. It is straightforward to perform the integral in the Sudakov factor and, using the variables $x \equiv \alpha_0 \kappa$ and $y \equiv \alpha_0 L$, we obtain

$$dP = g(y)dx(y - x)x^{y-1}\exp(-x) \tag{18.61}$$

Here $g(y)$ is a normalisation factor. We may once again use the same methods as we used to find the maximum of the Lund fragmentation distribution in Chapter 9, cf. Eq. (9.6), to prove that there is a pronounced maximum in the distribution for $x_{max} = y - \sqrt{y}$; in terms of L, κ it occurs for $\kappa_{max} = L - 1/\sqrt{\bar{\alpha}(L)}$.

There is a simple explanation for this result. The step $\delta\kappa$ from the starting point L to κ_{max} is evidently equal to the virtual size of the dipole at κ_{max}. The probability for decay inside this region is then

$$\bar{\alpha}\delta\kappa\delta y = \bar{\alpha}(\delta\kappa)^2 \simeq C \qquad (18.62)$$

where C is a constant of order unity. We will now consider the general case. If a dipole tends to have the same size as its 'survival time' this means in the phase-space triangle that a region free of gluons is as wide (in the generalised y-variable) as it is high (in the κ-direction). This means that *the step $\delta\kappa$ between two 'generations'* (meaning the decay-times of the typical dipoles) *of gluons behaves as the square root of the virtuality κ*.

Then the first generation (the typical 'hardest' single gluon when we start at the virtuality L) will occur after a step

$$\delta\kappa_1 \sim C'\sqrt{L}, \quad C' = C/\sqrt{\alpha_0} \qquad (18.63)$$

and we are left at the virtuality $L - C'\sqrt{L}$. The next generation (containing two 'typical' gluons) occurs after a second step

$$C'\sqrt{L - C'\sqrt{L}} \simeq C'\sqrt{L} - (C')^2/2 \qquad (18.64)$$

where we have expanded the square root under the assumption that $C'\sqrt{L} \ll L$. Counting downwards we obtain after $n \gg 1$ generations (in which there are 2^{n-1} gluons produced) a remaining virtuality

$$\simeq L - nC'\sqrt{L} + n^2(C')^2/4 \qquad (18.65)$$

If this is the endpoint κ we obtain

$$n \simeq 2(\sqrt{\alpha_0 L} - \sqrt{\alpha_0 \kappa})/C \qquad (18.66)$$

and the multiplicity of dipoles will be 2^n, so that using $C = \log 2$ we have

$$M_1 \sim \exp 2(\sqrt{\alpha_0 L} - \sqrt{\alpha_0 \kappa}) \qquad (18.67)$$

To obtain M_2 we just multiply by the typical dipole size at κ; we have thus in a simple way obtained good approximations for $\langle \lambda \rangle$ and its first κ-derivative, i.e. the average dipole multiplicity.

It is worthwhile to ponder the immensely slow development of a QCD cascade. Suppose that we consider an e^+e^- annihilation event with $\sqrt{s} = 1000$ GeV and $\Lambda_{QCD} = 250$ MeV. Then the first generation will occur, according to the calculation above, at around $k_\perp \sim 130$ GeV, the second at around $k_\perp \sim 15$ GeV, the third at around 5 GeV, etc.

Actually we are nevertheless exaggerating the rate. When we take the recoil corrections into account within the so-called modified leading-log scenario the average rate will become even slower.

5   The master equation

In [6] several other distributions are derived and investigated. We will here be satisfied to discuss the combined distribution in the number of dipoles, $n_d \equiv n$, and their total length $\sum_{j=1}^{n} \mu_j \equiv \lambda$, which we will call $P(n, \lambda, \kappa)$ (once again suppressing the dependence on L).

The master equation for the distribution $P(n, \lambda, \kappa)$ is

$$\frac{\partial P}{\partial \kappa} - n \frac{\partial P}{\partial \lambda} = \frac{\alpha_0}{\kappa} \lambda [P(n, \lambda, \kappa) - P(n-1, \lambda, \kappa)] \tag{18.68}$$

The result stems from the fact that the change in λ from n independent dipoles is $\lambda \to \lambda + n \delta \kappa$ when $\kappa \to \kappa - \delta \kappa$. Further there is a loss for $P(n)$ and a gain from $P(n-1)$ when any of the dipoles decays.

We will briefly consider this equation and its solutions before we show how to incorporate corrections due to recoils, the phase-space size and the neglect of the polarisation sum in the decay process.

We may make Eq. (18.68) into a linear partial differential equation in all the three variables n, λ, κ in the approximation when n is considered a continuous variable. Then Eq. (18.68) becomes

$$\frac{\partial P}{\partial \kappa} - n \frac{\partial P}{\partial \lambda} \simeq \frac{\alpha_0}{\kappa} \lambda \frac{\partial P}{\partial n} \tag{18.69}$$

We may again use the method of rays and look for $n = n(\kappa)$ and $\lambda = \lambda(\kappa)$ with the properties that

$$\frac{dn}{d\kappa} = -\frac{\alpha_0}{\kappa} \lambda, \quad \frac{d\lambda}{d\kappa} = -n \tag{18.70}$$

Given the solutions to these equations we then have that

$$\frac{\partial P}{\partial \kappa} + \frac{\partial P}{\partial \mu} \frac{d\lambda}{d\kappa} + \frac{\partial P}{\partial n} \frac{dn}{d\kappa} = \frac{d}{d\kappa} P(n(\kappa), \lambda(\kappa), \kappa) = 0 \tag{18.71}$$

i.e. any function which is constant along the rays will work!

We note that Eqs. (18.70) are just the equations we had for the quantities $M_j, j = 1, 2$ or $j = 0, 3$ (Eq. (18.53)) and they can consequently be solved in terms of these functions. We assume that we know the distribution in n_0, λ_0, to be called $F(n_0, \lambda_0)$, for a certain value of $\kappa = l_0$. It is straightforward to describe the values of λ, n for an arbitrary $\kappa \leq l_0$ from the results of Eqs. (18.53) with the boundary values n_0, λ_0

$$n = n_0 M_1(l_0, \kappa) + \lambda_0 M_0(l_0, \kappa)$$
$$\lambda = \lambda_0 M_3(l_0, \kappa) + n_0 M_2(l_0, \kappa) \tag{18.72}$$

We note in particular that we obtain the correct boundary values $n = n_0, \lambda = \lambda_0$ when $\kappa = l_0$ because the functions $M_1 = M_3 = 1$ and $M_2 = M_0 = 0$ when the arguments coincide, according to their definitions (Eq. (18.53)). We may then write a simple formula for the total function P in terms of the boundary-value distribution F:

$$P(n, \lambda; \kappa)$$
$$= \int d\lambda_0 dn_0 F(n_0, \lambda_0) \delta(n - n_0 M_1 + \lambda_0 M_0) \delta(\lambda - \lambda_0 M_3 + n_0 M_2). \quad (18.73)$$

The integral is easily solvable, once again using the properties of the functions M_j:

$$P(n, \lambda, \kappa) = F(nM_3 - \lambda M_0, \ \lambda M_1 - nM_2) \qquad (18.74)$$

It is useful to convince oneself that in this way *the properties of the M-functions lead to a transfer property for the solutions* (which, of course, is just the content of the original partial differential equation). By this we mean that if the distribution F at l_0 is described in terms of the distribution G at another $\kappa = l_1 \geq l_0$ then the formulas are identical if we exchange F for G and the argument l_0 for l_1. The changes in the system are all the time evidently occurring in the average way, according to the relevant average values of λ and n, as we can see from Eqs. (18.53) and (18.70).

The conclusion is that in this approximation the ensemble of states described by F at $\kappa = l_0$ will move on towards different (n, λ)-values at other κ-values *but a constant distribution occurs along the rays described by Eq. (18.72)*. The system corresponds to a hamiltonian flow in a space where the coordinate corresponds to λ and the momentum corresponds to the dipole multiplicity n and the hamiltonian is

$$H = +\frac{n^2}{2} - \frac{\bar{\alpha}\lambda^2}{2} \qquad (18.75)$$

At first sight the function P would seem to provide a possible tool to investigate the running of the QCD coupling constant. Thus it is possible to concentrate on an event sample in which each event contains a number of jets (related to n), the combined logarithmic phase space (related to λ) having a certain cut in the transverse jet energies (related to κ). Then one would continue with the same event sample using smaller transverse jet energy cuts and study the changes in the distribution in the number of jets and phase-space size. To perform a reliable such comparison it is, however, necessary to make a better approximation than $P(n) - P(n-1) = \partial P/\partial n$ and also to correct for recoils along the cascade.

In order to investigate the difference between a continuous and a discrete treatment of the multiplicity we will briefly consider a simplified model.

Suppose that we consider e.g. a population of bacteria and assume that as time t passes the population increases at a constant rate γ. This leads to the following differential equation for the bacteria multiplicity $P(n,t)$:

$$\frac{dP}{dt}(n,t) = \gamma[P(n-1,t) - P(n,t)] \tag{18.76}$$

i.e. there is a gain per time unit γ from $n-1$ and a corresponding loss to $n+1$. Using the generating function $\mathscr{P}(z,t) = \sum P(n,t)z^n$ we obtain the differential equation

$$\frac{d\mathscr{P}}{dt} = \gamma(z-1)\mathscr{P} \tag{18.77}$$

with the obvious solution $\mathscr{P} = z\exp[\gamma(z-1)t]$ if we start with a single bacterium at $t = 0$. Expanding the generating function we obtain as expected a Poissonian distribution (truncated at $n = 1$):

$$P(n,t) = \frac{(\gamma t)^{n-1}}{(n-1)!}\exp(-\gamma t), \quad n \geq 1 \tag{18.78}$$

If we use the continuous approximation, i.e. put $P(n) - P(n-1) = \partial P/\partial n$, we obtain by the method of rays that $n(t) = \gamma t + 1$, which obviously corresponds to the mean value of the distribution in Eq. (18.78). Although the mean value is the only size parameter in a Poissonian distribution it is obviously not a good approximation to write the distribution in Eq. (18.78) as a δ-distribution and neglect the width. It is, however, possible to include the width if we also take the second derivative into account. In this simple model we obtain

$$\frac{\partial \mathscr{P}}{\partial t} + \gamma\frac{\partial \mathscr{P}}{\partial n} = \frac{\gamma}{2}\frac{\partial^2 \mathscr{P}}{\partial n^2} \tag{18.79}$$

This corresponds to a diffusion equation and we are then very close to the considerations in Chapter 10 on the Brownian motion in impact space in multiperipheral ladder diagrams and in Chapter 12 on the transverse momentum generation. We obtain immediately the well-known (normalised) gaussian solution,

$$\frac{1}{\sqrt{2\pi\gamma t}}\exp\left[-\frac{(n-1-\gamma t)^2}{2\gamma t}\right] \tag{18.80}$$

and this coincides (as it should of course) with a stationary-phase approximation to the Poissonian distribution in Eq. (18.78). It is useful to carry through the calculations, using the Stirling approximation to the factorial.

We conclude that the QCD cascade evolutions do not only correspond to simple 'laminar flow along the mean streamlines', which is what Eq. (18.73) implies, if we use a hydrodynamical analogy. There is also diffusion among the streamlines because of the discreteness in the dipole multiplicity. When we take the second derivative into account we obtain what is known as the Fokker-Planck equations and there is a very general mathematical method to treat equations of this kind. We have already discussed the Langevin equation in Chapter 12 and we may apply it again, this time coupled to the ray equations which we obtained in the approximate treatment in connection with Eq. (18.69). If we again introduce a gaussian noise term R with properties in accordance with Eq. (12.19) we may write the following system of coupled stochastical equations:

$$\frac{d\lambda}{d\kappa} = -n$$
$$\frac{dn}{d\kappa} = -\bar{a}\lambda + R\sqrt{\bar{a}\lambda/2} \tag{18.81}$$

The square root in front of the gaussian noise term is this time not a constant but it is nevertheless possible to prove that the resulting equations actually converge to the solution of the equation

$$\frac{\partial P}{\partial \kappa} - n\frac{\partial P}{\partial \lambda} = \frac{\alpha_0}{\kappa}\lambda\left(\frac{\partial P}{\partial n} - \frac{1}{2}\frac{\partial^2 P}{\partial \kappa^2}\right) \tag{18.82}$$

(note that the variable κ decreases!). We will, however, end the investigation at this point.

18.4   The next-to-leading-order corrections

In order to obtain a better approximation it is not only necessary to go back to the discrete n-distribution, but it is also necessary to improve upon the master equation. We note in particular that we have up to now considered the phase space to be given by the triangular approximation

$$|y| \leq \log(\sqrt{s}/k_\perp) \quad \text{instead of} \quad k_\perp \cosh y \leq \frac{\sqrt{s}}{2} \tag{18.83}$$

which would e.g. imply that the maximum $k_{\perp max}^2 = s/4$ instead of s. We have also neglected the polarisation sums in the cross section, i.e. that the emission of a g from a $q\bar{q}$ should be weighted with the factor $x_1^2 + x_3^2$. This is again a factor which starts to play a role close to the triangle boundary, because at the boundary either x_1 or x_3 is small. The polarisation sum has all the time been approximated by 2 and to check on the approximation we will consider the integral

$$I \equiv \int_{y_{min}}^{y_{max}} dy(x_1^2 + x_3^2) \tag{18.84}$$

for a fixed value of $k_\perp$. According to our approximation $I = 2\log(s/k_\perp^2)$ but if we introduce (cf. the definitions in Eq. (16.39))

$$x_1 = 1 - k_\perp \exp(-y)/\sqrt{s}, \quad x_3 = 1 - k_\perp \exp y/\sqrt{s} \quad (18.85)$$

then we obtain a correction term, i.e.

$$I = 2[\log(s/k_\perp^2) - \delta_q]$$

$$\delta_q = \frac{3}{2}\sqrt{1 - \frac{4k_\perp^2}{s}} - 2\log\left[\frac{1}{2}\left(1 + \sqrt{1 - \frac{4k_\perp^2}{s}}\right)\right]$$

(18.86)

It turns out that the correction term varies very little as a function of $k_\perp$ and that if we cut off a strip $c_q/2 = 3/4$ on each side of the triangle we obtain a very good approximation to the suppression from the polarisation sum as well as to the neglected hyperbolic cutoff, Eq. (18.83).

It is possible [72] to subdivide the phase space for gluon emission into regions relevant to $q\bar{q}$-, qg-, $g\bar{q}$- and gg-dipoles. They all have different polarisation sums according to the dipole cascade model, Eq. (17.2), and for all of them one can calculate the decrease in the triangles. The corresponding decrease factor for gluon corners is e.g. $c_g = 11/6$. It is also possible to include gluon splitting as implemented in the dipole cascade model [10] and finally also to take into account the effective coupling constants, $3\alpha_s/2\pi$ and $4\alpha_s/3\pi$, respectively.

The result is a rather complex set of equations, [72], which are close to the so-called modified leading-log approximation in QCD, [52]. Using the correction terms in the L-method the result becomes a very good approximation to the Monte Carlo results for the multiplicity moments.

We will not go into the details but will briefly consider the simplest and actually also the largest correction to the master equation, Eq. (18.68). If we decrease all the n dipoles by a common factor c then we obtain

$$\frac{\partial P}{\partial \kappa} - n\frac{\partial P}{\partial \mu} = \frac{\alpha_0}{\kappa}\{(\lambda - nc)P(n, \lambda, \kappa) - [\lambda - (n-1)c]P(n-1, \kappa, \lambda)\}.$$

(18.87)

From this equation we can easily calculate the equations connecting the (modified) mean multiplicity M_{m1} and mean λ-phase space size, M_{m2}. We obtain

$$\frac{dM_{m1}}{d\kappa} = -\frac{\alpha_0}{\kappa}(M_{m2} - cM_{m1}), \quad \frac{dM_{m2}}{d\kappa} = -M_{m1} \quad (18.88)$$

which should be compared with Eqs. (18.53). There is then a second-order differential equation for M_{m2}:

$$\frac{d^2M_{m2}}{d\kappa^2} = c\frac{\alpha_0}{\kappa}\frac{dM_{m2}}{d\kappa} + \frac{\alpha_0}{\kappa}M_{m2} \quad (18.89)$$

This equation again has solutions in terms of the modified Bessel functions. We may define the functions

$$\mathscr{I}_\beta(\kappa) = \sqrt{2}\kappa^{\beta/2}I_\beta(2\sqrt{\alpha_0\kappa})$$

$$\frac{d\mathscr{I}_\beta}{d\kappa} \equiv \mathscr{I}_{\beta-1} = \sqrt{2\alpha_0}\kappa^{(\beta-1)/2}I_{\beta-1}(2\sqrt{\alpha_0\kappa}) \tag{18.90}$$

for $\beta = 1 + \alpha_0 c$ (we note the similarities to the functions $\mathscr{I}_j$, $j = 0, 1$, in Eqs. (18.21)) and similarly the functions $\mathscr{K}_\beta$ with respect to the modified Bessel functions K, with a minus sign in the derivative. Then the solution for M_2 is

$$M_{m2} = L^{1-\beta}[\mathscr{I}_\beta(L)\mathscr{K}_\beta(\kappa) - \mathscr{I}_\beta(\kappa)\mathscr{K}_\beta(L)]$$

$$M_{m1} = L^{1-\beta}[\mathscr{I}_\beta(L)\mathscr{K}_{\beta-1}(\kappa) + \mathscr{I}_{\beta-1}(\kappa)\mathscr{K}_\beta(L)] \tag{18.91}$$

if we start at $\kappa = L$ with $M_{m2} = 0$ and $M_{m1} = 1$. The asymptotic behaviour for $L \gg \kappa \gg 0$ is

$$M_{m2} \sim \left(\frac{L}{\kappa}\right)^{-\alpha_0 c/2} (L\kappa)^{1/4}\exp[2(\sqrt{\alpha_0 L} - \sqrt{\alpha_0\kappa})] \tag{18.92}$$

Evidently the introduction of a factor that diminishes the phase space is reflected directly in the power in front of the exponents, which has changed from $1/4$ to $1/4 - \alpha_0 c/2$ as compared to Eqs. (18.17) and (18.24). This factor directly reduces the L-dependence and increases the κ-dependence. The result is an even slower development of the cascades than in the LLA, described earlier.

18.5   On the running coupling in QCD

The equations for the λ- and n- variations with κ have a great similarity to the Callan-Symanzik equations, considered in Chapter 4. The background is, however, completely different. Equations (18.68) and (18.87) are derived from classical probability concepts. They are really classical gain-loss equations and there is not the renormalisation group background which may motivate a relationship to a Callan-Symanzik investigation. We have changed the scale from κ to $\kappa - d\kappa$ and then there is in this step a possibility that the number of dipoles n increases to $n + 1$, which for the equations mentioned above corresponds to a loss. But there is also a gain if the state changes from $n - 1$ to n.

The probability is governed by the relevant coupling, which mostly corresponds to the situation in gluon dipoles and therefore equals $3\alpha_s/2\pi$. This effective coupling is multiplied by the available phase space λ and the probability of finding just this number of dipoles at λ. The running of the coupling is introduced by hand, i.e. we use

$$\alpha_s = \left(\frac{12\pi}{33 - 2n_f}\right)\frac{1}{\kappa} \tag{18.93}$$

Now suppose that (without any further motivation) we introduce the running of the coupling in our equations along the same lines as in the Callan-Symanzik equations. That would mean that we would have an accompanying scale change of $\partial/\partial\kappa$ to $\partial/\partial\kappa + \beta\partial/\partial\alpha_s$ in the equations (at the same time, of course, we leave out the running of α_s in Eq. (18.93) and introduce this property through the β-dependence).

This means in the language of gain-loss equations that we introduce a loss term which looks like

$$-\beta(\alpha_s)\frac{\partial\mathscr{P}}{\partial\alpha_s} = \left(\frac{11}{6}\frac{3\alpha_s}{2\pi} - \frac{2}{3}\frac{n_f\alpha_s}{4\pi}\right)\alpha_s\frac{\partial\mathscr{P}}{\partial\alpha_s}$$
$$\equiv (\delta y_g \alpha_{effg} - \delta y_q \alpha_{effq})\mathscr{N}(\mathscr{P}) \tag{18.94}$$

The results in the first line correspond to straightforward algebra, using the expressions for the β-function and the running coupling constant of QCD. The symbol $\mathscr{N}(\mathscr{P})$ indicates a number operator in the sense that for any well-behaved function $f = f(\alpha)$ we obtain an average n with

$$f = \sum \alpha^n a_n, \quad \alpha\frac{\partial f}{\partial\alpha_s} = \sum n\alpha^n a_n \equiv \mathscr{N}(f) \tag{18.95}$$

The two effective couplings

$$\alpha_{effg} = \frac{3\alpha_s}{2\pi}, \quad \alpha_{effq} = \frac{n_f\alpha_s}{4\pi} \tag{18.96}$$

correspond to the gluon emission process $g \to gg$ and the gluon splitting process $g \to q\bar{q}$, respectively. Finally the two quantities δy are, respectively,

$$\delta y_g = \frac{11}{6} \to 2\log(s/k_\perp^2) - \int_{y_{min}}^{y_{max}} dy(x_1^3 + x_3^3)$$
$$\delta y_q = \frac{2}{3} = \int_0^1 dz[z^2 + (1-z)^2] \tag{18.97}$$

We met the quantity δy_g above as the decrease in phase space for emission of new gluons in the neighborhood of a gluon corner. It can most reasonably be considered as *a typical collinearity size in rapidity* for a fixed value of $k_\perp$ in the limit when $s \to \infty$.

The quantity δy_q, however, is for a fixed $k_\perp$ and the same limit in s the total rapidity phase space for a gluon to split into a $q\bar{q}$-pair.

Therefore such a variation added to the coupling constant in the master equation, and for that matter to any of the partial differential equations used in the κ-method, may be considered as

- a loss term proportional to the probability that the gluon fluctuates into a gg-pair within a small collinearity region. Intuitively that region should be equal to the region in phase space where such virtual fluctuations may occur in this approximation;

- a gain term proportional to the probability that the gluon may fluctuate into a $q\bar{q}$-pair. The rapidity size allowed for such a fluctuation is essentially smaller because there is no pole in z, which means that it exponentially falls off in rapidity from the gluon corner.

This interpretation of the running coupling constant is certainly somewhat imaginative, in particular the interpretation that the collinearity region is equal to the region of virtual fluctuations for the gluon.

We nevertheless note that if a gluon fluctuates into a gg-pair then there is a color flow across the region. But if it fluctuates into a $q\bar{q}$-pair then the color flow is broken over the corresponding region. Thus a gluon is never in the first case able to get away from the influence of its own Coulomb color field, but in the second case there is nothing to stop it moving around as a free asymptotic pair.

Another way to understand the gain-loss nature of the gluon emission and gluon splitting, respectively, is to consider the case when a gluon just above κ decays into two gluons (which then are counted at κ). Just below κ, however, the two gluons are reabsorbed into a single gluon again, thereby causing a loss in multiplicity. But if the gluon instead splits into a $q\bar{q}$-pair there is no gluon at κ, but if the pair reassembles to a gluon there will then be a gain in the gluon multiplicity. Note that the possibility of a loss term only occurs in a nonabelian gauge theory, because the abelian photons do not interact.

18.6   Discrete QCD, another approximation method

1   The method

We will in this section make explicit use of the properties of the running coupling, discussed in section 18.5, to present another analytical approximation method for the perturbative QCD parton cascades. It is called *discrete QCD*, [15], for reasons easily understood when it is demonstrated. To that end we start by using the Webber-Marchesini method to search through the triangular phase space, cf. Section 17.7.

This means that we will take steps in the rapidity y and consider the probability of obtaining a gluon emission in each step. We will in particular choose to make these steps *finite* in size, and equal to δ. Consider as an illustration the rapidity bin $\delta y_1 \equiv \delta$ around y_1 in Fig. 18.1 and assume that there is one emission for κ and none between κ and the maximum $\kappa_{max} \equiv \ell_1$ in this bin. The resulting probability is given by (note, see subsection 3 of section 17.3, the Sudakov factor!)

$$dP(\kappa, \ell_1) = \frac{(\alpha_0 \delta) d\kappa}{\kappa} \exp(-\alpha_0 \delta) \int_{\kappa}^{\ell_1} \frac{d\kappa'}{\kappa'} = d\kappa (d\kappa^{d-1} \ell_1^{-d}) \quad (18.98)$$

with $d = \alpha_0 \delta$. As yet we have not decided upon the size of δ but it is perfectly feasible to choose

$$\delta = \frac{1}{\alpha_0} \simeq \delta y_g = \frac{11}{6} \quad (18.99)$$

Here we have made the approximation of neglecting the contribution from the gluon splitting into $q\bar{q}$-pairs, or equivalently we have put the number of flavors $n_f = 0$ in the running coupling, cf. Eq. (18.97). In this way the power d in Eq. (18.98) becomes 1 and we obtain the simple result that *there is no κ-dependence left* in the probability

$$dP(\kappa, \ell_1) = \frac{d\kappa}{\ell_1} \quad (18.100)$$

This evidently goes for all steps of size $\delta \simeq \delta y_g$ and in particular for a (discrete) value of $L = \log(s/s_0) = 2N\delta$ the whole 'original' dipole phase-space triangle will contain $2N$ possible emissions.

We now make the further assumption that inside each δ-bin in rapidity the κ-variable is discretised so that $\ell_1 = n_1(2\delta)$ with n_1 an integer (the index 1 means the bin with the height ℓ_1). The result is that there may be any n-emission, $1 \le n \le n_1$, i.e. an emission in the κ-box n, with the same probability $P_n(n_1)$ for all integer n, where

$$P_n(n_1) = \frac{1}{n_1} \quad (18.101)$$

Each emission will produce a subtriangle, cf. section 17.5, which again contains $2n$ discrete steps in rapidity along the projecting folds. This means n steps on each of the two sides, see Fig. 18.1. The whole procedure can evidently be continued with new discretised (sub-)triangular folds, projecting folds, and so on. It is easy to convince oneself that (apart from the very outskirts of the phase-space triangles) the construction is consistent and that the above-mentioned κ-boxes fit in. Note that it is necessary to take 2δ-steps in κ for this consistency!

So what is the physics behind this seemingly simple but up to now purely mathematical scheme? We are actually doing exactly what the running coupling in QCD indicates to us, i.e. we are sending out the gluons with a distance at least $\delta y_g = 11/6$ between them. If they are closer than this distance in rapidity then they will in practice be reabsorbed, i.e. with this method we emit *effective gluons*, which are not reabsorbed in the next step in perturbation theory!

This is also an instructive example of the difference between *exclusive distributions*, with a probability normalised to one event having certain properties and *inclusive distributions*, which instead correspond to the average behaviour of many events. Therefore the results should in the latter case rather be called densities. The running coupling is characteristic for the inclusive density of gluons, but if we concentrate on the largest gluon then the probability is the same constant for all emissions!

There are two regions which need particular attention. The first is the region corresponding to $n = 1$, i.e. the lowest κ-box in each δy_g-step. We will interpret this box to correspond to no emission, i.e. the effective gluons in such a box are too soft to be noticeable. This means that we are actually bringing in a precise cutoff in the cascades with respect to κ, i.e. all 'observable' gluons have $\kappa \geq 2\delta y_g$.

The second comprises the regions close to the triangular border in each phase-space triangle, where the boxes are distorted. We have actually already discussed these regions repeatedly because the triangular border corresponds to gluons collinear with the parton emitting them. Just as in connection with the modified leading-log approximation, e.g. in Eq. (18.87), we will cut away a region equal to $\delta y_g/2$ along each border of the triangles. (We note that we are in this way making a small error for dipoles containing a q- or $\bar{q}$-particle as endpoint. For this case we should cut away the slice $\delta y_q/2 = 3/4$, which is less than the value of $\delta y_g/2$ according to Eq. (18.86)).

2   Some results in discrete QCD

We will now consider in some detail the structure of the scheme developed in the last subsection. The original dipole triangle will be called an *N-forest* (actually containing $2N$ bins) because in each δy_g rapidity bin in the original dipole there will be a tree-like structure, which we will call an n_1-*tree* (with the maximum κ in the bin equal to $n_1 2\delta y_g$).

Such an n_1-tree will, however, not necessarily have height n_1. But it will have two sides, each with length $n\delta y_g$ if there is an emission at $\kappa = n2\delta y_g$ (with $n \leq n_1$). This is illustrated in Fig. 18.5, which corresponds to the triangular phase space with its projecting folds looked upon from below.

The two sides of the tree correspond to the two sides of the projecting emission triangle according to the construction in subsection 17.5 (remember that each fold has length $n\delta y_g$). We will call this a *true n-tree*. We note that in this way the n_1-tree is a true n-tree, with each n occurring, according to Eq. (18.101), with the same probability, i.e. $1/n_1$.

It is worthwhile noting the following two features.

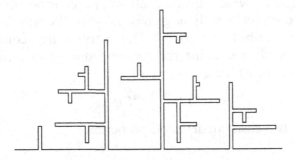

Fig. 18.5. The projecting folds of the phase-space triangle together with a set of subfolds and subsubfolds, etc.

- The two sides of the true tree correspond to the two color flows at a gluon corner. A gluon with e.g. the color combination r$\bar{\text{g}}$ contains one r-color flowing towards the gluon tip and one g-color starting at the gluon tip and flowing 'inwards' (being an anticolor its flow is oppositely directed to the color). The connection point between them is the 'top' of the true tree and they are each color-connected to partons on both sides of the emitted gluon.

It may seem for the projecting triangular fold in e.g. Fig. 18.1 that the two parts of the folds adjacent to the original background triangle are close in the rapidity variable of this original dipole. This is not true; they are actually far apart in true rapidity and in particular also distant with respect to color flow. The distance in true rapidity is just the distance from the bottom of the true n-tree up to the top and down again i.e. $n2\delta y_g$.

- A true n-tree actually has the same structure as an n-forest defined above (to see this cut it up along the centre line, i.e. along the gluon corner, see Fig. 18.5). Therefore all statements for one can be taken over to the other.

Each side of a true n-tree is now subdivided into n bins, containing 'n'-trees' (in realistic tree-language it may possibly be better to refer to them as branches) with $n' = 1, \ldots, n$. Each such tree can be treated just as the first one, i.e. it will contain a true tree with two sides each of a length at most equal to the length of the original tree. Further, all integer values have the same probability according to Eq. (18.101). Everything ends at the place where the probability is just 1 of obtaining a 1-tree everywhere, i.e. the situation which we have defined above to correspond to no further gluons.

As an example of what we can easily obtain consider the probability that an n-tree contains exactly m gluons, $P(m;n)$. We will also introduce the probability distribution, $P^t(m;n)$, that a true n-tree contains exactly m gluons and finally we define the corresponding generating functions, $\mathscr{P}(z;n)$ and $\mathscr{P}^t(z;n)$, with e.g.

$$\mathscr{P}(z;n) = \sum z^m P(m;n) \qquad (18.102)$$

Then there are two easily understood properties.

1 An $(n+1)$-tree is either a true $(n+1)$-tree with probability $x_{n+1} = 1/(n+1)$ or else an n-tree with probability $y_{n+1} = 1 - x_{n+1} = n/(n+1)$, i.e.

$$\mathscr{P}^t(z;n+1) = \frac{\mathscr{P}(z;n+1) - y_{n+1}\mathscr{P}(z;n)}{x_{n+1}}$$
$$= (n+1)\mathscr{P}(z;n+1) - n\mathscr{P}(z;n) \qquad (18.103)$$

2 The difference between a true $(n+1)$-tree and a true n-tree is that one gets in the first case contributions also from the two largest subtrees, i.e. one n-tree on each side:

$$P^t(m;n+1) = \sum_{\sum m_j = m} P(m_1;n)P^t(m_2;n)P(m_3;n) \qquad (18.104)$$

or equivalently in terms of the generating functions

$$\mathscr{P}^t(z;n+1) = [\mathscr{P}(z;n)]^2 \mathscr{P}^t(z;n) \qquad (18.105)$$

It is as always necessary to supply the boundary conditions and in this case we obtain easily $\mathscr{P}^t(z;1) \equiv \mathscr{P}(z;1) = 1$ and $\mathscr{P}(z;2) = (1+z)/2$. It is then straightforward to construct all the distributions.

Actually the formulas we have derived are discretised versions of the L-method formulas described in subsection 1 (although we have written them for the multiplicities and not for the λ-measure as there). This is shown in [15] and in this reference there are also a number of other applications mentioned, i.e. the λ-measure distributions, the combined distributions of the λ-measure and the multiplicities, how to go to the formulas of the κ-method and finally also a way to translate the results from this lattice description in an abstract space to the observable energy-momentum vectors of the emitted gluons.

As an example we consider the average multiplicity for an N-tree, $\bar{n}(N)$. This can as usual be obtained from the generating function, $\mathscr{P}(z;N)$, by means of the first derivative evaluated for $z = 1$; note that $\mathscr{P}(z=1;n) = \mathscr{P}^t(z=1;n) = 1$ always. We obtain from Eqs. (18.103) and (18.105) after some small manipulations:

$$\bar{n}(N+1) - 2\bar{n}(N) + \bar{n}(N-1) = \frac{2}{N+1}\bar{n}(N-1) \qquad (18.106)$$

which as expected is a discretised version of e.g. Eqs. (18.54) and (18.89). The result is a very good description of the jet multiplicities as obtained from Monte Carlo simulations by ARIADNE, see [15].

We end with a few remarks on the results of discrete QCD:

- Despite the discretisation approximation the resulting formulas for the jet multiplicities and hadron multiplicities are even better approximations to the Monte Carlo simulation results (including all the kinematics) than any modified leading-log result.

This may seem surprising but it is related to the fact that the 'reabsorption' length in rapidity we have introduced, i.e. δy_g, is also a good description of the rapidity region where the recoils from earlier emissions are noticeable.

- It turns out that if we consider the zeroes of the generating function, i.e. the values $z_j(n)$ with

$$\mathscr{P}(z; n) = A_n \prod [z - z_j(n)] \qquad (18.107)$$

where A_n is a suitable constant, then these quantities exhibit some surprising properties. They are all finite and stay in a region close to the origin in a structure called a Julia set by the mathematicians. Its properties are just as beautiful as those of the nowadays well-known Mandelbaum set. Each root of the generating function will according to Eq. (18.105) change into three when $n \rightarrow n+1$ and they are also closely located, albeit occurring in an irregular fashion nowadays known as a fractal curve.

Instead of going into the details (which are still under investigation) we will exhibit these irregularities in a different way in the next section.

It is finally of some interest to connect the formulas obtained in the discrete QCD model to the ordering procedure used in the dipole cascade model, i.e. to investigate how the particular basic property of discrete QCD, occurring in Eqs. (18.98) to (18.101), will come out of our ordinary treatment of the Sudakov factor. Then we return to Eq. (18.59) (cf. also Eq. (18.49)), which describes the first decay of a dipole with logarithmic squared mass L into two dipoles by a gluon emission at κ_1:

$$dP_1 = \frac{\alpha_0}{\kappa_1} d\kappa_1 (L - \kappa_1) f_{Sud}(L, \kappa_1) \qquad (18.108)$$

This formula is valid in the leading-log approximation and it is easy to extend it to the modified leading-log approximation by the exchange $L \to L - 1/\alpha_0$ (assuming that we neglect the flavored dipoles). This means that the rapidity space factor is changed from $L - \kappa_1$ to $L - \kappa_1 - 1/\alpha_0$ and the Sudakov factor will change to $f_{Sud}(L - 1/\alpha_0, \kappa_1)$. We now introduce the variables $2\ell \equiv \alpha_0 L$ and $2k_1 \equiv \alpha_0 \kappa_1$ and perform the Sudakov integral, to obtain in the modified leading-log approximation

$$
\begin{aligned}
f_{Sud}(L - 1/\alpha_0, \kappa_1) &\simeq \frac{k_1^{2\ell - 2k_1} \Gamma^2(k_1)}{\Gamma^2(\ell)} \\
dP_1 &= dk_1 [2(\ell - k_1) - 1] \frac{k_1^{2\ell - 2k_1 - 1} \Gamma^2(k_1)}{\Gamma(\ell)^2}
\end{aligned}
\tag{18.109}
$$

We have used the Stirling approximation for the Γ-functions:

$$
\Gamma(\ell) \simeq C \exp[(\ell - 1/2) \log \ell - \ell]
\tag{18.110}
$$

with C a normalisation constant.

In order to understand the result we return to the distributions of discrete QCD and interpret ℓ, k_1 as integers.

Consider the probability of emitting no gluon in an ℓ-forest with a 'tree-height' above k_1. Then for a j-tree with $j \le k_1$ the probability is obviously 1 because there can be no true k_1-tree in this case. For a $(k_1 + m)$-tree, when the probability is $1/(k_1 + m)$ for each integer 'height', there are m possibilities for making a true tree above k_1 and thus the probability of using none of them is $1 - m/(k_1 + m) = k_1/(k_1 + m)$.

The probability dP_1/dk in Eq. (18.109) contains two factors. The first corresponds to the number of ways that one can choose any one of the 'central' integer y-bins of height $k_1 + 1, \ldots, \ell - 1, \ell - 1, \ldots, k_1 + 1$. The second is the probability of making an effective gluon at k_1 in one of these bins and only gluons below k_1 in the rest of the bins.

The observant reader will note that there seems to be a mismatch, i.e. there is a factor $2(\ell - k_1) - 1$ for the number of bins but only $2(\ell - k_1) - 2$ central integers. A closer examination tells us, however, that the two k_1-bins, one on each side, should be incorporated in the possibility of making a true k_1-tree. Due to the triangular shape, however, the surface related to them is only half the surface related to those called central. Therefore the problem is solved if we incorporate them with unit probability and phase space size $1/2$ each.

Next we consider a second gluon emission at $\kappa_2 < \kappa_1 < L$. It is straightforward to prove (and the reader is strongly invited to think it through, in particular the factorisation property of the Sudakov factors!) that with proper Sudakov factors this probability is, in the modified leading-log approximation,

$$dP_2 = (\kappa_0 - \kappa_1 - 1/\alpha_0)(\kappa_0 + \kappa_1 - 2\kappa_2 - 2/\alpha_0)$$
$$\times \prod_{j=1}^{2} d\kappa_j \frac{\alpha_0}{\kappa_j} f_{Sud}(\kappa_{j-1} - 1/\alpha_0, \kappa_2) \quad . \tag{18.111}$$

where we have introduced the symmetric notation $L \equiv \kappa_0$. The two factors in front of the product sign correspond to the size of the original dipole when it decays at κ_1 and the sum of the sizes of the two emerging dipoles at κ_2. We again leave it to the reader to prove in terms of the $k_j = \alpha_0 \kappa_j/2$ variables that Eq. (18.111) can be written as

$$dP_2 = (2k_0 - 2k_1 - 1)dk_1(2k_0 + 2k_1 - 2k_2 - 2)dk_2$$
$$\times \prod_{j=0}^{1} \frac{\Gamma^2(k_2)k_2^{2k_j - 2k_2}}{\Gamma^2(k_j)} \prod_{j=1}^{2} \frac{1}{k_j} \tag{18.112}$$

and from this result it is obvious how to generalise the formulas to any number of gluon emissions. In terms of the notions of discrete QCD we conclude that if n gluon folds are produced at the integers $k_1, \ldots, k_n$ in an original k_0-forest then we have the following.

- For the emission of a gluon at $\kappa_j \equiv 2k_j/\alpha_0$ one should multiply by the number of possibilities for choosing a (generalised) rapidity bin (including the outer two with a common size 1). This is equal to the sum of all the virtual dipole sizes, i.e. to the size of the λ-measure at the virtuality κ_j, including the modified leading-log correction. For the case $n = 2$ this corresponds to the first two terms in Eq. (18.112).

- For the triangles that correspond to such a gluon emission one should multiply by the probability of making no true tree above k_n. This is the first product in Eq. (18.112)

- The running coupling, which in this notation corresponds to the last product term, then contains the probability of making a true k_j-tree in the central bin of each triangular fold, i.e. of the gluon being produced at $\kappa_j = 2k_j/\alpha_0$ (note that there is no $1/k_0$-factor and also note that the last factor $1/k_n$ can be interpreted as discussed in connection with the first gluon emission in Eq. (18.109). The emitted gluons can evidently be attached at any integer value along the relevant λ-size.

If this result is integrated over all the κ_j-variables we actually obtain a general formula for the distribution $P(\lambda, n, \kappa)$. It is a solution to the differential equation in Eq. (18.87) (with the parameter $c = 1/\alpha_0$), as the reader can readily verify by iteration from $n = 1$ upwards.

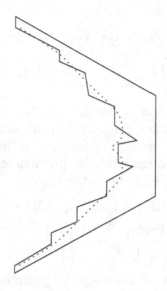

Fig. 18.6. The particles produced in connection with the fragmentation in Fig. 15.17 are redrawn along the directrix, together with the two connected hyperbolas (dotted).

18.7 The *x*-curve and an infrared-stable λ-measure

1 *Definitions*

An undesirable feature in the present definition of the generalised rapidity space region λ is that it can only be defined as long as the dipole (squared) masses are above a certain value s_0. We will in this section introduce an infrared-stable definition of λ and also a well-defined curve, [20], the *x*-curve, which describes the average energy-momentum-space behaviour of the final-state particles.

In order to understand the idea behind the *x*-curve consider Fig. 18.6. In this picture the yoyo-hadrons, which are produced around the two hyperbolas in Fig. 15.12, are drawn instead as a series of connected line-segments close to the two hyperbolas spanned along the directrix. They are of course the same hadrons but while their production points are emphasised in Fig. 15.17 it is instead their energy-momentum vectors which play the role of connectors in Fig. 18.6. The curve exhibited in this way is (approximately) the *x*-curve.

For a mathematical description we define a function $T(\xi) \equiv \exp[\lambda(\xi)]$ and a vector $q(\xi)$ along the directrix $A(\xi)$ by means of differential equations (we imagine that the directrix is parametrised by the parameter ξ and that it is differentiable so that dA has a meaning as a four-vector):

$$dT = \frac{(qdA)T}{m_0^2}, \quad dq = dA - \frac{(qdA)}{m_0^2}q \qquad (18.113)$$

For the vector q we obtain formally, with boundary value $q(\xi = 0) = 0$, that it is a weighted mean of the partonic energy-momentum vectors which describe the directrix according to e.g. subsection 17.6:

$$q(\xi) = \frac{1}{T(\xi)} \int_0^\xi dA(\xi')T(\xi') \qquad (18.114)$$

Similarly we obtain for T with the boundary value $T(\xi = 0) = 1$ that it is the exponent of an area:

$$T(\xi) = \exp\left(\frac{1}{m_0^2} \int_0^\xi q(\xi')dA(\xi')\right) \qquad (18.115)$$

Note that the area element spanned by the vectors q and (the lightlike) dA is $d\Sigma = \sqrt{(qdA)^2 - q^2dA^2} = qdA$. If we multiply the second of equations (18.113) by q we also find that q becomes timelike and its invariant length quickly approaches the value m_0:

$$dq^2 = 2\left(1 - \frac{q^2}{m_0^2}\right)qdA \quad \Rightarrow \quad q^2(\xi) = m_0^2\left[1 - T^{-2}(\xi)\right] \qquad (18.116)$$

If we introduce the case when the directrix is built up by finite lightlike parton energy-momentum vectors, then we can construct the quantities T and q recursively by

$$T_{j+1} = \frac{T_j}{\gamma_{j+1}}, \quad q_{j+1} = \gamma_{j+1}q_j + \tfrac{1}{2}(1 + \gamma_{j+1})k_{j+1} \qquad (18.117)$$

$$\gamma_{j+1} = \frac{1}{1 + (q_j k_{j+1})/m_0^2}$$

In this way we obtain

$$q_0 = 0, \quad q_1 = k_1, \quad q_2 = \frac{k_2}{2} + \left(k_1 + \frac{k_2}{2}\right) \Big/ \left(1 + \frac{k_1 k_2}{m_0^2}\right) \qquad (18.118)$$

etc. Similarly for T we have

$$T_0 = 1, \quad T_1 = 1 + \frac{s_{12}}{2m_0^2}, \quad T_2 = 1 + 2\left(\frac{s_{123}}{4m_0^2} + \frac{s_{12}s_{23}}{16m_0^4}\right) \qquad (18.119)$$

etc. The largest power in $T = T_n$ always has the generic form

$$2\frac{s_{12}}{4m_0^2}\frac{s_{23}}{4m_0^2}\cdots\frac{s_{n,n+1}}{4m_0^2} \qquad (18.120)$$

This means that

T1 in general $\log T$ is a good approximation to the λ-measure and the parameter $\log(4m_0^2)$ corresponds to the virtuality κ or the resolution power;

T2 if any of the partons become collinear or soft then the next-order term in T will take over so that λ, defined in this way, is infrared stable;

T3 the result in Eq. (18.115) that $\log T$ is the area between the x-curve and the directrix provides an intuitive understanding of the relationship between the fragmentation process and the partonic state as described by the directrix. *The string state has a (mean local) lifetime proportional to the region between the directrix and the x-curve.*

It is possible to find a solution for the vector q, which we will call $\hat{q}$, which is periodic in the same sense as the directrix is periodic. This means that $\hat{q}_j = \hat{q}_{j+2(n+2)}$. In this case $\hat{q}^2 = m_0^2$. The vector $\hat{q}$ can to a good approximation be constructed by iterating Eqs. (18.117) a few periods around the string directrix.

We will from now on only work with this periodic q-vector function and therefore we drop the circumflex notation. The x-curve is then defined in terms of this periodic q as

$$x(\xi) = A(\xi) - q(\xi) \tag{18.121}$$

It is not difficult to see that with this definition the x-curve is everywhere a timelike curve in the sense that its tangent is everywhere timelike.

From Eq. (18.121) we obtain that the vector q is the tangent of the x-curve at every point with a length such that it reaches from the x-curve to the directrix:

$$dx(\xi) = dA(\xi) - dq(\xi) = \frac{q\,dA}{m_0^2}q \tag{18.122}$$

Finally, it is possible to do exactly the same construction as we have done from the q-end also from the $\bar{q}$-end and to define the corresponding q-vector and x-curve in that case.

2   *Local parton-hadron duality in the Lund model*

The x-curve provides an interesting possibility for describing the average behaviour of the final-state hadron energy-momentum distributions, [48].

Suppose that we use the ordinary Lund fragmentation probabilities to decide upon a distribution of the rank-ordered group of hadrons in the fragmentation process so that we know the behaviour of the first, the second etc. hadron in rank. In this way we obtain the distributions of an ordered set of (transverse) mass variables.

After that we proceed to obtain an ensemble of partonic states by means of the dipole cascade model as implemented by ARIADNE and we may fragment these states in accordance with the Lund model prescriptions in JETSET. But we may also partition the x-curve (defined above) for each state into pieces, each with an invariant size corresponding to the mass distribution of the final-state hadrons.

Comparing the results of the two procedures we find that the partitioning of the x-curves provides a very good description of all inclusive (single-particle, i.e. average) features of the Lund model. In other words, the partitioning of the x-curves for the multigluon states provides the same inclusive distributions as the production of the multigluon states with Lund fragmentation added in for each state. This is true even if we decide upon a subdivision of all (partonic) states into states with a particular value of λ, a particular value of sphericity etc, [48].

The theory group from Gatchina in the present St Petersburg, [52], have introduced the *hypothesis of local parton-hadron duality* in order to be able to relate their analytical calculations of the partonic-state features (in the LLA and modified LLA) to the final observable hadronic states. They have obtained quite good descriptions of many (inclusive) features of the final states in this way, [52].

The results described above means that there is a direct correspondence in the Lund model. The Lund model results, [48], go even further because as far as we know all inclusive (single-particle) features can be derived from considerations of the x-curves, which correspond to the properties of the chosen partonic state ensemble. We note, however, that in this scheme with partitioning of the x-curve the result is not a property of the individual partons. All hadron formations involve at least two neighboring partons so that the final-state hadron makes use of the energy-momentum from at least two partons in order to come onto the mass shell.

The stretching of the curve stems from the color connection between the neighboring partons. The relationship between the x-curve and the parton energy-momenta is in that way similar to the relationship between a hyperbola and its asymptotes in the form of two lightlike vectors. If any of the partons in a state is collinear or soft the x-curve in a well-defined sense ignores that parton direction and just continues onwards along the main partonic directions.

It is necessary, in connection with the partitioning above, to decide upon the value of the parameter m_0 which occurs in the defining formulas. We find in [48] that m_0 actually corresponds to a *resolution parameter* along the directrix state. Small values of m_0 correspond to moving close to the

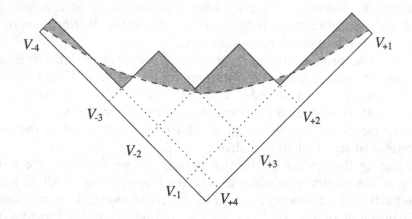

Fig. 18.7. The breakup in a Lund string segment, with the area, the typical hyperbola and the area in between exhibited, together with the characteristic coordinates for the decay.

lightcones while larger values mean that the hyperbolas are less sensitive to the many small fluctuations in a multiparton final state.

The best results are, not surprisingly, obtained if we choose the m_0-parameter so that it is close to the mean hyperbola in the breakup situation. If, as a theoretical experiment, we change the a-, b- and σ-parameters in the Lund model, it turns out that we still obtain a very good description of the final-state hadronic energy-momentum distributions if we adjust the parameter m_0 accordingly.

It is also possible within the same scenario to investigate the fluctuations along the fragmentation cascade in the Lund model. To that end we return to the breakup situation as depicted in Fig. 18.7 for a string segment between two partonic excitations. If we turn the figures discussed in Chapters 7–11 through 90 degrees we obtain the corresponding x-curve description. We have exhibited both the breakup area and the typical hyperbola (parametrised $V_+ V_- = B$), i.e. the correspondence to the x-curve, together with the area between the two; this 'in between' area is shaded in the figure.

In order to calculate the size of the shaded area we note that the part of it spanned by the region (V_{+1}, V_{-1}) is

$$V_{+1} V_{-1} - \int_{B/V_{-1}}^{V_{+1}} \frac{B\,dx}{x} - B = B(\rho_{11} - \log(\rho_{11}) - 1) \qquad (18.123)$$

where we have used that the hyperbola cuts the line $V_\pm = V_{\pm 1}$ at $V_\pm = B/V_{\mp 1}$. We have also defined $\rho_{11} = V_{+1} V_{-1}/B$.

The next part of the shaded area is below the hyperbola and its size is easily found to be

$$-\left(\frac{B}{V_{+2}} - V_{-1}\right) V_{+2} + \int_{V_{+2}}^{B/V_{-1}} \frac{B dx}{x} = B(\rho_{21} - \log \rho_{21} - 1) \quad (18.124)$$

where we have used the same methods as before and defined $\rho_{21} = V_{+2}V_{-1}/B$. This area should be subtracted because it corresponds to something lacking from the total. The procedure is now evident and if we define $\log \rho_{ij} = y_{ij}$ and

$$G(y) = \exp[-bB(\exp y - y - 1)] \quad (18.125)$$

we note that the negative exponential of the area AB can be written as

$$\frac{G_{11} G_{22} G_{33} \cdots}{G_{21} G_{32} \cdots} \quad (18.126)$$

with $G_{ij} \equiv G(y_{ij})$. The meaning of the hyperbolic angles is evident: they correspond to the length along the hyperbola which is spanned by the corresponding coordinates. For small values of y the function G may be approximated by a gaussian. We are evidently describing the Lund model breakup process as something rather similar to a Brownian motion along the typical hyperbola. The parameter of this hyperbola is typically $B = a/b$ in terms of the Lund a- and b- parameters.

If we go back to the process for transverse momentum generation (the Ornstein-Uhlenbeck process) which was discussed in Chapter 12, in particular to the Langevin equation

$$\frac{dv}{dt} = -\rho v + R, \quad \frac{dx}{dt} = v \quad (18.127)$$

and compare this to Eqs. (18.113) and (18.121),

$$\frac{dq}{d\lambda} = -q + \frac{dA}{d\lambda}, \quad \frac{dx}{d\lambda} = q \quad (18.128)$$

we notice the strong similarity. In both cases there is a 'friction term', corresponding to the fact that it takes some time to turn from one direction to another in the process. The particles in the Lund model are produced one after another, neighbors to some extent keeping close in phase space. We may further identify the 'time' variable in the transverse momentum with the λ-measure in the 'longitudinal' process. Further, while the longitudinal process is governed by the given directrix A, the transverse momentum is driven by the stochastic noise term R, which we may intuitively identify as describing the noise of the soft gluons which drown in the longitudinal process.

Let us finally mention that a correspondence to the Sjöstrand treatment of the Lund model fragmentation has been investigated, [21], as a process along the x-curve, producing the noise mentioned above. We will not go

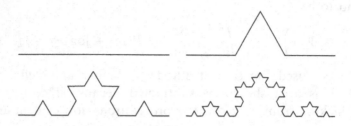

Fig. 18.8.   Three generations of the self-similar construction process for the straight-line snow star by von Koch.

into details because the results, although in principal interesting, to a very large degree coincide with the ones obtained from Sjöstrand's gluon fragmentation model described in Chapter 15.

18.8   The fractal properties of the QCD cascades

We will end this chapter by exhibiting an intuitively appealing way to describe the properties of the x-curve and the λ-measure.

We start by considering a so-called fractal curve, the snow star construction of von Koch, Fig. 18.8. This is an example of the 'mathematical monsters' invented and investigated in connection with the mathematical crisis at the end of the last century. The basic question was then to what extent the intuitive notions of continuity and connectedness also meant regularity. The uncomfortable answer is that it is possible in a straightforward way to recursively construct any number of seemingly 'nice' objects, like e.g. one-dimensional continuous curves. As one continues the iterations these curves, nevertheless, tend to 'fill up' the regions around them to the extent that they should no longer be considered as one-dimensional objects. The mathematicians were also able to construct single-point clusters, called Cantor dust, which similarly must be considered as one-dimensional curves, as well as two-dimensional surfaces, which tend to cover parts of the three-dimensional space.

One way to produce such objects is to make use of the notion of *self-similarity*. The construction by von Koch is only one of the simplest and most elegant. You take a continuous line of length 1 and subdivide it into three parts. Then on the middle part you change the straight line segment into an equilateral triangle. From then on the 'new' curve is defined to include the outskirts of the triangle. This means that an object of length 1 has changed into one of length 4/3. Then you repeat the procedure for the four different parts and obtain four new projecting triangles. The new curve will now have the total length $(4/3)^2$. This construction can go on as long as you are able to visualise the procedure and at the nth step you have a 'measuring rod' (i.e. the length of the elementary segments) of size $\ell_n = (1/3)^n$ to measure the curve length $L_n = (4/3)^n$. Note that every part is by construction the same as all others and related to the earlier steps only by a scale change.

Nowadays one defines the *fractal dimension*, D, as [80]

$$D = 1 - \frac{dL_n}{dl_n} = \frac{\log 4}{\log 3} \tag{18.129}$$

Therefore D is a number between 1 and 2, i.e. the construction leads to something which, intuitively, fills the plane 'partially'.

When we go back to Fig. 18.1 and the general construction we have presented for the λ-measure and the x-curve we note that the λ-measure depends upon the size of the 'measuring rod', i.e. the size of the $k_{\perp c}$ at which we stop the construction.

In Fig. 18.5 we have shown what Fig. 18.1 looks like from below. It is easy to see that the many out-sticking branches, twigs and subtwigs which occur in Fig. 18.5 have clear similarities to the construction by von Koch shown in Fig. 18.8. The main difference is that while the von Koch construction is a deterministic process, i.e. every step is completely fixed, the λ-structure is stochastic in nature. Thus every step is determined by a probabilistic scenario. If we use exactly the same considerations that led to Eq. (18.129) in Eq. (18.92), then for the mean value of λ using κ as the measuring rod we obtain

$$D = 1 + \sqrt{\frac{3\alpha_s\left(k_{\perp}^2\right)}{2\pi} - \left(\frac{1}{4\kappa} + \frac{3c\alpha_s\left(k_{\perp}^2\right)}{4\pi\kappa} + \cdots\right)} \tag{18.130}$$

These results would mean that it is possible, using today's fashionable language to call the quantity $\langle \lambda \rangle$ a *multifractal* with dimension equal to $1 + \epsilon$, where ϵ *is the anomalous dimension* of the QCD multiplicity distributions, [52]. The word multifractal, [80], is used in order to stress that the dimension is changing with the size of the measuring rod. The result of comparing the first term in Eq. (18.130), $\sqrt{3\alpha_s/2\pi}$, to the ARIADNE Monte Carlo, [73], is not a good approximation. The second (negative) term makes it into an essentially better approximation.

19

The parton model and QCD

19.1 Introduction

In this chapter we will provide the parton model, the PM, with a QCD field theoretical structure according to the conventional method; for more details see e.g. [52]. In the next chapter we continue the discussion and present the Lund model version of the properties of deep inelastic scattering (DIS) events, both the treatment of the fragmentation and, in particular, the use of the newly developed linked dipole chain model, [16] to provide the fragmenting string state.

The method of virtual quanta (MVQ) in Chapter 2 describes the electromagnetic field from a fast-moving charge in terms of the photon flux from the bremsstrahlung spectrum, and we will make use of this as an analogy. It is evident that Feynman picked up the basic features of the MVQ to make the PM into a description of the corresponding flux of the hadronic field quanta. In that way he made the PM into a useful tool to describe the cross sections for DIS events. Those we consider in this book are initiated by an electromagnetic probe, i.e. they correspond to inelastic electron-baryon (or muon-baryon) scatterings. But it is also possible to use the PM to describe e.g. inelastic neutrino-baryon scattering events as well as to consider the interactions between the partons themselves.

Feynman assumed that the partons can be treated as a stream of free elastic scatterers with respect to the probe. However, at that time there was no known field theory, besides that of non-interacting fields, in which the quanta could even approximately be treated in this way.

All the hadron-hadron cross sections are in the range of tens of millibarns, corresponding to a surface with a radius in the fm region. This is comparable to the size of the hadrons themselves, i.e. their form-factor extensions, cf. section 5.5. In a precise way we may say that within this region the forces are very strong. A hadron is almost black from the point of view of absorption, which means that (almost) all hadronic probes that penetrate inside the hadronic region are scattered. But we remember from earlier chapters that an electromagnetic cross section behaves as α/Q^2, i.e. it is proportional to the squared wavelength of the probing field and it is consequently small for large Q^2.

We have then two facts which it does not seem possible to explain within the same framework. On the one hand we know that the hadrons interact strongly inside their size radius. On the other hand when they are probed with a wavelength much smaller than this the hadronic wave function can, according to the PM, be projected into a stream of non-interacting partons.

Nevertheless, there is an ingenious answer inside QCD and we will consider it using several different methods. We will discuss in some detail the leading log approximation (LLA) to the relevant Feynman diagrams and we will also consider the lightcone singularities of the current matrix elements we met in Chapter 5. We will show that the two methods are equivalent and can be reformulated into the celebrated DGLAP equations. We will end with a discussion of several suggested corrections.

In section 19.3 there is a brief description of the contents of this chapter to provide the reader with a birds-eye view of the subjects to be covered. But we start in section 19.2 with the general field theoretical method to calculate the cross sections for DIS events. In particular we will clarify the partitioning of the radiation in these states into *initial-state bremsstrahlung* (ISB) and *final-state bremsstrahlung* (FSB). We will make use of these notions repeatedly in this and the next chapter.

19.2 The DIS cross sections, initial- and final-state bremsstrahlung

Until now we have in this book been mostly concerned with the production probabilities in e^+e^- annihilation events. The cross sections for DIS are different and in particular it is not sufficient to know a few low-order perturbative terms in order to describe them. The reason is, of course, the *parton flux factors*, i.e. the hadronic structure functions which we discussed in Chapter 5.

In the MVQ in Chapter 2 the electromagnetic fields of a moving charge are described. These fields, which can be considered as the 'wave functions' of the radiation states connected to the charge, are projected onto states with a fixed frequency ω and impact parameter b (later redefined into the canonically conjugate variable $k_\perp$). Finally the size of the field pulse, or flux, as seen by the measuring setup is described (this corresponds to the squared wave functions). We find that *it only depends upon the number of quanta with quantum numbers* $\omega, k_\perp$.

In Chapter 5 the corresponding flux factors are described as the current-current matrix elements for the probed hadron, cf. e.g. Eqs. (5.49) and (5.66). In this case we sum over all the states which can be reached from the hadron by the application of the current. This result, that there is a close connection between the MVQ radiation wave functions and the current matrix elements, can be inferred from Eq. (5.35). Asking for a particular frequency, sensitive to the probe, means that only the current-current matrix elements which end on this frequency should be included. This corresponds to the Fourier transform of the matrix elements with respect to the probe frequency, in this case given by the momentum transfer q. It is useful to subdivide the radiation into the primary emissions from the currents, called the initial-state bremsstrahlung (ISB), and the remaining radiation, the final-state bremsstrahlung (FSB).

We will, from now on, in general use partonic language and assume that the wave function of the original hadron can be projected onto a set of wave functions with a well-defined number of partons at some observational level. Thus there is, according to the original SLAC experiments, briefly described in Chapter 5, a distribution in x_B, measured at a momentum transfer scale $Q_0^2 \simeq 1 \ (\text{GeV}/c)^2$. This entity is nothing other than the (squared) wave function of the hadron, projected onto the partonic base states. We will now, using this input, construct the (squared) wave function corresponding to a larger resolution scale, as probed by smaller wavelengths $\lambda \simeq 1/Q$. This will be done by an analysis of the Feynman graphs corresponding to multiple gluon emission.

Let us consider, as we shall do more than once later on, the 'fan diagram' in Fig. 19.1. This is not meant as a single Feynman diagram, but rather corresponds to a set of such diagrams. A fan diagram contains a connection, in particular a color flow, from an *incoming parton* (included in the distribution at Q_0^2 and described by a massless energy-momentum vector P) to a *parton scattered by the probe* with energy-momentum q. At this point we will not consider the color connection along the fan diagram.

There is a set of *emissions* along the ladder, described by the energy-momentum vectors p_j (which are always taken to be on-shell and massless). There is also a set of connector lines, to be called *propagators*, described by the energy-momentum vectors q_j, which are all spacelike being the momentum transfers between on-shell lightlike vectors, cf. Chapter 2. At every *vertex* there is energy-momentum conservation,

$$q_j = P - \sum_{m=1}^{j} p_m, \quad \text{i.e.} \quad q_j = q_{j-1} - p_j \qquad (19.1)$$

Besides the emitted partons p_j we show a set of further parton emissions, which are (for each index j gathered into a set) called $(h)_j$. We will

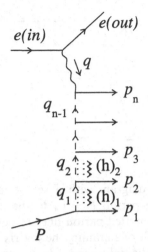

Fig. 19.1. A fan diagram, corresponding to a collection of QCD Feynman graphs between a parton with energy-momentum vector P and a probe with energy-momentum vector q, stemming from the momentum transfer from a leptonic probe e. Notation according to text.

assume that it is possible to emit the set $(h)_j$ according to the rules of a perturbative QCD cascade, i.e. in a coherent way and with negligible recoils, *if we already have emitted the set* (p_j). (Such cascades are the ones we have already encountered in the earlier chapters). The set (p_j) is known as the ISB while the corresponding sets $(h)_j$ are the FSB.

Before we clarify the precise partitioning between ISB and FSB we will exhibit how the cross sections for the radiation states can be described by means of field theoretical Feynman graphs. Consider the configuration in Fig. 19.2, where we again show a set of primary emissions (p_j) along a chain, together with the *radiative corrections* connected to this emission process. Note that we are, as always for the cross sections, considering the square of the production matrix (this time the above-mentioned current matrix) elements, called $\mathscr{J}$ (equal to $\mathscr{J}^*$ because the currents are real), summed over the final-state particles.

While the production matrix elements correspond to the diagrams in Fig. 19.1, the cross sections correspond to the symmetrised graphs in Fig. 19.2 (containing an implicit sum over all the final-state (on-shell) p_j-vectors). It is only the lines along the ladder sides, which are 'true' propagators, carrying the off-shell q_j-vectors. Such diagrams were referred to as *cut diagrams* in Chapter 4 (i.e. cut across the p_j-vectors, which means that the p_j are on the mass shell). Remember that by the renormalisation

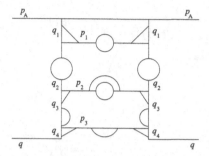

Fig. 19.2. An LLA ladder diagram with a set of radiative corrections along the chain from the incoming parton $P = p_A$ to the momentum transfer q and on the connecting (on-shell) emitted parton lines in between. Note the symmetry between the two sides, a result of summing the matrix elements $\mathscr{J}$ (the left-hand side) multiplied by $\mathscr{J}^* = \mathscr{J}$ (the right-hand side, the same for a real current) over the intermediate states $|X\rangle$ containing the partons p_j.

process the propagators are arranged to have a pole at the mass value of the observed particles, and at the same time a normalisation and a charge value are defined at some (arbitrary) mass scale. The particular diagrammatic contribution in Fig. 19.2 contains such radiative corrections to the normalisation and the charge of the emitted p_j and these contributions can be associated with the sets $(h)_j$, i.e. in the cross section these FSB emissions correspond to radiative corrections, cf. Chapter 4.

We note the close correspondence to the way in which the ladder diagrams occurring in Chapters 9 and 10 describe the states that can be reached by the hadronic scattering operator T (and from the other side by T^*). We may take over from these discussions the fact that in order to obtain a large contribution from such diagrams the momentum flowing along the lines should not be (too) large. Therefore, if the external probe's energy-momentum $q^2 = -Q^2$ increases then it is necessary to include more and more rungs in the ladder diagram.

The problem is to distinguish the two sets, i.e. those gluon emissions that are included in the set (p_j), and those in the sets $(h)_j$. Actually there is no clear distinction apart from the two features mentioned above, that in order to be able to 'sum away' the set $(h)_j$ as virtual corrections to the main p_j-emissions they should be coherent and leave small recoils. We are not allowed to make statements about time-ordering, as we found in Chapter 16 in connection with Figs. 16.1 and 16.2. None of the contributions from the single graphs is gauge-invariant by itself and although it may seem natural to partition them into a 'before' and 'after' scenario such a partitioning is not consistent.

The answer is that any choice of the q_j-emissions is allowed. But a particular choice will also contain a corresponding set of virtual corrections of the Sudakov kind, i.e. there will be a Sudakov factor for each choice, in accordance with the discussion in section 17.3. The cross sections of DIS are then given by the formula

$$d\sigma_{DIS} = \sum_I dw(I)Sud(I) \tag{19.2}$$

where $dw(I)$ is the inclusive weight for emission of a state I, i.e. an inelastic state included in the initial-state bremsstrahlung. If we change the content of the set of I-states this will be compensated by a corresponding change in the Sudakov factor $Sud(I)$ so that the sum is unchanged. At this point it is worthwhile to be a little more specific because the cross section is an observable quantity. It would be rather puzzling if a particular state provide different contributions solely because of our ISB choice.

Consider an exclusive partonic state in a DIS event, i.e. consider all the gluonic radiation emitted in a state. Then we may subdivide this radiation into the ISB part, defined by some rule or another, and the rest, the FSB. The total weight for the state is then given by the weight for the ISB choice, denoted $dw(I)Sud(I)$ in Eq. (19.2), *together with a Sudakov factor $dw(F)Sud(F)$, corresponding to the probability of emission the particular FSB group(s) from the chosen ISB state.* Remembering the properties of the Sudakov factors, cf. section 17.3, *we conclude that $dw(I)Sud(I)$ corresponds to the contribution from the sum over all states with the same ISB choice and all possible FSB gluons resulting from them.* If we change the ISB set then the weight for the exclusive state must be rearranged:

$$dw(I)Sud(I)dw(F)Sud(F) \rightarrow dw'(I)Sud'(I)dw'(F)Sud'(F) \tag{19.3}$$

and we obtain the primed ISB contribution to the cross section after summing over all the states defined by the new ISB gluons and all allowed (primed) FSB in these states. But the total result in Eq. (19.2) is the same!

19.3 A bird's-eye view of the features of deep inelastic scattering

1 Generalities on the leading-log approximation

There were, rather soon after the PM was suggested and the original SLAC experiments were completed, serious attempts by Gribov and his collaborators to provide a consistent method of re-summing the higher-order terms in perturbation theory. For DIS events they arrived, with due care to the Sudakov corrections discussed above, cf. [52], at the results which are known as the leading-log approximation (LLA), valid for field theories with a dimensionless coupling constant.

They found that in every order of perturbation theory there are new (in general squared) logarithmic contributions in the large variables such as the squared cms energy s. The sum of such contributions tends to grow exponentially so that the logarithms become powers in s. (The reason that there are two logarithmic powers is that both the transverse momentum and the rapidity variations provide contributions, although both of them will at the nth order be limited due to the iterations by inverse factors of $n!$, cf. Eq. (19.6) below.)

In DIS, where there are two basic dynamical variables Q^2 and v, there are problems when these variables are not of the same order, i.e. when x_B is very small. Then it is necessary to sample the double logarithms in Q^2 and $1/x_B$, but the power correction results are still true. This means a serious disagreement with the scaling results from SLAC and in particular that the PM could not be motivated within such a framework.

A major advancement at the theoretical level started when it was recognised, [68], that nonabelian gauge theories exhibit *asymptotic freedom*. The coupling constant for these theories does not behave as in other theories; cf. Chapter 4. Instead *the nonabelian coupling constants effectively become smaller when the momentum transfers increase*.

This is partly sufficient, although not enough, to obtain the original scaling arguments of the PM. A typical result might be if one calculates the multiplicity from the contributions of the ladder diagrams,

$$\sum_n \frac{[C\alpha_s \log^2(s/s_0)]^n}{(n!)^2} \tag{19.4}$$

In order to understand the behaviour of this sum we make use of the Stirling approximation for large values of n,

$$n! \simeq \exp[(n + 1/2)\log(n + 1) - (n + 1)] \tag{19.5}$$

and maximise the expression with respect to n. For positive values of d a sum of the following kind

$$\sum \frac{a^{dn}}{(n!)^d} \simeq \exp(ad) \tag{19.6}$$

is strongly governed by the term corresponding to this maximum, i.e. by the term with $n_{max} \simeq a$. Therefore the sum in Eq. (19.4) will behave like a power in s for a constant coupling α_s. But if the coupling behaves, as in QCD, effectively like $1/\log s$, the result in Eq. (19.4) will behave, as in our calculations of λ and the multiplicities in Chapter 18, like $\exp(c\sqrt{\log s})$. There will be scale-breaking logarithmic behaviour but there will not be power corrections within this framework.

We have also seen that, in accordance with the Callan-Symanzik equations in Chapter 4, there will be logarithmic power corrections to some quantities. This is a reflection of the same feature, i.e. that *asymptotically free theories involve some scale-breaking 'on the way to freedom'*.

Thus even inside QCD the partonic flux factors will contain a Q^2-dependence so that we should write, for the parton distributions in Chapter 5, $f(x_B, Q^2)$. In this and the next chapter we will consider this Q^2-dependence, which actually occurs as a dependence on

$$\tau = \log(Q^2/\Lambda_{QCD}^2) \tag{19.7}$$

We will investigate it from several different points of view but we note that such logarithmic corrections generally are slowly changing when $Q \gg \Lambda$.

2   Generalities on the moment method and the operator product expansion

We have already in Chapter 5, subsection 1, presented the reasons to go to a lightcone dynamical treatment of the matrix elements which occur in DIS. The structure functions f are given by the Fourier transforms of the squared current matrix elements but this result can be reformulated in terms of commutator matrix elements, which should vanish (according to causality) outside the lightcone. Therefore we found that the structure functions should in limiting situations be dominated by the current behaviour along the lightcones.

After a brief discussion of the kinematics we will use this lightcone dynamical treatment in terms of the moment method (MM) combined with the Wilson lightcone operator product expansion (OPE) in a way invented by Christ, Hasslacher and Mueller, [43]. (The historical and intellectual dependence of the results is outside the scope of this book. It is, however, evident that the original participants in the LLA adventure very early noted the simplicity of their results in terms of the moments of the structure functions.)

This treatment will lead to a description of QCD scale-breaking, [69], for the moments of the structure function f:

$$\mathscr{F}(j, Q^2) = \int_0^1 x^j dx f(x, Q^2) \sim \mathscr{F}(j, Q_0^2) \left(\frac{\alpha_s(Q_0^2)}{\alpha_s(Q^2)} \right)^{a_j} \tag{19.8}$$

Here Q_0^2 is a fixed scale (introduced above), $\alpha_s(Q^2)$ is the running coupling of QCD and the a_j are numbers that can be computed by means of the MM and the OPE.

We will present the physical arguments within a scalar field theoretical framework. The situation for the real world is somewhat more complex because there are vector indices as well as dimensional differences in connection with electromagnetic currents. There are also several flavor- and color-dependent contributions to the different parts of the parton distributions but the result in Eq. (19.8) is true for each part.

The MM, combined with the OPE, corresponds to a very neat method. It relates the moments of the structure functions to the behaviour of the matrix elements of space-time operators. Then the renormalisation group, in terms of the Callan-Symanzik equations [108] (cf. Chapter 5), is applied to the operator matrix elements to obtain the results in Eq. (19.8).

The method turns out to be equivalent to the LLA. If we invert the results for the moments in Eq. (19.8) we obtain a set of integro-differential equations, nowadays known as the DGLAP equations, for the structure functions $f(x_B, Q^2)$ (DGLAP is short for Dokshitzer-Gribov-Lipatov-Altarelli-Parisi). These equations are equivalent to the results derived within the LLA, [52], which means that the whole setup is consistently connected. It turns out that the Sudakov factors in this case simply correspond to a subtraction in the occurring splitting functions to fulfil the energy-momentum conservation constraints.

In this way a 'conventional' scenario emerges, which will be called the *ISB scenario*. Viewed from the lightcone point of view, larger values of Q^2 will probe regions closer and closer to the lightcone, cf. Chapter 5. Remembering that the variable x_B is the Fourier inverse of the variable px, we conclude that small values of j in Eq. (19.8) correspond to probing large distances along the lightcone direction px and large values of j correspond to probing small regions close to the origin.

For small values of j the numbers a_j in Eq. (19.8) are generally positive, meaning that the small-x_B region will increase in the structure functions. The a_j's turn round and become negative for larger values of j; then the main contributions to the moment integrals come from the large-x_B part of the distributions. Viewed from the LLA perspective, an increase in Q^2 corresponds to the possibility that a parton at Q_0^2 may split up into smaller-x_B partons at a higher scale of resolution. This is of course the same dynamics as before, namely that the small-x_B region obtains more and more contributions, as Q^2 increases, from the partons which decay along the fan diagrams, thereby depopulating the larger x_B-values.

3   Some problems in the ISB scenario

There are nevertheless a set of problems. Some of these are addressed in the work by Gribov jr, Levin and Ruskin (GLR) [67]. They are related to the uncomfortably large numbers of partons which may emerge at small-x_B and moderate-to-large Q^2 values from the ISB scenario.

GLR re-sum a set of Feynman graph contributions to calculate the probability that some of the already emitted partonic 'chains' reinteract, thereby decreasing the total partonic multiplicity, cf. also [95]. But this so-called *shadowing* method (where one emitted chain is in front of another emission) is only applicable inside certain regions of phase space and outside these there are more complex multiparton interactions.

The correction terms contain an unknown scale corresponding to the (transverse) region effectively inside which a parton chain is emitted. If this scale is determined by the expected hadronic size, around 1 fm, then the correction terms are rather small and the multiplicity growth of the partons at small x_B is not inhibited at the presently available energies.

It is, however, possible to imagine that the hadronic wave functions contain large- and small-density regions in a complex way, so that there are 'hotspots' of a small size, which will then provide large GLR corrections.

There are also other reasons for concern about the ISB scenario. To begin with, the MM and OPE results coincide with the LLA because both of them pick out only the leading contributions and neglect all corrections. Thus in the MM and OPE all non-scaling contributions are neglected and only the leading singularities on the lightcone, corrected by the logarithms from the renormalisation group equations, are retained. For the LLA, to all orders only the terms with the largest logarithmic factor are retained (it is, however, possible to use a modified leading-log scenario such as described in Chapter 18).

There have been efforts by Lipatov and his coworkers, [29] to take account also of (some of) the non-leading contributions. The result of their effort is, however, that the number of small-x partons increases even more, although it then tends to stabilise for the evolution equations.

The Lipatov results are that e.g. the gluon structure function will, for small x behave like a power in x (there is also some gaussian $\log Q^2$-behaviour, due to the projection on an eigenfunction):

$$g(x) \sim x^{-1-\lambda_L} \tag{19.9}$$

where λ_L is a number of order 0.5, stemming from the largest eigenvalue of an integral equation.

These results should be valid for medium to small Q^2. They imply that the ocean $q\bar{q}$-content of the nucleon structure functions, which is directly coupled to the gluon density, will make the cross sections very large indeed for increasing energies (which means that unitarity must be invoked and/or shadowing à la GLR). We will call this effect the *BFKL mechanism* (for Balitsky-Fadin-Kuraev-Lipatov). We will also point out that there are large corrections to the results in Eq. (19.9) both from energy-momentum conservation and from the QCD coherence conditions.

Note that the coherence conditions of QCD bremsstrahlung are not necessarily applied within the DGLAP and BFKL approaches. In section 19.6 we will consider the approach of Marchesini *et al.*, [44], in order to show the implications of a more sophisticated approach, which contains both the DGLAP and the BFKL contributions but nevertheless retains the coherence conditions. This is also the starting point of the linked dipole chain model, [16] to be described in the next chapter.

19.4 The moment method and the DGLAP mechanism

1 Kinematical preliminaries

There are several coordinate systems of interest used for the description of DIS events. One is called the *probe-hadron cms*. We will in this chapter mostly make use of this system or rather of a system which is somewhat more general, called 'equivalent to the hadron-probe cms'. This means that we boost along the momentum direction (conventionally the 3-axis) between the probe and the hadron. Then the probe q will have light-cone energy-momentum components ($\mathbf{0}_\perp$ stands for vanishing transverse momentum)

$$q = (-Q_+, \mathbf{0}_\perp, Q_-) \tag{19.10}$$

The hadron is in this frame described by a (large) lightcone energy-momentum $P_+ = E + P_3$. We neglect its transverse (i.e. along the $(1,2)$-axes) and negative lightcone components. We will assume that the hadron is described by P_+ and by its space-time component $x_{+3} = t + x_3$. This is compatible with quantum mechanical considerations since the quantities P_+ and x_{+3} commute. Thus the hadron is described by a wave function depending upon P_+, x_{+3}.

We may then consider the interaction as a *measuring process* in which the probe determines the hadron's x_{+3}-coordinate to a precision given by the 'interaction time', $\delta x_{+3} \simeq 1/Q_-$. The hadronic state can then be in any of its eigenstates within the energy-momentum range (P_+, P_-) with $P_- \leq Q_-$. These are the quantum states which live sufficiently long (at least as long as the interaction time) for a measurement to take place.

The measuring process is defined by an interaction with a parton with $x_B P_+ = Q_+$; the parton is then turned around by the momentum transfer so that the final state corresponds to a hadronic state in the energy-momentum range $(P_+ - Q_+ \equiv P_+(1 - x_B), Q_-)$. The phase space for gluon emission is evidently described by the triangular region in Fig. 19.3 in terms of the parton variables $\kappa = \log(k_\perp^2/s_0)$ and rapidity y.

The various useful kinematical variables are exhibited in Fig. 19.3. We note that a fixed value of the fractional energy-momentum $x = k_+/P_+$, with $k_+ = k_\perp \exp y$, corresponds to a straight line across the triangle. In particular, for $x = x_B$ we obtain a triangle corresponding to Q^2 on the left-hand side of the total phase-space triangle. We also note that the length of the baseline of the triangle corresponds to $\log W^2 = \log(P_+Q_- - Q^2) \simeq \log P_+Q_-$ (the approximation is valid unless $x_B \sim 1$). All partonic emission,

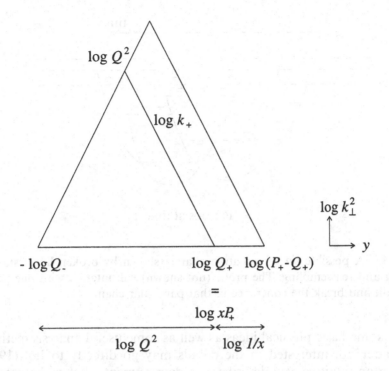

Fig. 19.3. The phase space for emission of gluons in a DIS event.

whether it should be referred to as ISB or FSB radiation, must occur inside this triangle in an energy-momentum conserving theory.

There are various intuitive pictures that can be used to imagine the hadronic state. In the ISB scenario it is useful to think of a virtual excitation living for a long time due to time dilation (cf. Chapter 2). Such an excitation may then be described as a cascade chain, which in a coherent way develops and then reassembles.

In this way there may be many chains available, each starting on a 'permanent' member of the wave function. The interaction probe will pick out one parton with fractional energy-momentum x_B, thereby breaking the coherence in that particular chain and realising the corresponding radiation state, see Fig. 19.4. This diagram is taken from [67] where a particularly lucid description is given of the ideology behind the ISB scenario.

2   The moment method based upon Wilson's operator product expansion

This section contains many formal notions and, although the mathematics will be rather informal, this is a worthwhile approach since we can then

time

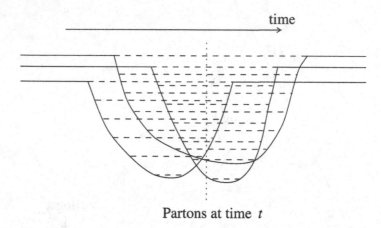

Partons at time t

Fig. 19.4.   A possible set of radiation chains (shown by broken lines) starting, evolving and reassembling. The probe (not shown) will interact with one parton in a chain and break the coherence in that particular chain.

present some basic physical ideas as well as some useful analysis methods. The reader not interested in the details may go directly to Eq. (19.20), which in an intuitive way describes the developments that go before.

We would like to isolate the major contributions to the current matrix element in Eq. (5.49), which describes the cross section in DIS reactions. This is done by an expansion around the lightcone singularities of the current commutator. A field operator is distribution-valued but we may nevertheless (with care) use a pointlike notation both for the free-field operators and also for the perturbed-field operators in an interacting theory, [36], although in that case after renormalisation.

The relationship between the time-ordered and normal-ordered operator products, which was derived in Chapter 3, implies for the current $j(x) = \lim_{y \to x} :\Phi(x)\Phi(y):$

$$j(x)j(0) \overset{x_\mu \to 0}{\longrightarrow} c_0 \frac{1}{(x^2)^2} I + c_1 \frac{1}{x^2} j(0)$$
$$+ c_2 \frac{1}{x^2} x_\mu :\Phi \partial^\mu \Phi:(0) + c_3 :jj:(0) \qquad (19.11)$$

The numbers c_j are constants for free fields and I stands for the unit operator. This is the original *Wilson short-distance expansion of an operator product*, in which one only takes the singular terms into account. As mentioned above, [36], it is well defined also in perturbation theory but then the numbers c_j become logarithmic functions of x.

We would now like to go over to the lightcone scenario described in

Chapter 5. We note that for a short-distance expansion such as Eq. (19.11), when each component of the vector x_μ tends to zero, then $1/x^2$ is one power more singular than x_μ/x^2. But when we consider the approach to the lightcone, $x^2 \to 0$, then these expressions are both equally singular. In this limit it turns out that one needs an infinite number of local-operator terms:

$$j(x)j(0) \overset{LC}{\to} c_0 \frac{1}{(x^2)^2} I + \frac{1}{x^2} \sum_{m=0}^{\infty} f_m x_{\mu_1} \cdots x_{\mu_m} O^m_{\mu_1,\ldots,\mu_m}(0) + \cdots \qquad (19.12)$$

The limit notation LC means the lightcone limit $x^2 \to 0$. The operators O^m are all the (local and symmetric) operators in the field theory that carry m Lorentz indices (we use the summation convention for repeated indices μ_j). As an example, in a theory with a scalar field Φ there will be, for the corresponding currents, an $O^m_{\mu_1,\ldots,\mu_m} = :\Phi \bar{\partial}^{\mu_1} \cdots \bar{\partial}^{\mu_m} \Phi:(0)$ where the notation $\Phi \bar{\partial}^\mu \Phi = \Phi(\partial/\partial x^\mu)\Phi - [(\partial/\partial x^\mu)\Phi]\Phi$ has been used repeatedly.

It is obvious that along the lightcone all the quantities in the sum have the same singularity, i.e. $1/x^2$. The final ellipses refer to less singular terms in the expansion. The idea behind the partitioning in Eq. (19.12) is *to find for each field theory the basic operators O^m, express them in terms of the free-field correspondences and then to include all the perturbation theoretical modifications in the coefficients f_m*. From perturbation theory with non-dimensional coupling it is possible to prove, [36], that the f_m are functions of $\log x^2$ expressible as power series in the coupling g:

$$f_m(x^2) = \sum_{j=0}^{\infty} \sum_{r=0}^{r=j+1} f_m(j,r) g^{2j} \log^r x^2 \qquad (19.13)$$

From the results in Eqs. (19.12) and (19.13) we have a method of analysing the current matrix elements in Eq. (5.49). We firstly note that if we evaluate the tensor O^m in a state with a well-defined energy-momentum p we will, due to Lorentz covariance, obtain

$$\langle p| O^m_{\mu_1,\ldots,\mu_m} |p \rangle = p^{\mu 1} \cdots p^{\mu m} C_m + \cdots \qquad (19.14)$$

The reason is that p is the only Lorentz vector available in a scalar theory. The so-called 'reduced matrix element' C_m is, in a free-field theory, a plain number and in this way we have been able to extract the 'trivial' Lorentz covariance requirement.

If we consider the Fourier transform of the current matrix element itself we obtain from Eqs. (19.12) and (19.14)

$$\mathcal{W} \equiv \int dx \exp(iqx) \langle p|j(x)j(0)|p \rangle$$

$$= \int dx \exp(iqx) \sum_m f_m(x^2)[(px)^m C_m + \cdots] \frac{1}{x^2 + i\epsilon x_0} \qquad (19.15)$$

The power $(px)^m$ is, using the notation $(P_+, \simeq 0, \mathbf{0})$ for p, given by the expression $[(P_+ x_-)/2]^m$ and can, at least formally, be written as

$$(px)^m \exp(iqx) = (i2pq)^m \left(\frac{\partial}{Q_- \partial Q_+} \right)^m \exp(iqx) \qquad (19.16)$$

This means that the whole expression can be rewritten as

$$\mathscr{W} = \sum_m \frac{1}{x_B^{m+1}} C_m E_m,$$

$$iv E_m = (iQ^2)^{m+1} \left(\frac{\partial}{\partial Q^2} \right)^m \int dx \exp(iqx) f_m(x^2) \frac{1}{x^2 + i\epsilon x_0} \qquad (19.17)$$

In this derivation we have freely interchanged sums and integrals and performed a set of formal operations such as the differentiations in Eq. (19.16). What we have obtained is an approximate expression for the quantity $\mathscr{W}$ in terms of a power series in (the inverse of) x_B multiplied by the matrix element functions C_m and the c-number functions E_m. In this way we have been able to rewrite the power series in px, in Eq. (19.15), as an inverse power series in x_B, which once again reminds us of the reciprocal relationship between these variables.

In order to relate the quantities C_m and E_m to measurables, it is necessary to make an assumption on analyticity for the quantity $\mathscr{W}$ in respect of the variables x_B and Q^2. In [43] the authors assume that $\mathscr{W}$ is, *for large values of* Q^2, *an analytic function of* x_B *apart from branch cuts for* $-1 \leq x_B \leq 1$. It is not possible to prove this statement outside perturbation theory so we are thus in the same situation as for the elastic form factors in Eq. (5.47).

The authors of [43] also assume that $\mathscr{W}$ is even in x_B (which corresponds to the property of crossing, in a field theory). Therefore we can use Cauchy's formula for the line integral around a curve c of an analytic function to write the function $x_B^n \mathscr{W}$ as follows:

$$\frac{1}{2i\pi} \oint_c dx_B x_B^n \mathscr{W} = C_n E_n \qquad (19.18)$$

$$= \frac{1}{2i\pi} \int_0^1 dx_B x_B^n [\mathscr{W}(x_B + i\epsilon) - \mathscr{W}(x_B - i\epsilon)] \qquad (19.19)$$

Now the integrand on the right-hand side is $2i \operatorname{Im}(\mathscr{W})$ and can be identified with the physically measurable quantity $W = 2\pi \tilde{f}(x_B, Q^2)/v$ discussed in Chapter 5. In Eq. (19.18) we have used the residue calculus and have diminished the Cauchy curve to include only an integral along the singularities lying on the cuts, using the symmetry $\mathscr{W}(-x_B) = \mathscr{W}(x_B)$ mentioned above.

Consequently, the nth moment of the structure function $\tilde{f}$ in this scalar theory can be identified with the (reduced) matrix element C_n multiplied by the c-number function E_n. This in turn can, according to Eq. (19.17), be expressed as the Q^2-variation of the (energy-momentum space) matrix element of the operator O^m, evaluated in perturbation theory. This should remind us of the Callan-Symanzik equation, discussed in Chapter 4, which describes just this, i.e. the effect on a matrix element stemming from scale changes in the renormalisation. Therefore if we perform the renormalisation just at the point Q^2 (which is allowed according to the assumptions on analyticity) we may apply the Callan-Symanzik formalism to derive the behaviour of the moments of the structure function!

We can thus summarise our results in the following simple statement (although it contains some subtle relations)

$$\int_0^1 dx_B x_B^m \tilde{f}(x_B, Q^2) \propto \frac{1}{(P_+)^m} \langle p| O^m_{++\cdots+} |p\rangle_{Q^2} \qquad (19.20)$$

The matrix element on the right-hand side is then evaluated in energy-momentum space and renormalised at the scale Q^2 according to Chapter 4, [52]. Within perturbation theory, it will coincide with the product $C_m E_m$ obtained in Eq. (19.18). It is, however, necessary to understand that there are at least two important aspects of this result. Firstly there is the assumption that the approximation of keeping only the most singular terms from perturbation theory in the lightcone expansion is a good one. Secondly it is necessary to invoke analyticity for the function $\mathcal{W}$ in order to derive the relationship of the moments to the derivatives of the matrix element.

3   *The Callan-Symanzik equation and its implications for the moments*

We will now use the *renormalisation group* of field theory, Chapter 4, to calculate the behaviour of the quantities E_m in Eq. (19.17) when Q^2 varies. The tool will be the Callan-Symanzik equation and we will extend it outside the scalar field theory scenario we have considered up to now.

We recall that the β-function of QCD is negative,

$$\beta(\alpha) = -b\alpha^2 - \cdots \qquad (19.21)$$

where the ellipses refer to higher-order terms, some of which have been calculated; but they do not play a major role in our argument. This implies that the QCD running coupling vanishes as the inverse of the log of the scale at which we perform our renormalisation. We may choose this scale at Q^2 (it is allowed according to the analyticity assumptions in the MM and OPE) and consider the large-Q^2 limit just as in the treatment of the Callan-Symanzik equation in Chapter 4. Because of the properties of the running coupling we need only the lowest-order perturbation theory results to calculate the anomalous-dimension functions γ.

In [69] the anomalous dimensions of the operator matrix elements $C_m E_m$ are calculated for QCD. They are unfortunately not as simple as the ones we encountered in Chapter 4, where $\gamma_m = d_m \alpha_s$, the d_m being plain numbers and α_s the QCD coupling. There are two reasons. The first is the tensor structure and the dimensions of the electromagnetic currents, but those cause only minor complications in comparison with the scalar version of the MM discussed in the last subsection.

The major reason for the complications is that in this case there are contributions to the current matrix elements not only from quark and antiquark intermediate states but also from the gluon states that can be reached by applying the gluon field operator A. This means that both of the matrix elements

$$\langle q| AA |q\rangle, \quad \langle q| j |q\rangle \tag{19.22}$$

are nonvanishing; thus a quark can be absorbed not only by the fermion current j but also by the gluon 'current' AA stemming from the three-gluon interaction of QCD. In Eq. (19.22) we have neglected all vector and color indices.

The fact that both the matrix elements are nonvanishing can be understood from our considerations relating the scattering from a potential to the scattering from protons in Chapter 5, cf. Eq. (5.35). There we used that

$$A^\mu(x) \to \int dx_1 \Delta_F(x - x_1) j_B^\mu(x_1) \tag{19.23}$$

This means that the matrix elements of the potential A behave like the corresponding (color) current ones *and they are nonvanishing*. Therefore the OPE couples any current to both the quark and the gluon contributions. This feature is called *operator mixing*. Then we obtain a *matrix form of the Callan–Symanzik equation* in this case, written for a matrix $E(m)$ instead of the plain function $E(m)$, in Eq. (19.17):

$$E(m) = \begin{pmatrix} E_{NS}(m) \\ E_q(m) \\ E_g(m) \end{pmatrix} \tag{19.24}$$

The indices refer to non-singlet, i.e. the valence flavor parts, cf. Chapter 5, and to quark (q) and gluon (g), respectively, and we get

$$\left[\mu \frac{\partial}{\partial \mu} + \beta \frac{\partial}{\partial \alpha} - \gamma(m) \right] E(m) = 0 \tag{19.25}$$

Here the β-contribution is diagonal and m-independent, $\beta = -b\alpha^2$, but

$\gamma(m)$ is non-diagonal with the numbers $d = d(m)$ plain numbers:

$$\gamma = \alpha \begin{pmatrix} d_q & 0 & 0 \\ 0 & d_q & 2n_f d_{qg} \\ 0 & d_{gq} & d_g \end{pmatrix} \qquad (19.26)$$

This is a linear equation and, like any linear equation, can be diagonalised by taking combinations of the quark and the gluon components to obtain 'eigenstates' and in particular 'eigenfrequencies' from the diagonalised γ-matrix elements. We will not do this, nor will give the formulas for the numbers d, because it is done in detail in the original papers, [69], as well as in [52].

The main point is that the different elements of the matrix E in Eq. (19.24) can be written as linear combinations of the following kind:

$$E_j(m) = \sum_i (\xi)^{\delta_{ji}(m)} \mathscr{E}_i(m), \quad \xi = \frac{\alpha_1}{\alpha_s(Q^2)} \qquad (19.27)$$

(this means that the E_j behave as powers in $\log Q^2$!). The powers δ are derivable from the matrix elements in Eq. (19.26) and the coefficient in the QCD runnning coupling and the quantities $\mathscr{E}_i(m)$ are the initial values of the moments at the scale where the running coupling is α_1. Thus *the moments of the structure functions for both quarks and gluons will, according to this result, contain computable logarithmic power corrections in the large-Q^2 limit.* This behaviour is very well confirmed experimentally, at least inside the presently available Q^2-region.

For the non-singlet moments, in particular, there is only one term in the sum and it corresponds for each moment m to $\delta_m = d_q^{(m)}/b$. Therefore we have in this case the simple differential equation

$$\frac{dE_{NS}^{(m)}}{d\tau} = d_q^{(m)} \alpha_s E_{NS}^{(m)} \qquad (19.28)$$

where τ is defined in Eq. (19.7). The general results can be reformulated into a relation between τ-derivatives of the matrix E and the γ-matrix:

$$\frac{dE(m)}{d\tau} = \alpha_s \gamma(m) E(m) \qquad (19.29)$$

4   The DGLAP equations

As we have said before this is not the place to discuss the historical and intellectual developments with respect to 'who did what first'. But it is evident that many different contributions did occur independently.

One major contribution to the understanding of the physics is given in [5], where it is proved that the differential equations for the moments in Eq. (19.29) can be rearranged into equations for the parton structure functions themselves, nowadays known as the DGLAP (Dokshitzer-Gribov-Lipatov-Altarelli-Parisi) equations:

$$\frac{dq_j}{d\tau} = \alpha_s \int_x^1 \frac{dz}{z} \left[\mathscr{P}_q^q(z) q_j \left(\frac{x}{z}, \tau \right) + \mathscr{P}_{q_j}^g(z) g \left(\frac{x}{z}, \tau \right) \right] \qquad (19.30)$$

$$\frac{dg}{d\tau} = \alpha_s \int_x^1 \frac{dz}{z} \left[\sum_j^{2n_f} \mathscr{P}_g^{q_j}(z) q_j \left(\frac{x}{z}, \tau \right) + \mathscr{P}_g^g(z) g \left(\frac{x}{z}, \tau \right) \right]$$

The index j corresponds in this case to different q- and $\bar{q}$-flavors and we obtain back the splitting functions which were derived in Chapter 17, $\mathscr{P}_b^a$, for the splitting of a parton a into a parton b (cf. below for the behaviour of $\mathscr{P}$ when its argument approaches unity).

The main part of the proof in [5] is to show that the anomalous dimension matrix $\gamma(m)$ fulfils the identity

$$\gamma(m) = \begin{pmatrix} d_q(m) & 2n_f d_{qg}(m) \\ d_{gq}(m) & d_g(m) \end{pmatrix} = \int_0^1 dz \, z^m \begin{pmatrix} \mathscr{P}_q^q(z) & 2n_f \mathscr{P}_g^q(z) \\ \mathscr{P}_q^g(z) & \mathscr{P}_g^g(z) \end{pmatrix} \qquad (19.31)$$

This is straightforward if we use the formulas for the splitting functions and for the anomalous-dimension matrix. After that one can rely on a mathematical theorem which tells us that a moment equation can be inverted in a unique way. (The observant reader may note that the first equation of (19.30) has been summed over the different q-flavors.)

The even more observant reader will note that some of the splitting functions are singular for $z = 1$ and therefore the integrals in Eqs. (19.30) and (19.31) are not well defined. A closer examination tells us, however, that this singularity is closely related to energy-momentum conservation. Formally it turns out that the singular behaviour of the splitting functions (this is shown in detail in [52]) is cancelled by a proper account of the virtual corrections to the emissions.

The result for the non-singlet is obtained by taking the difference between the equations for the derivatives of two quark (or antiquark) species. In that way the gluon term in the first equation of (19.30) vanishes and we obtain a diagonal contribution from the same difference between the structure functions integrated over the $q \rightarrow qq$ splitting function.

There is a direct connection between the results using the MM and the OPE, as in [43] and [69], and the LLA results of Gribov and collaborators, [52]. For the latter case one follows the emission lines in high-order perturbation theory and rewrites the results as exactly the integro-differential equations (19.30).

In order to understand the physics we consider again the phase-space triangle; see Fig. 19.5. Suppose that we increase Q^2 for a fixed value of P_+, i.e. of the hadron energy, and for a fixed value of $x = x_B = Q_+/P_+$. This means that the left-hand side of the triangle, $-\log(Q_-)$, will move to

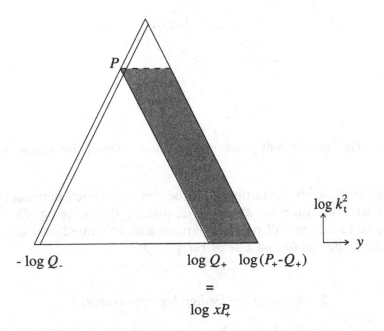

Fig. 19.5. The phase space corresponding to the emission of new partons at x, following an increase in Q^2, is shown, together with the region (shaded) inside which one must in an ISB model know which partons already exist.

the left. Then the DGLAP equations (19.30) describe the change of the structure function at the point P, corresponding to $k_\perp^2 = Q^2$ and x. This is the left-hand side of Eq. (19.30). On the right-hand side the change is related to the number of partons with values $x' > x$, each of which, due to the QCD processes that are possible, may decay into a parton at x. The region inside which we may sample such partons is shaded in Fig. 19.5.

Suppose that we consider such a parton decay, Fig. 19.6. The incoming parton will have a fraction x' of the hadron P_+, and we will assume for simplicity that it is massless. It will emit a massless gluon with $x_g = (1-z)x'$ thereby becoming a virtual parton with lightcone fraction $x = zx'$. Its virtual squared mass, which is usually related to the value $-Q^2$, can be calculated from the transverse momentum, $k_\perp$, in the emission by

$$Q^2 = \frac{k_\perp^2}{1 - z} \tag{19.32}$$

The transverse momentum variable, $k_\perp$, is compensated between the two partons emitted, so that they have $\pm k_\perp$ respectively, and the result of Eq. (19.32) stems from the conservation of the negative lightcone component.

The probability for the emission shown in Fig. 19.6 is given by the

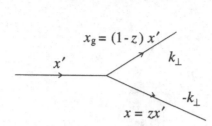

Fig. 19.6. The kinematics of parton decay with notation as discussed in the text.

splitting function, $\mathscr{P}(z)$, multiplied by the density of such partons, i.e. the relevant structure function at $x' = x/z$, and by the coupling. The result must be summed over all possible partons and integrated over all values of x', giving the right-hand side of Eq. (19.30).

5   *More on the leading-log approximation*

In Chapter 5 we derived the hadronic tensor $W_{\mu\nu}$, Eq. (5.49). In Fig. 19.2 we have shown a contribution with intermediate state $|X\rangle = |p_1, \ldots, p_n\rangle$; here $n = 4$. The state is produced by iteration (including the radiative corrections) along a main chain with propagators redistributing the large momentum transfer q into many rungs of the ladder. Although there are obvious similarities to the ladder diagrams for the unitarity equations in Chapter 10, the kinematics is different in this case, in which the virtuality is building up and the energy-momentum fraction is decreasing down the main chain.

The result in the LLA is that the main contributions stem from diagrams having the kinematical variables of the q_j-emissions *strongly ordered*:

$$q_{\perp 1} < q_{\perp 2} < \cdots < q_{\perp n} < Q; \quad 1 > z_1 > z_1 z_2 \cdots > x_B \equiv \prod z_j \quad (19.33)$$

The motivation is that to avoid strong damping from the propagators the large momentum transfer Q^2 must be partitioned over many steps. The larger is Q^2, the more steps are necessary for the energy-momentum flows in each rung of the diagram to be reasonably small.

If we use the approximate weight dw_j for every step (putting the gluon splitting function $\mathscr{P} \sim 1/z$ and $\log q_\perp^2 = \kappa$),

$$dw_j \sim \bar{\alpha}(\kappa) d\kappa_j \frac{dz_j}{z_j} \quad (19.34)$$

then, using the ordering given in Eq. (19.33) and the symmetrical resummation we have used before,

$$\int \prod_{j=1}^{n+1} dy_j \Theta(y_{j-1} - y_j) = \frac{Y^n}{(n)!}, \quad \chi(Q^2) = \int_{\log Q_0^2}^{\log Q^2} \frac{\alpha_0}{\kappa} d\kappa \quad (19.35)$$

we obtain the sum of the main contributions:

$$xg(x, Q^2) \sim \sum_n \frac{[\chi(Q^2) \log(1/x_B)]^n}{(n!)^2} \sim \exp\left[2\sqrt{\chi(Q^2) \log(1/x_B)}\right] \quad (19.36)$$

The result in Eq. (19.36) stems from a well-known formula for the modified Bessel function I_0, which we also encountered in connection with the λ-measure, in Chapter 18.

The upshot is that in the structure functions we have a very slow Q^2-dependence $\sim \log\log Q^2$ but there is an increase for small values of x_B. From numerical studies of the DGLAP equations this increase for small x_B is confirmed. These results are quite different from the ones we obtained from the method of virtual quanta in Chapters 2 and 5. The equations do not allow a stable constant behaviour for small-x values of the combination xg and therefore neither the gluon nor the ocean $q\bar{q}$ content will behave in accordance with Feynman's speculations on the wee parton spectrum. We are evidently in a different dynamical situation!

19.5 The Lipatov results and a critique on the stability

We will show that the situation may be even worse with respect to the small-x_B behaviour in an ISB scenario investigated by Lipatov and collaborators [29] (although in the end we present some doubts on the stability of the results, which will be further enhanced in section 20.8). They show that if we keep to the leading contributions in $\log(1/x)$ then there are many subleading contributions, neglected in connection with the transverse momentum generation, which may be essential for moderate and small Q^2-values and for very small x-values. In somewhat loose language we may say that Lipatov *et al.* have investigated the possibility that the transverse momenta are not ordered as in Eq. (19.33) but may go up and down in size along the ladder. This may happen many times if the ladder is very long counted in $\log(1/x)$ units.

The results of the DGLAP as well as the BFKL approach can be reformulated (cf. Eq. (19.36)) into an integral equation relating the contribution at the nth order, G_n, to the contribution at the $(n-1)$th order, G_{n-1}:

$$G_n(x, k_\perp^2) = \int_x^1 \frac{dz}{z} \int dk_{\perp 1}^2 K(k_\perp^2, k_{\perp 1}^2) G_{n-1}(z, k_{\perp 1}^2) \quad (19.37)$$

where the kernel K in the DGLAP case is very simple,

$$K(k_\perp^2, k_{\perp 1}^2) = \frac{\alpha_0}{k_\perp^2 \log(k_\perp^2/\Lambda_{QCD}^2)} \Theta(k_\perp^2 - k_{\perp 1}^2) \tag{19.38}$$

It is shown in [29] that the kernel should be more complex in order to contain the possibility that the new transverse momentum, $k_\perp$, is smaller than the one before, $k_{\perp 1}$. Although the kernel is not described here (cf. Section 20.8), there is a very general mathematical result for the case when the kernel in Eq. (19.37) is symmetric.

The way to solve integral equations of this kind is to assume that it is possible to write the kernel as a (sum of) factorisable contribution(s)

$$K(k_\perp^2, k_{\perp 1}^2) = u(k_\perp^2) v(k_{\perp 1}^2) \tag{19.39}$$

If K in Eq. (19.37) is of this form then we obtain a correspondingly factorised result for G:

$$G_n(x, k_\perp^2) = u(k_\perp^2) t_n(x) \tag{19.40}$$

where t_n is defined by an integral (containing the eigenvalue λ):

$$t_n(x) = \lambda \int_x^1 \frac{dz}{z} t_{n-1}(z), \quad \lambda = \int dk_{\perp 1}^2 u(k_{\perp 1}^2) v(k_{\perp 1}^2) \tag{19.41}$$

We may then iterate the equations and so obtain at the nth order of iteration

$$t_n \simeq \frac{[\lambda \log(1/x)]^n}{n!} \tag{19.42}$$

which leads to the following small-x behaviour:

$$xg \simeq \sum t_n \sim x^{-\lambda} \tag{19.43}$$

The kernel K in Eq. (19.37) is not factorisable in the BFKL case but there is a mathematical theorem that the eigenvalue λ and the eigenfunction u can be obtained as solutions to the following integral equation:

$$\int dk_{\perp 1}^2 K(k_\perp^2, k_{\perp 1}^2) u(k_{\perp 1}^2) = \lambda u(k_\perp^2) \tag{19.44}$$

There are in general many such solutions but the main behaviour will stem from the largest eigenvalue, which for a constant coupling α_s is for the BFKL kernel, [29] given by

$$\lambda_{max} \equiv \lambda_L = \frac{4N_c \alpha_s \log 2}{\pi} = \frac{12\alpha_s \log 2}{\pi} \tag{19.45}$$

(N_c is the number of colors). For $\alpha_s \simeq 0.2$ one obtains a value of $\lambda_L \simeq 0.5$.

It is of some interest to note that if we use as input a gluon structure function with $xg \sim x^{-0.5}$ then the DGLAP equations stabilise around such a behaviour also when $\log(1/x)$ becomes very large.

We also note that it is necessary to project the boundary conditions (i.e. the assumptions about the original parton wave function), onto the eigenfunction corresponding to this largest eigenvalue, λ_L. This provides a gaussian $\kappa = \log k_\perp^2$-contribution, i.e. there are some (logarithmic) transverse momentum fluctuations in the gluon emissions along the ladder that are of a gaussian character. As the length in the cascades corresponds to $\log(1/x)$ this *Brownian motion contribution* will have a width, according to BFKL, proportional to $\log(1/x)$. This is of interest with respect to the predictions of the transverse energy behaviour in the observable states in DIS. Unfortunately it seems as if the introduction of a running coupling will destroy this diffusion scenario, cf. Section 20.8.

In practice, what is done in the Lipatov treatment is to exchange a contribution stemming from the iterated integral in Eq. (19.36) for a plain number obtained in every iteration:

$$\frac{\chi(Q^2)^n}{n!} \to \lambda^n \qquad (19.46)$$

The DGLAP iteration is always directed towards larger $k_\perp$. Due to the finite available integration space it will then decrease with the number of iterations for a given top value, i.e. this contribution must diminish by a factorial. This upwards integration will always win out in the asymptotic limit when Q^2 is sufficiently large but the power may be relevant for smaller values of Q^2.

The reason for the increase in the number of small-x partons may seem rather obvious if we consider Fig. 19.5. According to the equations we are supposed to move from the right-hand lower corner in the phase-space triangle towards the point P, sampling all possible decays. In the ordinary DGLAP approach we are then supposed to move only upwards and leftwards in the shaded region. In the Lipatov treatment we are allowed to go both upwards and downwards, i.e. towards larger or smaller $k_\perp$ values, as we move to the left. This means that there are inherently more paths available in this case if we increase $\log(1/x_B)$ for a given Q^2.

The Lipatov mechanism and the BFKL effect are consistent ways to take into account some non-leading contributions but there should be corrections of the order α^2. The BFKL kernel, K, and its eigenvalues turn out to be very stable, for a constant coupling, against perturbations of the procedure. One may imagine that the (logarithmic) steps in the integration variable should be made into discrete steps (for a motivation, cf. [15] as described in section 18.6) so that Eq. (19.37) becomes a sum. This is easy to do but the results only correspond to tiny changes in the value of λ. Mueller, [4], has also considered the production process in the transverse coordinate space and again obtains the BFKL eigenvalues from the impact parameter distributions. We will, however, end this subsection with a remark to indicate that the BFKL results are rather unstable *with respect to the non-singular terms in the z-dependence of the iteration.*

There is one feature used in the BFKL approach, i.e. that in every splitting $q' \rightarrow pq$ the virtual (gluon) propagator q contains only a *small fraction z of the energy-momentum*, while most of this energy-momentum, the fraction $1 - z$, is carried away by the emitted gluon p (we have conventionally followed the z-pole contributions and Mueller's treatment uses a corresponding motion in the rapidity $dy = dz/z$). It is then necessary for consistency to demand that the major contributions stem from regions where z is actually small, i.e. in this convention that $z < \exp(-a)$ for some real number a (which must satisfy $a > \log 2$ in order that $z < 1 - z$). If we introduce such a simple restriction into the integrations then

$$\int \prod_{j=1}^{n} dy_j \delta \left(\sum_{j=1}^{n} y_j - Y \right) \rightarrow \int \prod_{j=1}^{n} dy_j \delta \left(\sum_{j=1}^{n} y_j - (Y - na) \right)$$
$$= \frac{(Y - na)^{n-1}}{(n-1)!} \qquad (19.47)$$

(keeping to the notation in Eqs. (19.36) and (19.35)). We have introduced the domain restriction in the expression following the arrow; the final expression summed over all values of n will no longer provide the BFKL exponential. It is straightforward, using the Stirling approximation to the factorial, to obtain the change to Eq. (19.43) as a power in $1/x$ with $\lambda_L \rightarrow \rho$, where ρ is determined by $\log(\lambda_L/\rho) = a\rho$. Thus the power in $1/x$ will be diminished so that $\lambda_L \rightarrow \rho \simeq \lambda_L(1 - a\lambda_L)$.

We conclude that *the BFKL mechanism obtains a large part of its contributions from the possibility of emitting the gluons along the ladder with moderate-to-small values of* $1 - z$, i.e. with moderate-to-large z-values. Actually this implies that one must take very many steps in order to obtain small x_B-values. One may then seriously doubt that it is allowable to neglect interference in the emissions and we will find in Section 20.8 that the QCD coherence properties are not fulfilled. We note, however, that the correction exhibited above is of order α^2 (which is expected in the BFKL treatment, and it has been repeatedly pointed out by the original authors that there should be such corrections). But it should also be noted that the correction is very large! It changes the (negative) x-power from 0.5 to about 0.3.

19.6 The CCMF model, interpolating between the DGLAP and the BFKL contributions

In this section we will consider how the DIS contributions appear in the formalism developed by Marchesini and his collaborators [44]. The ensuing model will be called the Ciafaloni-Catani-Marchesini-Fiorani (CCMF) model. This is one of the major efforts that anybody has undertaken in perturbative QCD. It was also pursued to a successful end (which has not been the case with most of the valiant efforts based upon 'good dynamics', which my generation have pursued!).

In the CCMF model there is a clever choice of the initial-state bremsstrahlung (ISB) set (p_j), which we discussed in section 19.2 in connection with the fan diagram in Fig. 19.1. (Remember also the notation, (q_j), for the propagators which connect the emission points along the fan diagram.) This choice can be described as the most general possible that is compatible with

- the QCD coherence conditions (the strong angular ordering, as described in Chapter 17)

- energy-momentum conservation, as implied by Eq. (19.1), and the possibility of keeping the (p_j) massless.

All emissions are ordered in rapidity, which (due to the relation between angle and rapidity, i.e. for a massless particle $y = \log \cot(\theta/2)$ with θ the ordering angle) means *strong angular ordering* along the chain, i.e. that the QCD coherence conditions are fulfilled. The CCMF model then picks the ISB set (p_j), from the set of all emissions, as those emissions each of which is *not* followed (in the rapidity ordering variable) by another one with a larger lightcone energy-momentum $p_+ (= p_0 + p_3 \equiv p_\perp \exp y)$. In this way the chosen p_j has a larger 'energy' than the rest and one may, in the leading-log approximation (LLA), neglect the recoils from the emission of the final-state bremsstrahlung (FSB).

More precisely, in terms of the ordinary variables $z_j, \mathbf{p}_{\perp j}$ with $q_{+j} = z_j q_{+(j-1)}$ and $\mathbf{q}_{\perp j} = \mathbf{q}_{\perp(j-1)} - \mathbf{p}_{\perp j}$, the CCMF choice for the q_+ implies (in the LLA) that $q_{+j} \ll q_{+(j-1)}$. Therefore the splitting function is again approximated as $\mathscr{P}(z) \propto 1/z$ so that z is small enough for the approximate relation $1-z \simeq 1$ to hold, which means that $p_{+j} = q_{+(j-1)}(1-z_j) \simeq q_{+(j-1)}$. Further, the gluons in the sets $(h)_j$ are, in accordance with the LLA, treated as soft enough that the p-vectors can be taken as on-shell and massless but the propagator vectors q are all spacelike. The transverse momenta of the propagators $q_\perp$ are dominated by the $p_\perp$-emissions in the neighborhood, see below. A major kinematical constraint is

$$q_{\perp j}^2 > z_j p_{\perp j}^2 \tag{19.48}$$

If this is not fulfilled then the virtuality of the propagator will, in the LLA, fulfil $|q^2| \gg q_\perp^2$, which implies strong suppression. Each step in the emission chain is, in the CCMF model, described by the weight

$$\bar{\alpha} \left(\frac{dz_j}{z_j} \right) \frac{dq_{\perp j}^2}{q_{\perp j}^2} \Delta_{NE}(z_j, q_{\perp j}, p_{\perp j}) \tag{19.49}$$

Here $\bar{\alpha}$ is the effective coupling (including color factors) and Δ_{NE} is the so-called 'non-eikonal form factor', with

$$\Delta_{NE}(z_j, q_{\perp j}, p_{\perp j}) = \exp[-\bar{\alpha} \log(1/z_j) \log(q_{\perp j}^2 / z_j p_{\perp j}^2)] \tag{19.50}$$

The first major result in the CCMF model is this non-eikonal form factor, corresponding to the radiative corrections for the choice of the ISB set defined above (for the second, i.e. the fact that there are no FSB emissions with $p_\perp^2 > -q^2$, see below). We note in particular that due to the properties of this form factor small values of z_j and $p_{\perp j}$ in Eq. (19.49) are effectively cut off if we assume that $q_{\perp j}$ is finite.

The negative exponential in the non-eikonal form factor corresponds to an area multiplied by the effective coupling $\bar{\alpha}$. We will end this section with a description of this area (and some associated ones) and show that we may interpret the occurrence of the non-eikonal form factor just as an ordinary Sudakov factor, i.e. there is a region excluded for gluon emissions because of the particular choice of ISB in the CCMF model.

In Fig. 19.7 a set of gluon emissions is shown, denoted from the hadron front end $a, b, 1, c, 2, d, 3$. The gluons denoted by the numbers $1, 2, 3$ fulfil the requirements for the ISB gluons in the CCMF model and in each case there are surfaces Aj, Bj, Cj exhibited (in between the consecutive gluons). The gluons denoted by letters, however, are all FSB gluons, i.e. they do not fulfil the CCMF conditions of rapidity and p_+-ordering necessary for ISB gluons. Note that the gluons denoted a, b are followed in rapidity by $p_{+1} > p_{+a}, p_{+b}$, gluon c by $p_{+2} > p_{+c}$ and gluon d by $p_{+3} > p_{+d}$. Actually all possible gluons inside the three regions denoted Aj, $j = 1, 2, 3$, in the figure are FSB gluons in the CCMF model, i.e. the gluons occurring inside the regions Aj may, according to the rules of the CCMF model, be emitted in connection with the ISB gluon j.

To understand the relationship between these surfaces and the non-eikonal form factor we start with the transverse momentum properties of the emissions. From the relationship $\mathbf{q}_{\perp j} = \mathbf{q}_{\perp j-1} - \mathbf{p}_{\perp j}$ we obtain in the leading-log approximation that there are three possible situations:

T1 $\mathbf{p}_{\perp j}^2 \simeq \mathbf{q}_{\perp j}^2 \gg \mathbf{q}_{\perp j-1}^2$, i.e. the propagator transverse momentum increases owing to the emission;

T2 $\mathbf{q}_{\perp j}^2 \simeq \mathbf{q}_{\perp j-1}^2 \gg \mathbf{p}_{\perp j}^2$, i.e. the emitted gluon momentum is much smaller so that the propagator retains its momentum in such a step;

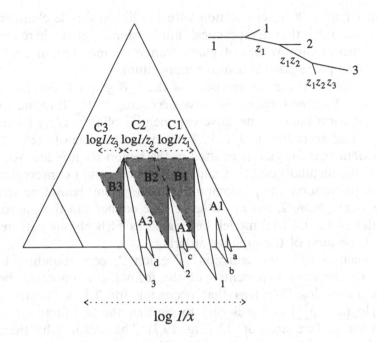

Fig. 19.7. Examples of gluon emissions in the CCMF model with the associated areas Aj, Bj, Cj. The notation is defined in the text.

T3 $\mathbf{q}_{\perp j-1}^2 \simeq \mathbf{p}_{\perp j}^2 \gg \mathbf{q}_{\perp j}^2$, i.e. as the emitted gluon picks up most of the momentum of the incoming propagator (index $j-1$) the momentum of the outgoing propagator (index j) goes down.

In Fig. 19.7 the examples are chosen so that emission j corresponds to case Tj above. There are, besides the surfaces Bj, also regions Aj and Cj and we note that the upper boundaries of the regions Bj (corresponding to the lower boundary of the regions Cj) each correspond to a measure of the relevant propagator transverse momentum, to be precise to $\log q_{\perp j}^2$. We have also indicated the distances $\log(1/z_j)$ in each step. We will now investigate the negative exponential of the non-eikonal form factor using these examples. We will find that it corresponds to (besides the effective coupling $\bar{\alpha}$) *the size of the regions Bj in phase space which are excluded due to this particular choice of ISB.*

We firstly note that there are no emissions inside the regions denoted Cj. It is shown in the CCMF model that *inside these regions there can be no emissions because the real emissions are just cancelled by the virtual corrections.* Although this statement is hardly noticeable for the results in Eq. (19.50) it is arguably the major result of the CCMF model (and is very difficult to prove!). We will provide a dynamical reason for this feature in Chapter 20 in connection with the linked dipole chain model, [16]. But we note that its practical implication is that there can be no FSB emissions for values of gluon transverse momentum above the corresponding propagator transverse momentum.

We now note that for the emission of the ISB gluon 1, corresponding to the case $T1$ defined above, we have according to the formula for the non-eikonal form factor a negative exponential of $\log^2(1/z_1)$ (note that $\mathbf{p}_{\perp 1}^2 \simeq \mathbf{q}_{\perp 1}^2$) and according to Fig. 19.7 this is just the size of area $B1$.

It is evident that for gluons emitted in the region $B1$ it is impossible to fulfil both the angular condition and energy-momentum conservation. In particular, to conserve the p_+-component such gluons cannot be sent out from the 'next' gluon, 2, but at the same time neither can they, according to the rules of the CCMF model, be treated as FSB gluons with respect to gluon 1, because of the rapidity ordering.

In connection with the emission of gluon 2, corresponding to the case $T2$, the negative exponential of the form factor contains, besides the same factor $\log^2(1/z_2)$ as that occurring for $T1$, a further factor $\log(1/z_2)\log(q_{\perp 2}^2/p_{\perp 2}^2)$ and it is easy to see that the two factors together make up the surface area of $B2$ (Fig. 19.7). The reason why there can be no gluon emission inside the region $B2$ is essentially the same as for $B1$, i.e. the rapidity ordering of the CCMF model forbids the region 'behind' gluon 2, while energy-momentum conservation does not allow these gluons to be emitted by the next ISB gluon, 3.

Finally, for the emission of gluon 3, we note the combined effects of the constraint in Eq. (19.48) and the fact mentioned above that the region $C3$ is (in the LLA) a strictly forbidden emission region. This means that due to Eq (19.48) there can be no gluon emission in front of the negative lightcone line of p_{-3} and above the new propagator transverse momentum $k_{\perp 3}$. The size of the area $B3$ again equals the (negative) logarithm of the non-eikonal form factor.

We may remark explicitly that although the total state weight in the CCMF model, given by the allowed ISB gluon phase space multiplied by the non-eikonal form factor, contains recognition of only the surface areas B_j (which must be empty due to the particular choice of ISB in the model) the regions A_j are not forgotten. Inside these regions any number of FSB gluons may be emitted in a state defined by the ISB gluons. All these emissions can be summed up so that the weight becomes 1. For any particular exclusive state, however, there will of course be a Sudakov factor, corresponding to the regions not used in that state.

It is possible, in accordance with [44], to write out integral equations for the structure functions in the CCMF model and, as we may expect, these equations will have solutions with behaviour in between the DGLAP and the BFKL results. The equations are, however, somewhat complicated and it is difficult to use them in connection with a Monte Carlo simulation. The reason is that to make it into a consistent stochastical process it is necessary to keep track of the constraints in z, rapidity (with respect to the earlier emission), $p_\perp$ and $q_\perp$. In Chapter 20 we will present a generalisation of the CCMF model, the linked dipole chain model, in which both the weight distributions and the Sudakov factors are simpler so that the implementation in terms of a Monte Carlo simulation process is straightforward.

19.7 The GLR model of reinteraction of partons

The basic idea in the GLR model is that if the number of gluons becomes very large then the partons will be very closely packed inside the proton. There will then be a correspondingly large probability for them to reinteract.

In order to find when this starts to happen we will use the arguments of Gribov, Levin and Ruskin, [67]. They noted that the number of gluons per unit rapidity is given by $dn_g/dy = xg$. If all these gluons are inside a transverse (impact parameter) region πR^2 then the average surface density is $xg/(\pi R^2)$. Further, the gluonic cross section at a given value of Q^2 is $\sigma_g \sim \alpha_s(Q^2)/Q^2$ and therefore it was concluded in [67] that the crucial parameter for a possible reinteraction is

$$\Omega(x, Q^2) = \frac{\alpha_s(Q^2)xg}{Q^2 \pi R^2} \qquad (19.51)$$

As long as the parameter Ω is very small the ordinary DGLAP equations (provided with the proper angular ordering) are expected to work. But, for sufficiently small values of x, when xg becomes large two gluons from different cascade chains may interact thereby fusing the different ladders and decreasing the total multiplicity.

The authors of [67] have been able to take into account such two-body interactions, cf. also [95]. The result is that the DGLAP equation for the gluon distribution obtains a negative contribution

$$\frac{dg}{d\tau} = \alpha_s \int_x^1 \frac{dz}{z} \left[\sum_i^{2n_f} \mathscr{P}_g^{q_i}\left(\frac{x}{z}\right) q_i(z, \tau) + \mathscr{P}_g^g\left(\frac{x}{z}\right) g(z, \tau) \right] - I$$

$$I = \frac{81\alpha_s^2(Q^2)}{16R^2Q^2x} \int_x^1 \frac{dz}{z} [zg(z, Q^2)]^2 \qquad (19.52)$$

This contains a non-linear contribution in which the square of the gluon structure function occurs together with a set of color factors and finally an unknown size parameter, R, with the dimension of length. The meaning of R is that it comes from the integral over the transverse region inside which the interaction takes place.

It is pointed out in [95] that if $R \sim 1$ fm, the approximate proton radius, then the correction term is very tiny indeed and will not play any role for the HERA region.

This has lead to some speculation that inside the proton there may be more or less dense subregions and that the correction term may play a large role in such a dense and small subregion, a 'hot spot'. We will, however, not pursue the question any further in this book.

20

Inelastic lepto-production in the Lund model, the soft radiation model and the linked dipole chain model

20.1 Introduction

As usual in connection with the Lund model we will start with a semi-classical string scenario to describe a deep inelastic scattering (DIS) event. We will show that if the target is a meson state of the yoyo type then the final state obtained, after a large momentum transfer to one of the endpoints q_0 or $\bar{q}_0$, is very similar to the state obtained in an e^+e^- annihilation event as long as we neglect gluon emission.

Depending upon the Lorentz frame used to describe the string motion, we obtain different shapes of the final state. If we use the final-state cms frame we obtain a longitudinally stretched string, which, if it does not break up, will have a length $\simeq W/\kappa$ just like the flat e^+e^- annihilation $q_0\bar{q}_0$-strings of the same cms energy W. This time the state will, however, contain a small bend. But the transverse extension of the state is always of order m/κ, with m the original meson mass. Therefore the transverse dimensions are in general negligible compared to the longitudinal size.

We will show that the properties related to the breakup of such a string state only depend upon *projections onto the momentum transfer direction*. Therefore it is easy to generalise the Lund model fragmentation formulas to such final states.

After that we consider a corresponding model for DIS from a baryon. We will use a simple but nontrivial string model for the baryon. We show that even if the baryon is not a yoyo string state, the final-state string will nevertheless look very similar to a stretched-out yoyo, almost independently of how it is hit.

This will also lead us to some considerations of baryon fragmentation. We will show how the baryon number is conserved in the Lund model breakup process. We note, however, that there are many question marks in connection with baryon fragmentation, partly because there are so few experimental data available at present.

Then we discuss the way in which the Lund model treats situations when the momentum transfer Q^2 acts on an ocean quark or antiquark. This is a part of the cross section which is expected to grow fast with the energy and therefore it will be more and more important for the future.

In the original Lund treatment as it is implemented in the Monte Carlo simulation program LEPTO, [63], the final state is treated as a two-string situation with a rather *ad hoc* parametrisation of the energy sharing between the three original valence quarks and the left-over ocean q ($\bar{q}$), if it is the ocean $\bar{q}$ (q) that is struck.

It is possible to make a case for a more precise structure from considerations of the time development of the final state, at least if the ocean partons are intrinsic parts of the wave function of the hadron. We present these ideas as they are implemented in the Monte Carlo program ARIADNE, [92].

Next we go over to gluon radiation for a DIS event in the Lund model. We introduce a model, the soft radiation model (SRM), [11], of a different kind from the one we considered in the last chapter in the context of the conventional ISB scenario.

The basic ideas are the following. Even if the final state in a DIS event develops on a long time scale in a way similar to the corresponding state in an e^+e^- annihilation event, there is one major difference on the short time scale relevant to gluon emission. In e^+e^- annihilation the produced q_0 and $\bar{q}_0$ are both expected to be essentially pointlike. This means that all the energy is readily available for gluon emission when they start to separate, forming the original dipole.

In DIS events, the struck-out parton is expected to be pointlike in the same sense. If it is a q_0, i.e. a color-3, *there is no reason why the corresponding color-$\bar{3}$ charge should be localised in the same way*. It is, in fact, probably spread out over all the remainder state. Similarly, while the struck parton's energy-momentum after the collision, in the notation of the last chapter Q_-, is strongly concentrated, the hadron remainder will contain the total energy-momentum $P_+ - Q_+$ in a (space-)extended form.

This means that the radiation in this case occurs from something similar to an extended 'antenna source'. It is well known that coherent emission of wavelengths much smaller than the antenna size is strongly suppressed. We have already discussed the notion of a *form factor* to describe extended charge distributions, cf. the size parameter in Eq. (5.47). For a wavelength larger than this size there is no difference between a pointlike and an extended charge. But for a smaller wavelength, corresponding to momentum transfers larger than e.g. the parameter $M_0 \simeq 0.7$ GeV in the elastic baryon form factors of Eq. (5.47), there is power suppression.

The SRM suggests one possible method to treat the extension. It contains two basic parameters, which have been investigated using the experimental data available at present. One of the parameters, corresponding to the inverse of the transverse size of a hadron, turns out from the experimental data to be of the order 0.5–1 GeV. The other parameter corresponds to the (space) dimensions of the extended system and for it we obtain in a very stable way the number 1. This would evidently be typical of a string or, remembering a motivation for the string presented in Chapter 6, of a vortex-line force field.

Since the SRM was suggested several years ago, [11], based upon arguments like those presented above, it is quite surprising that its implications are similar to those obtained from the CCMF model, [44], cf. Section 19.6. We will show that the so-called non-local form factor in that approach actually on average cuts off the gluon radiation along the same lines in phase space as the SRM.

After that we will introduce a different approach, which is very natural within the Lund model, where *all QCD properties are treated in accordance with dipole properties*. The Lund dipole cascade model describes the partonic states in timelike cascades (occurring in particular in e^+e^- annihilation events) in terms of the decay of color dipoles, as discussed in Chapters 16–18. Then the fragmentation process converts the ensuing string states into 'ultimate dipoles', i.e. hadrons modelled by means of $q\bar{q}$-states, with the charges stemming from different vertices connected by string-field pieces.

In section 20.7 we will show that if one probes such a hadronic dipole entity then the virtual states, i.e. the states encountered by a short-time probe, can also be most easily described in terms of dipoles, in particular in terms of *chains of linked dipoles*. To be more precise we will show that the CCMF model described in section 19.6 can be generalised and simplified into such a statement. As this is a very recent result, [16], within the Lund Group we will be content to describe the general ideas incorporated in the *linked dipole chain model* (LDC) and only briefly consider the consequences (in particular with respect to the ongoing measurements in HERA).

20.2 The classical motion of a yoyo-string exposed to a large momentum transfer at an endpoint

We start by considering the motion of a yoyo-hadron the constituents of which are originally moving in and out, as shown in Fig. 20.1 in the lab frame, in which the state is at rest before the momentum transfer. We have

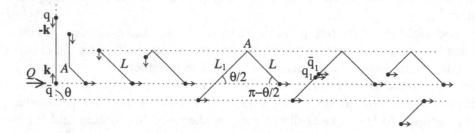

Fig. 20.1.   The development of a yoyo-state, originally at rest, after a large momentum transfer. The notation is explained in the text.

already discussed string motion in detail several times and will therefore be brief.

The two endpoint particles will have momenta $\pm \mathbf{k}$ along the string direction at the beginning. The main phases in the development after the string is hit at an endpoint with a momentum transfer $\mathbf{Q}$ are as follows.

- Suppose that the endpoint $\bar{q} \equiv \bar{q}_0$ is struck and moves along the direction $\mathbf{Q} + \mathbf{k}$ (at an angle θ with respect to the original string direction) with constantly decreasing energy-momentum. Its partner at the opposite endpoint, q_0, is as yet unaware of this and moves downwards gaining energy-momentum from the string.

- A straight string section L (with angle $(\pi - \theta)/2$ with respect to the string) is formed and a disturbance 'corner' A moves along the string (but does not carry any energy-momentum). The transverse velocity of L is $v_\perp = \cos(\theta/2)$ since the corner and $\bar{q}_0$ move with the velocity of light.

- The q_0 meets the corner A. The string is 'soft' and affects q_0 only with the finite force κ so that q_0 just continues downwards. A new segment L_1 is formed while the q_0 is losing energy, until it stops (at the same point where it would have stopped if there had been no momentum transfer to $\bar{q}_0$). The transverse velocity of L_1 is $v_{\perp 1} = \sin(\theta/2)$.

- The q_0 is then dragged along by the string and it will move along a line parallel to $\mathbf{Q} + \mathbf{k}$. If we boost to a system in which L_1 is at rest then q_0 actually moves along the string segment. In the lab frame the angle between the string segments L and L_1 is always $\pi/2$.

- From now on the string segments L and L_1 both serve as 'transporters' of energy-momentum from $\bar{q}_0$ to q_0.

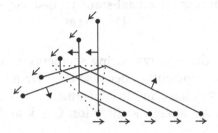

Fig. 20.2. The development of a yoyo-state exposed to a large momentum transfer, shown in the final-state cms.

We also show in Fig. 20.1 a breakup of the string; its characteristics will be discussed further below. But before this the string state looks like a rigid body moving forward in a triangular shape along the $(\mathbf{Q}+\mathbf{k})$-direction. In the lab frame the state evidently has an extension of the same order both transversely and longitudinally with respect to the momentum transfer. However, the state is moving very fast longitudinally if the momentum transfer is large.

The total momentum is evidently $\mathbf{Q}$ and the total energy is $|\mathbf{Q}+\mathbf{k}| + M - |\mathbf{k}| \simeq |\mathbf{Q}| + M - |\mathbf{k}|(1 - \cos\theta)$, where we have developed the square root $|\mathbf{Q}+\mathbf{k}|$ as $\sqrt{Q^2 + k^2 + 2\mathbf{k}\cdot\mathbf{Q}}$ to lowest order in $|\mathbf{k}|$. We conclude that the velocity is $\simeq 1 - [M - 2|\mathbf{k}|\sin^2(\theta/2)]/Q$.

It is also useful to consider the state in the cms after the momentum transfer, i.e. in this frame the struck $\bar{q}_0$ will move away with the cms energy $W/2$ along the $(\mathbf{Q}+\mathbf{k})$-direction and the remainder state will move as a whole with energy $W/2$ in the opposite direction. An approximate formula for W is $W \simeq \sqrt{2Q[M - 2|\mathbf{k}|\sin^2(\theta/2)]}$, which is easy to derive from the results in the last paragraph.

This situation is shown, for simplicity for the choice $\theta = \pi$, in Fig. 20.2 and we leave it to the reader to go through the development. The most noticeable thing in this case is that the state has become much longer longitudinally, i.e. if it does not break it will now be of order W/κ, while it is still of order m/κ transversely.

The lab-frame size is evidently an effect of Lorentz contraction, cf. Chapter 2. In the cms frame the bend around the corner is hardly noticeable and the situation is very similar to the flat e^+e^- annihilation $q\bar{q}$-strings we have treated before.

20.3   The fragmentation of a final-state yoyo-string stemming from a DIS event

We start by noticing that the two string segments L and L_1 in Fig. 20.1 both carry energy and momentum because they are moving. Thus a small element, dl, along L_1 will have, in the lab frame, energy dE and momentum $d\mathbf{p}$, with components dp_ℓ along the direction $\mathbf{Q} + \mathbf{k}$ and $dp_\perp$ transverse to it:

$$dE = \frac{\kappa dl}{\cos(\theta/2)}, \quad dp_\perp = \kappa dl \sin(\theta/2), \quad dp_l = \kappa dl \tan(\theta/2) \sin(\theta/2)$$

(20.1)

If the string breaks up into small parts then all of these, besides one, will be plain yoyo-states. The exception is the part containing the bend. Just as in connection with the breakup of strings with internal excitation gluons, cf. Chapter 15, we assume that the occurrence of such a bend still allows the same projection onto the hadronic states of a given mass.

We will next show that all properties related to the decay of the string depend only upon the longitudinal projections along $\mathbf{Q} + \mathbf{k}$. We start by considering the production of a $q_1\bar{q}_1$-pair along the segment L_1 in accordance with Fig. 20.1.

It is easy to convince oneself that, if the q_1 and $\bar{q}_1$ start out as massless and without energy, the $\bar{q}_1$ will move along the dotted line and together with the q_0 will form a yoyo-state. To prove this note that in the rest frame of the string piece L_1 the q_1 and $\bar{q}_1$ will move along the string with the velocity of light in opposite directions. Therefore in the frame where L_1 moves with the transverse velocity $\sin(\theta/2)$ they will both move at an angle $\theta/2$ with respect to the L_1-direction.

If we assume that the break occurs at a distance δ from the q_0 then we may calculate the following energy-momentum fraction,

$$z_- = \frac{E - p_\ell}{(E - p_\ell)_{tot}}$$

(20.2)

for the string piece. The variable z_- is Lorentz-invariant and it is the relevant variable in the target fragmentation region. For large energies it coincides with the Feynman scaling variable x_F, which is often used in hadronic collisions.

We note that $(E - p_\ell)_{tot}$ corresponds to the total longitudinal size of the system and that for the system $(\bar{q}_1, \delta, q_0)$ we have

$$E - p_\ell = \kappa\delta \cos(\theta/2) = \kappa\delta_\ell$$

(20.3)

according to the formulas derived above. Note that the energy-momentum of the $\bar{q}_0$ does not occur in the difference. This means that the variable

z_- only depends upon the longitudinal projection, $\delta_\ell = \delta \cos(\theta/2)$, of the string size. Further, the remainder system after the $q_1\bar{q}_1$-break will move as a rigid triangle with a final-state mass $M_1^2 \simeq (1 - z_-)W^2$, which again only depends upon the longitudinal projection.

Finally, if the production probability per unit time for a $q\bar{q}$-pair is proportional to the tension in the rest system of the relevant string piece then, owing to time dilation, it will be proportional to $\sqrt{1 - \mathbf{v}_\perp^2}$ in a frame where the string piece moves with transverse velocity $\mathbf{v}_\perp$, as was shown in Chapter 15:

$$\mathcal{T} = \kappa\sqrt{1 - \mathbf{v}_\perp^2} \tag{20.4}$$

In this way the probability of producing a pair in the string element $d\sigma$ in L_1 is again proportional to the longitudinal projection:

$$d\sigma\sqrt{1 - \mathbf{v}_\perp^2} = d\sigma \cos(\theta/2) = d\sigma_\ell \tag{20.5}$$

If, however, we consider the breakup properties in the segment L we find a corresponding factor $(\sqrt{1 - \mathbf{v}_\perp^2})_L = \sin(\theta/2)$. This will be the longitudinal projection factor for all elements of L. In this case the relevant fragmentation variable is z_+ and we note that we will have the relation

$$z_+ z_- = \frac{m_\perp^2}{W^2} \tag{20.6}$$

Thus for particles with large z_+ we will have tiny z_- components, corresponding to (seemingly) small longitudinal string elements.

We conclude that Lund model fragmentation can be performed just as for an e^+e^- annihilation event and that the process is Lorentz-invariant if we use the longitudinal projections of the string state. In this case there is also some intrinsic transverse momentum, which stems from the original motion of the q_0 and the $\bar{q}_0$ before the momentum transfer. We note that the transverse momentum component of the $\bar{q}_0$, $\mathbf{k}$, is carried forward to the so-called *quark fragmentation region* or *current fragmentation region*, while the component carried by the q_0, $-\mathbf{k}$, will be subdivided among the final-state particles in the target fragmentation region.

This means that there is a long-range compensation (i.e. the compensation occurs in regions with large relative rapidity) of this transverse momentum. This original motion is usually termed the *Fermi motion* inside the hadronic state (in accordance with the nomenclature for the motion of the nucleons in a bound nucleus). It turns out, however, that this effect is hardly noticeable at large values of Q, compared to the noise stemming from the gluon emission.

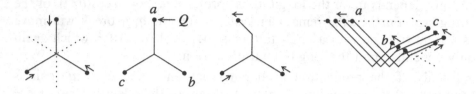

Fig. 20.3. A simple string model for a baryon and the way the state responds to a large momentum transfer.

20.4   A model for baryon fragmentation

1   A valence, i.e. endpoint, q is struck

In order to obtain a nontrivial model for an initial baryon state we assume that the three q-particles are all connected via strings to a common point, which we will call the *junction*, in accordance with Fig. 20.3. The junction does not carry any energy-momentum. It is merely a device which moves in such a way that the total tension at rest for the three connected string pieces will vanish, i.e. the strings are in equilibrium.

The baryon model has in the same way as the yoyo-states the property that, in the mean, half the energy will be kinetic energy of the q-particles while the other half is potential energy in the string's constituent gluons. We consider the motion of this state when one of the q's, called a in the figure, is struck thereby undergoing a large momentum transfer. The other two are denoted b and c and the direction along the momentum transfer is **n**.

When a moves away along **n** there will be a bend on the adjoining a-string, which will move inwards towards the junction. Assuming that at the start of the motion all three q's are moving inwards, the bend will reach the junction at the same time as b and c arrive. After this the junction will start to move with a velocity determined by **n**, i.e. the a-direction of motion. The reason is that in the rest frame of the junction there must always be an angle $2\pi/3$ between the three string pieces.

We will not trace the rest of the motion in detail but we note that both b and c will continue to move towards the places where they would have stopped if there had been no momentum transfer. Only after that will they start to respond to the momentum transfer and move after a. This is due to causality, i.e. that there is a finite transmission velocity of information along the force field.

It is a remarkable fact that, in almost all cases, in whichever direction **n** we hit a, roughly speaking one of b or c will go in the opposite direction

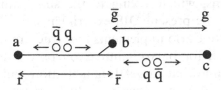

Fig. 20.4. The general appearance of a baryon which has been exposed to a large momentum transfer. The colors are exhibited together with the color fields between the forward quark *a* and the two backwards quarks (*b* and *c*) with the field $r\bar{r}$ and similarly between *a*, *b* and *c* with the field $\bar{g}g$. The drag force on produced pairs in the two segments is also exhibited.

while the other will go along **n**. This means as shown in Fig. 20.3 that *b* will end up with a rather short piece of string connecting it to the junction, while *c* will move away and end up far back before it is turned around. The size of the string segment around *b* will in practice serve as a kind of effective mass for *b*. The parton *b* will move repeatedly up and down around the junction but the effective mass will always be a rather small fraction of the original baryon mass.

The emerging picture is then of an essentially longitudinally stretched object with one of the valence *q*'s moving out along the momentum transfer, the second moving initially away and the third staying in the middle. A simplified version of such a state is shown in Fig. 20.4, and from this picture we may deduce a consistent way to treat the fragmentation in the Lund model.

We note that the whole state is a color singlet and consequently if *a* is colored r, *b* colored b and *c* colored g we will have at the forward end a color-3, r, joining a color-$\bar{3}$, composed of bg = $\bar{r}$ at the backward end. Similarly the segment between *c* and the forward-moving parts will behave like a $\bar{g}g$-string.

This means that if the string breaks up between *a* and *b* then the $\bar{q}$ from the break will be dragged towards *a*, and the *q* towards the color-$\bar{3}$ *bc*. Similarly, a break in the backward-moving string part will cause the $\bar{q}$ to move towards *c* and the *q* to move towards *ab*. In this way *there will always be a baryon produced around the junction*, while the rest fragments like ordinary Lund string pieces, including baryon-antibaryon production (possible at least in the high-energy forward part).

It is possible to make many further semi-classical remarks, [8], based upon this simple model. We will be content to say that there is no really well understood picture for the fragmentation of a baryon. Within the JETSET Monte Carlo scenario there are several phenomenologically reasonable predictions, which, owing to the small amount of data, are rather little tested in the present DIS experiments.

The JETSET Monte Carlo implements the structure we have introduced in the popcorn model for baryon-antibaryon production in Chapter 13. To that end we assume that target fragmentation will act just as in the $B\bar{B}$ production case after a $\bar{B}$ has already been produced. Then there is a given probability that the first-rank hadron is a particular baryon or that it is a particular meson and the baryon is then the second-rank particle. Then the remainder of the state is treated as an ordinary $q\bar{q}$-string fragmentation event.

2   An ocean q or $\bar{q}$ is struck

There is one kind of event upon which we have not touched as of yet, although it corresponds to a large cross section in DIS and the cross section grows with the energy. In these events an ocean quark or antiquark is struck, in contrast to the situation we have discussed above when we have always struck a valence q or $\bar{q}$, an endpoint of the string.

Here the target region will contain either three valence q's and one ocean q (if it is the ocean $\bar{q} \equiv \bar{q}_o$ that is struck) or else there will be three valence q's and an ocean $\bar{q}$ left (the ocean $q \equiv q_o$ is struck). We will call these situations cases A and B, respectively. There is no straightforward unique way to treat them dynamically and within the Lund model there have been two different suggestions, one which is used in the LEPTO Monte Carlo, [63], and one in ARIADNE, [10].

In the LEPTO treatment the final state contains the fragmentation of two distinct strings. For case A there will be a small baryonic string, between the ocean q_o and two of the valence q's, and a large mesonic one between the remaining valence q and the struck ocean $\bar{q}_o$. For case B there will be the opposite situation with the struck ocean q_o joined to a large baryonic string with two valence q's and the remaining valence q joined to a small mesonic string with the remaining ocean $\bar{q}_o$. 'Large' and 'small' here describe the energies and masses of the strings. There is a rather *ad hoc* parametrisation of the energy sharing: the treatment is only meant to provide a possible parametrisation of data.

It is possible to be a bit more sophisticated and to introduce some-what more structure, based upon the expected time development of the state. The ocean components of the structure functions must be consid-ered as long-lived parts of the fluctuations. Such parts, which we have discussed within the conventional ISB scenario in Chapter 19, stem from the DGLAP evolution equations. They are usually called 'intrinsic'.

There is also a possibility, which can be calculated in perturbation the-ory, that the process proceeds through the channel $\gamma^* g \rightarrow \gamma^*(q\bar{q})$ (usually named *boson-gluon fusion*). Then the lepto-production probe interacts with the short-lived fluctuation of an intrinsic gluon into a $q\bar{q}$-pair. It is difficult to distinguish between the two situations because the DGLAP mechanism stems from the same perturbative contribution.

In perturbation theory it is always the main momentum transfer behaviour which decides whether one diagrammatic contribution will dominate over other possible contributions. Later in this chapter, in connection with the linked dipole chain model, we will exhibit this feature in much more detail. For the present we make the following reasonable partitioning.

An intrinsic component in the hadron wave function ought to 'thermalise' in the sense that the ocean q_o and $\bar{q}_o$ should no longer be directly connected. If the interaction picks up an ocean q_o then we expect that the (anti-)color charge of its partner is distributed within the hadronic radius. Therefore there should be few (color-dynamical) differences between case B and the case when a valence quark is struck. In both cases the target is effectively in a color-$\bar{3}$ state.

There is nevertheless the difference that there are extra flavor numbers in the target region. In accordance with Lund model ideas we expect that when the struck q_o moves away the vacuum will compress the color field into a thin vortex-like string. When sufficient energy is stored in the field it will break up but this cannot occur until the string (in the rest system of the produced hadron) is larger than a hadronic radius.

This kind of out-moving string should be little affected by the finer details of the target charge distribution and we will therefore assume in accordance with the ARIADNE ansatz, [10], that *the momentum distributions of the final-state hadrons are the ordinary ones encountered in the Lund model*. Thus in the ocean-quark situation we will use the same breakup probabilities as before. We treat case A as in Fig. 20.5(a), i.e. we first produce a baryon at the end containing the q_o and two valence quarks and let the remaining energy go into a final string state between the struck $\bar{q}_o$ and the remaining valence q. Similarly, for case B, shown in Fig. 20.5(b), we choose to produce first a meson between the $\bar{q}_o$ and one valence q; the remaining energy-momentum then goes into a final string between the remaining valence constituents and the struck q_o.

The difference from the LEPTO case is that now there is no a priori energy sharing. The endpoint particle will have the same spectrum as if it were produced as the endpoint particle in any Lund string. Thus the whole momentum transfer is taken by the struck ocean component and we peel off one hadron at the backward end in a stochastical way, thereby defining the energy-momentum of the final remaining string.

In practice when one builds a Monte Carlo simulation it is necessary to include all possibilities. Therefore given a cross section for DIS events

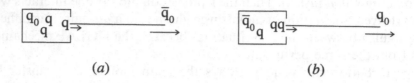

$$(a) \qquad\qquad\qquad\qquad\qquad (b)$$

Fig. 20.5.   The string breakup in the target fragmentation region when the momentum transfer acts on an ocean component that moves away leaving a state which is dragged apart, a particle being produced behind it. The cases (a) and (b) are explained in the text.

in accordance with Chapters 5 and 19 it is possible to subdivide it into the valence and the ocean parts. There is also the possibility of having a $\gamma^* g$ event (the boson-gluon-fusion (BGF) events described above) and then we use the energy-momentum partitioning prescribed by the perturbative expressions. In order not to double-count the contributions, we require in the BGF case that *the squared energy-momentum transfer between the q_o and the $\bar{q}_o$, $q_{\perp n}^2$, should be larger than Q^2*. If $q_{\perp n}^2 < Q^2$ then the contribution is already included in the conventional structure function cross section, i.e. we have the intrinsic ocean contributions mentioned above.

20.5   The soft radiation model

1   Preliminaries

We start by considering a simple model of DIS from a state with two charged particles bound together as in positronium, see Fig. 20.6. In this figure the state is assumed to move very fast and we also assume that the momentum transfer acts on the e^-. There are then two different but dynamically equivalent ways to look at the situation if we assume that the state is *loosely bound*.

P1 We may say that the e^+ is completely unaffected by the momentum transfer and therefore does not radiate at all. Then all the radiation stems from the e^--current, which comes in and is suddenly changed.

P2 We may alternatively say that there is no radiation from a bound state and therefore all the radiation stems from the dipole which is produced between the e^+ and e^- when the momentum transfer strikes.

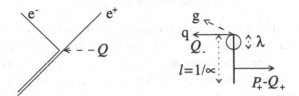

Fig. 20.6. The situation when a loosely bound system is affected by a large momentum transfer (left-hand part) together with the kinematics for the SRM (right-hand part).

We remember from the considerations on the bremsstrahlung cross section in Chapter 16 that these two descriptions will give the same result. They correspond to adding the two diagrams in Fig. 16.1 for case P1 and those in Fig. 16.2 for case P2. The sum of the two diagrams is in both cases gauge-invariant and also provides the same total radiation.

The situation is much more difficult to disentangle if we consider a *strongly bound system*. In that case both the e^+ and e^- are accelerated in the bound system throughout, but in accordance with quantum mechanics there is no radiation as long as the system is in its ground state before the interaction. For the case P1 it is necessary not only to describe the way e^- is localised in the wave function, i.e. to describe the transition from the bound state to a free e^- with a precise value of its energy-momentum. It is also necessary to describe what happens to the remainder system, in particular to the other charge(s), here e^+.

The basic proposal in the soft radiation model (SRM) is that the situation may be more easily described by case P2. This is in particular plausible for QCD because then the field itself is also color charged, so that there may be many charges accelerated during the interaction. We note that in the ordered dipole chains of the dipole cascade model, which directly mirrors the Lund string, these accelerations already occur in a coherent way.

It is also the case that the force field of a bound state has the energy distributed over a region, which for a vortex line or string would be one-dimensional. Any emission would only involve a part of the system and the ordinary radiation conditions mean a size of the order of a fraction of the wavelength. Therefore only that part of the energy-momentum is available for the emission.

In summary, in e^+e^- annihilation reactions the emitters can be considered as essentially pointlike objects but this may no longer be the case for DIS events. The probe, i.e. the field pulse, is well defined in size but the target hadron is extended.

2 *The details of the model*

If a q-parton, for example, is affected by a momentum transfer such that it obtains a lightcone energy-momentum Q_- then the rest of the state contains the corresponding energy, $P_+ - Q_+$, and the color-$\bar{3}$. Thus the initial dipole contains one pointlike object and one that is extended, in particular with respect to the carrying of energy-momentum (see the right-hand part of Fig. 20.6).

Then a gluon emitted from the phase-space element $(k_\perp, y)$ will need both a positive and a negative energy-momentum lightcone component:

$$k_\pm = k_\perp \exp(\pm y) \tag{20.7}$$

While the negative lightcone component is easily available from the large energy concentration in the struck parton, the positive one is spread over some region. It is a well-known property that coherent radiation of wavelength λ from an emitter of size l, where $\lambda \ll l$, stems only from a fraction of the emitter comparable to λ (shown as a 'bubble' in Fig. 20.6).

Then for a large-$k_\perp$ gluon (which has a small $\lambda \sim 1/k_\perp$) the total positive lightcone component will stem from a fraction of the emitter P_{+r} (the index r stands for the remainder after the struck parton leaves). This means that there will be strong damping of the radiation in the forward direction, i.e. in the target region.

In the SRM it is assumed that the phase-space limits are changed into

$$k_+ < \left(\frac{\mu}{k_\perp}\right)^d P_{+r}$$
$$k_- < Q_- \tag{20.8}$$

where the parameter μ corresponds to the inverse size and d is a number describing the dimensionality of the source-remnant.

The data from the EMC collaboration prefer a value of $\mu \approx 0.6$ GeV/c and $d \approx 1$, according to [11]. This corresponds to a mean transverse extension $\sim \pi/\mu \approx 1$ fm and an essentially one-dimensional energy density as in a string. The data covers a rather small (Q^2, x) range but we will nevertheless, perhaps rashly, assume that the result is valid for all available energies.

We will make one adjustment, however. In the case when there is a large x-interaction the remnant only contains energy-momentum-$P_{+r} = (1 - x_B)P_+$, which is less than that of the incoming hadronic state P_+. Assuming that it is *the energy density in the rest system that is constant* we are lead to expect that the effective emitter size l is correspondingly smaller and thus that μ behaves as

$$\mu \equiv \mu(x) = \frac{\mu_0}{1 - x} \tag{20.9}$$

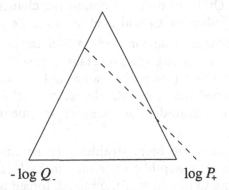

Fig. 20.7. The available phase space for gluon emission in the SRM shown in the κy-plane, where $\kappa = \log k_\perp^2$, with the region above the broken line forbidden (or exponentially suppressed) as described in the text.

This will mean that *the radiation in the target fragmentation region is the same, independently of the interaction*, i.e. the damping is governed by $(\mu_0/k_\perp)P_+$, independently of x and Q^2. The available energy will, however, only permit an emission if the energy-momentum component of the emitted gluon, k_+, is smaller than the available $P_{+r} = (1-x)P_+$.

The allowed emission region in the triangular phase space of the dipole is in this way changed (sometimes very much) in the target fragmentation region of a DIS event (cf. Fig. 20.7). From Eqs. (20.8), (20.9) we conclude that the new phase-space boundaries are

$$\log\left(\frac{k_\perp}{Q_-}\right) < y < \min\left[\log\left(\frac{\mu_0 P_+}{k_\perp^2}\right), \log\left(\frac{P_{+r}}{k_\perp}\right)\right] \quad (20.10)$$

Note that these curves correspond to straight lines in the κy-plane. The damping is, in the present case, of a step function character. If there were to be power suppression in the transverse momentum, as in a form factor, this would correspond to an exponential damping in the κ-variable. The results of the model are, however, rather insensitive to the possibility of such an exponential tail.

20.6 The relationship between the SRM and the non-local form factor of the CCMF model

In the CCMF model Marchesini *et al.*, [44], have used a very general technique to re-sum higher-order perturbative contributions to the basic ladder diagrams in QCD, in order to obtain the changes in the structure functions for small values of x_B and medium-to-large values of Q^2.

We have described their results in Section 19.6. In the CCMF model one follows an emission line along the fan diagrams, see Fig. 19.1 and there are gluon splittings into z (the main line) and $1 - z$ (the ladder gluons). It is implicitly assumed that the pole character of the splitting function implies $z \simeq 0$, i.e. a fast degrading of the energy-momentum fraction along the main line.

Therefore the initial-state bremsstrahlung (ISB) emissions result in a stochastical process of a step-like character in $\log(1/z)$: the next gluon emission is, according to the rules of the model, forced to be behind (i.e. in the dipole phase-space triangle to the left of) the line xP_+ where $x = \prod z_j$ from the previous emissions. Marchesini *et al.* have calculated the virtual corrections to these emissions and obtain a form factor, called non-eikonal, in the weights for producing such states. We have shown in section 19.6 that the non-eikonal form factor actually has the same properties as an 'ordinary' Sudakov factor, i.e. it corresponds to the negative exponential of the regions in which there is no emission, owing to the CCMF choice of ISB gluons. We will now show that *the average boundary of the area forbidden by CCMF is equal to the simple suggestion obtained from the SRM.*

We will thus assume that there is a set of steps in $\log(1/z)$, each one bringing the later emissions backwards a distance $l_j = \log(1/z_j)$. The emitted gluon j also contains transverse momentum, $\mathbf{p}_{\perp j}$, and the main line, i.e. the 'virtual propagator' line, then obtains a recoil $\mathbf{q}_{\perp n} = -\sum^n \mathbf{p}_{\perp j}$. We start with the case when the $p_\perp$'s are ordered such that $p_{\perp 1} \ll p_{\perp 2}$ etc. which also means (in the leading-log approximation) that $\mathbf{q}_{\perp n} \simeq -\mathbf{p}_{\perp n}$. Afterwards we will consider what happens if there is a different ordering (as the Lipatov approach allows, cf. Section 19.5).

In order to move, as we have decided, backwards and upwards in the phase-space triangle we will have to sum over all possibilities to reach the point $l = \sum_{j=1}^n l_j, h = \sum_{j=1}^n h_j$, with

$$h_j = \log(p_{\perp j}^2/p_{\perp j-1}^2) \tag{20.11}$$

We note that all transverse momenta are in accordance with the case T1 in Section 19.6 and we therefore use the relevant shape of the non-eikonal form factor, i.e. $\exp[-\bar{\alpha}\log^2(1/z_j)]$ with $\bar{\alpha} = 3\alpha_s/\pi$, to obtain

$$N(l,h) = \sum \int \prod \bar{\alpha} dl_j dh_j \exp(-\bar{\alpha} l_j^2) \delta\left(\sum l_j - l\right) \delta\left(\sum h_j - h\right) \tag{20.12}$$

All the quantities l_j and h_j are then positive and the result can be re-summed if we take the Laplace transform with respect to l, i.e.

$$\mathcal{N}(L, h) = \int dl \exp(-lL) N(l, h) = \exp[hI(L)],$$

$$I(L) = \int \bar{\alpha} dl \exp(-lL) \exp(-\bar{\alpha} l^2)$$

(20.13)

This means that for a fixed value of $h = \log p_\perp^2$ we may by expansion in L obtain both the average value, $\langle l \rangle$, and the width, $\sigma(l) = \sqrt{\langle l^2 \rangle - \langle l \rangle^2}$ reached by $l = \log(1/x)$ after an arbitrary number of steps, starting from $x \sim 1$. These results are similar to the ones obtained for the λ-measure in Chapter 18:

$$\langle l \rangle = \frac{h}{2}, \quad \frac{\sigma(l)}{\langle l \rangle} = \sqrt{\frac{1}{h} \left(\frac{\pi}{\bar{\alpha}} \right)}$$

(20.14)

This means that the average rapidity (remember that $l = \log(1/x) = \log P_+ - y - \log q_\perp$) will in this way behave as

$$\langle y \rangle + \log q_\perp^2 = C$$

(20.15)

where C is a constant, $\log(\mu P_+)$, and μ is some length scale which cannot be determined by the present method. *This is just the cutoff line occurring in the SRM*, as we have shown above. We also note that the relative width decreases with h (although at present energies the width is still rather large).

From an investigation of Fig. 19.7 we conclude that the average line described by Eq. (20.15) is always at the top of the region forbidden by the non-local form factor of Marchesini *et al.* We have used a constant value of α_s but we note that a running coupling will bring us down from the SRM cutoff line.

Evidently this result depends upon the fact that we have taken all the $p_{\perp j}$ as increasing with j, i.e. all the h_j as positive. If some of the numbers h_j are positive and some are negative (which means that the main-line transverse recoil is no longer dominated by the last emission) then there are further contributions to the integrals. It is, however, easily seen that for a fixed value of h such contributions will produce a larger value of l, i.e. we will then be even more below the SRM suggestion.

The intention of this subsection is not to prove that the SRM results are identical to the CCMF results but instead to show that there must be necessary damping properties in connection with the gluon emissions in the DIS states. In the next section we will show that the CCMF model can be reformulated into a set of linked dipoles instead of a single 'extended' dipole with SRM damping.

20.7   The linked dipole chain model

1   Introductory remarks

In the last section we have seen that the CCMF model, described in detail in section 19.6, implies results similar to those of the soft radiation model (SRM). The CCMF model contains a consistent description of the coherence conditions for QCD bremsstrahlung and includes the virtual corrections to the choice of ISB radiation in the model, cf. Eq. (19.2), in terms of the non-eikonal form factor. The SRM, however, stems from considerations of the behaviour of the radiation from an extended dipole. In this section we will show that the emerging 'extended dipole' picture of the radiation in the DIS states can be made even more precise, [16], i.e. *it is the radiation from a set of linked (color-connected) dipoles produced in a stochastical ISB scenario.* In order to obtain this result, the linked dipole chain (LDC) model, it turns out that the choice of ISB gluons must be restricted compared to the CCMF model.

To define the LDC model we recall that in the CCMF model, [44], the way to define the ISB is (i) to order all radiation in rapidity and (ii) to choose the ISB emissions as those that are not followed by a larger p_+-emission in this rapidity ordering (this is from the 'target' side, see Figs. 19.1 and 19.7). All the remaining radiation is included in the final-state bremsstrahlung (FSB). We also recall the particular requirement in Eq. (19.48) involving the fractional variable z_j and the transverse momenta of the jth emitted gluon, $p_{\perp j}$, and the ensuing propagator $q_{\perp j}$:

$$q_{\perp j}^2 > z_j p_{\perp j}^2 \tag{20.16}$$

The rapidity ordering is introduced to fulfil the strong angular ordering, i.e. the QCD coherence conditions for the bremsstrahlung emission. Increasing rapidity from the target side corresponds to decreasing angles along the emission line. It is, however, equally possible to order the emissions in rapidity from the probe side (we have already used this in connection with the description of the Webber–Marchesini model in section 17.7). Then the opposite rapidity ordering must be used and for every exclusive (i.e. fully defined) partonic final state the ISB should be chosen from a p_--ordering in the CCMF model formalism. A particular exclusive state will then contain a different set of ISB gluons and this results in a different non-eikonal form factor and a (seemingly) different contribution from the state to the cross section, cf. the discussion after Eq. (19.2).

We have already considered this question in detail in section 19.2 and we note that if the ISB gluons are more restricted then the $Sud(I)$ factors in Eq. (19.2) will generally be larger. It is a challenge to be able to partition the total state weight into a simple weight factor for the ISB, in accordance with QCD coherence, and at the same time obtain a simple description of the corresponding FSB gluons, given this ISB choice. One such approach is the LDC model, which exhibits the following features.

LDCa The final-state bremsstrahlung (FSB) correspond to emission from a set of color dipoles, spanned by the chosen gluons in the ISB set. Therefore the FSB can be treated by means of the Lund dipole cascade model (the DCM) (implemented in the Monte Carlo simulation program ARIADNE and described in detail in Chapter 17).

LDCb The inclusive weights for the ISB set of states chosen in the LDC model are simpler than the results for the CCMF model. The stochastical process obtained is, further, explicitly *local* (Markovian) and (in the leading-log approximation) symmetric with respect to emissions from the hadron and the probe end. In this way the predictions of the linked dipole chain model can be easily implemented in Monte Carlo simulation programs to study the particular ISB sets of the model.

LDCc It is possible to incorporate into the formalism both the ordinary perturbative QCD parton interactions, the boson-gluon fusion interactions and also the resolved (virtual) probe structure functions, including Rutherford interactions between the probe and the hadron ends. One consequence is that in the linked dipole chain (LDC) model there is no need of a cutoff (besides energy-momentum conservation) for large transverse momenta in the ISB gluon emissions, because such situations pass over in a well-defined way into Rutherford scattering. Correspondingly the gluonic bremsstrahlung also in a well-defined way defines a cutoff for small-transverse-momentum Rutherford scattering.

These statements will be clarified below. The model is defined in subsection 2 and in subsections 3 and 4 there is a description of the states in the triangular phase space as well as a discussion of how to generalise the model outside the leading-log approximation. Then the different channels are described in subsection 5 and finally, in Section 20.8, some features of the resulting structure functions are derived, in particular the relationship to the BFKL and DGLAP mechanisms, which we described in Chapter 19.

2   The definition of the LDC model

In [16] the LDC model is defined by the following restriction of the ISB gluons, compared to the CCMF model:

$$p_{\perp j}^2 \geq \min(q_{\perp j}^2, q_{\perp j-1}^2)$$
$$\simeq \min(-q_j^2, -q_{j-1}^2) \tag{20.17}$$

(The first line is defined by the Lorentz frame under consideration; the second is a Lorentz-invariant definition, which is approximately the same in the coordinate frames we have called 'equivalent to the hadron-probe cms' in Section 19.4.) Then it is possible to re-sum the weight in the emission step j of the CCMF model, Eq. 19.49, into

$$\bar{\alpha}\left(\frac{dz_j}{z_j}\right)\frac{dp_{\perp j}^2}{p_{\perp j}^2} = \begin{cases} \bar{\alpha}g_j d[\ln(1/z_j)]d\kappa_j & \text{if } \kappa_j > \kappa_{j-1} \\ \bar{\alpha}g_j d[\ln(1/z_j)]d\kappa_j \exp(\kappa_j - \kappa_{j-1}) & \text{otherwise} \end{cases} \quad (20.18)$$

Here on the right-hand side the constraint in Eq. (20.17) is introduced and we use $\kappa = \log q_\perp^2$. The quantity g_j corresponds to the azimuthal-angle (ϕ) average of the pole term $p_{\perp j}^2 = (\mathbf{q}_{\perp j} - \mathbf{q}_{\perp j-1})^2 = q_{\perp j}^2 + q_{\perp j-1}^2 - 2q_{\perp j}q_{\perp j-1}\cos\phi$, with the constraint in Eq. (20.17), using $a_j = \min(q_{\perp j}, q_{\perp j-1})/\max(q_{\perp j}, q_{\perp j-1})$:

$$g_j(a_j) = \frac{1}{2\pi}\int \frac{d\phi}{1 + a_j^2 - 2a_j\cos\phi}\Theta(a_j - 2\cos\phi) \quad (20.19)$$

It is straightforward to calculate the integral and we obtain

$$(1 - a_j^2)g_j(a_j) = \begin{cases} 1 - \frac{2}{\pi}\arctan\left[\frac{(1+a_j)\sqrt{2a_j-1}}{(1-a_j)\sqrt{2a_j+1}}\right] & 1 > a_j > 0.5 \\ 1 & 0.5 > a_j > 0 \end{cases} \quad (20.20)$$

To prove these statements it is necessary to carefully disentangle the contributions from the gluon emission in the CCMF model and sum over the non-eikonal form factor, cf. Eqs. (19.49), (19.50), for those which do not fulfil Eq. (20.17). Further it is necessary to convince oneself that it is possible to emit all the remaining gluons as FSB radiation from the dipoles between the chosen ISB gluons. We will consider these results in subsections 3 and 4, but in the remainder of this subsection we will investigate the consequences of the local stochastical process, which is symmetric with respect to the target and probe side defined by Eqs. (20.18) and (20.20).

'Local' means that the dipoles are determined by a Markovian stochastical process in the variables z, κ. With the $q_{\perp n}$-values and $x_n (= \prod^n z_j)$-values already obtained the next value of $q_\perp$ and a value of z can be chosen according to Eq. (20.18), e.g. by Monte Carlo simulation routines. The gluon $p_\perp$ is then defined by Eq. (20.17), its value of $(1-z)\prod^n z_j$ computed and we may then easily generate a dipole chain as shown in Fig. 20.8.

There is complete symmetry with respect to the target and the probe side, i.e. the values of the splitting variable z may be chosen along the positive or negative lightcones as $z_\pm$ (for the definition, see below and subsection 3 and for a discussion of this choice, subsection 4); the variables $z_\pm$ define the 'steps', $\log(1/z_\pm)$. The value of $\log q_\perp^2$ then defines the 'height' of the propagator virtuality and finally the emitted gluons (extended folds) are placed in an almost obvious way (for details, see the discussion in subsection 3).

There is a subtle but necessary change when we consider the production process from the probe side towards the target. The energy-momentum conservation equations at every vertex are written as $q_{j-1} = q_j + p_j$ and the index is increased from the target side. From the probe side we should, according to our convention, decrease the (propagator) indices, i.e. we must rewrite the relations as $-q_j = -q_{j-1} + p_j$ and generate 'negative' propagator vectors, but this is of course no problem (as they are all spacelike).

In Section 20.6 we noted the stepwise character of the process, i.e. that the splitting variable $z \ll 1$, in general. Then the positive and negative lightcone components fulfil $q_{+(j-1)} \gg z_{+j} q_{+(j-1)} = q_{+j} \simeq p_{+(j+1)}$ and $-q_{-(j+1)} \gg -z_{-j} q_{-(j+1)} = -q_{-j} \simeq p_{-j}$. Therefore the off-shell propagator vectors q_j are shown in Fig. 20.8 as horizontal lines between the two color-adjacent gluon emissions which form the dipole, with $-q_j^2 = -q_{+j} q_{-j} + q_{\perp j}^2 \simeq q_{\perp j}^2$.

In the next subsection we will consider the geometry of the triangular phase space for the emerging dipole chain. Although we have considered gluon emission within this phase space before, the situation for the bremsstrahlung in deep inelastic scattering is kinematically more complex.

3   The geometry of the triangular phase space for DIS events

In this subsection we will consider in some detail part of Fig. 20.8, in order to get acquainted with the way in which an emerging state in DIS is described in the triangular phase space of the LDC model. We concentrate on a dipole spanned between the emitted gluons p_1, p_2, with propagators q_1, q_0, q_2 in accordance with Fig. 20.9(a).

We note that the propagator q_0 is in between the massless gluons and we have chosen to exhibit the state in such a way that the propagator sizes are ordered as $-q_1^2 < -q_0^2 < -q_2^2$. (Exchanging the indices 1 and 2 would correspond to exchanging the probe and hadron side in the following arguments.) As the propagator sizes are dominated by the transverse momenta in most Lorentz frames (see below) we obtain, according to Eq. (20.17), $p_{\perp 1} = q_{\perp 0}$ and $p_{\perp 2} = q_{\perp 2}$. This is the most general situation possible (with obvious changes if we exchange the probe and the target sides) apart from the case when $p_{\perp 1} \simeq p_{\perp 2} \simeq q_{\perp 0} > \sqrt{-q_2^2}$, which, as we will later see, corresponds to a Rutherford scattering contribution (for the remainder of the discussion it is useful to note that in the LLA the inequalities $<$ can be exchanged for $\ll$, and $\simeq$ for equality).

The first observation is that there is a simple relationship between the variables $z_\pm$ corresponding to the splitting variables in the two different

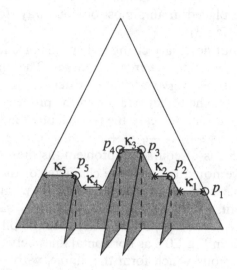

Fig. 20.8. A fan diagram in the LDC model as described in the triangular phase space. The extended folds correspond to on-the-mass-shell gluons and the arrows to the connector propagators. The front (back) borderline of the total triangle corresponds to the phase space boundaries, i.e. $\log P_+$ ($\log Q_-$) for the parton (probe).

directions, since for z_+, $q_1 \to q_0 p_1$ and for z_-, $-q_2 \to -q_0 p_1$. From the figure we may read off

$$\log(1/z_-) - \log(1/z_+) = \log(-q_2^2/-q_0^2) \tag{20.21}$$

(it is useful to relate this result to the requirement in Eq. (20.16)!).

According to perturbative QCD we may consider the state as resulting from the scattering of the two propagators $q_1, -q_2$, considered as the incoming entities, with the exchange of q_0 to obtain in the final state the two emitted gluons p_1, p_2, according to Fig. 20.9(b). (The minus sign in $-q_2$ is introduced to keep to energy-momentum conservation, as mentioned at the end of the last subsection, i.e. $q_1 - q_2 = p_1 + p_2$.)

While the triangular dipole phase space is manifestly invariant under Lorentz boosts along the chosen axis (i.e. the rapidity $y \to y + y_b$ with y_b, the boost rapidity, such that all rapidity differences stay constant) it is not so with respect to transverse boosts. We will therefore consider the scattering in two different, transversely boosted, Lorentz frames and also construct an approximate transformation between the frames in order to be able to understand the distribution in the triangular phase space.

As the dominating virtuality is $-q_2^2$ we will make the approximation $-q_1^2 \simeq q_{\perp 1}^2 \simeq 0$, using the LLA. We have then a dynamical situation

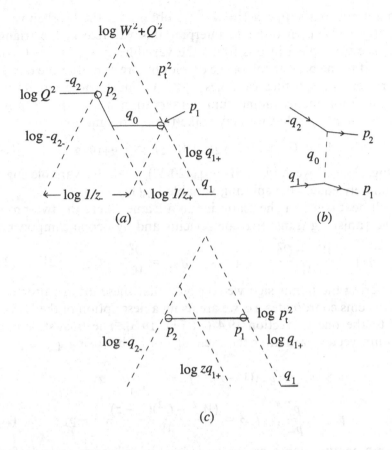

Fig. 20.9. A description of part of the fan diagram in the triangular phase space: (a) the three propagators q_1, q_0, q_2, denoted by lines, and the two emitted gluons p_1, p_2 of the dipole, for simplicity denoted by circles (although they correspond to triangular folds); (b) the corresponding scattering situation; (c) the resulting configuration in the probe$(-q_2)$-parton(q_1) cms frame.

similar to deep inelastic scattering with $-q_2$ as the probe and the other three entities as massless particles. In the first frame the probe virtuality $-q_2^2 \equiv Q^2$ will correspond to a strong transverse momentum pulse $Q^2 \simeq q_{\perp 2}^2 = p_{\perp 2}^2 \gg p_{\perp 1}^2 = q_{\perp 0}^2$. *Then the situation is similar to the description of a DIS event in the lepton-parton cms frame.* In this frame the lepton and hadron approach each other along an axis. After the encounter the lepton recoils with a large transverse momentum $-\mathbf{p}_{\perp 2}$. The field probe (emitted by the lepton), $-q_2$, then transfers the final-state parton, indexed 2, to $\simeq \mathbf{p}_{\perp 2}$. (The notation $\simeq$ is used because there is also some transverse momentum flowing through the propagator q_0 to p_1.)

With a dipole mass $(p_1 + p_2)^2 = W^2$ we obtain, for the Bjorken variable, $x = Q^2/(Q^2 + W^2)$. In order to interpret this variable in the triangular phase space we neglect in this frame the variables $q_{+2} \simeq q_{-1} \simeq 0$ (i.e. we assume that in the production of the dipoles before and after the one under consideration the splitting variables $z \ll 1$). Under those circumstances we obtain from energy-momentum conservation $q_{+1} \simeq p_{+1} + p_{+2}$ and $-q_{-2} \simeq p_{-1} + p_{-2}$ so that we may make the approximation

$$W^2 + Q^2 \simeq (p_1 + p_2)^2 + (\mathbf{p}_{\perp 2})^2 \simeq -q_{+1} q_{-2} \qquad (20.22)$$

From Fig. 20.9(a) we find, using Eq. (20.22), that the variable $\log(1/x)$ approximately equals the splitting variable $\log(1/z_+)$.

We will next consider the scattering in a frame where the two propagators have vanishing transverse components and lightcone components

$$q_{+1} = \frac{W^2 + Q^2}{W}, \quad q_{-1} = 0, \quad q_{+2} = \frac{Q^2}{W}, \quad q_{-2} = -W \qquad (20.23)$$

Remembering the minus sign we conclude that these are the probe$(-q_2)$-parton(q_1) cms coordinates, i.e. we are using a description of the DIS event similar to the one in Section 19.4 (cf. Fig. 19.3). The final-state energy-momentum vectors will after the exchange of q_0 be (using $\mathbf{q}_{\perp 0} \equiv \mathbf{p}_\perp$)

$$p_{+1} = q_{+1}(1 - z), \quad p_{-1} = \frac{p_\perp^2}{p_{+1}}, \quad \mathbf{p}_\perp$$

$$p_{+2} = \frac{p_\perp^2}{p_{-2}}, \quad p_{-2} = \frac{(W^2 + Q^2)(1 - z)}{W}, \quad -\mathbf{p}_\perp \qquad (20.24)$$

Energy-momentum conservation provides the following formula for the exchanged transverse momentum $p_\perp$:

$$p_\perp^2 = \frac{[zW^2 - (1-z)Q^2](W^2 + Q^2)(1-z)}{W^2} \qquad (20.25)$$

and we also note that the total virtuality of the q_0-propagator is

$$-q_0^2 = \frac{p_\perp^2}{(1-z)} = \frac{zW^2 - (1-z)Q^2(W^2 + Q^2)}{W^2}$$

$$\equiv \frac{(W^2 + Q^2)(z - x)}{(1 - x)} \qquad (20.26)$$

where we have introduced the value of x defined before.

There are a few conclusions to be drawn immediately.

I From the expressions for $p_\perp^2$ and $-q_0^2$ we conclude that the splitting variable actually must be $z_+ = z > x$, and also that for fixed values of W and Q, z must grow with the value of the propagator $-q_0^2$. A closer analysis of the components of the momentum transfer q_0 tells us that $-q_{+0}q_{-0} = zp_{\perp}^2/(1-z)$ so that unless $z < 1-z$ this part of the propagator size $-q_0^2$ will dominate over the transverse momentum part $-p_{\perp}^2$. We will analyse the occurrence of large values of z in the next subsection. Remember that the possibility of large z-values implies problems for the BFKL mechanism in the x-development of the structure functions, cf. Section 19.5.

II When we compare the configuration in Fig. 20.9(a), where $-q_2$ brings p_2 to the same transverse momentum, $p_{\perp 2}^2 \simeq Q^2 > -q_0^2$, as that in the second frame, cf. Fig. 20.9(c), there are some differences. The emitted gluon vectors in the final state, p_1, p_2, are both placed, in the second frame, at the same $\log p_{\perp}^2$-level, corresponding to the exchanged transverse momentum in the propagator q_0. While p_1 has basically the same position as before, p_2 has moved down along the lightcone line $\log |q_{-2}|$, i.e. it has a smaller transverse momentum. (It is useful to consider the appearance of the dipole having q_2 as the propagator in the new frame!)

III While the second frame is a rest frame for the dipole the main axis is not along the dipole axis. The vectors $\mathbf{p}_1 = -\mathbf{p}_2$ form an angle with the main axis and also have an azimuthal angular difference π; note that in the triangular phase space each point in general corresponds to all azimuthal angles around the main axis! It is instructive to perform the necessary rotation to make the dipole axis the main axis. We will find that while the dipole axis in the first situation is a smooth curve from the positions of p_1 to p_2, although with a dip at the centre corresponding to approximately vanishing transverse momentum, in the new frame the two axes are exchanged.

Next we will construct an approximate Lorentz transformation between the dipole rest frame and the lepton-hadron cms in Fig. 20.9(a). To that end we start in the cms with the axis along the dipole axis so that $p_{+1} = p_{-2} = W$; we then boost along the positive axis to obtain $p_{+1}^{(1)} = \exp(-y_1)p_{+1}$, $p_{-2}^{(1)} = \exp(y_1)p_{-2}$, with a large boost rapidity y_1; see the treatment of such boosts in Chapter 2.

Then we perform a transverse boost to obtain a large $|\mathbf{p}_{\perp 2}^{(2)}| \equiv Q$ along some azimuthal direction. Finally we perform (in this new frame) a boost along the original dipole axis to obtain the vector $p_2^{(3)} \simeq (Q^2/W, W, \mathbf{Q}_\perp)$ while $p_1^{(3)} \simeq (W, 0, \mathbf{0}_\perp)$; in both cases we use lightcone coordinates and there are corrections of order $\exp(-2y_1)$.

It is now of interest to investigate what the transformations defined above will do to the points in the triangular phase space. Suppose that we take a (lightlike) vector approximately equal to p_2; the simplest example would be $p_2' = \zeta p_2, 0 < \zeta < 1$. The three Lorentz transformations we have

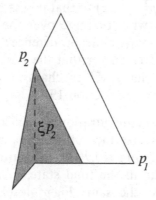

Fig. 20.10. The result of Lorentz transformation between the p_1p_2-dipole rest frame and a transversely moving frame, as described in the text.

described above will evidently transform it into $\zeta p_2^{(3)}$, i.e. with increasing ζ-values it will move from the baseline up along the triangle corresponding to p_2, see Fig. 20.10. Similarly the lightlike vectors (close to p_2) with small but nonvanishing transverse momenta (and azimuthal angles along and opposite to the boost direction) will fill in the region around this middle line (the shaded region in Fig. 20.10). Thus the shaded area, which comes to an apex at p_2, corresponds to the vectors which are collinear to p_2.

The reason for doing this transformation exercise for the points in the triangular phase space is to be able to describe the role of the virtuality $-q_0^2$ for the dipole spanned by p_1, p_2. In the dipole chain, shown in Fig. 20.8, we note that only final-state bremsstrahlung (FSB) is allowed within the shadowed region below the corresponding propagator. For the cases denoted by the indices $1, 2, 4$ the regions are bounded either by broken (logarithmic lightcone) lines, starting on the emitted gluons, or by the line corresponding to $\log(-q_j^2) \simeq \log(q_{\perp j}^2)$.

According to the results above it is the FSB emission, which is collinear to an original ISB gluon, that will cover the first kind of region, i.e. those FSB gluons that are produced close to the ISB gluon.

Actually it is straightforward to convince oneself that

FSB1 if we allow all possible bremsstrahlung in the rest frame of the dipole with $p_{\perp FSB} \leq \sqrt{-q^2}$ (q being the energy-momentum of the propagator in the relevant dipole) and

FSB2 if we then transform back to the original, i.e. the externally defined, frame then the FSB gluons will cover just the region below the broken lines and the logarithmic virtuality $\log(-q^2) \simeq \log q_\perp^2$.

4 The role of the virtuality and the possible extensions

According to the subsections above, in the LDC model the FSB is the radiation which can be emitted from the dipoles spanned by two adjacent initial state bremsstrahlung (ISB) partons. Each dipole can be characterised by two parameters, its mass, describing the size of the related triangular phase space, and its virtuality, i.e. the $-q^2$-size of the propagator which defines the largest transverse momentum allowed for FSB in the rest frame of the dipole. We will in this subsection provide a physical argument for the ocurrence of the virtuality cutoff and after that consider both the occurrence of large z-values and the possibility of extending the model outside the leading-log approximation (LLA).

In the description of timelike partonic cascades, which we encountered in e^+e^- annihilation events through the dipole cascade model (DCM), see section 17.2, there is a corresponding notion, i.e. no new emission is allowed with a transverse momentum above the earlier emission (the earlier emission may often define the dipole as such and the $p_\perp$-ordering corresponds to the coherence conditions of the radiation). The result that there can be no emission above the propagator $-q^2$ is, in connection with DIS events, a major result in the CCMF model, see section 19.6, and we will now outline a dynamical argument for this feature.

If we go back to the definition of a propagator in the Feynman way, cf. section 3.3, in particular subsection 3, we find that it describes the field activity inside a region compatible with Heisenberg's indeterminacy relations. If we use a coordinate system such that the energy-momentum-space propagator size $-q^2$ is essentially transverse, i.e. $-q^2 \simeq q_\perp^2$, then the size of the corresponding coordinate-space region is given by the canonically conjugate variable, i.e. the impact parameter b.

The occurrence of a particular $-q^2$-value consequently corresponds to a coordinate-space transverse distance $b \simeq 1/\sqrt{-q^2}$. Therefore the propagator size implies that *the two partons forming the dipole do not stem from a pointlike region*, i.e. there is a transverse distance between the emerging parton currents. In the soft radiation model in Section 20.5 we have already considered the radiation from an extended 'antenna'. *The result, that there is no emission with transverse momentum above the propagator* $-q^2 \geq p_{\perp FSB}^2$, *is due to the same extension property*, cf. Eq. (20.8). All radiation with a wavelength smaller than the typical size of the emitter is (form factor) suppressed. In the logarithmic phase space this implies at least exponential suppression, which in the LLA corresponds to a vanishing result. In the language of the CCMF model, [44], the same property is phrased as follows: above the propagator virtuality real emission is cancelled by the virtual corrections.

We note that the present results are only valid in the leading-log approximation (LLA). It is tempting to extend the model beyond the LLA by the use of the 'true' splitting functions $\mathscr{P}(z)$, see Eq. (19.30), instead of the $1/z$-pole. There are then a set of problems to be faced. In order to resolve them we will make use of the scattering results of subsection 3 above and the derivation of the splitting functions in Section 17.7.

The first problem we encounter is whether we should interpret the splitting variable z as z_+ or z_-, i.e. in the notation above whether we should consider the process from the target or from the probe side. To resolve that question we consider Fig. 19.6 and note that in the derivation of the splitting functions, see e.g. Eq. (17.25), the incoming parton is assumed to be massless. In Fig. 19.6 it is denoted by its lightcone energy-momentum fraction $x' \equiv q'_+/P_+$, with P_+ the (target) parton energy-momentum. It is split up into a parton, with $x \equiv q_+/P_+ = zx'$, assumed to have a (large) virtual mass, $-Q^2$, and another massless parton with $x_g = (1 - z)x'$. Then the variable z equals $z_+ \equiv q_+/q'_+$, i.e. the positive lightcone energy-momentum ratio of the consecutive propagator vectors.

Thus if we increase the virtuality from the target side, which in the notation above means that $\kappa_{j-1} < \kappa_j$ (examples in Fig. 20.8 are $\kappa_1 < \kappa_2 < \kappa_3$ and $\kappa_4 < \kappa_5$) – note that this means going 'down' in virtuality from the probe side! – then we should use z_+. In the opposite case, i.e. when we go down from the target side, which means that we go up from the probe side (the example in Fig. 20.8 is $\kappa_4 < \kappa_3$) then we should evidently use $z = z_-$ in order to keep to this virtuality ordering.

But there is another and more difficult problem. Besides the z-pole the splitting functions, $\mathscr{P}(z)$, contain some finite corrections and even a $(1 - z)$-pole, cf. section 17.7. We have already discussed the $(1 - z)$-pole in connection with the DGLAP mechanism (where it is formally regularised by means of the Sudakov factor or physically by energy-momentum conservation). Within the BFKL mechanism there is a problem if the bremsstrahlung emissions stem from large z-values, cf. section 19.5. In order to investigate the occurrence of large z-values we return to the results of subsection 3.

From Eqs. (20.23)–(20.26) we may read off the behaviour of the emitted gluon vectors p_1, p_2 for all values of z, see Fig. 20.11. The size of $-q_0^2$ will increase with z (for fixed W, Q) and be equal to $-q_2^2 = Q^2$ for $z = z_1 = x(2 - x)$ (which satisfies $1 \geq z_1 \geq x$ as it should). Further, the two vectors p_1, p_2 will be described by the same point in the triangular phase space when $z = z_2 = (1 + x)/2$ (note that $1 \geq z_2 \geq z_1$). At this point the scattering situation corresponds in the cms to a scattering angle $\theta = \pi/2$, (it is straightforward to see that $p_{+j} = p_{-j} = p_{\perp j}$, $j =$

1, 2), i.e. the vectors $\mathbf{p}_j$ are transversely directed, with opposite azimuthal angles.

If z increases further *the two vectors p_1, p_2 change place with respect to rapidity ordering in the triangular phase space.* We obtain a situation similar to that discussed in Section 15.4, in particular in subsection 1, and in Section 17.8, i.e. that the color-field flow is not stretched along the 'simplest' direction between the emitted partons in a cascade, see Figs. 15.8 and 20.11(b). *This is a forbidden configuration in a strictly strong-angular ordering scenario* but, although in general strongly suppressed, it is often an allowed configuration when QCD coherence is not taken approximately; we are after all working in a three-space-dimensional world!

When we continue, for values of $z > z_2$, the situation corresponds to backwards scattering with a cms scattering angle $\theta > \pi/2$. The transverse momenta of the two p_j-vectors decrease *but the propagator size $-q_0^2$ increases to $-q_{0max}^2 = Q^2 + W^2$.* The development for increasing z-values is shown in Fig. 20.11(a). Within our present knowledge, which comprises the results from the analysis in the CCMF model, there is no indication of the way to treat this situation. We only know that it must be suppressed owing to the difficulty of fulfilling the QCD coherence conditions.

We should also be aware, however, that color coherence may not be only a question of angular ordering when we consider the development along a line with gluon quantum numbers, i.e. when there is both a color and an anticolor line along the propagator chain and the emitted partons are gluons. It is straightforward to convince oneself that in half the cases the two (adjacent) gluons p_1, p_2, which we have considered repeatedly, will have a color charge in common but in the remaining cases they correspond to emissions from the two independent color lines of the propagator; the two situations are shown in Fig. 20.11(c) but at the present time we will have to leave the above-mentioned question as an open problem.

There is one situation which we have up to now completely neglected and that is the gluon splitting process $g \to q\bar{q}$. As always it is only at a small percentage level compared to gluon emission $g \to gg$ and therefore is of minor interest along the emission chains. But it is of direct interest at the end of any fan diagram because an electromagnetic (as well as a weak-interaction) probe can only interact with the q- and $\bar{q}$-partons. We note that in this case there is no pole for the splitting variable so that z and $1 - z$ are in general of the same size, cf. Eq. (17.26).

We finally note another consequence of the scattering behaviour described above when $z_1 < z < z_2$ (which for small values of $x \simeq Q^2/W^2$ is essentially the whole available z-region $0 < z < 1$). In this case the two ISB gluons p_1, p_2 are both placed at the same $\log p_\perp^2 \simeq \log(-q_0^2)$ level, i.e. at the two edges of the largest propagator q_0 above $-q_2$. In this

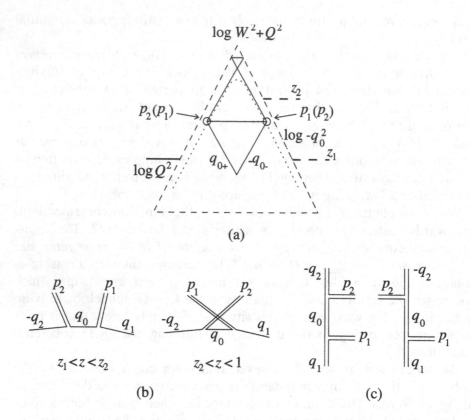

Fig. 20.11. (*a*) For small values of the splitting variable z the dipole virtuality size fulfils $-q_0^2 \simeq q_{\perp 0}^2 \gg -q_{+0}q_{-0}$ (the lower 'crossed' configuration) but for large z, $\log(-q_0^2)$ increases to the turning point $z = z_2$. For $1 > z > z_2$, $\log(-q_0^2)$ increases further but then $-q_{+0}q_{-0} > q_{\perp 0}^2$ (the upper 'crossed' configuration). The two cases correspond to the two color-line descriptions in (*b*). In (*c*) we have the two color-flow situations when there are gluons as 'in', 'out' and propagator partons.

situation the chain moves up and down in virtuality in a symmetric way from both the target and the probe side. *It corresponds to a Rutherford scattering interaction* like the one described in detail in Section 5.4. From the weight distribution in Eq. (20.18) we find that in the present case there will be a pole of the kind $(-q_0^2)^{-2} \simeq p_\perp^{-4}$ (there is one $p_\perp^{-2}$-factor from each side), in accordance with the results of Eq. (5.40)! Thus, as we will further discuss in the following subsections, *the ISB bremsstrahlung in the LDC model goes over to Rutherford scattering at the largest virtuality in the chain.*

5 The different channels in deep inelastic scattering

In the analysis of the structure functions it is in general tacitly assumed that the major momentum transfer stems from the external probe, $-q^2 = Q^2$. As we have seen in the subsections above the LDC model contains the possibility of considering also situations where there is some virtuality along the chain which exceeds Q^2. Such situations are of interest in particular for small and moderate Q^2-values (with large cross sections at HERA) and we will in this subsection subdivide the DIS cross section into three channels with different properties in this respect:

  I the usual quark-parton model interaction, in which the largest virtuality along the chain is given by Q^2,

  II a boson-gluon fusion (BGF) event, in which the (final) propagator virtuality exceeds the probe virtuality $-q^2 \simeq q_{\perp n}^2 > Q^2$,

  III a Rutherford parton-scattering event, in which there is one virtuality further down the chain, $q_{\perp max}^2$, exceeding all the remaining ones (note that there may in general be several 'local' maxima if the chains are sufficiently long, counted in $\log(1/x)$ units, but as such situations are very rare within the presently obtainable energy regimes we will neglect them).

In Fig. 20.12 the triangular phase space is again shown (we will subsequently use the probe-parton cms according to Section 19.4 and subsection 3 above) with the variables $\log(1/x_B)$ and $\log Q^2$ and with three 'chain-roads' as examples of the cases I–III. On the right-hand side of the figure we show the conventional Feynman diagrams for the three cases and the main momentum transfer is particularly emphasised.

The structure function f is for case I conventionally obtained from an integral over the last propagator $q_\perp \leq Q$ (here $x = \prod z_j = x_B$ and $\mathscr{F}$ is the so-called non-integrated structure function):

$$f(x, Q^2) = \int^{Q^2} \frac{dq_\perp^2}{q_\perp^2} \mathscr{F}(x, q_\perp^2) \tag{20.27}$$

Therefore in the probe-parton cms the chain will end somewhere along the (positive lightcone) line AB. Just like the emitted parton p_2 in the scattering discussion of subsection 3, the final-state parton, after absorption of the probe momentum, will keep the transverse momentum from the chain, i.e. $p_{\perp n+1} \simeq q_{\perp n}$, and end up at point E along the (negative lightcone) line CD, corresponding to the Q_- of the probe.

For the boson-gluon fusion event in case II we note that the last propagator will have $q_{\perp n}^2 > Q^2$ and also $q_{+n} = x_n P_+$ with $x_n = \prod z_j > x_B$. This last relation can be read off from Fig. 20.12 and can be understood

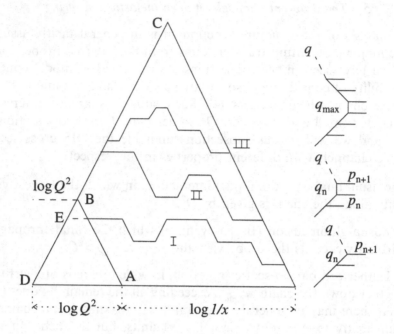

Fig. 20.12. Examples of the different kinds of fan diagram chain. Using $Q_+ = xP_+$ the baseline rapidity region is divided into $(-\log Q_-, \log Q_+)$ and $(\log Q_+, \log P_+)$ with lengths $\log Q^2$ and $\log(1/x)$, respectively.

from the following simple calculation. The last parton in the chain, see the second chain path in Fig. 20.12, will again have $p_{\perp(n+1)} \simeq q_{\perp n}$ and pick up the negative lightcone fraction of the probe, i.e. $p_{-(n+1)} \simeq Q_-$. Therefore energy-momentum conservation at the vertex $q_n \to -Qp_{(n+1)}$ provides, cf. the kinematics discussed in subsection 3 and in Section 19.4,

$$q_{+n} = -Q_+ + p_{+(n+1)} = \frac{Q^2}{Q_-} + \frac{p^2_{\perp(n+1)}}{p_{-(n+1)}} \simeq \frac{Q^2 + q^2_{\perp n}}{Q_-} \qquad (20.28)$$

We conclude that, while the first term in the last expression is by definition $x_B P_+$, in the boson-gluon-fusion situation the second term will dominate, i.e. $q^2_{\perp n} > Q^2$. We then obtain $q_{+n} \simeq p_{+(n+1)} = \prod(z_j)P_+ = x_n P_+ = xP_+ + q^2_{\perp n}/Q^2$. We may go further and conclude that the definition of the non-integrated structure function can be extended from Eq. (20.27) to

$$f(x, Q^2) = \int \frac{dq^2_\perp}{q^2_\perp} \mathscr{F}(\hat{x}, q^2_\perp), \quad \hat{x} = \begin{cases} x_B & \text{if } Q^2 > q^2_\perp \\ x_B q^2_\perp/Q^2 & \text{otherwise} \end{cases} \qquad (20.29)$$

but we must then in the second equation remember that there is a factor $Q^2/q^2_\perp$ included in the last step for the non-integated $\mathscr{F}$, corresponding to the definition of the transverse momentum dependence in Eq. (20.17) (we must go downwards from $q_\perp^2$ to get to Q^2).

Finally for chain-road III with $q_{\perp max}$ exceeding all other momentum transfers the same argument provides for the non-integrated $\mathscr{F}$

$$\frac{Q^2}{q_{\perp n}^2} \times \frac{q_{\perp n}^2}{q_{\perp n-1}^2} \times \cdots \times \frac{q_{\perp j}^2}{q_{\perp max}^2} = \frac{Q^2}{q_{\perp max}^2} \qquad (20.30)$$

i.e. a direct generalisation of the results for the boson-gluon-fusion events. As each step is weighted by $d\kappa = dq_\perp^2/q_\perp^2$ the Rutherford scattering result $dq_{\perp max}^2/q_{\perp max}^4$ will occur as soon as we reach a maximum and would like to go downwards along the chain. (An observant reader may note that the contributions appear in a non-symmetrical way with respect to the target and probe sides but an even more observant one will note that the structure function as it is defined is not a symmetric notion, cf. the discussion in section 20.8.)

The main result is, however, that the LDC chains include Rutherford parton-scattering contributions, too, and that the largest momentum transfer subdivides the event chain into one part stemming from the (coherent) bremsstrahlung from the parton while the other part can be attributed to the resolved probe. Thus *the largest transverse momentum of the bremsstrahlung chain(s) provides the lowest cutoff for Rutherford scattering, and conversely Rutherford scattering along a chain provides the upper cutoff for the bremsstrahlung.*

It is consequently possible to define a structure function for both the target and the projectile and then the notions of 'probe' and 'parton' are a matter of convention, because the cross section for the interaction corresponds to the convolution of these two structure functions together with the relevant Rutherford parton interaction.

20.8 The structure function behavior of the LDC model

1 Introduction

In this section we will provide a more detailed analysis of the properties of the structure functions than we have done before and in particular exhibit the combined role of the DGLAP and BFKL mechanisms for the final result (although we will find that over the HERA energy region the DGLAP mechanism will be the dominant one, except for such small Q^2-values that we cannot trust the results of perturbative QCD). To avoid confusion we repeat some of the notation. We use the ordinary DIS variables $W^2, Q^2, x_B = Q^2/(W^2 + Q^2)$. For the emitted massless gluon and the (spacelike) propagator vectors we use p, q, for the splitting variable and the transverse momenta we use $z, q_\perp, p_\perp$ and for the natural logarithmic variables we use $\ell \equiv \log(1/x_B), \kappa \equiv \log q_\perp^2, L_Q \equiv \log Q^2$.

2   The integral equations for the non-integrated structure function

The results of the earlier subsections can in a straightforward way be formulated as an integral equation for the non-integrated structure function $\mathscr{F}$, cf. Eq. (20.27), which actually is a function only of ℓ and κ:

$$\mathscr{F}(x, q_\perp^2) \equiv \mathscr{F}(\ell, \kappa) = \sum_{j=1}^{3} \mathscr{I}_j \qquad (20.31)$$

The meanings of the the terms $\mathscr{I}_j$ are as follows.

$\mathscr{I}_1$ This is the possibility of taking a single step from a starting point (ℓ_0, L_{Q0}) (conventionally $\ell_0 = 0, L_{Q0} = 0$, which defines both the properties of the starting-point and the scale of the coupling), to the final point (ℓ, κ). It is a boundary term, $\mathscr{I}_1 = \bar{\alpha}$, in the integral equation.

$\mathscr{I}_2$ This is the possibility of being at a point (κ', ℓ') below κ, i.e. $L_{Q0} < \kappa' \equiv \log q_\perp'^2 < \kappa$ and $\ell_0 < \ell' \equiv \log(1/x') < \ell = \log(1/x_B)$, and taking the final step to ℓ upwards in transverse momentum (as in the DGLAP mechanism). We obtain

$$\mathscr{I}_2 = \int_{L_{Q0}=0}^{\kappa} \bar{\alpha} d\kappa' \int_{\ell_0=0}^{\ell} d\ell' g \mathscr{F}(\ell', \kappa') \qquad (20.32)$$

$\mathscr{I}_3$ This is the possibility of being at $\kappa' > \kappa$ and taking a step downwards to κ. Note the factor $q_\perp^2/q_\perp'^2 = \exp(\kappa - \kappa')$ in $\mathscr{F}$ and a compensating change $\ell' \to \ell' + \kappa - \kappa'$ according to Eqs. (20.18) and (20.29):

$$\mathscr{I}_3 = \int_{\kappa}^{\ell} \bar{\alpha} d\kappa' \exp(\kappa - \kappa') \int_{0}^{\ell+\kappa-\kappa'} d\ell' g \mathscr{F}(\ell', \kappa') \qquad (20.33)$$

This is a leading-log approximation (LLA) equation. If we go further and neglect the variations around the pole and put $g = 1$, cf. Eq. (20.18) and the discussion below, it will be a symmetric (in mathematical terms 'Hermitian') integral equation in terms of the left-right (i.e. from the probe and parton side) symmetric non-integrated $\mathscr{F}_s \equiv \mathscr{F} \exp(-\kappa/2)$. (This is the lack of symmetry we noted in connection with the Rutherford contributions to the structure functions in subsection 4 of the last section.) We immediately recognize the DGLAP contribution in the first two terms, cf. Eqs. (19.33)–(19.35). We will next exhibit the corresponding BFKL contribution, in particular the Lipatov kernel, which was mentioned in section 19.5.

Before we continue we note the starting point of the LDC model, i.e. that the transverse momentum generation stems from $d^2q'_\perp/[2\pi(\mathbf{q}'_\perp - \mathbf{q}_\perp)^2]$, the azimuthal angular average of the pole term $1/\mathbf{p}_\perp^2$ in the emission, as seen in Eqs. (20.18)–(20.20). The factor g contains the LDC requirement given in Eq. (20.17) but it is unity unless we are close to $\kappa \simeq \kappa'$, i.e. when the emitted gluon transverse momentum is very small, $p_\perp \simeq 0$.

In order to relate the pole term to the Lipatov kernel we use the following approximate relationship:

$$\frac{\mathscr{F}(\ell, \kappa')}{p_\perp^2}\Theta(p_\perp - q'_\perp) \simeq \frac{\mathscr{F}(\ell, \kappa') - \mathscr{F}(\ell, \kappa)\Theta(q_\perp - p_\perp)}{p_\perp^2} \quad (20.34)$$

If this is inserted into the integral equation Eq. (20.31) and the equation differentiated with respect to $\ell = \log(1/x)$ we obtain

$$\frac{\partial \mathscr{F}(\ell, \kappa)}{\partial \ell} = \int \frac{\bar{\alpha}d^2q'_\perp}{2\pi p_\perp^2}[\mathscr{F}(\hat{\ell}, \kappa') - \mathscr{F}(\ell, \kappa)\Theta(q_\perp - p_\perp)] \quad (20.35)$$

which is almost the BFKL equation. The difference is the occurrence of $\hat{\ell} = \max(\ell, \ell + \kappa' - \kappa)$, i.e. the compensation in the contribution $\mathscr{I}_3$, which corresponds to the use of the relevant splitting variable z_- instead of z_+, cf. Eq. (20.21) and the discussion in subsection 4 of the previous section.

3   *The solutions to the integral equations*

Using the conventional methods for analysis, we will now investigate the solutions of Eqs. (20.31) and (20.35). To that end we introduce the moments of the non-integrated $\mathscr{F}_N$:

$$\mathscr{F}_N = \int dx\, x^{N-1}\mathscr{F} \quad (20.36)$$

and the *anomalous dimensions*, γ_N of this moment function:

$$\mathscr{F}_N \propto (q_\perp)^{2\gamma_N} = \exp(\gamma_N \kappa) \quad (20.37)$$

Insertion into Eq. (20.35) provides the following result:

$$1 = \frac{\alpha}{N-1}\xi(\gamma_N) \equiv \frac{\bar{\alpha}}{N-1}[h(\gamma_N) - h(N - \gamma_N)] \quad (20.38)$$

The two h-terms in ξ stem from the the transverse momentum integrals in $\mathscr{I}_2$ and $\mathscr{I}_3$ from Eqs. (20.32) and (20.33). If we put $\hat{\ell} = \ell$ in Eq. (20.35), i.e. in practice neglect the difference between $z+$ and z_-, then the argument $N - \gamma_N$ in the second h-term of ξ becomes $1 - \gamma_N$. This is a consistent procedure in the BFKL approach (to be called *conventional BFKL*) because the intention is to take into account the sub-leading contributions in the transverse momentum fluctuations while keeping to the $1/z$ pole in the longitudinal generation.

It is straightforward to obtain the following expression for the function h (if we use Eq. (20.35), i.e. basically put $g = 1$ in Eqs. (20.18) and (20.31)):

$$h(\gamma) = \int_0^1 \frac{du(u^{\gamma-1} - 1)}{(u - 1)} = \psi(1) - \psi(\gamma) \qquad (20.39)$$

where ψ is the Euler function (the derivative of the logarithm of the Γ-function). There is a mathematical theorem by which we may invert the moment equation (20.36) to obtain a formula for $\mathscr{F}$ itself, including both the x_B- and the $q_\perp$-dependence. We will not write out the formulas but, as we have noted before, the major x_B-dependence will stem from the largest value of $N - 1$ which is a solution to Eq. (20.38). For conventional BFKL any useful mathematical table will show that this occurs for $\gamma_N = 1/2$ and that for this value of $N = N_L$ (the Lipatov case)

$$N_L - 1 = 2\bar{\alpha}h(1/2) = 4\bar{\alpha}\log 2 \equiv \lambda_L \qquad (20.40)$$

It is an interesting fact that this value of N_L corresponds to the place where the two solutions of Eq. (20.38) coincide, in accordance with the symmetry $\gamma \leftrightarrow 1 - \gamma$. In the inverse moment integral we then have a singularity (the two poles, corresponding to the solutions, will approach their common value from each side of the integration contour and provide a 'pinch singularity'). While $\lambda_L \simeq 0.5$ in Eq. (20.40) (for $\bar{\alpha} = N_c\alpha_s/\pi \simeq 3/\pi \times 0.2$) the corresponding general solution of Eq. (20.38) (when the difference between z_+ and z_- is not neglected) corresponds to a pinch singularity for $\gamma = N/2$. Then the largest λ-value is diminished to $\lambda \simeq 0.31$!

We have already seen in section 19.5 that there are very large corrections to the BFKL mechanism (stemming from the contributions to the integrals for large values of of the splitting variables z). The result just mentioned corresponds to a different mechanism. If we allow for the simplest corrections, those subleading in N, to the BFKL eigenvalue equation (which recognise the fact that $z_+ \neq z_-$ in general) then we again obtain very large changes in λ. And these corrections will again result in essentially smaller effective λ-values!

4   Further remarks on the solutions

It is possible to continue the analytical investigations of the structure functions and e.g. to introduce the LDC requirement, which effectively means reintroducing the factor g in Eq. (20.18). But it is then necessary to take recourse to numerical calculations because the integrals no longer correspond to elementary functions. The result of such investigations are that for a sufficiently large value of $\log(1/x_B)$ there is always an approximate power behaviour, i.e. the gluon structure function $x_B g$ behaves as $x_B^{-\lambda_e}$ with an effective power $\lambda_e \simeq 0.3 \ll \lambda_L$. But for values of $\log(1/x_B)$ of the same order as $\log Q^2$ (cf. Fig. 20.12)) the DGLAP mechanism, leading to a structure function $x_B g \propto \exp[2\sqrt{\chi(Q^2)\log(1/x_B)}]$, cf. Eqs. (19.35)–(19.36), will be dominant (with a very slowly varying function $\chi(Q^2)$ and also a slow dependence on the proportionality constants).

We will end this investigation with a few simple analytical calculations to exhibit these facts. In the first we will find that for small $\kappa = \log q_\perp^2$ values the very construction of the LDC chains leads to a simple result of the BFKL kind. In the second we will nevertheless find that the BFKL mechanism is only relevant for very large values of $\ell = \log(1/x)$. Finally we will consider the effect of a running coupling on the LDC equations and show that *the BFKL diffusion effect in transverse momentum*, i.e. the gaussian $\log Q^2$ behaviour with a width proportional to ℓ mentioned in Section 19.5, is not consistent.

In the LDC model it is according to the earlier subsections possible to go up and down in κ along the chains. Let us consider all the possibilities of a *combined step-motion*, cf. Fig. 20.13: we consider the situations when there is one set of up-steps followed by another set of down-steps. Thus we start at $\ell_{+0} \equiv \ell_0 = L_{Q0} = 0$ and only follow (in κ) upwards directed chains to reach a stochastically chosen maximum point κ_1 at $\ell_{+1} \equiv \ell_1$. Then we continue (still from the target side) downwards to reach the final point $\ell_+ \equiv \ell$ at $\kappa < \kappa_1$ (note that the second part corresponds to up-steps from the probe side!).

We would like to obtain the total weight in the LDC model from all possible chains with this property. Then we have (from the target side) a DGLAP motion $(0,0) \to (\ell_1, \kappa_1)$ and from the probe side a corresponding DGLAP motion $(\ell_{-0} = 0, \kappa) \to (\ell_{-2} \equiv \ell_2, \kappa_1)$ (where ℓ_{-0}, κ_1) corresponds to the endpoint $\ell_+ \equiv \ell, \kappa$ and $\ell_{-2} \equiv \ell_2, \kappa_1$) the maximum point $\ell_{+1} \equiv \ell_1, \kappa$)). From Fig. 20.13 we obtain

$$\kappa_1 = \kappa_2 + \kappa, \quad \ell_1 + \ell_2 + \kappa_1 = \ell + \kappa \tag{20.41}$$

We will use a fixed coupling $\bar{\alpha}$, which means that the DGLAP contributions for the two cases are $\exp(2\sqrt{\bar{\alpha}\ell_j \kappa_j})$, $j = 1, 2$ (note that, for fixed coupling, $\chi - \bar{\alpha}\kappa$ according to Eq. (19.35)). For the symmetric non-integrated $\mathscr{F}_s$, defined above, we have the factors $\exp(-\kappa_j/2)$ from the transverse momentum generation and we are then supposed to sum over all contributions with the constraints in Eq. (20.41). Thus we have an integral in two independent variables, which may be chosen as e.g. ℓ_2 and κ_2:

$$\mathscr{F}_s = \int d\ell_2 d\kappa_2 \exp\left[2\sqrt{\bar{\alpha}\ell_2\kappa_2} + 2\sqrt{\bar{\alpha}(\kappa + \kappa_2)(\ell - \ell_2 - \kappa_2)} - \kappa_2 - \kappa/2\right] \tag{20.42}$$

where we have introduced the constraints from Eq. (20.41). This integral can be solved by stationary-phase or equivalently saddle-point methods, i.e. we look for the maximum in the integrand exponent as we did when investigating the fragmentation function in the Lund model in Eq. (9.6). This time we have to consider the maximum with respect to two variables but after a little algebra we find the surprisingly simple result that

$$\mathscr{F}_s \simeq \exp[(R-1)(2\ell+\kappa)/4], \quad R = \sqrt{1+8\bar{\alpha}} \qquad (20.43)$$

Thus *the symmetrical structure function only depends upon the rapidity difference* between the starting point and the endpoint, i.e. $\delta y \equiv \ell + \kappa/2$, cf. Fig. 20.13(*a*). This means that the result can be easily generalised to any number of added going-up and going-down cells. A closer examination tells us, however, that the maximum is only obtained within the integration region if ℓ is large compared to κ, i.e. if there is a sufficiently strong suppression of large κ-values. There will be a dividing line with

$$\kappa = \frac{R-1}{R}\delta y \qquad (20.44)$$

(where R is defined in Eq. (20.43)) with the property that for smaller κ-values there is a maximum but for larger κ the main contribution is a single DGLAP motion always directed upwards in κ. It is interesting that we again meet a result very similar to the one obtained in the soft radiation model in section 20.5, i.e. a cutoff which can be formulated as in Eq. (20.10). (It is worthwhile to calculate the corresponding 'dimension' in the cutoff as a function of both $\lambda_e = (R-1)/2$ and $\bar{\alpha}$ and consider the consequences of the results!) Anyway for κ below the line the result is obviously of the BFKL kind, i.e. there is (besides the symmetrical κ-dependence) an effective $x^{-\lambda_e}$ behaviour, but this time with $\lambda_e = (R-1)/2 \simeq 0.3$ for our 'conventional' value of $\bar{\alpha} \simeq 3/\pi \times 0.2$.

The next calculation will provide a useful formula for the major contributions to the structure functions when we start at the point $\ell_0 = L_{Q0} = 0$ and make use of all possible paths that end on the point (ℓ, L_Q). It is actually only possible to perform the full calculation by means of numerical methods but the final result is sufficiently simple that the following considerations apply. Firstly it turns out that the major contribution also corresponds closely to the 'average' path. This average is obtained if, for a fixed value of the rapidity in the triangular phase space, we consider the average 'passage' κ-value; the averaging is done by means of the LDC weights in Eq. (20.18).

In Fig. 20.13 we show the results for the two cases when $L_Q \simeq \ell$ and when ℓ dominates L_Q (we also show the fluctuation bands around the average paths). The most noticeable property is that *if we use a running coupling* (which is the case in the figures) *then there is a preference for small $q_\perp$-values*. The major contribution stems from an average path that

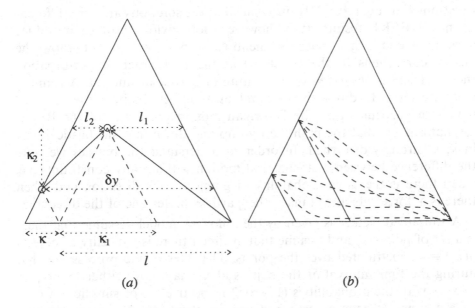

Fig. 20.13. (*a*) The combined DGLAP motion of one step (ℓ_1, κ_1) from the target and one step (ℓ_2, κ_2) from the probe, together forming $(\delta y, \kappa)$; (*b*) the average paths (together with the deviations) for a small and a large value of $L_Q = \log Q^2$.

stays close to $\kappa \simeq 0$ but, in the end, in order to reach the required L_Q goes up in κ. A simple assumption (which turns out to provide a very good approximation to the numerical results) is to subdivide the total $\ell = \log(1/x_B)$ into a 'first' BFKL contribution in ℓ_1, $\propto \exp(\lambda_e \ell_1)$ (close to the $\kappa = 0$ axis), times a DGLAP contribution in the 'final' ℓ_2-step upwards to L_Q, $\propto \exp[2\sqrt{\chi(Q^2)\ell_2}]$, and perform a convolution integral with the constraint $\ell_1 + \ell_2 = \ell$. It is easy (again using a stationary-phase method) to obtain that

$$\mathscr{F} \simeq \exp[\lambda_e \ell + \chi(Q^2)/\lambda_e] \tag{20.45}$$

if the saddle point $\ell_1 = \chi/\lambda_e^2$ is inside the integration region $0 < \ell_1 < \ell$. It is interesting to note that for a constant coupling $\bar{\alpha}$ we obtain back the division line in Eq. (20.44) if we use the value of λ_e defined by Eq. (20.43). For a truly running coupling, however, the corresponding requirement to obtain a consistent saddle-point approximation of $\mathscr{F}$ according to Eq. (20.45) will require a very large $\ell \equiv \log(1/x_B)$ compared to $L_Q = \log Q^2$. For the available ℓ-values in HERA we need such small Q^2-values that the results of perturbative QCD will no longer be valid.

It is worthwhile to further elaborate the use of a running coupling in connection with the integral equation for the non-integrated structure function $\mathscr{F}$ in Eq. (20.31). It turns out that the solutions are very different from the BFKL results. It is, however, not obvious how to introduce a running coupling because as mentioned above we do not know the virtual corrections to the equations at the next order of perturbation theory. There is nevertheless one feature of particular interest. A running coupling tends to diminish the κ-values along the main paths and this means in particular that the Brownian motion properties of the BFKL mechanism, i.e. that the paths can go up and down in a stochastical way in κ, is strongly disturbed. In order to understand the physics we note the difference between a stochastical motion without any constraints, i.e. when there is the same probability of going in each direction and when there is a 'force' which will make the 'particle' prefer one of the directions.

A simplified picture is given by the following model. Assume that there is a set of points x_j and assume that at time t there is a density of objects $\rho(x_j) \equiv \rho_j$ distributed over the points. The rules of the process are that during the time interval δt the objects at x_j may move either to x_{j+1} or to x_{j-1} with the probabilities $(1 \mp \alpha)/2$ respectively. For simplicity we let α be a constant between zero and one, i.e. there is a preference towards smaller j-values, which we will associate with smaller x. Then we obtain

$$\rho_j(t + \delta t) = \frac{1 - \alpha}{2} \rho_{j-1}(t) + \frac{1 + \alpha}{2} \rho_{j+1}(t) \qquad (20.46)$$

The equation can be rewritten

$$\rho_j(t + \delta t) - \rho_j(t) = \frac{1}{2}[\rho_{j+1}(t) - 2\rho_j(t) + \rho_{j-1}] + \frac{\alpha}{2}[\rho_{j+1}(t) - \rho_{j-1}(t)]$$

$$\Rightarrow \frac{\partial \rho}{\partial t} = a\frac{\partial^2 \rho}{\partial x^2} + b\frac{\partial \rho}{\partial x} \qquad (20.47)$$

where b is proportional to the parameter α in the process. In the second line we have gone to the limits $x_{j+1} \to x_j$ and $\delta t \to 0$. We have consequently derived the ordinary *diffusion equation*. It is well known that when b is zero (i.e. the case of symmetry where α vanishes) then the solution, which at time zero is centred at the origin $x = 0$ as a δ-distribution, is for finite t-values given by the (normalised) gaussian distribution

$$\frac{N}{\sqrt{\pi t}} \exp\left(-\frac{x^2}{4at}\right) \qquad (20.48)$$

This means that N objects, all starting at the origin, will perform a Brownian motion, i.e 'diffuse' away and after a time t on average reach the point $2\sqrt{at}$.

If b is different from zero, however, then *there are stationary, i.e. time-*

independent, solutions of an exponential kind:

$$C \exp \left(\frac{-bx}{a} \right) \qquad (20.49)$$

The distribution will settle into such a stationary state after a period which depends upon the boundary conditions. It is well known that this is just the density distribution of the atmosphere of the earth and the reason for the preference of small x-values is in that case the gravitational force.

The model described above is much too simple to describe fully the non-integrated structure function $\mathscr{F}$. It is, however, of interest that the integral equation (20.31) can under simplifying assumptions be rewritten as a second-order differential equation similar to the Schrödinger equation. Then $\mathscr{F}$ takes the role of the wave function and the coupling takes the role of a potential. In this way there will be a 'gravitational pull' towards small κ from a running coupling like $\alpha_s \propto 1/\kappa$. Unfortunately there is at present no infrared-stable solution, i.e. it is necessary to make use of some transverse momentum cutoff in the equations. Therefore we will not pursue this problem any longer.

Nevertheless, we note from the results of numerical solutions to the integral equation (20.31) that when the running-coupling solutions start to become of a BFKL kind, i.e. behave as a power in $1/x_B$, then the large-transverse-momentum tail of the distribution will exhibit an exponential falloff in κ with a slope independent of $\log(1/x_B)$. *Consequently there is no diffusion in κ leading to a gaussian distribution with a width proportional to the chain length $\log(1/x_B)$.*

In conclusion we have found that the two fundamental mechanisms, i.e. the DGLAP and the modified BFKL, are also relevant as basic analysis tools for the more complex interpolating linked dipole chain model equations ('modified BFKL' means that there is an effective λ_e essentially smaller than the original Lipatov index in Eq. (20.40)). The average paths, corresponding to the main contribution from the fan diagrams (Fig. 19.1) in the triangular phase space, are cigar-shaped when the main path is described together with the average fluctuations around it. This is nowadays known as the 'Bartel cigar', for a major contributor to the investigations. The Bartel cigar is situated along the small-κ region and contributes to the structure function behaviour in accordance with the modified BFKL mechanism. In the end the average path rises towards the required $\log Q^2$ value in accordance with the DGLAP mechanism. Owing to the limited $\log(1/x_B)$ range available, all our results inside the HERA region will be dominated by the latter mechanism.

There are, however, inside the presently available accelerator regions (referring to both the HERA and the FERMILAB facilities), the many investigations to be performed, both theoretically and experimentally, when it comes to the transverse momentum distributions and the variations, which will go over into jets (both of the Rutherford and the ordinary bremsstrahlung kind).

Consequently the phase space for deep inelastic scattering will most certainly contain many more degrees of freedom than we know of at present!

References

[1] D. Amati, A.Stanghellini and S. Fubini, *Nuovo Cimento* **26** 896, 1962.
[2] *Proceedings of the Workshop on Jet Studies at LEP and HERA, J. Phys.* **G17** 1441, 1991. S. Catani, Ettore Majorani Meeting QCD at 200 TeV (Plenum Press 1992).
[3] Y. Aharanov and D. Bohm, *Phys. Rev.* **115** 485, 1959. M. Peshkin and A. Tonomura, *The Aharanov-Bohm Effect* (Springer-Verlag 1989).
[4] A.H. Mueller, *Nucl. Phys.*, **B415** 373, 1994.
[5] G. Altarelli and G. Parisi, *Nucl. Phys.* **B126** 298, 1977.
[6] B. Andersson, *Nucl. Phys.* **B360** 109, 1991.
[7] B. Andersson and G. Gustafson, *Z. Phys.* **C3** 223, 1980.
[8] B. Andersson, G. Gustafson, I. Holgersson and O. Månsson, *Nucl. Phys.* **B178** 242, 1981.
[9] B. Andersson, G. Gustafson and G. Ingelman, *Phys. Lett.* **B85** 417, 1979.
[10] B. Andersson, G. Gustafson and L. Lönnblad, *Nucl Phys.* **B339** 393, 1990.
[11] B. Andersson, G. Gustafson L. Lönnblad, and U. Pettersson, *Z. Phys.*, **C43** 625, 1989.
[12] B. Andersson, G. Gustafson A. Nilsson, and C. Sjögren, *Z. Phys.*, **C49** 79, 1991.
[13] B. Andersson, G. Gustafson and C. Peterson, *Nucl. Phys* **B135** 273, 1978. R.D. Field and R.P. Feynman, *Nucl. Phys.* **B136** 1, (1978).
[14] B. Andersson, G. Gustafson and C. Peterson. *Z. Phys.*, **C1** 105, 1979.
[15] B. Andersson, G. Gustafson and J. Samuelsson, *Nucl. Phys.* **B463** 217, 1996.
[16] B. Andersson, G. Gustafson and J. Samuelsson, *Nucl. Phys.* **B467** 443, 1996.
[17] B. Andersson, G. Gustafson and T. Sjöstrand, *Z. Phys.* **C6** 235, 1980.
[18] B. Andersson, G. Gustafson and T. Sjöstrand, *Z. Phys.* **C6** 235, 1980. B. Andersson, G. Gustafson and T. Sjöstrand, *Phys. Lett.* **B94** 211, (1980).
[19] B. Andersson, G. Gustafson and B. Söderberg, *Z. Phys.* **C20** 317, 1983.
[20] B. Andersson, G. Gustafson and B. Söderberg, *Nucl. Phys.* **B264** 29, 1986.
[21] B. Andersson and A. Nilsson, Lund preprint, 1992.
[22] B. Anderssson and W. Hotmann, *Phys. Lett.* **B169** 364, 1986.
[23] A.G. Aranov and Yu.V. Sharvin, *Rev. Mod. Phys.* **59** 755, 1987. A review article containing early work.
[24] X. Artru, personal communication.
[25] X. Artru, *Z. Phys.* **C26** 83, 1984.
[26] X. Artru and G. Menessier, *Nucl. Phys.* **B70** 93, 1974.
[27] Ya.I. Azimov *et al. Phys. Lett.* **B165** 147, 1985. Yu.L. Dokshitzer, V.A. Khoze and S.I. Troyan, in A.H. Mueller ed., *Perturbative QCD*, p. 241, World Scientific.
[28] M. Baker and K.A. Ter-Martirosyan, *Phys. Rep.* **28C** 1, 1976.
[29] Ya.Ya. Balitzkij and L.N. Lipatov, *Sov. J. Nucl. Phys.* **28** 822, 1978. J.B. Bronzan and R.L. Sugar, *Phys. Rev.* **D17** 585, 1978; L.N. Lipatov, *Sov. Phys. JETP* **63** 904, 1986; V.S. Fadin, E.A. Kuraev and L.N. Lipatov, *Phys. Lett.* **B60** 50, 1975; L.N. Lipatov in *Perturbative QCD*, (World Scientific 1989).
[30] J.D. Bjorken and S.D. Drell, *Relativistic Quantum Fields* (McGraw-Hill 1965).

465

[31]   N. Bohr and L. Rosenfeld, *Dan. Mat. Fys. Medd.* **12(8)**, 1933.

[32]   M.G. Bowler, *Z. Phys.* **C11** 169, 1981.

[33]   M.G. Bowler, *Z. Phys.* **C29** 617, 1985.

[34]   M.G. Bowler, *Phys. Lett.* **B180** 299, 1986.

[35]   M.G. Bowler, *Phys. Lett.* **B185** 205, 1987.

[36]   R. Brandt, *Ann. Phys. (New York)* **44** 1967, 221. W. Zimmerman, in *Lectures on Elementary Particles*, S. Deser ed. (MIT Press 1971:I).

[37]   S.J. Brodsky, C. Peterson and N. Sakai, *Phys. Rev.* **D23** 2745, 1981.

[38]   C.D. Buchanan and S.-B. Chun, *Phys. Rev. Lett.* **59** 1997, 1987. C.D. Buchanan and S.-B. Chun, preprint UCLA-HEP-92-008.

[39]   A. Casher, J. Kogut and L. Susskind, *Phys. Rev.* **D10** 732, 1974.

[40]   A. Casher, H. Neuberger and S. Nussinov, *Phys. Rev.* **D20** 179, 1979.

[41]   H.B.G. Casimir, *Proc. Kon. Ned. Akad. Wetenschap Ser B* **51** 277, 1948.

[42]   S. Catani, F. Fiorani and G. Marchesini, *Phys. Lett.* **B234** 339, 1990; *Nucl. Phys.* 18, 1990.

[43]   N. Christ, B. Hasslacher and A.H. Mueller, *Phys. Rev.* **D6** 3543, 1972.

[44]   M. Ciafaloni, *Nucl. Phys.* **B296** 249, 1987. S. Catani, F. Fiorani and G. Marchesini, *Phys. Lett* **234B** 339, 1990; *Nucl. Phys.* **B336** 18, 1990.

[45]   S. Coleman, *Aspects of symmetry*, (Cambridge University Press 1985).

[46]   P.D.B. Collins and E.J. Squires, *Regge Poles in Particle Physics* (Springer Tracts in Modern Physics **45** 1968).

[47]   Per Dahlqvist, Lund preprint LU-TP 88-11.

[48]   Per Dahlqvist, *Perturbative and Nonperturbative Aspects of Multiparticle Production*, Ph.D. thesis, Lund University, Dept Theor. Physics, 1989. B. Andersson, P. Dahlqvist and G. Gustafson, *Phys. Lett.* **B214** 604, 1988; *Z. Phys.* **C44** 455, 1989; *Z. Phys.* **C44** 461, 1989.

[49]   B.S. Deaver and W.M. Fairbank, *Phys. Rev. Lett.* **7** 43, 1961. R. Doll and M. Näbauer, *Phys. Rev. Lett* **7** 51, 1961.

[50]   T.A. DeGrand and H.I. Miettinen *Phys. Rev.* **D23** 1227, 1981.

[51]   E. DeWolf, paper at 1993 Moriond Meeting, Les Arcs.

[52]   Yu.L. Dokshitzer, V.A. Khoze, A.H. Mueller and S.I. Troyan, *Basics of Perturbative QCD* (Edition Frontières 1991).

[53]   Yu.L. Dokshitzer and S.I. Troyan, *Proceedings of the XIX Winter School of the LNPI*, **1** 144, 1984. Ya.I. Azimov, Yu.L. Dokshitzer, V.A. Khoze and S.I. Troyan, *Z. Phys.* **C27** 6, 1985.

[54]   S.D. Drell and T.M. Yan, *Phys. Rev. Lett.* **24** 855, 1970. G.B. West, *Phys. Rev. Lett.* **24** 1206, 1970.

[55]   E. Eichten *et al.*, *Phys. Rev. Lett.* **34** 369, 1975.

[56]   R.K. Ellis, D.A. Ross and A.E. Terrano, *Nucl. Phys.* **B178** 421, 1981.

[57]   A. Erdelyi, W. Magnus, F. Oberhettinger and F.G. Tricomi, *Bateman Manuscript Project: Higher Transcendential Functions: 2*, (McGraw-Hill 1953).

[58]   S. Erhan *et al.*, *Phys. Lett.* **B85** 447, 1979. A.M. Smith *et al.*, *Phys. Lett.* **B185** 209, 1987; T. Henkes, in *Proceedings of XXIV Rencontre du Moriond*, J. TranThanhVan ed. (Editions Frontières 1989).

[59]   B.I. Ermolaev and V.S. Fadin, *JETP Lett.* **33** 161, 1981. A.H. Mueller, *Phys. Lett.* **B104** 161, 1981.

[60]   R.P. Feynman, *Photon-Hadron Interactions* (W.A. Benjamin 1972).

[61]   M. Fierz, *Helv. Physica Acta* **23** 731, 1950.

[62]   P.H. Frampton, *Dual Resonance Models and Superstrings* (World Scientific 1986). A general reference for formal string theory from Veneziano, Koba-Olesen and Nambu to Schwartz *et al.*

[63]   G. Ingelman, *Comp. Phys. Comm.* **46** 217, 1987. G. Ingelman and A. Weigend, *Comp. Phys. Comm.* **46** 241, 1987.

[64]   N.K. Glendenning and T. Matsui, *Phys. Rev.* **D28** 2890, 1983.

[65]   G. Goldhaber, S. Goldhaber, W. Lee and A. Pais, *Phys. Rev.* **120** 300, 1960.

[66]   K. Gottfried and F.E. Low, *Phys. Rev.* **D17** 2487, 1978.

[67]   L.V. Gribov, E.M. Levin and M.G. Ruskin. *Phys. Rep.* **100** 1, 1983.

[68]   D.J. Gross and F. Wilczek, *Phys. Rev. Lett.* **30** 1343, 1973. H.D. Politzer, *Phys. Rev. Lett.* **30** 1346, 1973.

[69] D.J. Gross and F. Wilczek, *Phys. Rev. Lett.* **30** 1343, 1973. H.D. Politzer, *Phys. Rev. Lett.* **30** 1346, 1973; D.J. Gross and F. Wilczek, *Phys. Rev.* **D8** 3633, 1963; H. Georgi and D.J. Politzer, *Phys. Rev.* **D9** 416, 1974.

[70] G. Gustafson, paper at QCD 20 Years After, Aachen meeting, 1992.

[71] G. Gustafson, *Z. Phys.* **C15** 155, 1982.

[72] G. Gustafson, *Nucl. Phys.* **B392** 251, 1993.

[73] G. Gustafson and A. Nilsson, *Z. Phys.* **C52** 553, 1991.

[74] G. Gustafson and M. Olsson, Lund preprint, 1992.

[75] G. Gustafson and U. Pettersson, *Nucl. Phys.* **B306** 746, 1988.

[76] R. Hagedorn, CERN Yellow Report 71-12.

[77] F. Halzen and A.D. Martin, *Quarks and Leptons*, John Wiley and Sons, 1984.

[78] R. Hanbury-Brown, *The Intensity Interferometer*, (Taylor and Francis, London 1974).

[79] W. Heisenberg and H. Euler, *Z. Phys.* **98** 714, 1936.

[80] H.G.E. Hentschel and I. Procachia, *Physica* **D8** 435, 1983. T. Halsey *et al.*, *Phys. Rev.* **A33** 1141, 1986.

[81] L. Van Hove, *Nucl. Phys.* **B9** 331, 1969.

[82] A. Jaffe, Ph.D. thesis, Princeton University, 1967.

[83] J. Glimm and A. Jaffe, *Quantum Physics, a Functional Integral Point of View* (Springer-Verlag 1981).

[84] G. Källén, *Elementary Particle Physics* (Addison-Wesley 1964).

[85] R. Kleiss, *Phys. Lett.* **B180** 400, 1986.

[86] Z. Koba, H.B. Nielsen and P. Olesen, *Nucl. Phys.* **B40** 317, 1972.

[87] J.B. Kogut and D.E. Soper, *Phys. Rev.* **D1** 2901, 1970.

[88] H.A. Kramers, in *Atti del Congress Internationale de Fisici Como*, 1927.

[89] R. Kronig, *J. Amer. Optical Soc.* **12** 547, 1926.

[90] A. Krzyvicki and B. Petersson, *Phys. Rev.* **D6** 924, 1972. F. Niedermayer, *Nucl. Phys.* **B79** 355, 1974.

[91] F. London and H. London, *Proc. Roy. Soc. (London)* **A147** 71, 1935.

[92] L. Lönnblad, DESY preprint 1992, ARIADNE version 4.

[93] B. Lörstad, *Int. J. Mod. Phys.* **A4** 2861, 1988. W.A. Zaic, in *Hadronic Multi-particle Production*, ed. P. Carruthers (World Scientific 1988); B.H. Boal, C.K. Gelbke and B.K. Jennings, *Rev. Mod. Phys.* **62** 553, 1990.

[94] G. Marchesini and B.R. Webber, *Nucl. Phys.* **B238** 1, 1984.

[95] A.H. Mueller and J. Qiu, *Nucl. Phys.* **B268** 427, 1986.

[96] A.H. Mueller, *Nucl. Phys.*, **B415** 373, 1994.

[97] Y. Nambu, *Phys. Rev.* **106** 1366, 1957.

[98] H.B. Nielsen and P. Olesen, *Nucl. Phys.* **B61** 45, 1973.

[99] C. Peterson, D. Schlatter, I. Schmitt and P.M. Zerwas, *Phys. Rev.* **D27** 105, 1983.

[100] J. Schwinger, *Phys. Rev.* **82** 664, 1951.

[101] J. Schwinger, *Phys. Rev.* **128** 2425, 1962.

[102] D.V. Shirkov, Mass and scheme effects in coupling constant evolution, preprint MPI-Ph/9-94.

[103] C. Sjögren, Parton cascades and multiparticle production, Ph.D. thesis, Lund University, Dept Theor. Physics, 1992.

[104] T. Sjöstrand, *Nucl. Phys.* **B248** 469, 1984.

[105] T. Sjöstrand, *Comp. Phys. Commun.* **39** 347, 1986. T. Sjöstrand and M. Bengtsson, *Comp. Phys. Commun.* **43** 367, 1987; T. Sjöstrand, CERN-TH.6488/92.

[106] A.M. Smith *et al.*, *Phys. Lett.* **B163** 267, 1985.

[107] A.E. Sudakov, *Izv. Akad. Nauk SSSR* **19** 650, 1955.

[108] K. Symanzik, *Comm. Mat. Phys.* **18** 227, 1970. C.G. Callan, *Phys. Rev.* **D2** 1541, 1970.

[109] G.E. Uhlenbeck and L.S. Ornstein, *Phys. Rev.* **36** 823, 1930. R. Kubo, M. Toda and N. Hashitsume, *Statistical Physics II* (Springer Series in Solid State Sciences **31**, 1983).

[110] B.R. Webber, *Nucl. Phys.* **B238** 492, 1984.

[111] S. Weinberg, *Phys. Rev.* **150** 1313, 1966.

[112] E.P. Wigner, *Ann. Math.* **40** 149, 1939. M.A. Naimark, *Uspekhi Mat. Nauk* **9** 19, 1954.

Index

Printed in the United States
by Baker & Taylor Publisher Services

Printed in the United States
by Baker & Taylor Publisher Services